COMBINATORIAL AND TORIC HOMOTOPY

Introductory Lectures

LECTURE NOTES SERIES
Institute for Mathematical Sciences, National University of Singapore

Series Editors: Chitat Chong and Kwok Pui Choi
Institute for Mathematical Sciences
National University of Singapore

ISSN: 1793-0758

Published

Vol. 26 Notes on Forcing Axioms
by Stevo Todorcevic, edited by Chitat Chong, Qi Feng, Theodore A Slaman, W Hugh Woodin & Yue Yang

Vol. 27 *E*-Recursion, Forcing and C*-Algebras
edited by Chitat Chong, Qi Feng, Theodore A Slaman, W Hugh Woodin & Yue Yang

Vol. 28 Slicing the Turth:
On the Computable and Reverse Mathematics of Combinatorial Principles
by Denis R Hirschfeldt, edited by Chitat Chong, Qi Feng, Theodore A Slaman, W Hugh Woodin & Yue Yang

Vol. 29 Forcing, Iterated Ultrapowers, and Turing Degrees
edited by Chitat Chong, Qi Feng, Theodore A Slaman, W Hugh Woodin & Yue Yang

Vol. 30 Modular Representation Theory of Finite and *p*-Adic Groups
edited by Wee Teck Gan & Kai Meng Tan

Vol. 31 Geometric Analysis Around Scalar Curvatures
edited by Fei Han, Xingwang Xu & Weiping Zhang

Vol. 32 Mathemusical Conversations: Mathematics and Computation in Music Performance and Composition
edited by Jordan B L Smith, Elaine Chew & Gérard Assayag

Vol. 33 Sets and Computations
edited by Dilip Raghavan, Sy-David Friedman & Yue Yang

Vol. 34 White Noise Analysis and Quantum Information
edited by Luigi Accardi, Louis H Y Chen, Takeyuki Hida, Masanori Ohya, Si Si & Noboru Watanabe

Vol. 35 Combinatorial and Toric Homotopy: Introductory Lectures
edited by Alastair Darby, Jelena Grbić, Zhi Lü & Jie Wu

*For the complete list of titles in this series, please go to
http://www.worldscientific.com/series/LNIMSNUS

Lecture Notes Series, Institute for Mathematical Sciences,
National University of Singapore Vol. 35

COMBINATORIAL AND TORIC HOMOTOPY

Introductory Lectures

Editors

Alastair Darby
Fudan University, China

Jelena Grbić
University of Southampton, UK

Zhi Lü
Fudan University, China

Jie Wu
National University of Singapore, Singapore

World Scientific

NEW JERSEY • LONDON • SINGAPORE • BEIJING • SHANGHAI • HONG KONG • TAIPEI • CHENNAI • TOKYO

Published by

World Scientific Publishing Co. Pte. Ltd.
5 Toh Tuck Link, Singapore 596224
USA office: 27 Warren Street, Suite 401-402, Hackensack, NJ 07601
UK office: 57 Shelton Street, Covent Garden, London WC2H 9HE

Library of Congress Cataloging-in-Publication Data
Names: Darby, Alastair, editor.
Title: Combinatorial and toric homotopy : introductory lectures / edited by
Alastair Darby (Fudan University, China) [and three others].
Description: New Jersey : World Scientific, 2017. | Series: Lecture notes series,
Institute for Mathematical Sciences, National University of Singapore ; volume 35 |
Includes bibliographical references.
Identifiers: LCCN 2017030713 | ISBN 9789813226562 (hardcover : alk. paper)
Subjects: LCSH: Combinatorial topology. | Toric varieties. | Geometry, Algebraic. | Homotopy theory.
Classification: LCC QA612.3 .C66 2017 | DDC 514/.22--dc23
LC record available at https://lccn.loc.gov/2017030713

British Library Cataloguing-in-Publication Data
A catalogue record for this book is available from the British Library.

Cover image: Adapted from the visualization of the Hopf fibration by Niles Johnson, taken from Wikimedia Commons.

Printed in Singapore

CONTENTS

FOREWORD

The Institute for Mathematical Sciences (IMS) at the National University of Singapore was established on 1 July 2000. Its mission is to foster mathematical research, both fundamental and multidisciplinary, particularly research that links mathematics to other efforts of human endeavor, and to nurture the growth of mathematical talent and expertise in research scientists, as well as to serve as a platform for research interaction between scientists in Singapore and the international scientific community.

The Institute organizes thematic programs of longer duration and mathematical activities including workshops and public lectures. The program or workshop themes are selected from among areas at the forefront of current research in the mathematical sciences and their applications.

Each volume of the *IMS Lecture Notes Series* is a compendium of papers based on lectures or tutorials delivered at a program/workshop. It brings to the international research community original results or expository articles on a subject of current interest. These volumes also serve as a record of activities that took place at the IMS.

We hope that through the regular publication of these *Lecture Notes* the Institute will achieve, in part, its objective of reaching out to the community of scholars in the promotion of research in the mathematical sciences.

April 2017

Chitat Chong
Kwok Pui Choi
Series Editors

PREFACE

These lecture notes arose from the program "Combinatorial and Toric Homotopy" which was held during 1–31 August 2015 at the National University of Singapore's Institute for Mathematical Sciences (IMS). The program aimed to explore toric homotopy theory and combinatorial homotopy theory as well as their connections with other areas of mathematics.

Sometime around the turn of the recent millennium, topological explorations of toric objects led to a new field on the boundary of equivariant topology, algebraic and symplectic geometry, combinatorics and commutative algebra under the name toric topology. Since then, the subject has rapidly developed. The methods from homotopy theory enriched the development of toric topology. On the other hand, combinatorial techniques coming from toric topology were applied to determine the homotopy type of certain configuration spaces. The interplay between homotopy theory and toric topology sets new directions in the development of homotopy theory, where closed connections between polyhedron products, Whitehead products and Hopf invariants provide a fundamental clue for the deep understanding of the homotopy theory of toric objects. The whole development further highlighted huge potential of applying homotopy theory to problems in various mathematical disciplines as well as to solving real-world problems.

Our program investigated interactions between toric topology and the closely related subject of combinatorial homotopy theory. Combinatorial homotopy theory was born in the 1950s and recent developments establish various connections between algebraic topology and other areas of mathematics. Ideas, methods and constructions of algebraic topology (especially the methodology from combinatorial topology) are finding applications in various domains stimulating the development of the subjects in new areas of applied and computational topology.

Our program had the distinction of being a part of Professor Frederick Cohen's celebration of his 70th birthday. Professor Cohen is one of the greatest topologists of our time and his insight and enthusiasm never cease to impress.

The program consisted of the following activities:

(1) Young Topologist Seminar, 11–19 August 2015.
(2) Workshop on Applied Topology, 20–21 August 2015.
(3) International Conference on Combinatorial and Toric Homotopy, 24–28 August 2015.

As a special perk of our program, Professor Vladimir Voevodsky gave a research seminar talk on univalent morphisms as a part of his univalent foundations of mathematics and homotopy type theory. Voevodsky's work in developing a homotopy theory for algebraic varieties and formulating motivic cohomology led to the award of a Fields Medal in 2002.

The Young Topologist Seminar focused on the training of young topologists. The lectures of this seminar form a basis of this book. In addition to lectures, various junior topologists contributed talks on their research. The workshop on applied topology continued in spirit of the activities of the Young Topologist Seminar focusing on applications. Various talks were given by active researchers in applied topology. The International Conference on Combinatorial and Toric Homotopy focused on recent developments in toric topology, combinatorial groups, geometric groups and algebraic topology with the talks contributed by eminent researchers in the subjects.

Professor Stephen Theriault gave an introduction to the toric homotopy theory describing homotopy theory surrounding Davis-Januszkiewicz spaces, moment-angle complexes and their generalizations to polyhedral products, and showing how the homotopy types of toric objects depend on a beautiful interplay between topology and combinatorics.

The lectures given by Professor Victor M. Buchstaber were devoted to a remarkable class of 3-dimensional polytopes, which are mathematical models of the important object of quantum physics, quantum chemistry and nanotechnology – fullerenes. His lectures showed how the results of toric topology help to build combinatorial invariants of fullerenes.

Professor Vladimir Vershinin's lectures gave various non-classical topics concerning braids. These topics enriched the study on the theory of braids and its connections with other areas of mathematics.

The lectures given by Professor Sergei O. Ivanov gave an introduction to the theory of limits over the category of presentations, with the aim to study the functors from the category of groups to the category of abelian groups using the theory of **fr**-codes.

Professor Michael Farber's lectures described several mathematical models producing random simplicial complexes and some recent results about topological properties of these spaces.

In addition to the lectures of the program, Professor Dai Tamaki contributed a chapter on cellular stratified spaces, which enriches the methodologies for studying the homotopy type of configuration spaces.

These were generally agreed to be great courses whose legacy will live on through these lecture notes. We would like to thank the lecturers most warmly for the effort that they have put into the preparation of the lecture notes, and the patience with which they have dealt with our editorial queries.

Besides us, the other members of the Organizing Committee were Alejandro Adem (University of British Columbia), Jon Berrick (National University of Singapore), Fengchun Lei (Dalian University of Technology), Ran Levi (The University of Aberdeen), Lek-Heng Lim (University of Chicago), Roman Mikhailov (St. Petersburg State University), Mikiya Masuda (Osaka City University), Dong Youp Suh (Korea Advanced Institute of Science and Technology), Dai Tamaki (Shinshu University), Vladimir Vershinin (University of Montpellier II), and Yan Loi Wong (National University of Singapore). We are very much grateful to their invaluable services. In addition to short term visitors and participants, there were two long term visitors (Sergei Ivanov and Djordje Baralic) and distinguished visitor (Vladimir Voevodsky) in the program. We would like to thank all the participants of this program for their support and stimulating interactions during the month.

We would like to take this opportunity to thank Professor Chi Tat Chong, Director of IMS, for his leadership in creating an exciting environment for mathematical research in the IMS and for his guidance throughout our program. The expertise and dedication of all IMS staff contributed essentially to the success of this program. Last but not least, we would like to acknowledge IMS, the Department of Mathematics of National University of Singapore and MOE Tier 1 Grant R-146-000-190-112 for providing financial support to the program.

October 2016

Alastair Darby
Fudan University, China

Jelena Grbić
University of Southampton, UK

Zhi Lü
Fudan University, China

Jie Wu
National University of Singapore
Singapore

Volume Editors

TORIC HOMOTOPY THEORY

Stephen Theriault

Mathematical Sciences
University of Southampton
Southampton SO17 1BJ, UK
S.D.Theriault@soton.ac.uk

These notes describe some of the homotopy theory surrounding Davis-Januszkiewicz spaces, moment-angle complexes and their generalizations to polyhedral products. These spaces are defined by gluing together products formed from pairs of spaces (X, A), where the gluing is determined by the faces of a simplicial complex K. The emphasis is on determining homotopy types — of the spaces themselves, their suspensions and their based loop spaces — and showing how these homotopy types depend on a beautiful interplay between the topology of the pairs (X, A) and the combinatorics of the simplicial complex K.

Contents

1. Introduction

Toric topology is at the nexus of a wide range of mathematical subjects: algebraic topology, symplectic and algebraic geometry, combinatorics and commutative algebra. Homotopy theory has proved to be very useful in analyzing the fundamental spaces of toric topology: Davis-Januszkiewicz spaces DJ_K, moment-angle complexes $\mathcal{Z}_K$, and their generalizations to polyhedral products $(\underline{X}, \underline{A})^K$. All three are defined in terms of products of topological spaces glued together at intersections as determined by the faces of a simplicial complex K. The topology of these spaces is fascinating, involving a beautiful interplay between homotopy theory and properties of simplicial complexes.

The purpose of these notes is to give an introduction to the homotopy theory surrounding Davis-Januszkiewicz spaces, moment-angle complexes and polyhedral products. In one direction, methods and tools are developed that give the reader a grounding in the techniques that have led to the main advances in the homotopy theory of toric topology over the past decade. In another direction, these methods are applied to address important questions and conjectures in homotopy theory.

The main goals are to identify the homotopy types of the spaces DJ_K, $\mathcal{Z}_K$ and $(\underline{X}, \underline{A})^K$, the homotopy types of their suspensions ΣDJ_K, $\Sigma\mathcal{Z}_K$ and $\Sigma(\underline{X}, \underline{A})^K$, and the homotopy types of their based loop spaces ΩDJ_K, $\Omega\mathcal{Z}_K$ and $\Omega(\underline{X}, \underline{A})^K$. General formulas for the homotopy type of $\Sigma(\underline{X}, \underline{A})^K$ and $\Omega(\underline{X}, \underline{A})^K$ will be discussed in Theorems 4.5 and 9.4 respectively. The homotopy type of $(\underline{X}, \underline{A})^K$ itself is more subtle, and only special cases will be considered. The loop space decomposition of $(\underline{X}, \underline{A})^K$ is used in the latter part of the notes to address problems related to major questions in homotopy theory: Barratt's conjecture, the elliptic-hyperbolic duality in rational homotopy theory, and Moore's conjecture. Along the way, various open problems will be pointed out in the hope of provoking interest.

2. Davis-Januszkiewicz spaces, moment-angle complexes and polyhedral products

The two fundamental objects in toric topology are Davis-Januszkiewicz spaces and moment-angle complexes. They were first constructed by Davis and Januszkiewicz [10] but were reformulated more conveniently by Buchstaber and Panov [5, 6]. For a positive integer m, let $[m] = \{1, \ldots, m\}$. Let K be a simplicial complex on the vertex set $[m]$. Let $T^m = \prod_{i=1}^{m} S^1$ be the m-fold torus and let $BT^m = \prod_{i=1}^{\infty} \mathbb{C}P^\infty$ be its classifying space. For a face $\sigma \in K$, let

$$DJ_\sigma = \prod_{i=1}^{m} Y_i \qquad \text{where} \qquad Y_i = \begin{cases} \mathbb{C}P^\infty & \text{if } i \in \sigma \\ * & \text{if } i \notin \sigma. \end{cases}$$

The *Davis-Januszkiewicz space* DJ_K is defined by

$$DJ_K = \bigcup_{\sigma \in K} DJ_\sigma. \tag{2.1}$$

Observe that DJ_K is a subspace of the product $\prod_{i=1}^{m} \mathbb{C}P^\infty$.

The moment-angle complex is constructed analogously. For a face $\sigma \in K$, let

$$\mathcal{Z}_\sigma = \prod_{i=1}^{m} Y_i \qquad \text{where} \qquad Y_i = \begin{cases} D^2 & \text{if } i \in \sigma \\ S^1 & \text{if } i \notin \sigma. \end{cases}$$

The *moment-angle complex* $\mathcal{Z}_K$ is defined by

$$\mathcal{Z}_K = \bigcup_{\sigma \in K} \mathcal{Z}_\sigma. \tag{2.2}$$

Observe that $\mathcal{Z}_K$ is a subspace of the product $\prod_{i=1}^{m} D^2$. There is a canonical action of T^m on $\prod_{i=1}^{m} D^2$ given by letting the circle in coordinate i of T^m act by rotation on the disc in coordinate i of $\prod_{i=1}^{m} D^2$. This induces an action of T^m on the subspace $\mathcal{Z}_K$. The orbit space of this action on $\mathcal{Z}_K$ is DJ_K. Thus there is a homotopy fibration sequence

$$T^m \longrightarrow \mathcal{Z}_K \longrightarrow DJ_K \longrightarrow BT^m. \tag{2.3}$$

The analogous definitions of DJ_K and $\mathcal{Z}_K$ suggest that there is something more functorial going on that ought to be investigated. The generalization was formulated independently in unpublished notes by Strickland

and by Buchstaber and Panov [5], and first seriously investigated by Bahri, Bendersky, Cohen and Gitler [1]. The generalization is called a polyhedral product.

Let K be a simplicial complex on m vertices. For $1 \leq i \leq m$, let (X_i, A_i) be a pair of pointed CW-complexes, where A_i is a pointed subspace of X_i. Let $(\underline{X}, \underline{A}) = \{(X_i, A_i)\}_{i=1}^m$ be the sequence of CW-pairs. Regard each simplex (face) $\sigma \in K$ as being identified with the sequence $(i_1, \ldots, i_k)$ where $i_1, \ldots, i_k$ are the vertices of σ and $i_1 < \cdots < i_k$. Let $(\underline{X}, \underline{A})^\sigma$ be the subspace of $\prod_{i=1}^m X_i$ defined by

$$(\underline{X}, \underline{A})^\sigma = \prod_{i=1}^{m} Y_i \qquad \text{where} \qquad Y_i = \begin{cases} X_i & \text{if } i \in \sigma \\ A_i & \text{if } i \notin \sigma. \end{cases}$$

The *polyhedral product* determined by $(\underline{X}, \underline{A})$ and K is

$$(\underline{X}, \underline{A})^K = \bigcup_{\sigma \in K} (\underline{X}, \underline{A})^\sigma \subseteq \prod_{i=1}^{m} X_i.$$

In particular, if each pair of spaces (X_i, A_i) equals $(\mathbb{C}P^\infty, *)$ then the polyhedral product $(\underline{X}, \underline{A})^K$ is the definition of the Davis-Januszkiewicz space DJ_K. If each pair of spaces (X_i, A_i) equals (D^2, S^1) then the polyhedral product $(\underline{X}, \underline{A})^K$ is the definition of the moment-angle complex $\mathcal{Z}_K$. Thus determining homotopy theoretic information about polyhedral products in general lets one deduce properties of Davis-Januszkiewicz spaces and moment-angle complexes.

To get a feeling for polyhedral products, we consider several examples. Let I be the unit interval and let A and B be pointed topological spaces. The *reduced join* of A and B is the quotient space

$$A * B = (A \times I \times B)/\sim$$

where $(a, 1, b) \sim (a', 1, b)$, $(a, 0, b) \sim (a, 0, b')$ and $(*, t, *) \sim *$ for all $a, a' \in A$, $b, b' \in B$ and $t \in I$. It is well known that there is a natural homotopy equivalence $A * B \simeq \Sigma A \wedge B$.

Example 2.1: Suppose that K is m disjoint points. Then the faces of K consist of the emptyset and the vertices. Since the emptyset has no vertices, from the definition of a polyhedral product we obtain

$$(\underline{X}, \underline{A})^\emptyset = A_1 \times \cdots \times A_m.$$

Since the face $\{i\}$ consists only of one vertex, from the definition of a polyhedral product we obtain

$$(\underline{X}, \underline{A})^{\{i\}} = A_1 \times \cdots \times A_{i-1} \times X_i \times A_{i+1} \times \cdots \times A_m.$$

As $A_1 \times \cdots \times A_m$ is a subspace of $(\underline{X}, \underline{A})^{\{i\}}$ for all $1 \leq i \leq m$, we obtain

$$(\underline{X}, \underline{A})^K = \bigcup_{i=1}^{m} A_1 \times \cdots \times A_{i-1} \times X_i \times A_{i+1} \times \cdots \times A_m.$$

In particular, if each A_i equals the basepoint, then $A_1 \times \cdots \times A_m = * \times \cdots \times *$, which equals the basepoint in the product. So

$$(\underline{X}, \underline{A})^K = \bigvee_{i=1}^{m} X_i.$$

Example 2.2: Changing the pairs of spaces (X_i, A_i) can have a dramatic effect on the homotopy type of the polyhedral product. For example, if K is two disjoint points then as in Example 2.1 we have

$$(\underline{X}, \underline{A})^K = X_1 \times A_2 \cup A_1 \times X_2$$

where the union is over $A_1 \times A_2$. This may be better represented as a pushout

$$\begin{array}{ccc} A_1 \times A_2 & \longrightarrow & A_1 \times X_2 \\ \downarrow & & \downarrow \\ X_1 \times A_2 & \longrightarrow & (\underline{X}, \underline{A})^K. \end{array}$$

If A_1 and A_2 are the basepoints in X_1 and X_2 respectively, then $(\underline{X}, \underline{*})^K \simeq X_1 \vee X_2$. But if X_1 and X_2 are the reduced cones on A_1 and A_2 respectively, then as we shall see in Section 6, the pushout becomes that for the join

$$\begin{array}{ccc} A_1 \times A_2 & \longrightarrow & A_1 \times CA_2 \\ \downarrow & & \downarrow \\ CA_1 \times A_2 & \longrightarrow & A_1 * A_2. \end{array}$$

That is $(\underline{CA}, \underline{A})^K \simeq A_1 * A_2$.

Example 2.3: At the other extreme, if $K = \Delta^{m-1}$ is the full $(m-1)$-simplex, then $\sigma = (1, \ldots, m)$ is a face of K, so from the definition of a polyhedral product we obtain

$$(\underline{X}, \underline{A})^\sigma = X_1 \times \cdots \times X_m.$$

As $(\underline{X}, \underline{A})^\sigma$ is a subspace of $(\underline{X}, \underline{A})^K$, and $(\underline{X}, \underline{A})^K$ is a subspace of $\prod_{i=1}^{m} X_i$, we obtain

$$(\underline{X}, \underline{A})^{\Delta^{m-1}} = X_1 \times \cdots \times X_m.$$

Note that this holds regardless of the subspaces A_i in the pairs (X_i, A_i).

Example 2.4: Seemingly small changes in the simplicial complex can also result in dramatic changes in the homotopy type of the polyhedral product. Consider the simplicial complexes

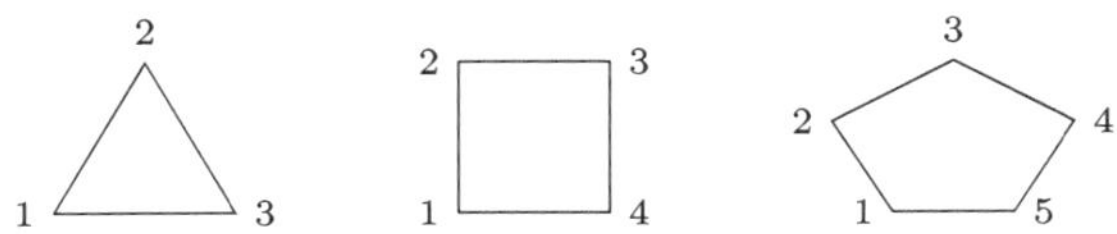

K_1 K_2 K_3

It is not obvious simply from the definition of a polyhedral product, but in these cases it turns out that $\mathcal{Z}_{K_1} \simeq S^5$, $\mathcal{Z}_{K_2} \simeq S^3 \times S^3$ and $\mathcal{Z}_{K_3} \simeq (S^3 \times S^4)^{\#5}$, where $\#$ is the connected sum operation.

3. Fundamental properties of polyhedral products

In this section five fundamental properties of polyhedral products are discussed: functorial properties, retractions of $(\underline{X}, \underline{A})^K$ induced by full subcomplexes of K, how pushouts of simplicial complexes induce pushouts of polyhedral products, fibrations and homotopy fibrations, and an elementary loop space decomposition.

3.1. *Functorial properties*

There are two ingredients in defining a polyhedral product, the sequence of pairs $(\underline{X}, \underline{A})$ and the simplicial complex K. Both have functorial properties.

Proposition 3.1: *The polyhedral product $(\underline{X}, \underline{A})^K$ satisfies the following:*

(a) if there are maps of pairs $(X_i, A_i) \longrightarrow (X'_i, A'_i)$, for $1 \le i \le m$ then for any simplicial complex K on the vertex set $[m]$ there is an induced map of polyhedral products $(\underline{X}, \underline{A})^K \longrightarrow (\underline{X'}, \underline{A'})^K$;

(b) if K and L are simplicial complexes on the vertex sets $[m]$ and $[n]$ respectively and there is an inclusion of simplicial complexes $K \longrightarrow L$, then for any sequence $(\underline{X}, \underline{A})$ of n pairs of pointed, path-connected CW-complexes there is an induced map of polyhedral products $(\underline{X}, \underline{A})^K \longrightarrow (\underline{X}, \underline{A})^L$.

Proof: Let σ be a face of K and recall that $(\underline{X}, \underline{A})^\sigma = \prod_{i=1}^m Y_i$ where $Y_i = X_i$ if $i \in \sigma$ and $Y_i = A_i$ if $i \notin \sigma$. The maps of pairs $(X_i, A_i) \longrightarrow (X'_i, A'_i)$

therefore induce a map of products

$$(\underline{X}, \underline{A})^\sigma = \prod_{i=1}^{m} Y_i \longrightarrow \prod_{i=1}^{m} Y_i' = (\underline{X}', \underline{A}')^\sigma.$$

Taking the union over all $\sigma \in K$ we obtain a map

$$(\underline{X}, \underline{A})^K = \bigcup_{\sigma \in K} (\underline{X}, \underline{A})^\sigma \longrightarrow \bigcup_{\sigma \in K} (\underline{X}', \underline{A}')^\sigma = (\underline{X}', \underline{A}')^K.$$

This proves part (a).

For part (b), let σ be a face of K. Since the map of simplicial complexes $f\colon K \longrightarrow L$ is an inclusion, σ is also a face of L. Write σ_K, σ_L for σ viewed as a face of K or L respectively. The map f sends the vertex $i \in [m]$ to $f(i) \in [n]$, so we obtain a map of products

$$(\underline{X}, \underline{A})^{\sigma_K} = \prod_{i=1}^{m} Y_i \longrightarrow \prod_{i=1}^{m} Y_{f(i)} \hookrightarrow \prod_{j=1}^{n} Y_j = (\underline{X}, \underline{A})^{\sigma_L}.$$

Taking the union over all the faces of K, we obtain

$$(\underline{X}, \underline{A})^K = \bigcup_{\sigma_K \in K} (\underline{X}, \underline{A})^{\sigma_K} \longrightarrow \bigcup_{\sigma_K \in K} (\underline{X}, \underline{A})^{\sigma_L} \hookrightarrow \bigcup_{\sigma_L \in L} (\underline{X}, \underline{A})^{\sigma_L} = (\underline{X}, \underline{A})^L.$$

□

Example 3.2: Regarding $\mathbb{C}P^1$ as S^2, there is an inclusion $S^2 \longrightarrow \mathbb{C}P^\infty$. This induces a map of pairs $(S^2, *) \longrightarrow (\mathbb{C}P^\infty, *)$. Taking m copies of the same pair, by Proposition 3.1 (a) and the definition of DJ_K, there is a map of polyhedral products $(\underline{S^2}, \underline{*})^K \longrightarrow (\underline{\mathbb{C}P^\infty}, \underline{*})^K = DJ_K$.

Example 3.3: Let L be a simplicial complex on the vertex set m and let K be the set of vertices of L. Then there is an inclusion of simplicial complexes $K \longrightarrow L$. In the case of $(\underline{X}, \underline{*})$, by Example 2.1, $(\underline{X}, \underline{*})^K = \bigvee_{i=1}^m X_i$. So by Proposition 3.1 (b) we obtain a map of polyhedral products $\bigvee_{i=1}^m X_i \longrightarrow (\underline{X}, \underline{*})^L$.

3.2. *Retractions of* $(\underline{X}, \underline{A})^K$ *induced by full subcomplexes of* K

There are distinguished subcomplexes of a simplicial complex K which play a special role in the homotopy theory of polyhedral products.

Definition 3.4: Let K be a simplicial complex on the vertex set $[m]$. If $I \subseteq [m]$ then the *full subcomplex* K_I of K is defined as the simplicial complex

$$K_I = \bigcup \{\sigma \in K \mid \text{the vertex set of } \sigma \text{ is in } I\}.$$

Example 3.5: Let K be the simplical complex on 4 vertices pictured as follows:

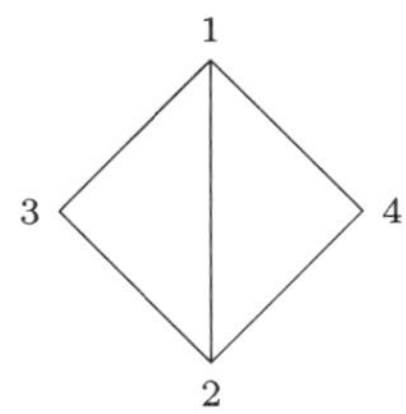

K

If $I_1 = \{1, 2, 3\}$, $I_2 = \{1, 3, 4\}$ and $I_3 = \{3, 4\}$ then the full subcomplexes K_{I_1}, K_{I_2} and K_{I_3} are:

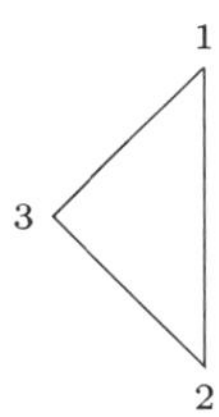

K_{I_1}

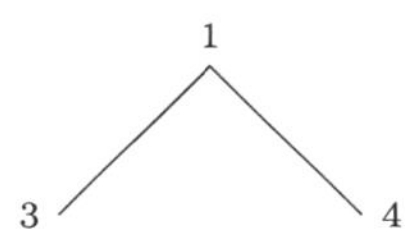

K_{I_2}

K_{I_3}

The definition of K_I implies that the inclusion of K_I into K is a map of simplicial complexes. So by Proposition 3.1 (b) there is a map of polyhedral products $(\underline{X}, \underline{A})^{K_I} \longrightarrow (\underline{X}, \underline{A})^K$. However, projecting from $[m]$ to I does not induce a map of simplicial complexes $K \longrightarrow K_I$. Yet on the level of polyhedral products, such a projection does exist. If $I = \{i_1, \ldots, i_k\}$ for $1 \leq i_1 < \cdots < i_k \leq m$, let $X^I = \prod_{j=1}^k X_{i_j}$.

Proposition 3.6: *Let K be a simplicial complex on the vertex set $[m]$ and let $(\underline{X}, \underline{A})$ be any sequence of pointed, path-connected CW-pairs. Let $I \subseteq [m]$. Then the following hold:*

(a) the inclusion $X^I \longrightarrow \prod_{i=1}^m X_i$ induces a map $(\underline{X}, \underline{A})^{K_I} \longrightarrow (\underline{X}, \underline{A})^K$ that is the same as that induced by the inclusion of simplicial complexes $K_I \longrightarrow K$;
(b) the projection $\prod_{i=1}^m X_i \longrightarrow X^I$ induces a map $(\underline{X}, \underline{A})^K \longrightarrow (\underline{X}, \underline{A})^{K_I}$;
(c) the composite $(\underline{X}, \underline{A})^{K_I} \longrightarrow (\underline{X}, \underline{A})^K \longrightarrow (\underline{X}, \underline{A})^{K_I}$ is the identity map.

Proof: Let $\bar{f}\colon I \longrightarrow [m]$ be the inclusion. Then $\bar{f}$ induces a map of products $\widetilde{f}\colon X^I \longrightarrow \prod_{i=1}^m X_i$ by sending the ith-factor of X^I to the $\bar{f}(i)$-factor of $\prod_{i=1}^m X_i$. For $\sigma \in K_I$, write σ_{K_I}, σ_K for σ viewed as a face of K_I or K respectively. Write $|I|$ for the cardinality of I. Define a map of products by

$$(\underline{X}, \underline{A})^{\sigma_{K_I}} = \prod_{i=1}^{|I|} Y_i \longrightarrow \prod_{i=1}^{|I|} Y_{\bar{f}(i)} \hookrightarrow \prod_{j=1}^{m} Y_j = (\underline{X}, \underline{A})^{\sigma_K}.$$

Taking the union over all faces in K_I, then exactly as in the proof of Proposition 3.1 (b), we obtain a map of polyhedral products

$$(\underline{X}, \underline{A})^{K_I} \longrightarrow (\underline{X}, \underline{A})^{K}.$$

Notice that this is exactly the same construction as in the proof of Proposition 3.1 (b), so the map just produced is the same as that induced by the inclusion of simplicial complexes $K_I \longrightarrow K$.

Next, for $I \subseteq [m]$ there is a projection

$$\prod_{i=1}^{m} X_i \longrightarrow \prod_{j=1}^{|I|} X_j.$$

For any subspace $(\underline{X}, \underline{A})^{\sigma} = \prod_{i=1}^m Y_i$ of $\prod_{i=1}^m X_i$, we obtain a projection onto $\prod_{j=1}^{|I|} Y_j$. Notice that if σ is a face of K_I then $\prod_{j=1}^{|I|} Y_i = (\underline{X}, \underline{A})^{\sigma_{K_I}}$, and if σ is not a face of K_I then $\tau = \sigma \cap K_I$ is a face of K_I, so the projection induces a map from $(\underline{X}, \underline{A})^{\sigma_K}$ to $(\underline{X}, \underline{A})^{\tau}$. Taking the union over all the faces in K we obtain a map

$$(\underline{X}, \underline{A})^{K} \longrightarrow (\underline{X}, \underline{A})^{K_I}.$$

Finally, as the composite $X^I \longrightarrow \prod_{i=1}^m X_i \longrightarrow X^I$ is the identity map, the constructions of the maps in the composite $(\underline{X}, \underline{A})^{\sigma_{K_I}} \longrightarrow (\underline{X}, \underline{A})^{\sigma_K} \longrightarrow (\underline{X}, \underline{A})^{\sigma_{K_I}}$ imply that this composite is the identity map. Taking the union over all $\sigma \in K_I$ shows that the composite $(\underline{X}, \underline{A})^{K_I} \longrightarrow (\underline{X}, \underline{A})^{K} \longrightarrow (\underline{X}, \underline{A})^{K_I}$ is the identity map. □

Proposition 3.6 is often very valuable in helping to determine features of the homotopy types of polyhedral products.

Example 3.7: Return to Example 3.5. The full subcomplex I_3 of K consists of two vertices. So by Example 2.2, $(\underline{CX}, \underline{X})^{I_3} \simeq X_3 * X_4$. Therefore, by Proposition 3.6 (c), $X_3 * X_4$ retracts off of $(\underline{CX}, \underline{X})^{K}$. In particular, in the case of the moment-angle complex $\mathcal{Z}_K$, we have $X_3 = X_4 = \Omega\mathbb{C}P^{\infty} \simeq S^1$, so we obtain a retraction of S^3 off of $\mathcal{Z}_K$.

3.3. *Pushouts of simplicial complexes induce pushouts of polyhedral products*

Let K be a simplicial complex on the vertex set $[m]$ and suppose that there is a pushout of simplicial complexes

$$\begin{array}{ccc} L & \longrightarrow & K_2 \\ \downarrow & & \downarrow \\ K_1 & \longrightarrow & K. \end{array}$$

That is, $K = K_1 \cup_L K_2$. To compare the polyhedral products for L, K_1, K_2 and K we should really consider each of the four simplicial complexes as being on the same vertex set $[m]$. This may introduce "ghost" vertices if, for example, L has fewer vertices than K. Let $\overline{L}$, $\overline{K}_1$ and $\overline{K}_2$ be the simplicial complexes L, K_1 and K_2, regarded as simplicial complexes on the vertex set $[m]$. Then $K = \overline{K}_1 \cup_{\overline{L}} \overline{K}_2$.

Proposition 3.8: *Let K be a simplicial complex on the vertex set $[m]$. Suppose that there is a pushout of simplicial complexes*

$$\begin{array}{ccc} L & \longrightarrow & K_2 \\ \downarrow & & \downarrow \\ K_1 & \longrightarrow & K \end{array}$$

Then there is a pushout of polyhedral products

$$\begin{array}{ccc} (\underline{X}, \underline{A})^{\overline{L}} & \longrightarrow & (\underline{X}, \underline{A})^{\overline{K}_2} \\ \downarrow & & \downarrow \\ (\underline{X}, \underline{A})^{\overline{K}_1} & \longrightarrow & (\underline{X}, \underline{A})^{K}. \end{array}$$

Proof: Since $K = K_1 \cup_L K_2$ and K is finite, the simplices in K can be put into three finite collections: $(\mathcal{A})$ the simplices in L, $(\mathcal{B})$ the simplices in K_1 that are not simplices of L and $(\mathcal{C})$ the simplices of K_2 that are not simplices of L. Then we have

$$L = \bigcup_{\sigma\in\mathcal{A}} \sigma$$

$$K_1 = \left(\bigcup_{\sigma\in\mathcal{A}} \sigma\right) \cup \left(\bigcup_{\sigma'\in\mathcal{B}} \sigma'\right)$$

$$K_2 = \left(\bigcup_{\sigma\in\mathcal{A}} \sigma\right) \cup \left(\bigcup_{\sigma''\in\mathcal{C}} \sigma''\right)$$

$$K = \left(\bigcup_{\sigma\in\mathcal{A}} \sigma\right) \cup \left(\bigcup_{\sigma'\in\mathcal{B}} \sigma'\right) \cup \left(\bigcup_{\sigma''\in\mathcal{C}} \sigma''\right).$$

By definition, for any simplicial complex M on the vertex set $[m]$, $(\underline{X},\underline{A})^M = \bigcup_{\sigma\in M}(\underline{X},\underline{A})^\sigma$. So in our case, we have

$$(\underline{X},\underline{A})^{\overline{L}} = \bigcup_{\sigma\in\mathcal{A}} (\underline{X},\underline{A})^\sigma$$

$$(\underline{X},\underline{A})^{\overline{K_1}} = \left(\bigcup_{\sigma\in\mathcal{A}} (\underline{X},\underline{A})^\sigma\right) \cup \left(\bigcup_{\sigma'\in\mathcal{B}} (\underline{X},\underline{A})^{\sigma'}\right)$$

$$(\underline{X},\underline{A})^{\overline{K_2}} = \left(\bigcup_{\sigma\in\mathcal{A}} (\underline{X},\underline{A})^\sigma\right) \cup \left(\bigcup_{\sigma''\in\mathcal{C}} (\underline{X},\underline{A})^{\sigma''}\right)$$

$$(\underline{X},\underline{A})^{K} = \left(\bigcup_{\sigma\in\mathcal{A}} (\underline{X},\underline{A})^\sigma\right) \cup \left(\bigcup_{\sigma'\in\mathcal{B}} (\underline{X},\underline{A})^{\sigma'}\right) \cup \left(\bigcup_{\sigma''\in\mathcal{C}} (\underline{X},\underline{A})^{\sigma''}\right).$$

In particular, since $(\underline{X},\underline{A})^{\overline{L}} = (\underline{X},\underline{A})^{\overline{K_1}} \cap (\underline{X},\underline{A})^{\overline{K_2}}$ we have

$$(\underline{X},\underline{A})^{K} = (\underline{X},\underline{A})^{\overline{K_1}} \cup_{(\underline{X},\underline{A})^{\overline{L}}} (\underline{X},\underline{A})^{\overline{K_2}}$$

which implies the existence of the asserted pushout. □

Example 3.9: Let $L = \emptyset$, and let $K_1 = \{1\}$ and $K_2 = \{2\}$ be single points. Then $K = K_1 \cup_L K_2$ is two disjoint points. By definition of the polyhedral product, $(\underline{X},\underline{A})^{\overline{L}} = A_1 \times A_2$, $(\underline{X},\underline{A})^{\overline{K_1}} = X_1 \times A_2$, $(\underline{X},\underline{A})^{\overline{K_2}} = A_1 \times X_2$, and $(\underline{X},\underline{A})^{K} = X_1 \times A_2 \cup_{A_1\times A_2} A_1 \times X_2$. So $(\underline{X},\underline{A})^{K}$ is the pushout of $(\underline{X},\underline{A})^{\overline{K_1}}$ and $(\underline{X},\underline{A})^{\overline{K_2}}$ over $(\underline{X},\underline{A})^{\overline{L}}$. Note that this is exactly the same as Example 2.2.

3.4. *Fibrations and homotopy fibrations*

The next fundamental property is how the polyhedral product behaves with respect to fibrations of pairs. Proposition 3.10 was proved in [11] in the special case of m copies of the same pairs, and stated in the more general case of m different pairs in [1].

Proposition 3.10: *Let K be a simplicial complex on the vertex set $[m]$. Suppose that for $1 \leq i \leq m$ there are maps of pairs $p_i\colon (E_i, E_i') \longrightarrow (B_i, B_i)$ such that the restrictions $E_i \longrightarrow B_i$ and $E_i' \longrightarrow B_i$ are fibrations with fibres F_i and F_i' respectively. Then there is a fibration*

$$(\underline{F}, \underline{F'})^K \longrightarrow (\underline{E}, \underline{E'})^K \longrightarrow \prod_{i=1}^{m} B_i.$$

Proof: By Proposition 3.1 (a), the maps of pairs $(E_i, E_i') \longrightarrow (F_i, F_i')$ induce a map of polyhedral products $f\colon (\underline{E}, \underline{E'})^K \longrightarrow (\underline{B}, \underline{B})^K$. The definition of the polyhedral product implies that $(\underline{B}, \underline{B})^K = B_1 \times \cdots \times B_m$.

First consider the fibre F of f, that is, the collection of points in $(\underline{E}, \underline{E'})^K$ that are sent to the basepoint $(*, \ldots, *) \in B_1 \times \cdots \times B_m$. Notice that for any $\sigma \in K$, f restricts to a product map

$$f^\sigma : (\underline{E}, \underline{E'})^\sigma = \prod_{i=1}^{m} Y_i \longrightarrow \prod_{i=1}^{m} B_i = (\underline{B}, \underline{B})^\sigma$$

where $Y_i = E_i$ if $i \in K$ and $Y_i = E_i'$ if $i \notin K$. Thus the fibre of f^σ is the product of the fibres of the maps in the m coordinates. That is, the fibre is $\prod_{i=1}^m Z_i$ where $Z_i = F_i$ if $i \in \sigma$ and $Z_i = F_i'$ if $i \notin \sigma$. In other words, the fibre is exactly $(\underline{F}, \underline{F'})^\sigma$. Now for the fibre of f as a whole we have:

$$\begin{aligned}
F &= \{(x_1, \ldots, x_m) \in (\underline{E}, \underline{E'})^K \mid f(x_1, \ldots, x_m) = (*, \ldots, *)\} \\
&= \{(x_1, \ldots, x_m) \in \bigcup_{\sigma \in K} (\underline{E}, \underline{E'})^\sigma \mid f(x_1, \ldots, x_m) = (*, \ldots, *)\} \\
&= \bigcup_{\sigma \in K} \{(x_1, \ldots, x_m) \in (\underline{E}, \underline{E'})^\sigma \mid f(x_1, \ldots, x_m) = (*, \ldots, *)\} \\
&= \bigcup_{\sigma \in K} (\underline{F}, \underline{F'})^\sigma \\
&= (\underline{F}, \underline{F'})^K.
\end{aligned}$$

It remains to show that the map $(\underline{E}, \underline{E'})^K \longrightarrow B_1 \times \cdots \times B_m$ has the homotopy lifting property. But each of the pairs $(E_i, E_i') \longrightarrow (B_i, B_i)$ is a fibration and so has the homotopy lifting property. A product of fibrations is a fibration, so the map $(\underline{E}, \underline{E'})^\sigma \xrightarrow{f^\sigma} B_1 \times \cdots \times B_m$ has the homotopy lifting property. Finally, since the map $(\underline{E}, \underline{E'})^K \longrightarrow B_1 \times \cdots \times B_m$ is the union of the maps f^σ, it too has the homotopy lifting property. $\square$

We will now use Proposition 3.10 to produce explicit fibrations between polyhedral products. The argument roughly follows [1]. For a pointed, path-connected space X let PX be the *path space* of X. That is, if x_0 is the basepoint of X and 0 is the basepoint of the unit interval $[0,1]$, then PX is the space of all pointed, continuous maps $f\colon [0,1] \longrightarrow X$ such that $f(0) = x_0$. The evaluation map $e_1\colon PX \longrightarrow X$ is defined by $e_1(f) = f(1)$. It is well known that there is a fibration

$$\Omega X \longrightarrow PX \xrightarrow{e_1} X.$$

Consider the pair $(X \times PX, PX)$ where PX is the subspace of $X \times PX$ given by pairs $(f(1), f)$. Observe that there is a fibration of pairs

$$(PX, \Omega X) \longrightarrow (X \times PX, PX) \xrightarrow{\pi_X \times e_1} (X \times X)$$

where π_X is the projection. Proposition 3.10 therefore implies the following.

Proposition 3.11: *Let K be a simplicial complex on the vertex set $[m]$ and let $X_1, \ldots, X_m$ be pointed, path-connected spaces. Then there is a fibration*

$$(\underline{PX}, \underline{\Omega X})^K \longrightarrow (\underline{X \times PX}, \underline{PX})^K \xrightarrow{(\pi_X \times e_1)^K} \prod_{i=1}^{m} X_i.$$

□

Next, observe that there is an equivalence of pairs

$$(X, *) \xrightarrow{(\kappa, *)} (X \times PX, PX)$$

where κ is defined by $\kappa(x) = (x, f_x)$ for $f_x\colon [0,1] \longrightarrow X$ the constant map $f_x(t) = x$ for all $t \in [0,1]$. Consequently, for any simplicial complex K, there is a homotopy equivalence $(\underline{X}, \underline{*})^K \simeq (\underline{X \times PX}, \underline{PX})^K$. Thus from the fibration in Proposition 3.11 we obtain the following.

Corollary 3.12: *Let K be a simplicial complex on the vertex set $[m]$ and let $X_1, \ldots, X_m$ be pointed, path-connected spaces. Then there is a homotopy fibration*

$$(\underline{PX}, \underline{\Omega X})^K \longrightarrow (\underline{X}, \underline{*})^K \longrightarrow \prod_{i=1}^{m} X_i.$$

□

Example 3.13: Take each $X_i = \mathbb{C}P^\infty$. Then $(\underline{X}, \underline{*})^K = DJ_K$, implying that $(\underline{PX}, \underline{X})^K \simeq \mathcal{Z}_K$.

Finally, it is common to replace path spaces with cones. For a pointed, path-connected space X, let CX be the *reduced cone* on X, defined by

$$CX = (X \times I)/\sim$$

where $(x,1) \sim (x',1)$ for all $x, x' \in X$ and $(x_0, t) \sim (x_0, 0)$ for all $t \in [0,1]$. Observe that there is a homotopy equivalence of pairs $(CPX, \Omega X) \longrightarrow (PX, \Omega X)$ given by dilating the cone on PX. Notice that the inclusion of ΩX into CPX factors through $C\Omega X$, and the dilation of the cone induces a homotopy equivalence of pairs $(C\Omega X, \Omega X) \longrightarrow (CPX, \Omega X)$. Thus there is a homotopy equivalence of pairs $(C\Omega X, \Omega X) \longrightarrow (PX, \Omega X)$. So Corollary 3.12 can be rewritten as follows.

Corollary 3.14: *Let K be a simplicial complex on the vertex set $[m]$ and let $X_1, \ldots, X_m$ be pointed, path-connected spaces. Then there is a homotopy fibration*

$$(\underline{C\Omega X}, \underline{\Omega X})^K \longrightarrow (\underline{X}, \underline{*})^K \longrightarrow \prod_{i=1}^{m} X_i.$$

□

Example 3.15: Including the wedge into the product, there is a homotopy fibration

$$F_m \longrightarrow \bigvee_{i=1}^{m} X_i \longrightarrow \prod_{i=1}^{m} X_i$$

which defines the space F_m. This homotopy fibration can be modelled by polyhedral products. Let K be m disjoint points. By Example 2.1, $(\underline{X}, \underline{*})^K = \bigvee_{i=1}^{m} X_i$, so the inclusion of the wedge into the product is modelled by the map $(\underline{X}, \underline{*})^K \longrightarrow \prod_{i=1}^{m} X_i$. By Corollary 3.14, the homotopy fibre of this inclusion is $(\underline{C\Omega X}, \underline{\Omega X})^K$. That is, $F_m \simeq (\underline{C\Omega X}, \underline{\Omega X})^K$. In Proposition 7.1, the homotopy type of $(\underline{C\Omega X}, \underline{\Omega X})^K$ will be determined.

3.5. *An elementary loop space decomposition*

Let i be a vertex in K. Including $\{i\}$ into K gives a map of simplicial complexes $\{i\} \longrightarrow K$. This induces a map of polyhedral products $(\underline{X}, \underline{*})^{\{i\}} \longrightarrow (\underline{X}, \underline{*})^K$. By the definition of the polyhedral product, $(\underline{X}, \underline{*})^{K_{\{i\}}} = X_i$. So the previous map of polyhedral products is just the inclusion of the ith-coordinate space, $X_i \longrightarrow (\underline{X}, \underline{*})^K$. Therefore, when included into $\prod_{i=1}^{m} X_i$ we obtain the inclusion of the ith factor. In general, $(\underline{X}, \underline{*})^K$ is not an

H-space so we cannot take the product of these coordinate inclusions. But after looping there is a product, so multiplying together the maps $\Omega X_i \longrightarrow \Omega(\underline{X},\underline{*})^K$ for $1 \leq i \leq m$ we obtain a right homotopy inverse for the map $\Omega(\underline{X},\underline{*})^K \longrightarrow \prod_{i=1}^{m} \Omega X_i$, implying the following.

Lemma 3.16: *The homotopy fibration in Corollary 3.14 splits after looping, resulting in a homotopy equivalence*

$$\Omega(\underline{X},\underline{*})^K \simeq (\prod_{i=1}^{m} \Omega X_i) \times \Omega(\underline{C\Omega X},\underline{\Omega X})^K.$$

□

Example 3.17: Let K be two disjoint points. By Example 2.1, $(\underline{X},\underline{*})^K = X_1 \vee X_2$, and by Example 2.2, $(\underline{C\Omega X},\underline{\Omega X})^K \simeq \Omega X_1 * \Omega X_2$. In this case Lemma 3.16 recovers a result of Ganea [14],

$$\Omega(X_1 \vee X_2) \simeq \Omega X_1 \times \Omega X_2 \times \Omega(\Omega X_1 * \Omega X_2).$$

An important special case is for Davis-Januszkiewicz spaces and moment-angle complexes. Noting that $\Omega\mathbb{C}P^\infty \simeq S^1$, Example 3.13 and Lemma 3.16 combine to immediately give the following.

Corollary 3.18: *Let K be a simplicial complex on the vertex set $[m]$. Then the homotopy fibration $\mathcal{Z}_K \longrightarrow DJ_K \longrightarrow \prod_{i=1}^{m} \mathbb{C}P^\infty$ splits after looping, giving a homotopy equivalence*

$$\Omega DJ_K \simeq (\prod_{i=1}^{m} S^1) \times \Omega\mathcal{Z}_K.$$

□

4. The homotopy type of $\Sigma(\underline{X},\underline{A})^K$

This section considers the homotopy type of $\Sigma(\underline{X},\underline{A})^K$. It turns out that this decomposes very neatly as a wedge of summands. The basic idea is to generalize the wedge decomposition for $\Sigma(X \times Y)$. Recall the following.

Lemma 4.1: *Let X and Y be path-connected, pointed CW-complexes. Then there is a homotopy equivalence*

$$\Sigma(X \times Y) \simeq \Sigma X \vee \Sigma Y \vee (\Sigma X \wedge Y).$$

Proof: Including the wedge into the product there is a cofibration

$$X \vee Y \longrightarrow X \times Y \longrightarrow X \wedge Y.$$

Notice that the composite $X \vee Y \longrightarrow X \times Y \xrightarrow{\text{proj}} X$ equals the pinch map onto the left wedge summand, and $X \vee Y \longrightarrow X \times Y \xrightarrow{\text{proj}} Y$ is the pinch map onto the right wedge summand. As $X \times Y$ is not a co-H-space, the two projections cannot be added to produce a left homotopy inverse for the inclusion $X \vee Y \longrightarrow X \times Y$, but after suspending this can be done. Thus the homotopy cofibration $\Sigma X \vee \Sigma Y \longrightarrow \Sigma(X \times Y) \longrightarrow \Sigma X \wedge Y$ splits to give a homotopy equivalence $\Sigma(X \times Y) \simeq \Sigma X \vee \Sigma Y \vee (\Sigma X \wedge Y)$. □

Iterating gives the following.

Lemma 4.2: *Let $X_1, \ldots, X_m$ be path-connected, pointed CW-complexes. Then there is a homotopy equivalence*

$$\Sigma(X_1 \times \cdots \times X_m) \simeq \bigvee_{1 \leq i_1 < \cdots i_k \leq m} \Sigma X_{i_1} \wedge \cdots \wedge X_{i_k}.$$

□

In terms of polyhedral products, $X_1 \times \cdots \times X_m = (\underline{X}, \underline{A})^{\Delta^{m-1}}$. Notice that the faces of Δ^{m-1} are in one-to-one correspondence with the sequences $(i_1, \ldots, i_k)$ for $1 \leq i_1 < \cdots < i_k \leq m$, and each such face corresponds to the polyhedral product $X_{i_1} \times \cdots \times X_{i_k}$. Thus, if $\sigma = (i_1, \ldots, i_k)$ is a face of K and we write

$$\underline{X}^{\wedge\sigma} = X_{i_1} \wedge \cdots \wedge X_{i_k},$$

then the homotopy equivalence in Lemma 4.2 can be rewritten as

$$\Sigma(X_1 \times \cdots \times X_m) \simeq \bigvee_{\sigma \in K} \Sigma \underline{X}^{\wedge\sigma}. \tag{4.1}$$

Therefore it could be expected that, at least in the case $(\underline{X}, \underline{*})^K$, we should have $\Sigma(\underline{X}, \underline{*})^K$ consisting of those wedge summands in $\Sigma(X_1 \times \cdots \times X_m)$ that correspond to the faces of K. In fact, this is exactly what happens.

Proposition 4.3: *Let K be a simplicial complex on the vertex set $[m]$ and let $X_1, \ldots, X_m$ be path-connected, pointed CW-complexes. Then there is a homotopy equivalence*

$$\Sigma(\underline{X}, \underline{*})^K \simeq \bigvee_{\sigma \in K} \Sigma \underline{X}^{\wedge\sigma}.$$

Proof: Let $\sigma = (i_1, \ldots, i_k)$ be a face of K. By definition of the polyhedral product

$$(\underline{X}, \underline{*})^\sigma = \prod_{i=1}^{m} Y_i$$

where $Y_i = X_i$ if $i \in \sigma$ and $Y_i = *$ if $i \notin \sigma$. That is,

$$(\underline{X}, \underline{*})^\sigma = \prod_{j=1}^{k} X_{i_j}.$$

Therefore, by (4.1) there is a homotopy equivalence

$$\Sigma(\underline{X}, \underline{*})^\sigma \simeq \bigvee_{\tau \in \sigma} \Sigma \underline{X}^{\wedge \tau}.$$

On the other hand, observe that any proper sub-product of $(\underline{X}, \underline{*})^\sigma = \prod_{j=1}^{k} X_{i_j}$ is the polyhedral product of a corresponding proper sub-face of K. Write $\partial\sigma$ for the union of proper sub-faces of σ. Then there is an identification of quotient spaces

$$\begin{aligned}(\underline{X}, \underline{*})^\sigma / (\underline{X}, \underline{*})^{\partial\sigma} &= (\prod_{j=1}^{k} X_{i_j}) / (\text{union of proper sub-products}) \\ &= X_{i_1} \wedge \cdots \wedge X_{i_k}.\end{aligned}$$

By Lemma 4.2, the quotient map $(\underline{X}, \underline{*})^\sigma = X_{i_1} \times \cdots \times X_{i_k} \longrightarrow X_{i_1} \wedge \cdots \wedge X_{i_k}$ has a right homotopy inverse after suspending. Thus there is a homotopy equivalence

$$\Sigma(\underline{X}, \underline{*})^\sigma \simeq \Sigma(\underline{X}, \underline{*})^{\partial\sigma} \vee (\Sigma X_{i_1} \wedge \cdots \wedge X_{i_k}) = \Sigma(\underline{X}, \underline{*})^{\partial\sigma} \vee \Sigma \underline{X}^{\wedge \sigma}. \quad (4.2)$$

Comparing the two homotopy equivalences for $\Sigma(\underline{X}, \underline{*})^\sigma$ we obtain

$$\Sigma(\underline{X}, \underline{*})^{\partial\sigma} \simeq \bigvee_{\tau \in \partial\sigma} \Sigma \underline{X}^{\wedge \tau}. \quad (4.3)$$

Now consider the string of equivalences

$$\begin{aligned}\Sigma(\underline{X},\underline{*})^K &= \Sigma\left(\bigcup_{\sigma\in K}(\underline{X},\underline{*})^\sigma\right)\\ &\simeq \bigcup_{\sigma\in K}\Sigma(\underline{X},\underline{*})^\sigma\\ &\simeq \bigcup_{\sigma\in K}\left(\Sigma(\underline{X},\underline{*})^{\partial\sigma}\vee(\Sigma X_{i_1}\wedge\cdots\wedge X_{i_k})\right)\\ &= \bigcup_{\sigma\in K}\Sigma X_{i_1}\wedge\cdots\wedge X_{i_k}\\ &\simeq \bigvee_{\sigma\in K}\Sigma X_{i_1}\wedge\cdots\wedge X_{i_k}\end{aligned}$$

The first equality holds by definition of the polyhedral product, the second holds because the reduced suspension over a union is homotopy equvalent to the union of the reduced suspensions over each term in the union, the third holds by (4.2), the fourth holds because the decomposition for $\Sigma(\underline{X},\underline{*})^{\partial\sigma}$ in (4.3) is in terms of proper sub-faces of σ which are otherwise accounted for in the union, and the fifth holds because the terms in the fourth line intersect only at the basepoint. The string of equivalences as a whole gives the asserted homotopy equivalence. □

Example 4.4: Let K be the following simplical complex on 4 vertices:

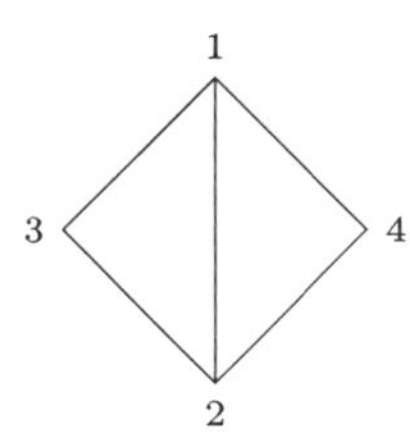

K

The faces of K are the vertices $\{1\}$, $\{2\}$, $\{3\}$, $\{4\}$ and the edges $(1,2)$, $(1,3)$, $(1,4)$, $(2,3)$ and $(2,4)$. For a vertex $\{i\}$ we have $\underline{X}^{\{i\}} = X_i$ and for an edge (i,j) we have $\underline{X}^{(i,j)} = X_i \wedge X_j$. So by Proposition 4.3 there is a homotopy equivalence

$$\begin{aligned}(\underline{X},\underline{*})^K \simeq{}& \Sigma X_1\vee\Sigma X_2\vee\Sigma X_3\vee\Sigma X_4\vee(\Sigma X_1\wedge X_2)\vee\\ &(\Sigma X_1\wedge X_3)\vee(\Sigma X_1\wedge X_4)\vee(\Sigma X_2\wedge X_3)\vee(\Sigma X_2\wedge X_4).\end{aligned}$$

Proving a decomposition for $\Sigma(\underline{X},\underline{A})^K$ in the general case is a bit more involved because the subspaces A_i of X_i need not be points. But the idea behind the argument remains the same. Instead of proving the general statement we will just state it. This requires some notation.

Let K be a simplicial complex on the vertex set $[m]$ and for $1 \le i \le m$ let (X_i, A_i) be pairs of pointed, path-connected CW-complexes. Recall that

$$(\underline{X},\underline{A})^\sigma = \prod_{i=1}^{m} Y_i \qquad \text{where} \qquad Y_i = \begin{cases} X_i \text{ if } i \in K \\ A_i \text{ if } i \notin K \end{cases}$$

and the polyhedral product is

$$(\underline{X},\underline{A})^K = \bigcup_{\sigma \in K} (\underline{X},\underline{A})^\sigma.$$

For $\sigma \in K$, let

$$\widehat{(\underline{X},\underline{A})}^\sigma = \bigwedge_{i=1}^{m} Y_i \qquad \text{where} \qquad Y_i = \begin{cases} X_i \text{ if } i \in K \\ A_i \text{ if } i \notin K \end{cases}$$

The *smash polyhedral product* is

$$\widehat{(\underline{X},\underline{A})}^K = \bigcup_{\sigma \in K} \widehat{(\underline{X},\underline{A})}^\sigma.$$

Another way to think of this is to recall that $(\underline{X},\underline{A})^K$ is a subspace of the product $\prod_{i=1}^m X_i$, and $\widehat{(\underline{X},\underline{A})}^K$ is the image of $(\underline{X},\underline{A})^K$ under the natural quotient map $\prod_{i=1}^m X_i \longrightarrow \bigwedge_{i=1}^m X_i$.

Let $I \subset [m]$. By Proposition 3.6, there is a map of polyhedral products $(\underline{X},\underline{A})^K \longrightarrow (\underline{X},\underline{A})^{K_I}$. Composing, we obtain a map into a smash polyhedral product,

$$(\underline{X},\underline{A})^K \longrightarrow (\underline{X},\underline{A})^{K_I} \longrightarrow \widehat{(\underline{X},\underline{A})}^{K_I}.$$

Such a composition exists for each $I \subseteq [m]$. Suspending, these can be added, giving a composition

$$\overline{H}\colon \Sigma(\underline{X},\underline{A})^K \longrightarrow \bigvee_{I \subseteq [m]} \Sigma(\underline{X},\underline{A})^{K_I} \longrightarrow \bigvee_{I \subseteq [m]} \Sigma\widehat{(\underline{X},\underline{A})}^{K_I}.$$

Bahri, Bendersky, Cohen and Gitler [1] proved the following.

Theorem 4.5: *Let K be a simplicial complex on the vertex set $[m]$ and for $1 \le i \le m$ let (X_i, A_i) be pairs of pointed, path-connected CW-complexes.*

The map

$$\overline{H}\colon \Sigma(\underline{X},\underline{A})^K \longrightarrow \bigvee_{I\subseteq[m]} \Sigma\widehat{(\underline{X},\underline{A})}^{K_I}$$

is a homotopy equivalence. □

Further, in the special case when each X_i is contractible, Bahri, Bendersky, Cohen and Gitler show that: (i) if $I \in K$ then the smash polyhedral product $\widehat{(\underline{X},\underline{A})}^{K_I}$ is contractible [1] ; and (ii) if $I \notin K$ and $I = (i_1, \ldots, i_k)$ then there is a homotopy equivalence $\widehat{(\underline{X},\underline{A})}^{K_I} \simeq \Sigma(|K_I| \wedge \widehat{A}^I)$ where $\widehat{A}^I = \bigwedge_{j=1}^k A_{i_j}$ [1] . Changing notation a bit, write the pairs as (CX_i, X_i), where CX_i is the reduced cone on X_i. Now $\widehat{(\underline{CX},\underline{X})}^{K_I} \simeq \Sigma(|K_I| \wedge \widehat{X}^I)$ and we obtain the following.

Theorem 4.6: *Let K be a simplicial complex on the vertex set $[m]$ and for $1 \le i \le m$ let X_i be a pointed, path-connected CW-complex. Then the map $\overline{H}$ specializes to a homotopy equivalence*

$$H\colon \Sigma(\underline{CX},\underline{X})^K \longrightarrow \bigvee_{I\notin K} \Sigma\widehat{(\underline{CX},\underline{X})}^{K_I} \xrightarrow{\simeq} \bigvee_{I\notin K} \Sigma^2(|K_I| \wedge \widehat{X}^I).$$

□

For the moment-angle complex $\mathcal{Z}_K$, each pair is (D^2, S^1), which is homeomorphic to the pair (CS^1, S^1). As each $X_i = S^1$, we have $\widehat{X}^I = \bigwedge_{j=1}^k (S^1)_{i_j} \simeq S^k$. If $I = \{i_1, \ldots, i_k\}$, let $|I| = k$. We obtain the following corollary.

Corollary 4.7: *Let K be a simplicial complex on the vertex set $[m]$. There is a homotopy equivalence*

$$\Sigma\mathcal{Z}_K \longrightarrow \bigvee_{I\notin K} \Sigma^{|I|+2}|K_I|.$$

□

Example 4.8: Return to the simplicial complex K in Example 4.4:

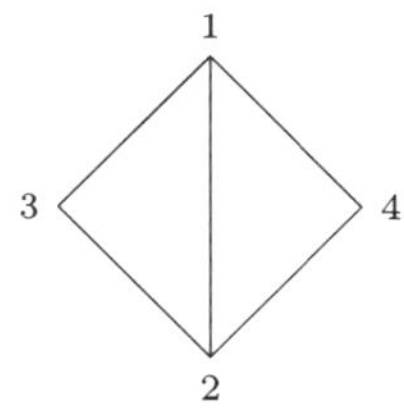

K

The missing faces of K are $(3,4)$, $(1,2,3)$, $(1,2,4)$, $(1,3,4)$, $(2,3,4)$ and $(1,2,3,4)$. The full subcomplexes K_I of K on these vertex sets are:

$K_{(3,4)}$

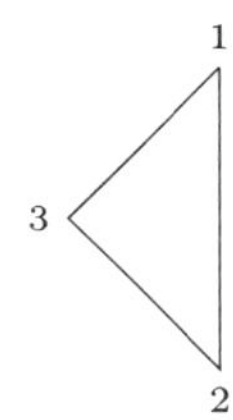

$K_{(1,2,3)}$

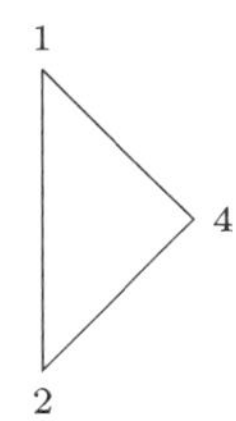

$K_{(1,2,4)}$

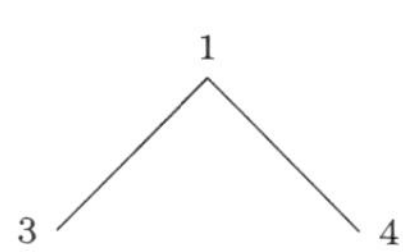

$K_{(1,3,4)}$

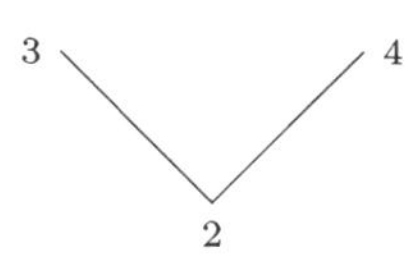

$K_{(1,3,4)}$

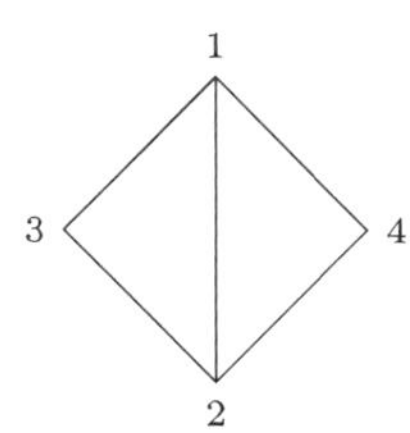

$K_{(1,2,3,4)} = K$

First, observe that $|K_{(3,4)}| = S^0$ and $|(3,4)| = 2$ so $\Sigma^{|I|+2}|K_I|$ in this case is S^4. Second, observe that $|K_{(1,2,3)}| = S^1$ and $|(1,2,3)| = 3$ so $\Sigma^{|I|+2}|K_I|$ in this case is S^6. Similarly, $K_{(1,2,4)}$ contributes an S^6 to $\Sigma\mathcal{Z}_K$. Third, observe that $|K_{(1,3,4)}|$ is contractible, so $\Sigma^{|I|+2}|K_I|$ in this case is $*$. Similarly, $K_{(2,3,4)}$ contributes a point to $\Sigma\mathcal{Z}_K$. Finally, observe that $|K_{(1,2,3,4)}| = S^1 \vee S^1$ and $|(1,2,3,4)| = 4$, so $\Sigma^{|I|+2}|K_I|$ in this case is $S^7 \vee S^7$. Hence there is a homotopy equivalence

$$\Sigma\mathcal{Z}_K \simeq S^4 \vee S^6 \vee S^6 \vee S^7 \vee S^7.$$

5. Building simplicial complexes and polyhedral products vertex by vertex

Having determined the homotopy type of $\Sigma(\underline{X},\underline{A})^K$, we now turn our attention to the homotopy type of $(\underline{X},\underline{A})^K$ itself. One approach to this is to build up the simplicial complex one vertex at a time and examine the affect on the homotopy type of the corresponding polyhedral product. This approach has had good success [17, 18, 19, 21, 22, 13].

5.1. *Building simplicial complexes*

Let K be a simplicial complex on the vertex set $[m]$. For a vertex $v \in K$, the *star*, *restriction* (or *deletion*) and *link* of v are the subcomplexes

$$\begin{aligned} \mathrm{star}_K(v) &= \{\tau \in K \mid \{v\} \cup \tau \in K\}; \\ K\backslash\{v\} = K_{[m]\backslash\{v\}} &= \{\tau \in K \mid \{v\} \cap \tau = \emptyset\}; \\ \mathrm{link}_K(v) &= \mathrm{star}_K(v) \cap K\backslash\{v\}. \end{aligned}$$

The *join* of two simplicial complexes K_1, K_2 on disjoint index sets is the simplicial complex

$$K_1 * K_2 = \{\sigma_1 \cup \sigma_2 \mid \sigma_i \in K_i\}.$$

From the definitions, it follows that $\mathrm{star}_K(v)$ is a join,

$$\mathrm{star}_K(v) = \{v\} * \mathrm{link}_K(v),$$

and there is a pushout of simplicial complexes

$$\begin{array}{ccc} \mathrm{link}_K(v) & \longrightarrow & \mathrm{star}_K(v) \\ \downarrow & & \downarrow \\ K\backslash\{v\} & \longrightarrow & K. \end{array} \tag{5.1}$$

Example 5.1: Let K be the simplical complex on 4 vertices:

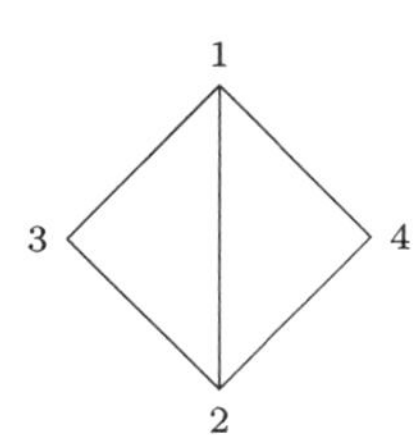

K

The star, restriction and link subcomplexes for the vertex 1 are:

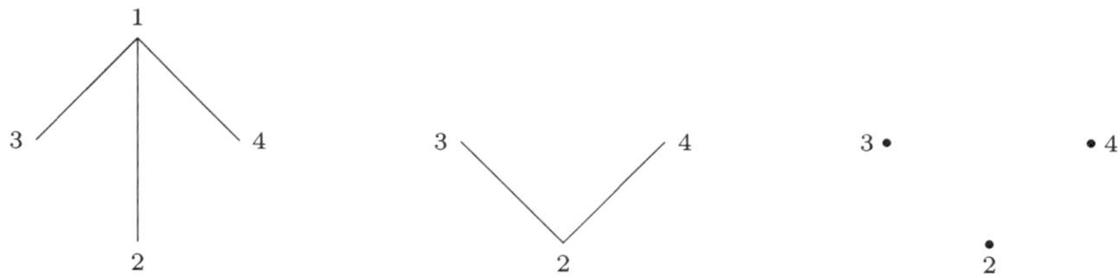

$\mathrm{star}_K(1)$ $\qquad\qquad$ $K\backslash\{1\}$ $\qquad\qquad$ $\mathrm{link}_K(1)$

Notice that $star_K(1)$ is clearly $\{1\}*\mathrm{link}_K(1)$, and K is the union of $\mathrm{star}_K(1)$ and $K\backslash\{1\}$ over $\mathrm{link}_K(1)$.

A family of simplicial complexes whose polyhedral products have been studied by this vertex-by-vertex approach is shifted complexes.

Definition 5.2: A simplicial complex K is *shifted* if there is an ordering on its vertices such that whenever $\sigma \in K$, ν is a vertex of σ, and $\nu' < \nu$, then $(\sigma - \nu) \cup \nu' \in K$.

Equivalently, regard the faces of K as sequences $(i_1, \ldots, i_k)$ where $1 \leq i_1 < \cdots < i_k \leq m$. Order the faces by saying that $(j_1, \ldots, j_t) \leq (i_1, \ldots, i_k)$ if $t \leq k$ and $j_1 \leq i_1, \cdots, j_t \leq i_t$. Then K is shifted if, whenever $\sigma = (i_1, \ldots, i_k) \in K$, then $\tau \in K$ for all $\tau \leq \sigma$.

Example 5.3: Consider the three simplicial complexes

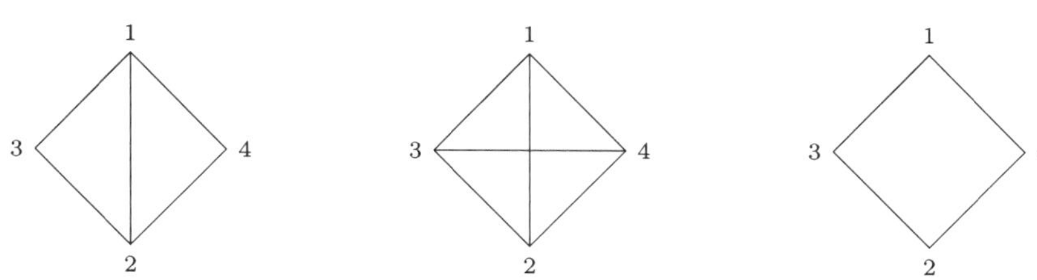

K_1 $\qquad\qquad$ K_2 $\qquad\qquad$ K_3

The simplicial complex K_1 is shifted using the given ordering of the vertices $(1 < 2 < 3 < 4)$, as is K_2. However, K_3 is not shifted under any ordering of the vertices. In the given ordering, as $(1, 3) \in K_3$, if K_3 were shifted then $(1, 2)$ would also have to be in K_3, which it is not. Under any reordering of the vertices, the vertex 4 would have to be connected to two other vertices by an edge: the only choices that could lead to a shifted complex are 1 and 2. But then 3 also has to be connected by an edge to 1 and 2, and now we are back to the description of K_3 with its given ordering of vertices, which is not shifted.

Another useful collection of examples is given by the full t-skeletons of the $(m-1)$-simplex.

Definition 5.4: Let K be a simplicial complex on the vertex set $[m]$. For $0 \leq t \leq m-1$, the *full t-skeleton* of K is the subcomplex K_t consisting of those faces $\sigma \in K$ on $t+1$ vertices for $t \leq k$.

For example, the 0-skeleton of K consists of the vertices, the 1-skeleton consists of the vertices and edges, and the $(m-1)$-skeleton is all of K.

Example 5.5: For $0 \leq t \leq m-1$, the full t-skeleton Δ_t^{m-1} of the $(m-1)$-simplex Δ^{m-1} is shifted. The shifted property is a consequence of the fact that the symmetric group on m letters acts freely on Δ_t^{m-1} by permuting the vertices.

5.2. *Building polyhedral products*

Let K be a simplicial complex on the vertex set $[m]$. We will fix the vertex m for convenience. By (5.1) there is a pushout describing K as the union of $\mathrm{star}_K(m)$ and $K\backslash\{m\}$ over $\mathrm{link}_K(m)$. As for Proposition 3.8, to compare the polyhedral products for the four simplicial complexes K, $\mathrm{link}_K(m)$, $\mathrm{star}_K(m)$ and $K\backslash\{m\}$, we should really consider each simplicial complex as being on the vertex set $[m]$. This changes nothing for K and $\mathrm{star}_K(m)$ as their vertex sets are already $[m]$ but it introduces a ghost vertex $\{m\}$ for $\mathrm{link}_K(m)$ and $K\backslash\{m\}$. Let $\overline{\mathrm{link}_K(m)}$ and $\overline{K\backslash\{m\}}$ be the simplicial complexes $\mathrm{link}_K(m)$ and $K\backslash\{m\}$, regarded as simplicial complexes on the vertex set $[m]$. By Proposition 3.8, the pushout of simplicial complexes in (5.1) induces a pushout of polyhedral products

$$\begin{array}{ccc} (\underline{X},\underline{A})^{\overline{\mathrm{link}_K(m)}} & \longrightarrow & (\underline{X},\underline{A})^{\mathrm{star}_K(m)} \\ \downarrow & & \downarrow \\ (\underline{X},\underline{A})^{\overline{K\backslash\{m\}}} & \longrightarrow & (\underline{X},\underline{A})^{K}. \end{array} \tag{5.2}$$

The spaces and maps in (5.2) can be better identified. First, the definition of the polyhedral product implies that

$$(\underline{X},\underline{A})^{\overline{\mathrm{link}_K(m)}} = (\underline{X},\underline{A})^{\mathrm{link}_K(m)} \times A_m$$

$$(\underline{X}, \underline{A})^{\overline{K\backslash\{m\}}} = (\underline{X}, \underline{A})^{K\backslash\{m\}} \times A_m.$$

By Proposition 3.1 (b), the inclusion of simplicial complexes $\mathrm{link}_K(m) \longrightarrow K\backslash\{m\}$ induces a map of polyhedral products

$$j\colon (\underline{X}, \underline{A})^{\mathrm{link}_K(m)} \longrightarrow (\underline{X}, \underline{A})^{K\backslash\{m\}}.$$

Including the ghost vertex at m, the map $\overline{\mathrm{link}_K(m)} \longrightarrow \overline{K\backslash\{m\}}$ therefore induces the map of polyhedral products

$$(\underline{X}, \underline{A})^{\overline{\mathrm{link}_K(m)}} = (\underline{X}, \underline{A})^{\mathrm{link}_K(m)} \times A_m \xrightarrow{j\times 1} (\underline{X}, \underline{A})^{K\backslash\{m\}} \times A_m = (\underline{X}, \underline{A})^{\overline{K\backslash\{m\}}}.$$

Second, the definition of the join of two simplicial complexes implies that if $K = K_1 * K_2$ then there is a homeomorphism

$$(\underline{X}, \underline{A})^K \cong (\underline{X}, \underline{A})^{K_1} \times (\underline{X}, \underline{A})^{K_2}.$$

In particular, since $\mathrm{star}_K(m) = \mathrm{link}_K(m) * \{m\}$, there is an identity

$$(\underline{X}, \underline{A})^{\mathrm{star}_K(m)} \cong (\underline{X}, \underline{A})^{\mathrm{link}_K(m)} \times X_m.$$

We may also regard $\overline{\mathrm{link}_K(m)}$ as $\mathrm{link}_K(m) * \emptyset$, where $\emptyset$ corresponds to the ghost vertex at $\{m\}$. The inclusion $\mathrm{link}_K(m) \longrightarrow \mathrm{star}_K(m)$ therefore induces a map of joins

$$\overline{\mathrm{link}_K(m)} = \mathrm{link}_K(m) * \emptyset \longrightarrow \mathrm{link}_K(m) * \{m\} = \mathrm{star}_K(m).$$

Thus $\mathrm{link}_K(m) \longrightarrow \mathrm{star}_K(m)$ induces a product map

$$(\underline{X}, \underline{A})^{\mathrm{link}_K(m)} \times A_m \xrightarrow{1\times i_m} (\underline{X}, \underline{A})^{\mathrm{link}_K(m)} \times X_m$$

where i_m is the inclusion of A_m into X_m.

Combining these two observations, (5.2) can be rewritten as follows.

Proposition 5.6: *Let K be a simplicial complex on the vertex set $[m]$. Let $(\underline{X}, \underline{A})$ be a sequence of pointed CW-pairs (X_i, A_i) for $1 \leq i \leq m$. Then there is a pushout of polyhedral products*

$$\begin{array}{ccc}
(\underline{X}, \underline{A})^{link_K(m)} \times A_m & \xrightarrow{1\times i_m} & (\underline{X}, \underline{A})^{link_K(m)} \times X_m \\
\downarrow{\scriptstyle j\times 1} & & \downarrow \\
(\underline{X}, \underline{A})^{K\backslash\{m\}} \times A_m & \longrightarrow & (\underline{X}, \underline{A})^{K}.
\end{array}$$

□

An important special case is that for $(\underline{CX},\underline{X})$. Then the map $(\underline{CX},\underline{X})^{\text{link}_K(m)}\times X \xrightarrow{1\times i_m} (\underline{CX},\underline{X})^{\text{link}_K(m)}\times CX_m$ is equivalent, up to homotopy, to the map $(\underline{CX},\underline{X})^{\text{link}_K(m)}\times X_m \xrightarrow{\pi_1} (\underline{CX},\underline{X})^{\text{link}_K(m)}$, where π_1 is the projection onto the first factor. Thus Proposition 5.6 specializes to the following.

Corollary 5.7: *Let K be a simplicial complex on the vertex set $[m]$. For $1 \leq i \leq m$ let X_i be a pointed CW-complex. Then there is a homotopy pushout*

$$\begin{array}{ccc} (\underline{CX},\underline{X})^{link_K(m)}\times X_m & \xrightarrow{\pi_1} & (\underline{CX},\underline{X})^{link_K(m)} \\ \big\downarrow{\scriptstyle j\times 1} & & \big\downarrow \\ (\underline{CX},\underline{X})^{K\backslash\{m\}}\times X_m & \longrightarrow & (\underline{CX},\underline{X})^{K}. \end{array}$$

□

In principle, if the homotopy class of the map j is known then it should be possible to identify the homotopy type of $(\underline{CX},\underline{X})^K$. This is essentially the approach taken in [18, 19, 21, 22] in order to identify the homotopy type of $(\underline{CX},\underline{X})^K$ when K is a shifted complex. In that case, the map j is null homotopic. The next section reviews the homotopy theory needed to handle the case when j is null homotopic. In cases when K is not shifted the map j may not be null homotopic, but the same principle should apply. However, this is largely unexplored territory.

6. Identifying the homotopy type of certain pushouts

We begin with some definitions. Let I be the unit interval. Let A and B be pointed topological spaces. Recall that the *reduced join* of A and B is the quotient space

$$A * B = (A \times I \times B)/\sim$$

where $(a,1,b) \sim (a',1,b)$, $(a,0,b) \sim (a,0,b')$ and $(*,t,*) \sim *$ for all $a,a' \in A$, $b,b' \in B$ and $t \in I$. Ignoring the relation $(*,t,*) \sim *$ for the moment, the unreduced join can be pictured as follows:

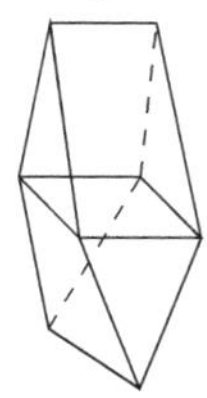

where the middle square is $A \times B$, with B running horizontally. Notice that the top part of $A * B$ is $\overline{CA} \times B$, where $\overline{CA}$ is the unreduced cone on A, and the bottom part of $A * B$ is $A \times \overline{CB}$. These are glued together in the middle at $A \times B$. The reduced join has the same description, using the reduced cone. Thus there is a topological pushout

$$\begin{array}{ccc} A \times B & \longrightarrow & A \times CB \\ \downarrow & & \downarrow \\ CA \times B & \longrightarrow & A * B. \end{array} \tag{6.1}$$

Since CA and CB are contractible, the maps $A \times B \longrightarrow A \times CB$ and $A \times B \longrightarrow CA \times B$ are equivalent, up to homotopy, to the projections $A \times B \xrightarrow{\pi_1} A$ and $A \times B \xrightarrow{\pi_2} B$ respectively. The contractibility of CB also implies that the homotopy class of the map $A \times CB \longrightarrow A * B$ is determined by the restriction to A. The commutativity of (6.1) implies that this restriction equals the composite $A \hookrightarrow A \times B \longrightarrow CA \times B \longrightarrow A * B$. The latter is null homotopic since it factors through CA. Therefore the map $A \times CB \longrightarrow A * B$ is null homotopic. Similarly, the map $CA \times B \longrightarrow A * B$ is null homotopic. Putting all this together we obtain a homotopy-theoretic version of (6.1).

Lemma 6.1: *There is a homotopy pushout*

$$\begin{array}{ccc} A \times B & \xrightarrow{\pi_1} & A \\ \downarrow{\scriptstyle \pi_2} & & \downarrow{\scriptstyle *} \\ B & \xrightarrow{*} & A * B \end{array}$$

where π_1 and π_2 are the projections onto the first and second factors respectively. □

The (right) *half-smash* of two pointed spaces C and B is the quotient space

$$C \rtimes B = C \times B / \sim$$

where $(*, b) \sim *$. In particular, if $i_2 \colon B \longrightarrow C \times B$ is the inclusion of the second factor, then there is a cofibration $B \xrightarrow{i_2} C \times B \longrightarrow C \rtimes B$. In what follows it is better to regard this cofibration as a pushout

$$\begin{array}{ccc} B & \longrightarrow & * \\ \downarrow{\scriptstyle i_2} & & \downarrow \\ C \times B & \longrightarrow & C \rtimes B. \end{array} \tag{6.2}$$

With Corollary 5.7 in mind, we now prove a key lemma that identifies the homotopy type of certain pushouts.

Lemma 6.2: *Suppose that there is a homotopy pushout*

$$\begin{array}{ccc} A\times B & \xrightarrow{\pi_1} & A \\ \big\downarrow{\scriptstyle *\times 1} & & \big\downarrow \\ C\times B & \longrightarrow & Q \end{array}$$

where π_1 *is the projection. Then there is a homotopy equivalence*

$$Q \simeq (A * B) \vee (C \rtimes B).$$

Proof: The map $A\times B \xrightarrow{*\times 1} C\times B$ is equal to the composite $A\times B \xrightarrow{\pi_2} B \xrightarrow{i_2} C\times B$. Thus, using Lemma 6.1, the homotopy pushout in the statement of the lemma may be written as an iterated homotopy pushout diagram

$$\begin{array}{ccc} A\times B & \xrightarrow{\pi_1} & A \\ \big\downarrow{\scriptstyle \pi_2} & & \big\downarrow{\scriptstyle *} \\ B & \xrightarrow{*} & A * B \\ \big\downarrow{\scriptstyle i_2} & & \big\downarrow \\ C\times B & \longrightarrow & Q. \end{array}$$

For the bottom square, since the map $B \xrightarrow{*} A*B$ factors as the composite $B \longrightarrow * \longrightarrow A*B$, using (6.2) we obtain an iterated homotopy pushout diagram

$$\begin{array}{ccccc} B & \longrightarrow & * & \longrightarrow & A * B \\ \big\downarrow{\scriptstyle i_2} & & \big\downarrow & & \big\downarrow \\ C\times B & \longrightarrow & C\rtimes B & \longrightarrow & Q. \end{array}$$

The right square immediately implies that $Q \simeq (A * B) \vee (C \rtimes B)$. □

7. A strategy for identifying the homotopy type of $(\underline{CX}, \underline{X})^K$ for certain K

Consider the homotopy pushout

$$\begin{array}{ccc} (\underline{CX}, \underline{X})^{\mathrm{link}_K(m)} \times X_m & \xrightarrow{\pi_1} & (\underline{CX}, \underline{X})^{\mathrm{link}_K(m)} \\ \Big\downarrow{\scriptstyle j\times 1} & & \Big\downarrow \\ (\underline{CX}, \underline{X})^{K\backslash\{m\}} \times X_m & \longrightarrow & (\underline{CX}, \underline{X})^{K} \end{array}$$

from Corollary 5.7. If the map j is null homotopic then Lemma 6.2 implies that there is a homotopy equivalence

$$(\underline{CX}, \underline{X})^K \simeq ((\underline{CX}, \underline{X})^{\mathrm{link}_K(m)} * X_m) \vee ((\underline{CX}, \underline{X})^{K\backslash\{m\}} \rtimes X_m). \quad (7.1)$$

Notice that on the right side both $\mathrm{link}_K(m)$ and $K\backslash\{m\}$ are simplicial complexes on the smaller vertex set $[m-1]$. This suggests an inductive procedure.

What is needed is a family $\mathcal{K}$ of simplicial complexes such that, if $K \in \mathcal{K}$ is on the vertex set $[m]$ then: (i) the vertices of K may be ordered so that the map $(\underline{CX}, \underline{X})^{\mathrm{link}_K(m)} \xrightarrow{j} (\underline{CX}, \underline{X})^{K\backslash\{m\}}$ is null homotopic, and (ii) both $\mathrm{link}_K(m)$ and $K\backslash\{m\}$ are in $\mathcal{K}$. In particular, property (ii) implies that both $\mathrm{link}_K(m)$ and $K\backslash\{m\}$ have property (i), so that they also satisfy homotopy decompositions as in (7.1), and these decompositions can then be substituted into that for $(\underline{CX}, \underline{X})^K$. Iterating, we would obtain a fine decomposition of $(\underline{CX}, \underline{X})^K$.

It would be even better if an extra condition were satisfied. It is well known that if C is a co-H-space then there is a homotopy equivalence $C \rtimes B \simeq C \vee (C \wedge B)$. So if $(\underline{CX}, \underline{X})^{K\backslash\{m\}}$ was a suspension, then (7.1) refines to a homotopy equivalence

$$(\underline{CX}, \underline{X})^K \simeq ((\underline{CX}, \underline{X})^{\mathrm{link}_K(m)} * X_m) \vee (\underline{CX}, \underline{X})^{K\backslash\{m\}} \vee ((\underline{CX}, \underline{X})^{K\backslash\{m\}} \wedge X_m). \quad (7.2)$$

In general, there is a homotopy equivalence $A * B \simeq \Sigma A \wedge B$. So if $(\underline{CX}, \underline{X})^{K\backslash\{m\}}$ was a suspension, then not only does (7.2) hold, but it implies that $(\underline{CX}, \underline{X})^K$ is also a suspension. Again, as K has one more vertex than $K\backslash\{m\}$, we are in the position of producing an inductive procedure.

Let's carry this through with an example to see the strategy in action. For $m \geq 1$, let K_m be the simplicial complex consisting of m disjoint vertices. There is a trivial case: as K_1 has 1 vertex, by definition of the polyhedral product $(\underline{CX}, \underline{X})^{K_1} = CX_1$ and so is contractible.

Proposition 7.1: *If $m \geq 2$ then there is a homotopy equivalence*

$$(\underline{CX}, \underline{X})^{K_m} \simeq \bigvee_{j=2}^{m} \left(\bigvee_{1 \leq i_1 < \cdots < i_j \leq m} (j-1)\Sigma X_{i_1} \wedge \cdots \wedge X_{i_j} \right).$$

Proof: The proof is by induction on the number of vertices. As K_2 has 2 vertices, by Example 2.2 there is a homotopy equivalence $(\underline{CX}, \underline{X})^{K_2} \simeq X_1 * X_2 \simeq \Sigma X_1 \wedge X_2$, as asserted. Suppose that the homotopy equivalence in the statement of the Proposition holds for K_{m-1}. Consider the pushout of simplicial complexes

$$\begin{array}{ccc} \text{link}_{K_m}(m) & \longrightarrow & \text{star}_{K_m}(m) \\ \big\downarrow & & \big\downarrow \\ K_m \backslash \{m\} & \longrightarrow & K_m. \end{array}$$

Since K_m is m disjoint vertices, $\text{star}_{K_m}(m) = \{m\}$ and $K_m \backslash \{m\} = K_{m-1}$. So $\text{link}_{K_m}(m) = \text{star}_{K_m}(m) \cap K_m \backslash \{m\} = \emptyset$ and the pushout of simplicial complexes above is the pushout

$$\begin{array}{ccc} \emptyset & \longrightarrow & \{m\} \\ \big\downarrow & & \big\downarrow \\ K_{m-1} & \longrightarrow & K_m. \end{array}$$

Note here that $\emptyset$ is regarded as a simplicial complex on the vertex set $[m-1]$. By Corollary 5.7 there is a homotopy pushout

$$\begin{array}{ccc} (\underline{CX}, \underline{X})^{\emptyset} \times X_m & \xrightarrow{\pi_1} & (\underline{CX}, \underline{X})^{\emptyset} \\ \big\downarrow{\scriptstyle j \times 1} & & \big\downarrow \\ (\underline{CX}, \underline{X})^{K_{m-1}} \times X_m & \longrightarrow & (\underline{CX}, \underline{X})^{K_m}. \end{array}$$

We aim to show that the map j is null homotopic. For the sequence of CW-complexes $(\underline{CX}, \underline{X})$ the definition of the polyhedral product implies that

$$(\underline{CX}, \underline{X})^{\emptyset} = X_1 \times \cdots \times X_{m-1}.$$

Since K_{m-1} is $m-1$ disjoint vertices, the definition of the polyhedral product implies that

$$(\underline{CX}, \underline{X})^{K_{m-1}} = \bigcup_{i=1}^{m-1} X_1 \times \cdots \times X_{i-1} \times CX_i \times X_{i+1} \times \cdots \times X_{m-1}.$$

Further, the definition of the polyhedral product implies that the map $(\underline{CX},\underline{X})^{\emptyset} \xrightarrow{j} (\underline{CX},\underline{X})^{K_{m-1}}$ is the inclusion. In particular, j factors as the composite of inclusions

$$\begin{aligned} j\colon X_1 \times \cdots \times X_{m-1} &\longrightarrow CX_1 \times X_2 \times \cdots \times X_{m-1} \\ &\hookrightarrow \bigcup_{i=1}^{m-1} X_1 \times \cdots \times X_{i-1} \times CX_i \times X_{i+1} \times \cdots \times X_{m-1}. \end{aligned}$$

Since CX_1 is contractible, the first map in the composite for j is null homotopic in the first coordinate, so j is homotopic to the composite

$$\begin{aligned} X_1 \times \cdots \times X_{m-1} &\xrightarrow{\text{proj}} X_2 \times \cdots \times X_{m-1} \\ &\hookrightarrow X_1 \times \cdots \times X_{m-1} \\ &\xrightarrow{j} \bigcup_{i=1}^{m-1} X_1 \times \cdots \times X_{i-1} \times CX_i \times X_{i+1} \times \cdots \times X_{m-1}. \end{aligned}$$

Repeating this argument for each coordinate $2 \leq i \leq m-1$ we obtain a factorization of j through the map $X_1 \times \cdots \times X_{m-1} \xrightarrow{\text{proj}} * \hookrightarrow X_1 \times \cdots \times X_m$, implying that j is null homotopic.

Thus by Lemma 6.2 there is a homotopy equivalence

$$(\underline{CX},\underline{X})^{K_m} \simeq ((\underline{CX},\underline{X})^{\emptyset} * X_m) \vee ((\underline{CX},\underline{X})^{K_{m-1}} \rtimes X_m).$$

Since $(\underline{CX},\underline{X})^{K_{m-1}}$ is a suspension by inductive hypothesis, there is a refined homotopy equivalence

$$(\underline{CX},\underline{X})^{K_m} \simeq ((\underline{CX},\underline{X})^{\emptyset} * X_m) \vee (\underline{CX},\underline{X})^{K_{m-1}} \vee ((\underline{CX},\underline{X})^{K_{m-1}} \wedge X_m).$$

Substituting in the inductively assumed homotopy equivalence for $(\underline{CX},\underline{X})^{K_{m-1}}$, using the fact that $(\underline{CX},\underline{X})^{\emptyset} = X_1 \times \cdots \times X_{m-1}$, and using Lemma 4.2 to decompose $(\underline{CX},\underline{X})^{\emptyset} * X_m \simeq \Sigma(\underline{CX},\underline{X})^{\emptyset} \wedge X_m$, we obtain the asserted homotopy equivalence for $(\underline{CX},\underline{X})^{K_m}$. □

It may be worth noting that in terms of the strategy for identifying the homotopy type of $(\underline{CX},\underline{X})^K$, if $\mathcal{K}$ is the family of simplicial complexes consisting of disjoint vertices, then the argument for the proof of Proposition 7.1 shows that as $K_m \in \mathcal{K}$ we have: (i) the map $(\underline{CX},\underline{X})^{\text{link}_K(m)} \xrightarrow{j} (\underline{CX},\underline{X})^{K\backslash\{m\}}$ is null homotopic, (ii) $\text{link}_{K_m}(m)$ and $K_m\backslash\{m\}$ are both in $\mathcal{K}$, and (iii) $(\underline{CX},\underline{X})^{\text{link}_K(m)}$ is a suspension.

Proposition 7.1 generalizes a result of Porter [35]. His work long predates the notion of a polyhedral product but in modern language he proved

Proposition 7.1 in the case of $(\underline{C\Omega X}, \underline{\Omega X})^{K_m}$. It also desuspends the homotopy decomposition of $\Sigma(\underline{CX}, \underline{X})^{K_m}$ in Theorem 4.6.

This leads to a natural question. For which simplicial complexes does the homotopy decomposition of $\Sigma(\underline{CX}, \underline{X})^K$ in Theorem 4.6 desuspend? The case when K is shifted was done in special cases of $(\underline{CX}, \underline{X})$ in [18, 21] and in general in [19, 22]. In [23], a desuspension was proved for a strictly larger family of simplicial complexes called sequential Cohen-Macauley. In [3, 23] it was shown that a desuspension exists if and only if $(\underline{CX}, \underline{X})^K$ is a co-H-space, and near characterizations of when this occurs were given. The strategy for desuspending the decomposition in Theorem 4.6 in the shifted case was exactly the strategy that we have been working with: the shifted property implies - after some work - that there is a reordering of the vertices with the property that the map $(\underline{CX}, \underline{X})^{\text{link}_K(m)} \xrightarrow{j} (\underline{CX}, \underline{X})^{K\backslash\{m\}}$ is null homotopic, and it is not difficult to see that the shifted property is inherited by both $\text{link}_K(m)$ and $K\backslash\{m\}$. We record the outcome without proving it.

Theorem 7.2: *Let K be a shifted simplicial complex on the vertex set $[m]$ and let $X_1, \ldots, X_m$ be pointed, path-connected CW-complexes. Then the homotopy decomposition in Theorem 4.6 desuspends to give a homotopy equivalence*

$$(\underline{CX}, \underline{X})^K \simeq \bigvee_{I \notin K} \Sigma(|K_I| \wedge \widehat{X}^I).$$

□

In the special case of the moment-angle complex $\mathcal{Z}_K$, we obtain the following corollary.

Corollary 7.3: *Let K be a shifted simplicial complex on the vertex set $[m]$. There is a homotopy equivalence*

$$\mathcal{Z}_K \longrightarrow \bigvee_{I \notin K} \Sigma^{|I|+1}|K_I|.$$

□

In particular, as in [1], if K is shifted and $I \subseteq [m]$ then the geometric realization $|K_I|$ is homotopy equivalent to a wedge of spheres. So when K is shifted $\mathcal{Z}_K$ is homotopy equivalent to a wedge of spheres.

Example 7.4: Return to Example 4.8. It was shown that if K is the

simplicial complex

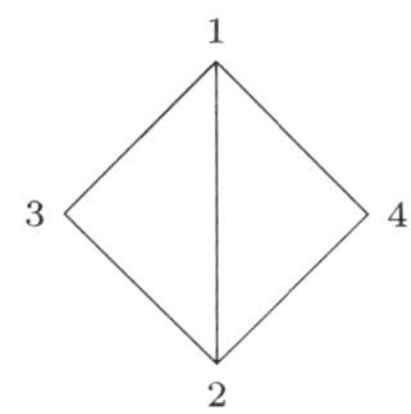

K

then there is a homotopy equivalence

$$\Sigma \mathcal{Z}_K \simeq S^4 \vee S^6 \vee S^6 \vee S^7 \vee S^7.$$

By Example 5.3, K is shifted. Let's go through the argument to desuspend the homotopy equivalence for $\Sigma \mathcal{Z}_K$. It is clearer to do the more general $(\underline{CX}, \underline{X})$ case as that more naturally keeps track of coordinate. Recall from Example 5.1 that the star, link and restriction of K for the vertex 1 are the simplicial complexes:

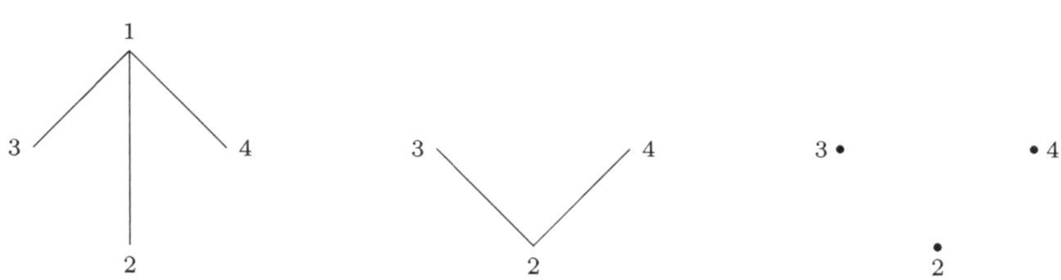

$\mathrm{star}_K(1)$ $K \backslash \{1\}$ $\mathrm{link}_K(1)$

However, as will be more clear by the end of the next section, the fact that the missing face $(3, 4)$ of $\mathrm{link}_K(1)$ is also a missing face of $K\backslash\{1\}$ implies that the map $(\underline{CX}, \underline{X})^{\mathrm{link}_K(1)} \xrightarrow{j} (\underline{CX}, \underline{X})^{K\backslash\{1\}}$ is not null homotopic. Instead, we consider a different vertex. Note that choosing the vertex 2 as the starting point gives star, link and restriction diagrams similar to those for 1. So instead let's start with vertex 3. Then we have:

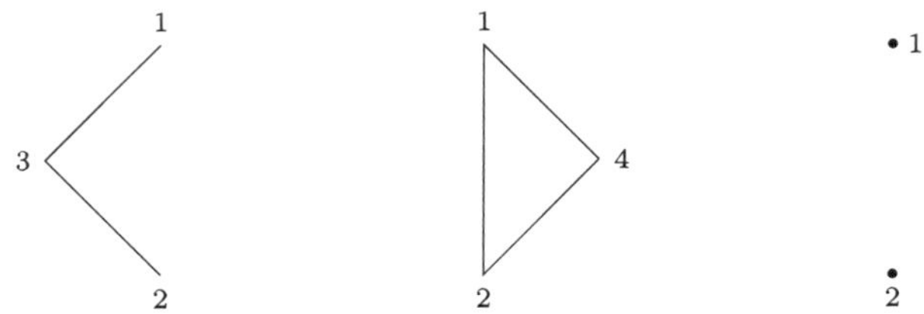

$\text{star}_K(3)$ $\quad$ $K\backslash\{3\}$ $\quad$ $\text{link}_K(3)$

Note here that $\text{link}_K(3)$ must be regarded as a simplicial complex on the vertex set $\{1,2,4\}$, the same vertex set as $K\backslash\{3\}$. Let L be $\text{link}_K(3)$ regarded as a simplicial complex on the vertex set $\{1,2\}$. Then $(\underline{CX},\underline{X})^{\text{link}_K(3)} = (\underline{CX},\underline{X})^L \times X_4$. As L is two disjoint points, by Example 2.2, $(\underline{CX},\underline{X})^L \simeq X_1 * X_2$. Therefore

$$(\underline{CX},\underline{X})^{\text{link}_K(3)} = (X_1 * X_2) \times X_4.$$

Observe next that the full subcomplex of $K\backslash\{3\}$ on the vertex set $\{1,2\}$ is Δ^1, so the map $\text{link}_K(3) \longrightarrow K\backslash\{3\}$ factors as the composite $\text{link}_K(3) \longrightarrow \Delta^1 \longrightarrow K\backslash\{3\}$, implying that the map $(\underline{CX},\underline{X})^{\text{link}_K(3)} \xrightarrow{j} (\underline{CX},\underline{X})^{K\backslash\{3\}}$ factors as the composite $(\underline{CX},\underline{X})^{\text{link}_K(3)} \longrightarrow (\underline{CX},\underline{X})^{\Delta^1} \longrightarrow (\underline{CX},\underline{X})^{K\backslash\{3\}}$. But by the definition of the polyhedral product, $(\underline{CX},\underline{X})^{\Delta^1} = CX_1 \times CX_2$ is contractible. Therefore j is null homotopic. Thus by Lemma 6.2 there is a homotopy equivalence

$$(\underline{CX},\underline{X})^K \simeq ((\underline{CX},\underline{X})^{\text{link}_K(3)} * X_3) \vee ((\underline{CX},\underline{X})^{K\backslash\{3\}} \rtimes X_3).$$

We have already mentioned that $(\underline{CX},\underline{X})^{\text{link}_K(3)} \simeq (X_1 * X_2) \times X_4$. Thus

$$(\underline{CX},\underline{X})^{\text{link}_K(3)} * X_3 \simeq [(X_1 * X_2) \times X_4] * X_3.$$

Using the fact that $A * B \simeq \Sigma A \wedge B$ and Lemma 4.1 to decompose the suspension of a product, we obtain homotopy equivalences

$$\begin{aligned}(\underline{CX},\underline{X})^{\text{link}_K(3)} * X_3 &\simeq \Sigma[(\Sigma X_1 \wedge X_2) \times X_4] \wedge X_3 \\ &\simeq [(\Sigma^2 X_1 \wedge X_4) \vee \Sigma X_4 \vee (\Sigma^2 X_1 \wedge X_2 \wedge X_4)] \wedge X_3 \\ &\simeq (\Sigma^2 X_1 \wedge X_2 \wedge X_3) \vee (\Sigma X_3 \wedge X_4) \\ &\qquad \vee (\Sigma^2 X_1 \wedge X_2 \wedge X_3 \wedge X_4).\end{aligned}$$

On the other hand, as we will see in the next section, $(\underline{CX},\underline{X})^{K\backslash\{3\}} \simeq X_1 * X_2 * X_4 \simeq \Sigma^2 X_1 \wedge X_2 \wedge X_4$. Thus there is a homotopy equivalence

$$(\underline{CX},\underline{X})^{K\backslash\{3\}} \rtimes X_3 \simeq (\Sigma^2 X_1 \wedge X_2 \wedge X_4) \vee (\Sigma^2 X_1 \wedge X_2 \wedge X_3 \wedge X_4).$$

Hence there is a homotopy equivalence

$$\begin{aligned}(\underline{CX},\underline{X})^{K} \simeq (\Sigma^2 X_1 \wedge X_2 \wedge X_4) \vee (\Sigma X_3 \wedge X_4) \vee (\Sigma^2 X_1 \wedge X_2 \wedge X_3 \wedge X_4)\\ \vee (\Sigma^2 X_1 \wedge X_2 \wedge X_3) \vee (\Sigma^2 X_1 \wedge X_2 \wedge X_3 \wedge X_4).\end{aligned}$$

In particular, in the case of $\mathcal{Z}_K$ each $X_i = S^1$, so we obtain a homotopy equivalence

$$\mathcal{Z}_K \simeq S^5 \vee S^3 \vee S^6 \vee S^5 \vee S^6,$$

as desired.

8. Higher Whitehead products and minimal missing faces

Consider the homotopy fibration

$$(\underline{C\Omega X},\underline{\Omega X})^K \longrightarrow (\underline{X},\underline{*})^K \longrightarrow \prod_{i=1}^{m} X_i$$

from Corollary 3.14. At this point the homotopy type of $(\underline{C\Omega X},\underline{\Omega X})^K$ has been identified for various simplicial complexes K. However, nothing has been said about the homotopy type of the map $(\underline{C\Omega X},\underline{\Omega X})^K \longrightarrow (\underline{X},\underline{*})^K$. This is the problem addressed in this section. In [20, 16, 24] it has been determined that Whitehead products and higher Whitehead products play an important role. In particular, the existence of nontrivial higher Whitehead products in $(\underline{X},\underline{*})^K$ corresponds to the existence of minimal missing faces in K. We begin with a discussion of the standard Whitehead product.

8.1. *Whitehead products*

One way of defining a Whitehead product is as the adjoint of a Samelson product. Let Y be an H-group, that is, Y is a homotopy associative H-space with a homotopy inverse. Let $\overline{c}\colon Y \times Y \longrightarrow Y$ be the commutator of Y, defined pointwise by $\overline{c}(x,y) = xyx^{-1}y^{-1}$. Observe that the restriction of $\overline{c}$ to the wedge is null homotopic, so $\overline{c}$ extends to a map

$$c\colon Y \wedge Y \longrightarrow Y.$$

By Lemma 4.1, the quotient map $Y \times Y \longrightarrow Y \wedge Y$ has a right homotopy inverse after suspending. Therefore the homotopy class of c is determined uniquely by the homotopy class of $\overline{c}$. Let $a\colon A \longrightarrow Y$ and $b\colon B \longrightarrow Y$ be maps. The composite

$$\langle a,b\rangle\colon A \wedge B \xrightarrow{a\wedge b} Y \wedge Y \xrightarrow{c} Y$$

is the *Samelson product* of a and b. In particular, c itself is the Samelson product of the identity map on Y with itself. The Samelson product has many useful properties, including the fact that if $x \in H_*(A)$ and $y \in H_*(B)$ are primitive then $(\langle a,b\rangle)_*(x,y) = [a(x), b(y)]$, where $[\,\cdot\,,\,\cdot\,]$ is the graded commutator in the homology algebra $H_*(Y)$.

Now suppose that X is a space and there are maps $f\colon \Sigma A \longrightarrow X$ and $g\colon \Sigma B \longrightarrow X$. Let $\tilde{f}\colon A \longrightarrow \Omega X$ and $\tilde{g}\colon B \longrightarrow \Omega X$ be the adjoints of f and g respectively. We may take the Samelson product $\langle \tilde{f},\tilde{g}\rangle\colon A \wedge B \longrightarrow \Omega X$. Its adjoint is a map

$$[f,g]\colon \Sigma A \wedge B \longrightarrow X$$

called the *Whitehead product* of f and g.

Whitehead products play a crucial role in homotopy theory. One key context in which they arise is the following. Let X and Y be pointed, simply-connected spaces and consider the inclusion $X \vee Y \longrightarrow X \times Y$. By Proposition 7.1, the homotopy fibre of this inclusion is homotopy equivalent to $\Sigma\Omega X \wedge \Omega Y$. Ganea [14] proved a much sharper result by identifying the homotopy class of the map from the fibre to $X \vee Y$. Let $i_L\colon X \longrightarrow X \vee Y$ and $i_R\colon Y \longrightarrow X \vee Y$ be the inclusions of the left and right wedge summands respectively. Let ev_L and ev_R be the composites $ev_L\colon \Sigma\Omega X \xrightarrow{ev} X \xrightarrow{i_L} X \vee Y$ and $ev_R\colon \Sigma\Omega Y \xrightarrow{ev} Y \xrightarrow{i_R} X \vee Y$ respectively, where ev is the canonical evaluation map.

Theorem 8.1: ***(Ganea)*** *Let X and Y be pointed, simply-connected spaces. Then there is a homotopy fibration*

$$\Sigma\Omega X \wedge \Omega Y \xrightarrow{[ev_L, ev_R]} X \vee Y \longrightarrow X \times Y.$$

□

One consequence is the following. In general, let $E\colon X \longrightarrow \Omega\Sigma X$ be the suspension map, which is adjoint to the identity map on ΣX. If $X = \Sigma A$ and $Y = \Sigma B$ then the composite $\Sigma A \wedge B \xrightarrow{\Sigma E \wedge E} \Sigma\Omega\Sigma A \wedge \Omega\Sigma B \xrightarrow{[ev_L, ev_R]} \Sigma A \vee \Sigma B$ is homotopic to the Whitehead product $[i_L, i_R]$, and the homotopy cofibre of this map is $\Sigma A \times \Sigma B$.

8.2. *Higher Whitehead products*

Attaching by a Whitehead product is what is needed to produce $\Sigma A \times \Sigma B$ from $\Sigma A \vee \Sigma B$. Theorem 8.1 says that there is a generalization when looking at the homotopy fibre of any wedge into a product. To get higher versions of Whitehead products, we should look at what is needed to produce $\prod_{i=1}^{m} \Sigma A_i$ from the fat wedge on $\Sigma A_1, \ldots, \Sigma A_m$, and then generalize by looking at the homotopy fibre of any fat wedge into a product. To make this precise some definitions are needed.

Definition 8.2: For spaces $X_1, \ldots, X_m$, let $FW(1, \ldots, m)$ be the subspace of $X_1 \times \cdots \times X_m$ defined by

$$FW(1, \ldots, m) = \bigcup_{i=1}^{m} X_1 \times \cdots \times X_{i-1} \times * \times X_{i+1} \times \cdots \times X_m.$$

The space $FW(1, \ldots, m)$ is the *fat wedge* of $X_1, \ldots, X_m$.

By analogy with Theorem 8.1, a higher Whitehead product should be defined via the homotopy fibre of the inclusion of $FW(1, \ldots, m)$ into $X_1 \times \cdots \times X_m$. To do so, it is easier to phrase this homotopy fibration in terms of polyhedral products. Let $\partial\Delta^{m-1}$ be the boundary of Δ^{m-1}. Note that $\partial\Delta^{m-1}$ is the full $m-2$-skeleton of Δ^{m-1}. Therefore, comparing the definition of the polyhedral product for $(\underline{X}, \underline{*})^{\Delta^{m-1}}$ and that for $FW(1, \ldots, m)$ we obtain

$$FW(1, \ldots, m) = (\underline{X}, \underline{*})^{\partial\Delta^{m-1}}.$$

So by Corollary 3.14 there is a homotopy fibration

$$(\underline{C\Omega X}, \underline{\Omega X})^{\partial\Delta^{m-1}} \longrightarrow FW(1, \ldots, m) \longrightarrow \prod_{i=1}^{m} X_i. \tag{8.1}$$

Thus we want to identity the homotopy type of $(\underline{C\Omega X}, \underline{\Omega X})^{\partial\Delta^{m-1}}$. We will do this in the more general case of $(\underline{CX}, \underline{X})^{\partial\Delta^{m-1}}$.

Lemma 8.3: *For $k \geq 2$, there is a pushout of simplicial complexes*

$$\begin{array}{ccc} \partial\Delta^{k-1} & \longrightarrow & \partial\Delta^{k-1} * \{k+1\} \\ \downarrow & & \downarrow \\ \Delta^{k-1} & \longrightarrow & \partial\Delta^{k}. \end{array}$$

Proof: If K is a simplicial complex on the index set $[k]$, regard the faces of K as subsequences $(i_1, \ldots, i_\ell)$ where $1 \le i_1 < \cdots < i_\ell \le k$. Observe that Δ^k has as faces all subseqences of $\{1, \ldots, k+1\}$, and $\partial\Delta^k$ has as faces all proper subsequences of $\{1, \ldots, k+1\}$. In particular, $\partial\Delta^k$ is the union of: (i) all subsequences of the form $(i_1, \ldots, i_\ell)$ where $1 \le i_1 < \cdots < i_\ell \le k$ – that is, those subsequences corresponding to Δ^{k-1}, and (ii) all the subsequences of the form $(i_1, \ldots, i_\ell, k+1)$ where $1 \le i_1 < \cdots < i_\ell \le k$ but at least one i_j is missing – that is, those subsequences corresponding to $\partial\Delta^{k-1} * \{k+1\}$. The intersection of the subsequences in (i) and (ii) consists of all subsequences of the form $(i_1, \ldots, i_\ell)$ where $1 \le i_1 < \cdots < i_\ell \le k$ and at least one i_j is missing – that is, those subsequences corresponding to $\partial\Delta^{k-1}$. This proves the lemma. □

Remark 8.4: The proof of Lemma 8.3 shows that the pushout in the statement of the lemma is the usual star-link-restriction pushout in the case of $K = \partial\Delta^k$. That is, $\text{star}_{\partial\Delta^k} = \partial\Delta^{k-1} * \{k+1\}$, $\text{link}_{\partial\Delta^k} = \partial\Delta^{k-1}$ and $(\partial\Delta^k)\backslash\{k+1\} = \Delta^{k-1}$.

Remark 8.4 implies that, by Proposition 5.6, there is a corresponding pushout of polyhedral products

$$\begin{array}{ccc} (\underline{X}, \underline{A})^{\partial\Delta^{k-1}} \times A_m & \xrightarrow{1 \times i_m} & (\underline{X}, \underline{A})^{\partial\Delta^{k-1}} \times X_m \\ \Big\downarrow{\scriptstyle j \times 1} & & \Big\downarrow \\ (\underline{X}, \underline{A})^{\Delta^{k-1}} \times A_m & \longrightarrow & (\underline{X}, \underline{A})^{\partial\Delta^k}. \end{array} \tag{8.2}$$

Consider the special case of $(\underline{X}, \underline{*})$. Then $(\underline{X}, \underline{*})^{\Delta^{k-1}} = X_1 \times \cdots \times X_k$, $(\underline{X}, \underline{*})^{\partial\Delta^{k-1}} = FW(1, \ldots, k)$, and $(\underline{X}, \underline{*})^{\partial\Delta^k} = FW(1, \ldots, k+1)$. Phrased in terms of a larger vertex set, diagram (8.2) implies the following.

Lemma 8.5: *Let $\{i_1, \ldots, i_k\}$ be a subset of $\{1, \ldots, m-1\}$ with the property that $1 \le i_1 < \cdots < i_k \le m-1$. Then there is a pushout*

$$\begin{array}{ccc} FW(i_1, \ldots, i_k) & \longrightarrow & FW(i_1, \ldots, i_k) \times X_m \\ \Big\downarrow & & \Big\downarrow \\ X_{i_1} \times \cdots \times X_{i_k} & \longrightarrow & FW(i_1, \ldots, i_k, m). \end{array}$$

□

Remark 8.6: Lemma 8.5 has a direct proof as well. By definition of the fat wedge,

$$FW(i_1,\ldots,i_k,m)=\left(\bigcup_{j=1}^{k} X_{i_1}\times\cdots\times X_{i_{j-1}}\times *\times X_{i_{j+1}}\times\cdots\times\cdots X_{i_k}\times X_m\right)$$
$$\cup\,(X_{i_1}\times\cdots\times X_{i_k}\times *)\,.$$

Notice that the first term in this union is $FW(i_1,\ldots,i_k)\times X_m$, the second term is $X_{i_1}\times\cdots\times X_{i_k}$, and the union is taken over the common subspace $FW(i_1,\ldots,i_k)$. The asserted pushout now follows immediately.

Next, consider the special case of $(\underline{CX},\underline{X})$. By (8.2) and Corollary 5.7, for $m\geq 2$ there is a pushout of polyhedral products

$$\begin{array}{ccc}(\underline{CX},\underline{X})^{\partial\Delta^{m-2}}\times X_m & \xrightarrow{\pi_1} & (\underline{CX},\underline{X})^{\partial\Delta^{m-2}}\\ \downarrow{\scriptstyle j\times 1} & & \downarrow\\ (\underline{CX},\underline{X})^{\Delta^{m-2}}\times X_m & \longrightarrow & (\underline{CX},\underline{X})^{\partial\Delta^{m-1}}.\end{array}\tag{8.3}$$

Lemma 8.7: *For $m\geq 2$ there is a homotopy equivalence*

$$(\underline{CX},\underline{X})^{\partial\Delta^{m-1}}\simeq\Sigma^{m-1}X_1\wedge\cdots\wedge X_m.$$

Proof: The proof is by induction on the number of vertices. When $m=2$ then $\partial\Delta^1$ is two disjoint vertices, so by Example 2.2 we have $(\underline{CX},\underline{X})^{\partial\Delta^1}\simeq\Sigma X_1\wedge X_2$. Assume inductively that $(\underline{CX},\underline{X})^{\partial\Delta^{m-2}}\simeq\Sigma^{m-2}X_1\wedge\cdots\wedge X_{m-1}$. By Example 2.3, $(\underline{CX},\underline{X})^{\Delta^{m-1}}=CX_1\times\cdots\times CX_{m-1}\simeq *$, so the homotopy pushout (8.3) is equivalent to the homotopy pushout

$$\begin{array}{ccc}(\Sigma^{m-2}X_1\wedge\cdots\wedge X_{m-1})\times X_m & \xrightarrow{\pi_1} & \Sigma^{m-2}X_1\wedge\cdots\wedge X_{m-1}\\ \downarrow{\scriptstyle \pi_2} & & \downarrow\\ X_m & \longrightarrow & (\underline{CX},\underline{X})^{\partial\Delta^{m-1}}.\end{array}$$

Therefore, by Lemma 6.1, there are homotopy equivalences

$$(\underline{CX},\underline{X})^{\partial\Delta^{m-1}}\simeq(\Sigma^{m-2}X_1\wedge\cdots\wedge X_{m-1})*X_m\simeq\Sigma^{m-1}X_1\wedge\cdots\wedge X_m.\ \square$$

By Lemma 8.7, the homotopy fibration (8.1) can be rewritten as a homotopy fibration

$$\Omega X_1*\cdots*\Omega X_m\xrightarrow{\phi_m}FW(1,\ldots,m)\longrightarrow X_1\times\cdots\times X_m$$

for some map ϕ_m. This lets us define higher Whitehead products.

Definition 8.8: For $1 \leq i \leq m$, suppose that there are maps $f_i \colon X_i \longrightarrow Y$ for some space Y. Suppose also that there is an extension of the wedge $\bigvee_{i=1}^m X_i \xrightarrow{\bigvee_{i=1}^m f_i} Y$ to a map $\overline{f} \colon FW(1,\dots,m) \longrightarrow Y$. Then the composite

$$\Omega X_1 * \cdots * \Omega X_m \xrightarrow{\phi_m} FW(1,\dots,m) \xrightarrow{\overline{f}} Y$$

is a *higher Whitehead product.*

Observe that if $m = 2$ then $\overline{f} = f_1 \vee f_2$ and ϕ_2 is the ordinary Whitehead product $[ev_L, ev_R]$. So the homotopy class of $\overline{f} \circ \phi_2$ is determined by the homotopy classes of f_1 and f_2. However, if $m \geq 3$ then there may be different choices of an extension $\overline{f}$ of $\bigvee_{i=1}^m f_i$, so the homotopy class of the higher Whitehead product $\overline{f} \circ \phi_m$ need not be determined only by the homotopy classes of $f_1, \dots, f_m$.

8.3. *Relating higher Whitehead products to minimal missing faces*

Let K be a simplicial complex on the vertex set $[m]$. We use the homotopy fibration

$$(\underline{C\Omega X}, \underline{\Omega X})^K \longrightarrow (\underline{X}, \underline{*})^K \longrightarrow \prod_{i=1}^m X_i$$

from Corollary 3.14 to establish a correspondence between the minimal missing faces in K and nontrivial higher Whitehead products on $(\underline{X}, \underline{*})^K$. This requires some definitions.

Definition 8.9: Let K be a simplicial complex on the vertex set $[m]$. A subset $\sigma = (i_1, \dots, i_k)$ of $[m]$ is a *missing face* of K if $\sigma \notin K$. It is a *minimal missing face* if any proper subset of σ is a face of K. Equivalently, σ is a minimal missing face of K if $\partial\sigma \subseteq K$ but $\sigma \notin K$.

Example 8.10: Let K be the following simplical complex:

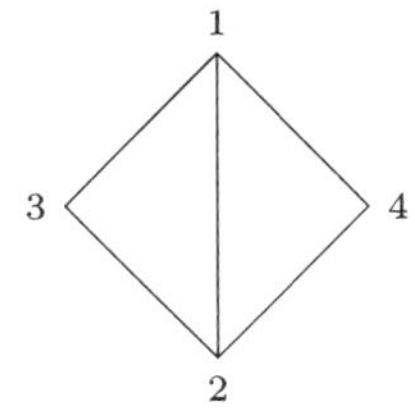

K

The missing faces of K are

$$MF(K) = \{(3,4),(1,2,3),(1,2,4),(1,3,4),(2,3,4),(1,2,3,4)\}.$$

As $(3,4)$ is a proper face of $(1,3,4)$, $(2,3,4)$ and $(1,2,3,4)$ which is missing from K, none of $(1,3,4)$, $(2,3,4)$ or $(1,2,3,4)$ are minimal missing faces. The complete list of minimal missing faces of K is

$$MMF(K) = \{(3,4),(1,2,3),(1,2,4)\}.$$

Let K be a simplicial complex on the vertex set $[m]$. Recall that K_m is the simplicial complex consisting of m disjoint points. Including the vertices into K gives a simplicial map $K_m \longrightarrow K$. This induces a map of polyhedral products $(\underline{X},\underline{*})^{K_m} \longrightarrow (\underline{X},\underline{*})^K$. By Example 2.1, $(\underline{X},\underline{*})^{K_m} = \bigvee_{i=1}^m X_i$, so the inclusion of K_m into K induces a map

$$F_m \colon \bigvee_{i=1}^{m} X_i \longrightarrow (\underline{X},\underline{*})^K.$$

Lemma 8.11: *Let K be a simplicial complex on the vertex set $[m]$. Then $\sigma = (i_1,\ldots,i_k)$ is a minimal missing face of K if and only if the composite $\bigvee_{j=1}^k X_{i_j} \hookrightarrow \bigvee_{i=1}^m X_i \xrightarrow{F_m} (\underline{X},\underline{*})^K$ extends to a map $FW(i_1,\ldots,i_k) \longrightarrow (\underline{X},\underline{*})^K$ but does not extend to a map $\prod_{j=1}^k X_{j_k} \longrightarrow (\underline{X},\underline{*})^K$.*

Proof: By definition, $\sigma = (i_1,\ldots,i_k)$ is a minimal missing face of K if and only if $\partial\sigma \subseteq K$ but $\sigma \notin K$. That is, if and only if there is a map of simplicial complexes $\partial\sigma \longrightarrow K$ which does not extend to a map $\sigma \longrightarrow K$. By definition of the polyhedral product, $(\underline{X},\underline{*})^K$ is the union of the coordinate subspaces formed from the faces of K. That is, $X_{j_1} \times \cdots \times X_{j_t}$ is a coordinate subspace of $(\underline{X},\underline{*})^K$ if and only if the face $\tau = (j_1,\ldots,j_t) \in K$. In our case, since σ is a $(k-1)$-simplex, the simplicial map $\partial\sigma \longrightarrow K$ exists if and only if every coordinate subspace of the fat wedge $FW(i_1,\ldots,i_k)$ is in $(\underline{X},\underline{*})^K$. That is, if and only if there is a map $FW(i_1,\ldots,i_k) \longrightarrow (\underline{X},\underline{*})^K$ extending the inclusion of the coordinates $\bigvee_{j=1}^k X_{i_j} \longrightarrow (\underline{X},\underline{*})^K$. Further, the simplicial map $\partial\sigma \longrightarrow K$ does not extend to a simplicial map $\sigma \longrightarrow K$ if and only if $X_{1_1} \times \cdots \times X_{i_k}$ is not a coordinate subspace of $(\underline{X},\underline{*})^K$. That is, if and only if the map $FW(i_1,\ldots,i_k) \longrightarrow (\underline{X},\underline{*})^K$ does not extend to a map $X_{i_1} \times \cdots \times X_{i_k} \longrightarrow (\underline{X},\underline{*})^K$. □

Let $\sigma = (i_1,\ldots,i_k)$ be a minimal missing face of K. By Lemma 8.11, there is an extension of the composite $\bigvee_{j=1}^k X_{i_j} \hookrightarrow \bigvee_{i=1}^m X_i \xrightarrow{F_m} (\underline{X},\underline{*})^K$ to a map $f\colon FW(i_1,\ldots,i_k) \longrightarrow (\underline{X},\underline{*})^K$. Define the higher Whitehead product $w(\sigma)$ by the composite

$$w(\sigma)\colon \Omega X_{i_1} * \cdots * \Omega X_{i_k} \xrightarrow{\phi_k} FW(i_1,\ldots,i_k) \xrightarrow{f} (\underline{X},\underline{*})^K.$$

In the next Proposition the equivalence in Lemma 8.11 is rephrased in terms of the existence of nontrivial higher Whitehead products.

Proposition 8.12: *Let K be a simplicial complex on the vertex set $[m]$ and let $(\underline{X},\underline{*})$ be a sequence of pointed CW-pairs. Then $\sigma = (i_1,\ldots,i_k)$ is a minimal missing face of K if and only if the higher Whitehead product $w(\sigma)\colon \Omega X_{i_1} * \cdots * \Omega X_{i_k} \xrightarrow{\phi_k} FW(i_1,\ldots,i_k) \xrightarrow{f} (\underline{X},\underline{*})^K$ is nontrivial.*

Proof: Suppose that $w(\sigma) = f\circ\phi_k$ is nontrivial. Since ϕ_k is the homotopy fibre of the inclusion $FW(i_1,\ldots,i_k) \longrightarrow X_{i_1}\times\cdots\times X_{i_k}$, the nontriviality of $f\circ\phi_k$ implies that there can be no extension of $FW(i_1,\ldots,i_k) \xrightarrow{f} (\underline{X},\underline{*})^K$ to a map $g\colon X_{i_1} \times \cdots \times X_{i_k} \longrightarrow (\underline{X},\underline{*})^K$. For otherwise $f\circ\phi_k$ would factor through two consecutive maps in a homotopy fibration and so be null homotopic. Therefore, as f does not extend to g, by Lemma 8.11 σ is a minimal missing face of K.

Conversely, suppose that $w(\sigma)$ is null homotopic. We wish to show that σ is not a minimal missing face of K. This case is trickier for the following reason. If each X_{i_j} is a suspension, $X_{i_j} = \Sigma X'_{i_j}$, then the composite

$$X'_{i_1} * \cdots * X'_{i_k} \xrightarrow{\overline{E}} \Omega\Sigma X'_{i_1} * \cdots * \Omega\Sigma X'_{i_k} \xrightarrow{\phi_k} FW(i_1,\ldots,i_k)$$

is the attaching map for the product $\Sigma X'_{i_1}\times\cdots\times\Sigma X'_{i_k}$, where $\overline{E} = E*\cdots*E$. Therefore, f extends to g in this case if and only if $f\circ\phi_k\circ\overline{E}$ is null homotopic. In particular, if $w(\sigma) = f\circ\phi_k$ is null homotopic then $f\circ\phi_k\circ\overline{E}$ is null homotopic, implying that f extends to g, and so by Lemma 8.11, σ is not a minimal missing face of K. However, in the general case X_{i_j} may not be a suspension, so the product $X_{i_1} \times \cdots \times X_{i_k}$ may not be the cofibre of a map into the $FW(i_1, \ldots, i_m)$.

Instead, for each $1 \leq j \leq k$, let $g_j \colon S^{n_j} \longrightarrow X_{i_j}$ be the inclusion of a bottom cell. Let $FW(\underline{S})$ be the fat wedge on $S^{n_1} \times \cdots \times S^{n_k}$. By Proposition 3.1 (a), the polyhedral product is natural with respect to the maps g_j, so we obtain a homotopy fibration diagram

$$\begin{array}{ccccc} \Omega S^{n_1} * \cdots * \Omega S^{n_j} & \xrightarrow{\phi_k} & FW(\underline{S}) & \longrightarrow & \prod_{j=1}^{k} S^{n_j} \\ \Big\downarrow{\scriptstyle G} & & \Big\downarrow{\scriptstyle FW(g_1,\ldots,g_k)} & & \Big\downarrow{\scriptstyle \prod_{j=1}^{k} g_j} \\ \Omega X_{i_1} * \cdots \wedge * \Omega X_{i_k} & \xrightarrow{\phi_k} & FW(i_1, \ldots, i_k) & \longrightarrow & \prod_{j=1}^{k} X_{i_k} \end{array} \qquad (8.4)$$

where $G = \Omega g_1 * \cdots * \Omega g_k$.

Now suppose that $w(\sigma) = f \circ \phi_k$ is null homotopic. Then the composite $f \circ \phi_k \circ G$ is null homotopic, implying by the homotopy commutativity of the left square in (8.4) that $f \circ FW(g_1, \ldots, g_k) \circ \phi_k$ is null homotopic. Notice that this is also a higher Whitehead product $\overline{f} \circ \phi_k$, where $\overline{f} = f \circ FW(g_1, \ldots, g_k)$. Since each S^{n_j} is a suspension, by the argument above, the null homotopy for $\overline{f} \circ \phi_k$ implies that σ is not a minimal missing face of K, and we are done. $\square$

Example 8.13: Return to Example 8.10. By Proposition 8.12, the minimal missing face $(3, 4)$ corresponds to a nontrivial Whitehead product $\Omega X_3 * \Omega X_4 \longrightarrow (\underline{X}, \underline{*})^K$. The minimal missing faces $(1, 2, 3)$ and $(1, 2, 4)$ correspond to nontrivial higher Whitehead products $\Omega X_1 * \Omega X_2 * \Omega X_3 \xrightarrow{\phi_3} FW(1, 2, 3) \longrightarrow (\underline{X}, \underline{*})^K$ and $\Omega X_1 * \Omega X_2 * \Omega X_4 \xrightarrow{\phi_3} FW(1, 2, 4) \longrightarrow (\underline{X}, \underline{*})^K$.

To close this section we return to the homotopy fibration $(\underline{C\Omega X}, \underline{\Omega X})^K \xrightarrow{\varphi} (\underline{X}, \underline{*})^K \longrightarrow \prod_{i=1}^{m} X_i$ and state some results. In [20] it is shown that if K is an "MF-complex" and each X_i is a suspension, then $(\underline{C\Omega X}, \underline{\Omega X})^K$ is homotopy equivalent to a wedge of iterated suspensions of the spaces ΩX_i and the map φ is a wedge sum of: higher Whitehead products, one for each minimal missing face of K; and iterated Whitehead products formed from one higher Whitehead product and possibly multiple copies of the coordiate inclusions $X_i \longrightarrow (\underline{X}, \underline{*})^K$. A similar statement is proved for the moment-angle complex $\mathcal{Z}_K$ even though $X_i = \mathbb{C}P^\infty$ is not a suspension. In [24] many of these results are generalized to the case when K is shifted.

9. A decomposition of the based loops on a polyhedral product

In this section we turn to investigating the homotopy type of $\Omega(\underline{X},\underline{A})^K$. We prove a new integral homotopy decomposition for $\Omega(\underline{X},\underline{A})^K$ that holds for any simplicial complex K. A different, much finer homotopy decomposition of $\Omega(\underline{X},\underline{A})^K$ can be found in [40]. The advantage of the cruder homotopy decomposition proved here is that the factors are easy to identify in terms of other spaces.

Let K be a simplicial complex on the vertex set $[m]$. Recall from (5.1) that there is a pushout of simplicial complexes

$$\begin{array}{ccc} \mathrm{link}_K(m) & \longrightarrow & \mathrm{star}_K(m) \\ \downarrow & & \downarrow \\ K\backslash\{m\} & \longrightarrow & K \end{array} \tag{9.1}$$

where $\mathrm{star}_K(m) = link_K(m) * \{m\}$. Define a new simplicial complex $\overline{K}$ on the vertex set $[m]$ by

$$\overline{K} = K\backslash\{m\} * \{m\}.$$

Observe that there is an inclusion of simplicial complexes $K\backslash\{m\} \longrightarrow \overline{K}$ given by including the join factor, so as $\mathrm{star}_K(m) = \mathrm{link}_K(m) * \{m\}$, there is a pushout map

$$K \longrightarrow \overline{K}.$$

Observe also that $K\backslash\{m\}$ is the full subcomplex of $\overline{K}$ on the vertex set $[m-1]$. That is, $\overline{K}\backslash\{m\} = K\backslash\{m\}$.

Now turn to polyhedral products. By Proposition 5.6, the pushout of simplicial complexes in (9.1) induces a pushout of polyhedral products

$$\begin{array}{ccc} (\underline{X},\underline{A})^{\mathrm{link}_K(m)} \times A_m & \xrightarrow{1\times i_m} & (\underline{X},\underline{A})^{\mathrm{link}_K(m)} \times X_m \\ \Big\downarrow{\scriptstyle j\times 1} & & \Big\downarrow \\ (\underline{X},\underline{A})^{K\backslash\{m\}} \times A_m & \longrightarrow & (\underline{X},\underline{A})^K \end{array} \tag{9.2}$$

where i_m is the inclusion. To relate this to $(\underline{X},\underline{A})^{\overline{K}}$, observe that the definition of the join of two simplicial complexes implies that if $K = K_1 * K_2$ then there is a homeomorphism

$$(\underline{X},\underline{A})^K \simeq (\underline{X},\underline{A})^{K_1} \times (\underline{X},\underline{A})^{K_2}.$$

In particular, as $\overline{K} = K\backslash\{m\} * \{m\}$ there is a homeomorphism

$$(\underline{X}, \underline{A})^{\overline{K}} \simeq (\underline{X}, \underline{A})^{K\backslash\{m\}} \times X_m.$$

We follow Félix and Tanré's idea [13] of mapping all four corners of the pushout (9.2) into $(\underline{X}, \underline{A})^{\overline{K}}$ and taking homotopy fibres. Mather's Cube Lemma [30] will help identify the homotopy types of the fibres.

Lemma 9.1: ***(Mather)*** *Suppose that there is a diagram of spaces and maps*

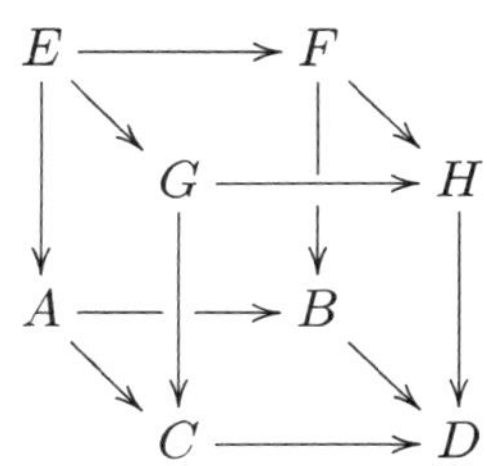

where the bottom face is a homotopy pushout and the four sides are obtained by pulling back with $H \longrightarrow D$. Then the top face is a homotopy pushout. □

Consider the map $(\underline{X}, \underline{A})^{K} \longrightarrow (\underline{X}, \underline{A})^{\overline{K}} \simeq (\underline{X}, \underline{A})^{K\backslash\{m\}} \times X_m$. Composing from all four corners of the pushout (9.2), we obtain homotopy fibrations

$$G_m \times Y_m \longrightarrow (\underline{X}, \underline{A})^{\mathrm{link}_K(m)} \times A_m \xrightarrow{j \times i_m} (\underline{X}, \underline{A})^{K\backslash\{m\}} \times X_m$$

$$Y_m \longrightarrow (\underline{X}, \underline{A})^{K\backslash\{m\}} \times A_m \xrightarrow{1 \times i_m} (\underline{X}, \underline{A})^{K\backslash\{m\}} \times X_m$$

$$G_m \longrightarrow (\underline{X}, \underline{A})^{\mathrm{link}_K(m)} \times X_m \xrightarrow{j \times 1} (\underline{X}, \underline{A})^{K\backslash\{m\}} \times X_m$$

$$F_m \longrightarrow (\underline{X}, \underline{A})^{K} \longrightarrow (\underline{X}, \underline{A})^{K\backslash\{m\}} \times X_m$$

where G_m is the homotopy fibre of $(\underline{X}, \underline{A})^{\mathrm{link}_K(m)} \xrightarrow{j} (\underline{X}, \underline{A})^{K\backslash\{m\}}$, Y_m is the homotopy fibre of the inclusion $A_m \xrightarrow{i_m} X_m$, and F_m is defined as the homotopy fibre of the map $(\underline{X}, \underline{A})^{K} \longrightarrow (\underline{X}, \underline{A})^{K\backslash\{m\}} \times X_m$. By Lemma 9.1, we obtain a homotopy pushout

$$\begin{array}{ccc} G_m \times Y_m & \longrightarrow & G_m \\ \downarrow & & \downarrow \\ Y_m & \longrightarrow & F_m. \end{array} \qquad (9.3)$$

We claim that the maps $G_m \times Y_m \longrightarrow G_m$ and $G_m \times Y_m \longrightarrow Y_m$ in (9.3) are projections. In the first case, the map is induced from Lemma 9.1 by the homotopy pullback diagram

$$\begin{array}{ccccc} G_m \times Y_m & \longrightarrow & (\underline{X},\underline{A})^{\mathrm{link}_K(m)} \times A_m & \longrightarrow & (\underline{X},\underline{A})^{K\backslash\{m\}} \times X_m \\ \big\downarrow & & \big\downarrow{\scriptstyle 1\times i_m} & & \Vert \\ G_m & \longrightarrow & (\underline{X},\underline{A})^{\mathrm{link}_K(m)} \times X_m & \longrightarrow & (\underline{X},\underline{A})^{K\backslash\{m\}} \times X_m. \end{array}$$

The maps in the right square are defined coordinate-wise, so the homotopy pullback is homotopy equivalent to the topological pullback obtained by turning both $(\underline{X},\underline{A})^{\mathrm{link}_K(m)} \longrightarrow (\underline{X},\underline{A})^{K\backslash\{m\}}$ and $A_m \longrightarrow X_m$ into (strict) fibrations. By doing so the map of fibres is also defined coordinate-wise, implying that $G_m \times Y_m \longrightarrow G_m$ is homotopic to the projection. Similarly, the map $G_m \times Y_m \longrightarrow Y_m$ is the projection. Therefore (9.3) has been more accurately identified as the homotopy pushout

$$\begin{array}{ccc} G_m \times Y_m & \xrightarrow{\pi_1} & G_m \\ \big\downarrow{\scriptstyle \pi_2} & & \big\downarrow \\ Y_m & \longrightarrow & F_m. \end{array}$$

Hence, Lemma 6.1 immediately implies the following.

Lemma 9.2: *There is an integral homotopy equivalence*

$$F_m \simeq G_m * Y_m.$$

□

We take this a step further. Consider the homotopy fibration

$$F_m \longrightarrow (\underline{X},\underline{A})^K \longrightarrow (\underline{X},\underline{A})^{K\backslash\{m\}} \times X_m.$$

Since $K\backslash\{m\}$ and $\{m\}$ are the full subcomplexes of K on the vertex sets $[m-1]$ and $\{m\}$ respectively, by Proposition 3.6, the polyhedral products $(\underline{X},\underline{A})^{K\backslash\{m\}}$ and $(\underline{X},\underline{A})^{\{m\}} = X_m$ both retract off $(\underline{X},\underline{A})^K$. In general, these retractions cannot be multiplied together in $(\underline{X},\underline{A})^K$, nor multiplied with the image of the map from F_m, because $(\underline{X},\underline{A})^K$ is rarely an H-space. But after looping this can be done. Thus, if μ is the loop multiplication on $(\underline{X},\underline{A})^K$, then the composite

$$\Omega X_m \times \Omega(\underline{X},\underline{A})^{K\backslash\{m\}} \times \Omega F_m \longrightarrow \Omega(\underline{X},\underline{A})^K \times \Omega(\underline{X},\underline{A})^K \times \Omega(\underline{X},\underline{A})^K \xrightarrow{\mu} \Omega(\underline{X},\underline{A})^K$$

is a homotopy equivalence. Combining this with Lemma 9.2, we obtain the following.

Proposition 9.3: *Let K be a simplicial complex on the index set $[m]$. Then there is an integral homotopy equivalence*

$$\Omega(\underline{X},\underline{A})^K \simeq \Omega X_m \times \Omega(\underline{X},\underline{A})^{K\backslash\{m\}} \times \Omega(G_m * Y_m).$$

□

Since $K\backslash\{m\}$ is a simplicial complex on the index set $[m-1]$, Proposition 9.3 can be applied again to decompose $\Omega(\underline{X},\underline{A})^{K\backslash\{m\}}$. Iterating, for $1 \leq i \leq m$, let R_{m-i} be the restriction of K to the index set $[m-i]$. Note that $R_m = K$. For $1 \leq i < m$, let $L_{m-i} = \text{link}_{R_{m-i+1}}(m-i+1)$. Then there is a star-link-restriction pushout of simplicial complexes

$$\begin{array}{ccc} L_{m-i} & \longrightarrow & L_{m-i} * \{m-i+1\} \\ \downarrow & & \downarrow \\ R_{m-i} & \longrightarrow & R_{m-i+1}. \end{array}$$

Let G_{m-i} be the homotopy fibre of $(\underline{X},\underline{A})^{L_{m-i}} \longrightarrow (\underline{X},\underline{A})^{R_{m-i}}$, and let Y_{m-i} be the homotopy fibre of the inclusion $A_{m-i} \longrightarrow X_{m-i}$. Then iterating Proposition 9.3 gives the following.

Theorem 9.4: *Let K be a simplicial complex on the vertex set $[m]$. Then there is an integral homotopy equivalence*

$$\Omega(\underline{X},\underline{A})^K \simeq \left(\prod_{i=1}^{m} \Omega X_i\right) \times \left(\prod_{i=1}^{m} \Omega(G_i * Y_i)\right).$$

□

Remark 9.5: Theorem 9.4 is completely general, and says that the homotopy type of $\Omega(\underline{X},\underline{A})^K$ depends only on the spaces X_i, the homotopy fibres of the inclusions $A_m \longrightarrow X_m$ and the homotopy fibres of the link-restriction maps $(\underline{X},\underline{A})^{L_i} \longrightarrow (\underline{X},\underline{A})^{R_i}$. So again we see the importance of knowing the homotopy class of the link-restriction maps.

Remark 9.6: Since $G_i * Y_i \simeq \Sigma G_i \wedge Y_i$, each of the factors $\Omega(G_i * Y_i)$ in the homotopy decomposition for $\Omega(\underline{X},\underline{A})^K$ is a loop suspension. A great deal is known about the homotopy theory of loop suspensions. In particular, the Bott-Samelson Theorem states that if R is a field then there is an algebra isomorphism $H_*(\Omega\Sigma Y; R) \cong T(\widetilde{H}_*(Y; R))$, where $T(\cdot)$ is the free tensor algebra functor. Thus, from Theorem 9.4 and the Kunneth isomorphism, there is an algebra isomorphism

$$H_*((\underline{X},\underline{A})^K;R)\cong\left(\bigotimes_{i=1}^m H_*(\Omega X_i;R)\right)\otimes\left(\bigotimes_{i=1}^m T(\widetilde{H}_*(G_i\wedge Y_i;R))\right).$$

Also, James [25] shows that there is a homotopy equivalence

$$\Sigma\Omega\Sigma Y\simeq\bigvee_{n=1}^{\infty}\Sigma Y^{(n)},$$

where $Y^{(n)}$ is the n-fold smash product of Y with itself. Thus, after suspending the homotopy equivalence for $\Omega(\underline{X},\underline{A})^K$ and using James' result together with an iteration of the fact that $\Sigma(X\times Y)\simeq\Sigma X\vee\Sigma Y\vee(\Sigma X\wedge Y)$, we obtain a wedge decomposition for $\Sigma\Omega(\underline{X},\underline{A})^K$ in terms of suspensions of iterated smash products of the spaces ΩX_i and $G_i\wedge Y_i$.

We mention two cases of Theorem 9.4. First, in the case $(\underline{X},\underline{*})^K$, the homotopy fibre Y_i of the inclusion $*\longrightarrow X_i$ is ΩX_i. Thus we obtain the following.

Corollary 9.7: *Let K be a simplicial complex on the vertex set $[m]$. Then there is an integral homotopy equivalence*

$$\Omega(\underline{X},\underline{*})^K\simeq\left(\prod_{i=1}^m\Omega X_i\right)\times\left(\prod_{i=1}^m\Omega(G_i*\Omega X_i)\right).$$

□

Second, in the case of $(\underline{CX},\underline{X})$, since CX_i is contractible, the term ΩCX_i in Theorem 9.4 is contractible, and the homotopy fibre of the inclusion $X_i\longrightarrow CX_i$ is homotopy equivalent to X_i. Thus we obtain the following.

Corollary 9.8: *Let K be a simplicial complex on the vertex set $[m]$. Then there is an integral homotopy equivalence*

$$\Omega(\underline{CX},\underline{X})^K\simeq\prod_{i=1}^m\Omega(G_i*X_i).$$

□

As special cases of Corollaries 9.7 and 9.8 we obtain homotopy decompositions for the loops on the Davis-Januszkiewicz space and the loops on the moment-angle complex. In the first case, each pair $(X_i,*)$ is $(\mathbb{C}P^\infty,*)$, and $\Omega\mathbb{C}P^\infty \simeq S^1$. In the second case, each pair (CX_i, X_i) is (D^2, S^1). In particular, in the first case we have $G_i * \Omega X_i \simeq G_i * S^1 \simeq \Sigma^2 G_i$ and in the second case we have $G_i * X_i \simeq G_i * S^1 \simeq \Sigma^2 G_i$.

Corollary 9.9: *Let K be a simplicial complex on the vertex set $[m]$. Then there are integral homotopy equivalences*

$$\Omega DJ_K \simeq \left(\prod_{i=1}^m S^1\right) \times \left(\prod_{i=1}^m \Omega\Sigma^2 G_i\right)$$

and

$$\Omega\mathcal{Z}_K \simeq \prod_{i=1}^m \Omega\Sigma^2 G_i.$$

□

Remark 9.10: Specializing Remark 9.6 to the case of $\Omega\mathcal{Z}_K$, for any field R there is an algebra isomorphism

$$H_*(\Omega\mathcal{Z}_K; R) \cong \bigotimes_{i=1}^m T(\widetilde{H}_*(\Sigma G_i; R))$$

and a homotopy equivalence for $\Sigma\Omega\mathcal{Z}_K$ in terms of suspensions of iterated smash products of the spaces ΣG_i.

10. Exponents and Barratt's conjecture

The homotopy decomposition in Theorem 9.4 is provocative. It is often the case in unstable homotopy theory that properties of a space Y are studied by considering properties of ΩY. For example, if Y is simply-connected then the homotopy groups of ΩY are the same as those for Y but shifted down one degree. One advantage of ΩY over Y is the loop multiplication, which better allows for the possibility of a homotopy decomposition into simpler factors. If $Y = \Sigma Y'$ then the situation is greatly improved, as in Remark 9.6. Thus Theorem 9.4 suggests that there may be properties of $(\underline{X}, \underline{A})^K$ that may be better discerned by looking at $\Omega(\underline{X}, \underline{A})^K$.

In this section and Section 13 we describe applications of the loop space decomposition, although none is fully explored and many open questions remain. In this section we discuss a connection with homotopy exponents and Barratt's conjecture, and in Section 13 we discuss connections with rational homotopy theory and Moore's conjecture. To consider homotopy exponents we begin with some definitions.

Definition 10.1: Let X be a path-connected, pointed topological space and let p be a prime. The *homotopy exponent* of X at p is the least power of p that annihilates the p-torsion in $\pi_*(X)$.

If the power of p is finite, say r, write $\exp_p(X) = p^r$. If there is no such least power of p, write $\exp_p(X) = \infty$.

Definition 10.2: Let X be a path-connected co-H-space and let p be a prime. The space X has *co-H-exponent* p^r if the degree p^r map on X is null homotopic but the degree p^{r-1} map is not.

For example, for $m \geq 2$, let $P^m(p^r)$ be the *Moore space*, defined by the homotopy cofibration

$$S^{m-1} \xrightarrow{p^r} S^{m-1} \longrightarrow P^m(p^r).$$

Observe that if $m \geq 3$ then $P^m(p^r) \simeq \Sigma P^{m-1}(p^r)$, implying that $P^m(p^r)$ is a co-H-space. By [32] the degree p^r map on $P^m(p^r)$ is null homotopic if $p^r \neq 2$, and if $p^r = 2$ then the degree 4 map on $P^m(2)$ is null homotopic but the degree 2 map is not. Therefore the co-H-exponent of $P^m(p^r)$ is p^r if $p^r \neq 2$, and is 4 if $p^r = 2$.

Definition 10.3: Let Y be a path-connected H-space and let p be a prime. The space Y has *H-exponent* p^r if the p^r-power map on Y is null homotopic but the p^{r-1}-power map is not.

Note that if Y has H-exponent p^r then $\exp_p(Y) \leq p^r$. For the p^r-power map on Y induces multiplication by p^r on homotopy groups. However, if X has co-H-exponent p^r then the p^r-power map on ΩX may not be null homotopic. This is due to the nontrivial difference between the loops on the degree p^r map and the p^r-power map; this difference exists by the distributivity law and may be difficult to control (see [9] for an excellent discussion of the distributivity law). Barratt [4] conjectured that, in fact, while the difference may be complex its effect on exponents is minimal.

Conjecture: (Barratt) Let X be a space with co-H-exponent p^r. Then the p^{r+1}-power map on $\Omega^2\Sigma X$ is null homotopic.

Barratt's conjecture is known to be true only for certain Moore spaces. If p is odd and $r \geq 1$ then the conjecture holds by [33]. If $p = 2$ and $r \geq 6$ the conjecture holds by [38]. For $p = 2$ and $2 \leq r \leq 5$, upper bounds on the homotopy exponent of $P^m(2^r)$ are known [38]. The most interesting case is $P^m(2)$. Barratt's conjecture predicts that $\exp_2(P^m(2)) = 8$. However, there is no known upper bound on the homotopy exponent of $P^m(2)$. That is, as far as anyone knows, there could be torsion of arbitrarily high order in $\pi_*(P^m(2))$.

We will use polyhedral products to extend the range of spaces to which a positive resolution of Barratt's conjecture would apply.

Proposition 10.4: *Let K be a simplicial complex on the vertex set $[m]$ and let p be a prime. Let $X_1, \ldots, X_m$ be co-H-spaces with co-H-exponent bounded above by p^r. If Barratt's conjecture holds, then the p^{r+1}-power map on $\Omega^2(\underline{X}, \underline{A})^K$ is null homotopic, and consequently* $\exp_p((\underline{X}, \underline{A})^K) \leq p^{r+1}$.

Proof: In general, if X is a co-H-space and Y is a space then $Y \wedge X$ is also a co-H-space with comultiplication given by the composite

$$Y \wedge X \xrightarrow{1\wedge\sigma} Y \wedge (X \vee X) \xrightarrow{\simeq} (Y \wedge X) \vee (Y \wedge X)$$

where σ is the co-multiplication on X. Therefore, if the degree p^r map on X is null homotopic then so is the degree p^r map on $Y \wedge X$.

In our case, by Corollary 9.8 there is a homotopy equivalence $\Omega(\underline{CX}, \underline{X})^K \simeq \prod_{i=1}^m \Omega(G_i * X_i)$. Looping, we obtain a homotopy equivalence of H-spaces $\Omega^2(\underline{CX}, \underline{X})^K \simeq \prod_{i=1}^m \Omega^2(G_i * X_i)$. Therefore the p^{r+1}-power map on $\Omega^2(\underline{CX}, \underline{X})^K$ is null homotopic if and only if the same is true for each factor $\Omega^2(G_i * X_i)$. Since each co-H-space X_i has co-H-exponent at most p^r, the same is therefore true for each $G_i \wedge X_i$. So Barratt's conjecture implies that the p^{r+1}-power map on $\Omega^2(G_i * X_i) \simeq \Omega^2(\Sigma G_i \wedge X_i)$ is null homotopic. □

Since Barratt's conjecture holds for Moore spaces $P^m(p^r)$ where p is odd or where $p = 2$ and $r \geq 6$, Proposition 10.4 also holds in these cases.

11. Elliptic and hyperbolic spaces

Félix, Halperin and Thomas [12] proved the remarkable fact that a finite dimensional path-connected, pointed space X either has finitely many rational homotopy groups or the total number of rational homotopy groups in dimensions $m \leq n$ grows exponentially with n. That is, there is no space whose rational homotopy groups have polynomial growth. Plainly put, either X has very few rational homotopy groups or it has an awful lot. There's no middle ground.

It is natural to ask which spaces fall into which category. It helps to give the two options names.

Definition 11.1: A finite dimensional space X is *elliptic* if it has finitely many rational homotopy groups. Otherwise X is *hyperbolic.*

The canonical example of an elliptic space is a product of spheres. The canonical example of a hyperbolic space is a wedge of two spheres (hyperbolicity is proved using the Hilton-Milnor Theorem). In what follows we will give a simple way of identifying a large number of hyperbolic polyhedral products.

Observe that if $K \neq \Delta^{m-1}$ then it has a missing face, and so it has at least one minimal missing face.

Lemma 11.2: *Let K be a simplicial complex on the vertex set $[m]$ and suppose that $K \neq \Delta^{m-1}$. Let $\sigma = (i_1, \ldots, i_k) \in MMF(K)$. Then $\Omega X_{i_1} * \cdots * \Omega X_{i_k}$ is an integral retract of $(\underline{C\Omega X}, \underline{\Omega X})^K$ and $\Omega(\Omega X_{i_1} * \cdots * \Omega X_{i_k})$ is an integral retract of $\Omega(\underline{X}, \underline{*})^K$.*

Proof: Since $\sigma = (i_1, \ldots, i_k)$ is a minimal missing face of K, we have $\sigma \notin K$ but $\partial\sigma \subseteq K$. In particular, the full subcomplex of K on the vertex set $I = \{i_1, \ldots, i_k\}$ is $K_I = \partial\sigma$. So by Proposition 3.6, $(\underline{C\Omega X}, \underline{\Omega X})^{K_I}$ retracts off of $(\underline{C\Omega X}, \underline{\Omega X})^K$. On the other hand, by Lemma 8.7, there is an integral homotopy equivalence $(\underline{C\Omega X}, \underline{\Omega X})^{K_I} = (\underline{C\Omega X}, \underline{\Omega X})^{\partial\sigma} \simeq \Omega X_{i_1} * \cdots * \Omega X_{i_k}$. Thus $\Omega X_{i_1} * \cdots * \Omega X_{i_k}$ is a retract of $(\underline{C\Omega X}, \underline{\Omega X})^K$.

By Lemma 3.16, there is an integral homotopy equivalence $\Omega(\underline{X}, \underline{*})^K \simeq \prod_{i=1}^m \Omega X_i \times \Omega(\underline{C\Omega X}, \underline{\Omega X})^K$. Therefore the second assertion of the lemma follows from the first. □

Theorem 11.3: *Let K be a simplicial complex on the vertex set $[m]$ and suppose that $K \neq \Delta^{m-1}$. Let $\sigma = (i_1, \ldots, i_k) \in MMF(K)$. Suppose that $(\underline{X}, \underline{*})$ is a sequence of pointed CW-pairs where each X_i is simply-connected and rationally nontrivial, and suppose that the rank of $\widetilde{H}_*(\Omega X_{i_j}; \mathbb{Q})$ is at least 2 for some $1 \leq j \leq k$. Then $(\underline{X}, \underline{*})^K$ is hyperbolic.*

Proof: To show that a simply-connected space Y is hyperbolic it is equivalent to show that ΩY is hyperbolic. By Lemma 3.16, there is a homotopy equivalence $\Omega(\underline{X}, \underline{*})^K \simeq \prod_{i=1}^m \Omega X_i \times \Omega(\underline{C\Omega X}, \underline{\Omega X})^K$. So to show that $(\underline{X}, \underline{*})^K$ is hyperbolic it suffices to show that $(\underline{C\Omega X}, \underline{\Omega X})^K$ is hyperbolic. Further, by Lemma 11.2, $\Omega X_{i_1} * \cdots * \Omega X_{i_k}$ is a retract of $(\underline{C\Omega X}, \underline{\Omega X})^K$, so it suffices to show that $\Omega X_{i_1} * \cdots * \Omega X_{i_k}$ is hyperbolic.

By hypothesis, each X_{i_j} is simply-connected and rationally nontrivial, so the rank of $\widetilde{H}_*(\Omega X_{i_j}; \mathbb{Q})$ is at least 1 for each $1 \leq j \leq k$. Therefore, as at least one of these ranks must be strictly greater than 1, the rank of $\widetilde{H}_*(\Omega X_{i_1} * \cdots * \Omega X_{i_k};\mathbb{Q})$ is at least 2. In general, a co-H-space is rationally homotopy equivalent to a wedge of spheres. As $\Omega X_{i_1} * \cdots * \Omega X_{i_j} \simeq \Sigma^{k-1}\Omega X_{i_1} \wedge \cdots \wedge \Omega X_{i_k}$, it follows that $\Omega X_{i_1} * \cdots * \Omega X_{i_k}$ is rationally homotopy equivalent to a wedge of spheres. As the rational homology of this space has rank at least 2, $\Omega X_{i_1} * \cdots * \Omega X_{i_k}$ is rationally homotopy equivalent to a wedge of at least 2 spheres. But a wedge of 2 or more spheres is hyperbolic, and so $\Omega X_{i_1} * \cdots * \Omega X_{i_k}$ is hyperbolic. □

Theorem 11.3 is a mild generalization of a result by Félix and Tanré [13], which considered the case when each X_i equals a common space X. However, the methods are very different. Félix and Tanré used a Sullivan minimal model argument which gives only rational information. The proof of Theorem 11.3 used a minimal missing face argument which gives integral information, from which rational information can be deduced. This distinction will be important in the next section when we consider Moore's conjecture.

Theorem 11.3 implies that if $(\underline{X},\underline{*})^K$ is elliptic for $K \neq \Delta^{m-1}$ then each ΩX_i must have at most one generator in reduced rational homology. That is, each ΩX_i is either rationally contractible or rationally a sphere. So while Theorem 11.3 identifies many hyperbolic spaces, it does not apply at all to the case of $DJ_K = (\mathbb{C}P^\infty, *)^K$ since each ΩX_i equals $\Omega\mathbb{C}P^\infty \simeq S^1$. Bahri, Bendersky, Cohen and Gitler [2] considered the case of Davis-Januszkiewicz spaces and showed that DJ_K is elliptic if and only if it is a product of spheres and discs.

12. Moore's conjecture

There is a major conjecture in homotopy theory which relates the notion of a space being elliptic or hyperbolic with homotopy exponents.

Conjecture: Moore Let X be a finite CW-complex. Then the following are equivalent:

(a) X is elliptic;
(b) X has a finite homotopy exponent at every prime p;
(c) X has a finite homotopy exponent at some prime p.

Moore's conjecture is known to hold in a limited number of cases. Elliptic spaces with finite exponents at all primes include spheres [26, 41], finite H-spaces [28], H-spaces with finitely generated cohomology [8], and odd primary Moore spaces [33]. Hyperbolic spaces with no exponent at any prime include $S^n \vee S^m$ by the Hilton-Milnor Theorem, torsion-free suspensions [36], and two-cell complexes [34]. There are also partial results. In [31] it was shown that if X is elliptic then it has an exponent at all but finitely many primes, and in [37] it was shown that if X is hyperbolic and $H_*(\Omega X;\mathbb{Z})$ is p-torsion free then, provided p is large enough, X has no exponent at p.

The methods used in Theorem 11.3 to show that $(\underline{X},\underline{*})^K$ is hyperbolic given certain conditions on the spaces X_i do more. We will use them to given examples when Moore's conjecture can be resolved. The statement is about hyperbolic spaces, so Moore's conjecture takes the form: if a space Y is hyperbolic then it has no exponent at p for any prime p. In this context the proof of the following lemma is immediate.

Lemma 12.1: *Given the hypotheses of Theorem 11.3, if there is a space Y which retracts off $\Omega X_{i_1} * \cdots * \Omega X_{i_k}$ and Y has no exponent at any prime p, then $(\underline{X},\underline{*})^K$ has no exponent any prime p.* □

For example, if $S^{n_1} \vee S^{n_2}$ retracts off $\Omega X_{i_1} * \cdots * \Omega X_{i_k}$ then $(\underline{X},\underline{*})^K$ has no exponent at any prime p. We give two examples where this occurs.

Example 12.2: Take $X_i = S^{n_i}$ for each $1 \leq i \leq m$, where $n_i \geq 2$, and consider the CW-pairs $(S^{n_i}, *)$. By [25], there is a homotopy equivalence

$$\Sigma\Omega S^{n+1} \simeq \bigvee_{i=1}^{\infty} S^{ni+1}. \tag{12.1}$$

Iterating this homotopy equivalence implies that $\Omega S^{n_{i_1}} * \cdots * \Omega S^{n_{i_k}} \simeq \Sigma^{k-1}\Omega S^{n_{i_1}} \wedge \cdots \wedge \Omega S^{n_{i_k}}$ is homotopy equivalent to an infinite wedge of spheres. Consequently, $\Omega S^{n_{i_1}} * \cdots * \Omega S^{n_{i_k}}$ has no exponent at any prime p and so the corresponding polyhedral product $(\underline{X},\underline{*})^K$ has no exponent at any prime p.

Example 12.3: Take $X_i = \mathbb{C}P^{n_i}$ for each $1 \leq i \leq m$, where $n_i \geq 1$, and consider the CW-pairs $(\mathbb{C}P^{n_i}, *)$. It is well known that there is a homotopy equivalence

$$\Omega\mathbb{C}P^n \simeq S^1 \times \Omega S^{2n+1}.$$

Therefore, by Lemma 4.1 and (12.1), there are homotopy equivalences

$$\begin{aligned}\Sigma\Omega\mathbb{C}P^n &\simeq \Sigma(S^1 \times \Omega S^{2n+1}) \\ &\simeq \Sigma S^1 \vee \Sigma\Omega S^{2n+1} \vee (\Sigma S^1 \wedge \Omega S^{2n+1}) \\ &\simeq S^2 \vee (\bigvee_{i=1}^{\infty} S^{2ni+1}) \vee (\bigvee_{j=1}^{\infty} S^{2ni+2}).\end{aligned}$$

Iterating this homotopy decomposition, $\Omega\mathbb{C}P^{n_{i_1}} * \cdots * \Omega\mathbb{C}P^{n_{i_k}}$ is homotopy equivalent to an infinite wedge of spheres, and so has no exponent at any prime p. Therefore the corresponding polyhedral product $(\underline{X}, \underline{*})^K$ also has no exponent at any prime p.

13. More on elliptic and hyperbolic spaces, and Moore's conjecture

In this section we use the homotopy equivalence

$$\Omega(\underline{X}, \underline{A})^K \simeq \left(\prod_{i=1}^{m} \Omega X_i\right) \times \left(\prod_{i=1}^{m} \Omega(G_i * Y_i)\right)$$

in Theorem 9.4 as a tool to further investigate whether polyhedral products are elliptic or hyperbolic, and whether Moore's conjecture holds. Two initial observations are as follows.

Lemma 13.1: *Let K be a simplicial complex on the vertex set $[m]$ and let $(\underline{X}, \underline{A})$ be a sequence of pointed CW-pairs. Then the following hold:*

(a) if any one of $X_1, \ldots, X_m$ is hyperbolic then so is $(\underline{X}, \underline{A})^K$;
(b) if any $G_i \wedge Y_i$ is rationally nontrivial and has the property that one of G_i or Y_i has two or more generators in rational homology then $(\underline{X}, \underline{A})^K$ is hyperbolic.

Proof: Part (a) follows immediately since each ΩX_i is a retract of $\Omega(\underline{X}, \underline{A})^K$, implying that the rational homotopy groups of X_i retract off those of $(\underline{X}, \underline{A})^K$.

For part (b), in general, if X is a co-H-space with n generators in rational homology then X is rationally equivalent to a wedge of spheres, one for each generator. In our case, each $G_i * Y_i \simeq \Sigma G_i \wedge Y_i$ is a co-H-space and so is rationally equivalent to a wedge of spheres. Rationally, there is a Kunneth isomorphism, so

$$\widetilde{H}_*(G_i \wedge Y_i; \mathbb{Q}) \cong \widetilde{H}_*(G_i; \mathbb{Q}) \otimes \widetilde{H}_*(Y_i; \mathbb{Q}).$$

By hypothesis, $G_i \wedge Y_i$ is rationally nontrivial, which implies that there is some nontrivial class in $\widetilde{H}_*(G_i \wedge Y_i; \mathbb{Q})$. The Kunneth isomorphism therefore implies that there is at least one nontrivial class in each of $\widetilde{H}_*(G_i; \mathbb{Q})$ and $\widetilde{H}_*(Y_i; \mathbb{Q})$. Thus if at least one of $\widetilde{H}_*(G_i; \mathbb{Q})$ or $\widetilde{H}_*(Y_i; \mathbb{Q})$ has two or more generators then so does $\widetilde{H}_*(G_i \wedge Y_i; \mathbb{Q})$. The same is then true for $\widetilde{H}_*(\Sigma G_i \wedge Y_i; \mathbb{Q})$, and so $\Sigma G_i \wedge Y_i$ is homotopy equivalent to a wedge of at least two spheres, implying that it is hyperbolic. □

Lemma 13.1 indicates that the property of being elliptic is pretty restrictive. If $(\underline{X}, \underline{A})^K$ is elliptic then each of $X_1, \ldots, X_m$ must be elliptic, and no $G_i \wedge Y_i$ can have more than one rational homology class. To have one rational homology class implies that $G_i \wedge Y_i$ is rationally a sphere, which further implies that both G_i and Y_i are rationally spheres. This restriction on G_i, which is the homotopy fibre of a link into a restriction, is particularly severe.

We will give a more explicit description of a large family of hyperbolic spaces in Propositions 13.7 and 13.10. The idea is to relate the minimal missing faces of $\text{link}_K(v)$ to $K \backslash \{v\}$ in order to identify nontrivial rational homotopy classes in one of the spaces G_i. Recall from Section 8 that $\text{MMF}(K)$ is the set of minimal missing faces of K.

Lemma 13.2: *Let K and L be simplicial complexes on the same vertex set $[m]$ and suppose that K is a subcomplex of L. If MMF(K) $\subseteq$ MMF(L) then MMF(K) = MMF(L).*

Proof: We aim for a contradiction. Suppose that $\sigma = (i_1, \ldots, i_k) \in \text{MMF}(L)$ but $\sigma \notin \text{MMF}(K)$. Since σ is a minimial missing face of L, by definition we have $\partial\sigma \subset L$. This implies that every proper face of σ is a face of L. On the other hand, since σ is a missing face of L it is also a missing face of the subcomplex K. Therefore some proper face τ of σ must be in $\text{MMF}(K)$. By hypothesis, $\text{MMF}(K) \subseteq \text{MMF}(L)$, implying that $\tau \in \text{MMF}(L)$. But we have already said that every proper face of σ is a face of L, and so cannot be a missing face, and therefore not in $\text{MMF}(L)$. This is a contradiction. □

Remark 13.3: In Lemma 13.2 it is important that K and L have the same vertex set. For if K is two disjoint vertices v_1 and v_2, and L is three disjoint vertices v_1, v_2 and v_3 then $(v_2, v_3) \in \text{MMF}(L)$ but it is not in $\text{MMF}(K)$ since 3 is not even a vertex of K.

Lemma 13.4: *Let K and L be simplicial complexes on the same vertex set $[m]$ and suppose that K is a subcomplex of L. If MMF$(K) =$ MMF(L) then $K = L$.*

Proof: Suppose that $K \neq L$. Then there is a face $\sigma \in L$ such that $\sigma \notin K$. Therefore σ is a missing face of K, and so either $\sigma \in \mathrm{MMF}(K)$ or some proper face τ of σ is in $\mathrm{MMF}(K)$. If $\sigma \in \mathrm{MMF}(K)$ then by hypothesis $\sigma \in \mathrm{MMF}(L)$. But then σ is a missing face of L, a contradiction. If $\tau \in \mathrm{MMF}(K)$ for some proper face τ of σ then, by hypothesis, $\tau \in \mathrm{MMF}(L)$ as well. But then τ is a missing face of L, implying that σ is a missing face of L, a contradiction. □

Suppose that K is a subcomplex of L. A minimal missing face σ of K is said to be filled in inside L if $\sigma \in L$. Specialize to the case of $(\underline{X}, \underline{*})$.

Lemma 13.5: *Let K be a simplicial complex on the vertex set $[m]$. Let v be a vertex of K. Then precisely one of the following hold:*

- *(a) $\mathrm{link}_K(v)$ is a full subcomplex of $K\backslash\{v\}$;*
- *(b) there is a minimal missing face in $\mathrm{link}_K(v)$ that is filled in inside $K\backslash\{v\}$.*

Proof: Suppose that $\mathrm{link}_K(v)$ is not a full subcomplex of $K\backslash\{v\}$. Then it must be shown that $\mathrm{link}_K(v)$ has a minimal missing face that is filled in inside $K\backslash\{v\}$. First we refine the problem. Let I be the vertex set of $\mathrm{link}_K(v)$. This may be a proper subset of the vertex set of $K\backslash\{v\}$. So let $R(v)$ be the full subcomplex of $K\backslash\{v\}$ on the vertex set I. Observe that a minimal missing face of σ is filled in inside $K\backslash\{v\}$ if and only if it is filled in inside $R(v)$. So it is equivalent to show that $\mathrm{link}_K(v)$ has a minimal missing face that is filled in inside $R(v)$.

Suppose not. Then every minimal missing face σ of $\mathrm{link}_K(v)$ is also a missing face of $R(v)$. But then σ is a minimal missing face of $R(v)$, for every proper face of σ is in $\mathrm{link}_K(v)$ and so must be in $R(v)$ since $\mathrm{link}_K(v)$ is a subcomplex. Therefore $\mathrm{MMF}(\mathrm{link}_K(v)) \subseteq \mathrm{MMF}(R(v))$. By definition of $R(v)$, both $R(v)$ and $\mathrm{link}_K(v)$ have the same vertex set, so by Lemma 13.2 $\mathrm{MMF}(\mathrm{link}_K(v)) = \mathrm{MMF}(R(v))$, and therefore by Lemma 13.4, $\mathrm{link}_K(v) = R(v)$. That is, $\mathrm{link}_K(v)$ is a full subcomplex of $K\backslash\{v\}$, a contradiction. □

Guided by the two options in Lemma 13.5, we consider conditions that guarantee that certain polyhedral products are hyperbolic by studying how the combinatorics of K influence the homotopy type of the homotopy fibre of the inclusion of $(\underline{X},\underline{A})^{\mathrm{link}_K(v)}$ into $(\underline{X},\underline{A})^{K\backslash\{v\}}$. Let v be a vertex of K. Define the space $G(v)$ and the map $g(v)$ by the homotopy fibration

$$G(v) \xrightarrow{g(v)} (\underline{X},\underline{A})^{\mathrm{link}_K(v)} \longrightarrow (\underline{X},\underline{A})^{K\backslash\{v\}}.$$

Lemma 13.6: *Let K be a simplicial complex on the vertex set $[m]$. Suppose that v is a vertex of K such that $link_K(v)$ is a full subcomplex of $K\backslash\{v\}$. Then the map $g(v)$ is null homotopic, implying that there is a homotopy equivalence*

$$\Omega(\underline{X},\underline{A})^{K\backslash\{v\}} \simeq \Omega(\underline{X},\underline{A})^{link_K(v)} \times G(v).$$

Further, if $link_K(v)$ is a proper full subcomplex of $K\backslash\{v\}$ and $u_{j_1},\ldots,u_{j_\ell}$ are vertices of $K\backslash\{v\}$ but not of $link_K(v)$, then $\prod_{t=1}^{\ell}\Omega X_{j_t}$ retracts off $G(v)$.

Proof: Since $\mathrm{link}_K(v)$ is a full subcomplex of $K\backslash\{v\}$, by Proposition 3.6, $(\underline{X},\underline{A})^{\mathrm{link}_K(v)}$ is a retract of $(\underline{X},\underline{A})^{K\backslash\{v\}}$. Therefore the map $g(v)$ is null homotopic, and the asserted homotopy equivalence for $\Omega(\underline{X},\underline{A})^{K\backslash\{v\}}$ follows immediately.

For the second statement, let u be a vertex of $K\backslash\{v\}$ that is not a vertex of $\mathrm{link}_K(v)$. The full subcomplex of $K\backslash\{v\}$ on the vertex set $\{u\}$ is just the single vertex u. So by Proposition 3.6, $(\underline{X},\underline{A})^{\{u\}}$ retracts off of $(\underline{X},\underline{A})^{K\backslash\{v\}}$. But by definition of the polyhedral product, $(\underline{X},\underline{A})^{\{u\}} = X_u$. Therefore, X_u retracts off $(\underline{X},\underline{A})^{K\backslash\{v\}}$, but it cannot retract off of $(\underline{X},\underline{A})^{\mathrm{link}_K(v)}$ since this is a subspace of $\prod_{\substack{i=1\\ i\neq u,v}}^{m} X_i$. Thus ΩX_u is a factor of $\Omega(\underline{X},\underline{A})^{K\backslash\{v\}}$ but not a factor of $\Omega(\underline{X},\underline{A})^{\mathrm{link}_K(v)}$, implying that ΩX_u is a factor of $G(v)$. As this holds for each of the vertices $u_{j_1},\ldots,u_{j_\ell}$, the product $\prod_{t=1}^{\ell}\Omega X_{j_t}$ retracts off $G(v)$. □

Proposition 13.7: *Let K be a simplicial complex on the vertex set $[m]$. Suppose that v is a vertex of K such that $link_K(v)$ is a full subcomplex of $K\backslash\{v\}$, and suppose that $u_{j_1},\ldots,u_{j_\ell}$ are vertices of $K\backslash\{v\}$ that are not in $link_K(v)$. If $X_1,\ldots,X_m$ are pointed, simply-connected CW-complexes such that the rank of $\widetilde{H}_*(\prod_{t=1}^{\ell}\Omega X_{j_t};\mathbb{Q})$ is at least 2 and ΩX_v is rationally nontrivial then $(\underline{X},\underline{*})^K$ is hyperbolic.*

Proof: By Lemma 13.6, $\prod_{t=1}^{\ell}\Omega X_{j_t}$ retracts off $G(v)$. So if the rank of $\widetilde{H}_*(\prod_{t=1}^{\ell}\Omega X_{j_t};\mathbb{Q})$ is at least 2 then the rank of $\widetilde{H}_*(G(v);\mathbb{Q})$ is at least 2.

Therefore, as ΩX_v is rationally nontrivial, the rank of $\widetilde{H}_*(G(v) * \Omega X_v; \mathbb{Q})$ is at least 2. Since $G(v) * \Omega X_v$ is a suspension, it is rationally homotopy equivalent to a wedge of spheres, and as the space has rank at least 2 in rational homology, this wedge is of at least 2 spheres. Therefore $G(v) * \Omega X_v$ is hyperbolic. By Theorem 9.4, $\Omega(G(v) * \Omega X_v)$ retracts off $\Omega(\underline{X}, \underline{*})^K$, implying that $\Omega(\underline{X}, \underline{*})^K$ is hyperbolic. $\square$

Remark 13.8: The proof of Proposition 13.7 implies more. Since $\prod_{t=1}^{\ell} \Omega X_{j_t}$ retracts off $G(v)$, there is a retraction of $\Sigma(\prod_{t=1}^{\ell} \Omega X_{j_t}) \wedge \Omega X_v$ off of $G(v) * \Omega X_v$. So if a wedge of spheres retracts off $\Sigma(\prod_{t=1}^{\ell} \Omega X_{j_t}) \wedge \Omega X_v$ then $G(v) * \Omega X_v$ has no exponent at any prime p and therefore neither does $(\underline{X}, \underline{*})^K$, implying that Moore's conjecture holds in this case.

The second case is when, for some vertex v of K, there is a minimal missing face of $\mathrm{link}_K(v)$ that is filled in inside $K \backslash \{v\}$. For this case we specialize to $(\underline{CX}, \underline{X})$ so that a product of cones, which is contractible, shows up at the right point. By definition, if $\sigma = (i_1, \ldots, i_k)$ is a minimal missing face of $\mathrm{link}_K(v)$ then $\partial\sigma \subseteq \mathrm{link}_K(v)$. The full subcomplex of $\mathrm{link}_K(v)$ on the vertex set $\{i_1, \ldots, i_k\}$ is therefore $\partial\sigma$. Therefore, by Proposition 3.6, $(\underline{CX}, \underline{X})^{\partial\sigma}$ is a retract of $(\underline{CX}, \underline{X})^{\mathrm{link}_K(v)}$. Since $\partial\sigma$ is the boundary of a simplex, by Lemma 8.7, $(\underline{CX}, \underline{X})^{\partial\sigma} \simeq X_{i_1} * \cdots * X_{i_k}$.

Lemma 13.9: *Let K be a simplicial complex on the vertex set $[m]$. Suppose that v is a vertex of K such that there is a minimal missing face $\sigma = (i_1, \ldots, i_k)$ of $\mathrm{link}_K(v)$ that is filled in inside $K \backslash \{v\}$. Then there is a lift*

$$\begin{array}{ccc} & & G(v) \\ & \overset{\lambda}{\nearrow} & \big\downarrow{\scriptstyle g(v)} \\ X_{i_1} * \cdots * X_{i_k} & \hookrightarrow & (\underline{CX}, \underline{X})^{\mathrm{link}_K(v)} \end{array}$$

for some map λ which has a left homotopy inverse.

Proof: Since σ is a minimal missing face of $\mathrm{link}_K(v)$ we have $\partial\sigma \subseteq \mathrm{link}_K(v)$. Since σ is filled in inside $K \backslash \{v\}$ there is a commutative diagram of simplicial complexes

$$\begin{array}{ccc} \partial\sigma & \longrightarrow & \mathrm{link}_K(v) \\ \big\downarrow & & \big\downarrow \\ \sigma & \longrightarrow & K \backslash \{v\}. \end{array}$$

Now take polyhedral products with respect to $(\underline{CX}, \underline{X})$ to obtain a commutative diagram of polyhedral products

$$\begin{array}{ccc} (\underline{CX}, \underline{X})^{\partial\sigma} & \longrightarrow & (\underline{CX}, \underline{X})^{\mathrm{link}_K(v)} \\ \downarrow & & \downarrow \\ (\underline{CX}, \underline{X})^{\sigma} & \longrightarrow & (\underline{CX}, \underline{X})^{K\backslash\{v\}}. \end{array} \tag{13.1}$$

As mentioned prior to the lemma, $(\underline{CX}, \underline{X})^{\partial\sigma} \simeq X_{i_1} * \cdots * X_{i_k}$. Since σ is a simplex, by Example 2.3 $(\underline{CX}, \underline{X})^{\sigma} = CX_{i_1} \times \cdots \times CX_{i_k}$, so $(\underline{CX}, \underline{X})^{\sigma}$ is contractible. Thus from (13.1) we obtain a homotopy commutative diagram

$$\begin{array}{ccc} X_{i_1} * \cdots * X_{i_k} & \longrightarrow & (\underline{CX}, \underline{X})^{\mathrm{link}_K(v)} \\ \downarrow & & \downarrow \\ * & \longrightarrow & (\underline{CX}, \underline{X})^{K\backslash\{v\}}. \end{array}$$

Therefore, as there is a homotopy fibration $G(v) \xrightarrow{g(v)} (\underline{CX}, \underline{X})^{\mathrm{link}_K(v)} \longrightarrow (\underline{CX}, \underline{X})^{K\backslash\{v\}}$, the map $X_{i_1} * \cdots * X_{i_k} \longrightarrow (\underline{CX}, \underline{X})^{\mathrm{link}_K(v)}$ lifts through $g(v)$ to a map $\lambda\colon X_{i_1} * \cdots * X_{i_k} \longrightarrow G(v)$. Since the map $X_{i_1} * \cdots * X_{i_k} \longrightarrow (\underline{CX}, \underline{X})^{\mathrm{link}_K(v)}$ is induced by the inclusion of the full subcomplex $\partial\sigma$ into $\mathrm{link}_K(v)$, by Proposition 3.6 it has a left homotopy inverse $t\colon (\underline{CX}, \underline{X})^{\mathrm{link}_K(v)} \longrightarrow X_{i_1} * \cdots * X_{i_k}$. Therefore $t \circ g(v)$ is a left homotopy inverse of λ. □

Proposition 13.10: *Let K be a simplicial complex on the vertex set $[m]$. Suppose that v is a vertex of K such that there is a minimal missing face $\sigma = (i_1, \ldots, i_k)$ of $\mathrm{link}_K(v)$ that is filled in inside $K\backslash\{v\}$. Suppose that $X_1, \ldots, X_m$ are pointed, connected CW-complexes such that each X_{i_j} for $1 \leq j \leq k$ and X_v are rationally nontrivial, and at least one of these spaces has rank 2 in rational homology. Then $(\underline{CX}, \underline{X})^K$ is hyperbolic.*

Proof: By Lemma 13.9, $X_{i_1} * \cdots * X_{i_k}$ retracts off $G(v)$, and so $X_{i_1} * \cdots * X_{i_k} * X_v$ retracts off $G(v) * X_v$. The hypotheses on the spaces X_{i_j} and X_v imply that $X_{i_1} * \cdots * X_{i_k} * X_v$ is rationally nontrivial and has rank at least 2 in rational homology. Therefore $G(v) * X_v$ has rank at least 2 in rational homology. As $G(v) * X_v$ is a suspension, it is rationally homotopy equivalent to a wedge of spheres. The fact that $\widetilde{H}_*(G(v) * X_v; \mathbb{Q})$ has rank at least 2 implies that this wedge is of at least 2 spheres. Therefore $G(v) * X_v$ is hyperbolic.

By Theorem 9.4, $\Omega(G(v) * X_v)$ retracts off $\Omega(\underline{CX}, \underline{X})^K$, so as $G(v) * X_v$ is hyperbolic, so is $(\underline{CX}, \underline{X})^K$. $\square$

Remark 13.11: As in Remark 13.8, the proof of Proposition 13.10 shows that if a wedge of spheres retracts off of $G(v) * X_v$ then $(\underline{CX}, \underline{X})^K$ has no exponent at any prime p, implying that Moore's conjecture holds in this case.

Propositions 13.7 and 13.10 should be regarded as a starting point for investigating whether polyhedral products are elliptic or hyperbolic. A key feature is that this is done by first determining integral results, and then deducing the rational information about ellipticity or hyperbolicity. The integral results have the advantage that they are well suited to going further to address Moore's conjecture. Much remains to be done in this direction, and we close with some questions.

Problem 13.12: *Determine which polyhedral products* $(\underline{X}, \underline{*})^K$ *or* $(\underline{CX}, \underline{X})^K$ *are elliptic and which are hyperbolic.*

One way forward in Problem 13.12 is to look for two rational classes in $\widetilde{H}_*(G(v) * \Omega X_v; \mathbb{Q})$ (or $\widetilde{H}_*(G(v) * X_v; \mathbb{Q})$) not by looking for one non-trivial class in $\widetilde{H}_*(G(v); \mathbb{Q})$ and two in $\widetilde{H}_*(\Omega X_v; \mathbb{Q})$ (or $\widetilde{H}_*(X_v; \mathbb{Q})$) but to look for two nontrivial classes in $\widetilde{H}_*(G(v); \mathbb{Q})$ and one in $\widetilde{H}_*(\Omega X_v; \mathbb{Q})$ (or $\widetilde{H}_*(X_v; \mathbb{Q})$). For example, if $\mathrm{link}_K(v)$ has at least two minimal missing faces filled in inside $K \backslash \{v\}$ then Lemma 13.9 gives two retractions off $G(v)$, and these are independent since they correspond to independent full sub-complex retractions off of $(\underline{CX}, \underline{X})^{\mathrm{link}_K(v)}$. So one is left having to deal with the cases where, for every vertex v in K, $\mathrm{link}_K(v)$ has 0 or 1 minimal missing faces filled in inside $K \backslash \{v\}$.

This perspective is particularly important because Propositions 13.7 and 13.10 do not give any information at all about the case of $DJ_K = (\mathbb{C}P^\infty, *)^K$ or $\mathcal{Z}_K = (D^2, S^1)^K$. In the latter case, for example, the join terms $X_{i_1} * \cdots * X_{i_k}$ are just single spheres. However, Bahri, Bendersky, Cohen and Gitler [2] considered the case of Davis-Januszkiewicz spaces and showed that DJ_K is elliptic if and only if it is a product of spheres and discs.

Another problem is to consider what happens in the more general case of $(\underline{X}, \underline{A})^K$.

Problem 13.13: *Determine which polyhedral products* $(\underline{X}, \underline{A})^K$ *are elliptic and which are hyperbolic.*

It may be helpful to approach Problem 13.13 by considering more families of special cases, such as when the CW-pairs (X_i, A_i) have the property that the homotopy fibre Y_i of the inclusion $A_i \longrightarrow X_i$ has at least two generators in reduced rational homology, or some other special property. Many interesting pairs fall into this case, such as $(\mathbb{C}P^n, \mathbb{C}P^k)$ for $1 \leq k < n \leq \infty$.

As for Moore's conjecture, the key idea behind showing the nonexistence of an exponent at any prime p in all the cases we considered was to find a wedge of two spheres retracting off $G_v * \Omega X_v$ or $G(v) * X_v$ for some vertex v. The same method should work if the spaces X_i are taken to be torsion-free suspensions by using Selick's result [36]. It is natural to ask for a solution more generally.

Problem 13.14: *Determine whether* $(\underline{X}, \underline{*})^K$, $(\underline{CX}, \underline{X})^K$, *or* $(\underline{X}, \underline{A})^K$ *satisfies Moore's conjecture.*

14. Further problems

In these notes three trends have emerged regarding the homotopy type of polyhedral products $(\underline{X}, \underline{A})^K$: (i) many of the results apply only for the cases of $(\underline{X}, \underline{*})$ and $(\underline{CX}, \underline{X})$, which happily gives results about the Davis-Januszkiewicz space DJ_K and the moment-angle complex $\mathcal{Z}_K$; (ii) knowing the homotopy class of the map $(\underline{X}, \underline{A})^{\mathrm{link}_K(v)} \xrightarrow{j} (\underline{X}, \underline{A})^{K \backslash \{v\}}$ is very helpful, and (iii) most of the results mentioned are of the form - for certain simplicial complexes K the space $(\underline{CX}, \underline{X})^K$ is homotopy equivalent to a wedge of suspensions of iterated smashes of the spaces X_i.

Let's address these trends in turn. First, identifying the homotopy types of DJ_K and $\mathcal{Z}_K$ are crucial, which is what motivates the generalization to identifying the homotopy types of $(\underline{X}, \underline{*})^K$ and $(\underline{CX}, \underline{X})^K$. However, there are other interesting polyhedral products and very little has been done to determine their homotopy types.

Problem 14.1: *Identify the homotopy type of* $(\underline{X}, \underline{A})^K$ *for pairs of spaces* (X_i, A_i) *where neither* X_i *nor* A_i *is contractible.*

Of course, this will probably involve specializing to certain pairs with special properties, and certain simplicial complexes. But at this point almost nothing is known so any progress would be interesting. A curious case to begin with may be pairs $(\mathbb{C}P^n, \mathbb{C}P^k)$ for $1 \leq k < n \leq \infty$.

Second, the homotopy class of the map $(\underline{X}, \underline{A})^{\mathrm{link}_K(v)} \xrightarrow{j} (\underline{X}, \underline{A})^{K \backslash \{v\}}$ played a pivotal role in the strategy for identifying the homotopy type of

$(\underline{CX}, \underline{X})^K$ for K shifted in Section 7, and it played an equally important role in identifying hyperbolic polyhedral products in Section 11. Moreover, knowing the homotopy class of j informs on the homotopy type of its fibre $G(v)$, which appears in the homotopy decomposition for $\Omega(\underline{X}, \underline{A})^K$.

Problem 14.2: *Find other sequences of pairs $(\underline{X}, \underline{A})$ and simplicial complexes K for which the homotopy class of the map $(\underline{X}, \underline{A})^{link_K(v)} \xrightarrow{j} (\underline{X}, \underline{A})^{K\backslash\{v\}}$ can be identified, and use this to determine information about $(\underline{X}, \underline{A})^K$ and $\Omega(\underline{X}, \underline{A})^K$.*

In particular, a better understanding of the map j would help resolve the problems already stated regarding the elliptic-hyperbolic dichotomy and Moore's conjecture.

Third, most moment-angle complexes $\mathcal{Z}_K$ and polyhedral products $(\underline{CX}, \underline{X})^K$ are not homotopy equivalent to wedges of suspensions of iterated smashes of the spaces X_i. In particular, it is known [3] that if K is a polytopal sphere then $\mathcal{Z}_K$ is a manifold, and so cannot be homotopy equivalent to a wedge of two or more spaces. In particular, if K is the boundary of an m-gon then by [7, 29] there is a diffeomorphism

$$\mathcal{Z}_K \simeq \#_{k=3}^{m-1}(S^k \times S^{m+2-k})^{\#(k-2)\binom{m-2}{k-1}} \tag{14.1}$$

where $\#$ is the connected sum operation. This generalizes to K being the dual of the boundary of certain higher dimensional polytopes. Further generalizations were proved in [15]. This raises several questions.

Problem 14.3: *The diffeomorphism with the connected sum in (14.1) was proved using geometric methods. Is there a homotopy-theoretic way of recognizing a space having the homotopy type of a connected sum?*

Problem 14.4: *If K is as in (14.1) then what is the homotopy type of $(\underline{CX}, \underline{X})^K$ in the case of more general pairs (CX_i, X_i)? While not a manifold in general, it would appear to be a useful generalization.*

Problem 14.5: *What other homotopy types can be realized by a moment-angle complex or a polyhedral product?*

Some initial progress has been made in Problem 14.3: in [16] it was shown that Whitehead products play an important role in the homotopy class of the map $\mathcal{Z}_K \longrightarrow DJ_K$, and in [39] an intriguing connection was made between the homotopy type of the moment-angle complex $\mathcal{Z}_K$

in (14.1) and the homotopy type of $m-2$ disjoint points. Nothing is really known about Problem 14.4.

For Problem 14.5, we mention two examples. In [16] it was shown that there is a 6-vertex triangulation K of $\mathbb{R}P^2$ with the property that $\mathcal{Z}_K$ has the Moore space $P^9(2)$ as a wedge summand. In particular, there is a non-trivial Sq^1 operation in the mod-2 cohomology of $\mathcal{Z}_K$. In [27] it was shown that there is a 9-vertex triangulation L of $\mathbb{C}P^2$ with the property that $\mathcal{Z}_L$ has $\Sigma^{10}\mathbb{C}P^2$ as a wedge summand. In particular, there is a nontrivial Sq^2 operation in the mod-2 cohomology of $\mathcal{Z}_L$. Connected to Problem 14.5 then, is the following problem.

Problem 14.6: *For which simplicial complexes K does $\mathcal{Z}_K$ have nontrivial Steenrod operations in mod-p cohomology? In particular, for which simplicial complexes K does $\mathcal{Z}_K$ have mod-p torsion in cohomology?*

Acknowledgments

These notes were prepared for a lecture series as part of the programme, "Combinatorial and Toric Homotopy", held at the IMS in Singapore in August 2015. I would like to thank the IMS for its support; the organizers, Jelena Grbić, Zhi Lü and Jie Wu, for inviting me to give these talks; and Alastair Darby for carefully proofreading the notes.

References

1. A. Bahri, M. Bendersky, F.R. Cohen and S. Gitler, The polyhedral product functor: a method of decomposition for moment-angle complexes, arrangements and related spaces, *Adv. Math.* **225** (2010), 1634–1668.
2. A. Bahri, M. Bendersky, F.R. Cohen and S. Gitler, On the rational homotopy type of moment-angle complexes, *Proc. Steklov Inst. Math.* **286** (2014), 219–223.
3. P. Beben and J. Grbić, Configuration spaces and polyhedral products, arXiv:1409.4462.
4. M.G. Barratt, Spaces of finite characteristic, *Quart. J. Math. Oxford* **11** (1960), 124-136.
5. V.M. Buchstaber and T.E. Panov, *Torus actions and their applications in topology and combinatorics*, University Lecture Series **24**, American Mathematical Society, 2002.
6. V.M. Buchstaber and T.E. Panov, *Toric topology*, Mathematical Surveys and Monographs **204**, American Mathematical Society, 2015.
7. F. Bosio and L. Meersseman, Real quadrics in $\mathbb{C}^n$, complex manifolds and convex polytopes, *Acta Math.* **197** (2006), 53–127.

8. W. Chachólski, W. Pitsch, J. Scherer and D. Stanley, Homotopy exponents for large H-spaces, *Int. Math. Res. Not. IMRN 2008*, **16**, Art. ID rnn061, 5pp.
9. F.R. Cohen, A course in some aspects of classical homotopy theory, Lecture Notes in Math. **1286**, Springer-Verlag (1987), 1–92.
10. M.W. Davis and T. Januszkiewicz, Convex polytopes, Coxeter orbifolds and torus actions, *Duke Mathematical Journal* **62** (1991), 417–452.
11. G. Denham and A.I. Suciu, Moment-angle complexex, monomial ideals, and Massey products, *Pure Appl. Math. Q.* **3** (2007), 25–60.
12. Y. Félix, S. Halperin and J.-C. Thomas, Elliptic spaces II. *Enseign. Math. (2)* **39** (1993), 25–32.
13. Y. Félix and D. Tanré, Rational homotopy of the polyhedral product functor, *Proc. Amer. Math. Soc.* **137** (2009), 891–898.
14. T. Ganea, A generalization of the homology and homotopy suspension, *Comment. Math. Helv.* **39** (1965), 295–322.
15. S. Gitler and S. Lopez de Medrano, Intersections of quadrics, moment-angle manifolds and connected sums, *Geom. Topol.* **17** (2013), 1497–1534.
16. J. Grbić, T. Panov, S. Theriault and J. Wu, The homotopy types of moment-angle complexes for flag complexes, *Trans. Amer. Math. Soc.* **368** (2016), 6663–6682.
17. J. Grbić and S. Theriault, The homotopy type of the complement of the codimension two coordinate subspace arrangement, *Russian Math. Surveys* **59** (2005), 203–204.
18. J. Grbić and S. Theriault, The homotopy type of the complement of a coordinate subspace arrangement, *Topology* **46** (2007), 357–396.
19. J. Grbić and S. Theriault, The homotopy type of the polyhedral product for shifted complexes, *Adv. Math.* **245** (2013), 690–715.
20. J. Grbić and S. Theriault, Homotopy theory in toric topology, *Russian Math. Surveys* **71** (2016), 185–251.
21. V. Grujić and V. Welker, Moment-angle complexes of pairs (D^n, S^{n-1}) and simplicial complexes with vertex-decomposable duals, *Monatsh. Math.* **176** (2015), 255–273.
22. K. Iriye and D. Kishimoto, Decompositions of polyhedral products, *Adv. Math.* **245** (2013), 716–736.
23. K. Iriye and D. Kishimoto, Fat wedge filtrations and decomposition of polyhedral products, arXiv:1412.4866.
24. K. Iriye and D. Kishimoto, Polyhedral products for shifted complexes and higher Whitehead products, arXiv:1505.04892.
25. I.M. James, Reduced product spaces, *Ann. of Math.* **62** (1955), 170–197.
26. I.M. James, The suspension triad of a sphere, *Ann. of Math.* **63** (1956), 407–429.
27. I. Limonchenko, *Combinatorial commutative algebra and topology of moment-angle complexes*, PhD thesis, Moscow State University, 2015.
28. J. Long, Thesis, Princeton University, 1978.
29. D. MacGavran, Adjacent connected sums and torus actions, *Trans. Amer. Math. Soc.* **251** (1979), 235–254.

30. M. Mather, Pull-backs in homotopy theory, *Canad. J. Math.* **28** (1976), 225–263.
31. C.A. McGibbon and C.W. Wilkerson, Loop spaces of finite complexes at large primes, *Proc. Amer. Math. Soc.* **96** (1986), 698–702.
32. J.A. Neisendorfer, *Primary Homotopy Theory*, Mem. Amer. Math. Soc. **232** (1980).
33. J.A. Neisendorfer, The exponent of a Moore space. *Algebraic topology and algebraic K-theory (Princeton, N.J., 1983)*, 35–71, Ann. of Math. Stud. **113**, Princeton Univ. Press, Princeton, NJ, 1987.
34. J. Neisendorfer and P. Selick, Some examples of spaces with and without homotopy exponents, *Current trends in algebraic topology, Part 1 (London, Ont., 1981)*, pp. 343–357, CMS Conf. Proc. **2**, Amer. Math. Soc., Providence, R.I., 1982.
35. G.J. Porter, The homotopy groups of wedges of suspensions, *Amer. J. Math.* **88** (1966), 655–663.
36. P. Selick, On conjectures of Moore and Serre in the case of torsion-free suspensions, *Math. Proc. Cambridge Philos. Soc.* **94** (1983), 53–60.
37. M. Stelzer, Hyperbolic spaces at large primes and a conjecture of Moore, *Topology* **43** (2004), 667–675.
38. S. Theriault, Homotopy exponents of mod 2^r Moore spaces, *Topology* **47** (2008), 369–398.
39. S. Theriault, Moment-angle manifolds and Panov's problem, *Int. Math. Res. Not.* (2015), doi:10.1093/imrn/rnu281.
40. S. Theriault, The dual polyhedral product, cocategory and nilpotence, arXiv:1506.05998.
41. H. Toda, On the double suspension E^2, *J. Inst. Polytech. Osaka City Univ., Ser. A* **7** (1956), 103–145.

FULLERENES, POLYTOPES AND TORIC TOPOLOGY

Victor M. Buchstaber
Steklov Mathematical Institute of Russian Academy of Sciences
Gubkina str. 8, 119991, Moscow, Russia
Department of Geometry and Topology
Faculty of Mechanics and Mathematics
Lomonosov Moscow State University
Leninskie Gory 1, 119991, Moscow, Russia
buchstab@mi.ras.ru

Nikolay Yu. Erokhovets
Department of Geometry and Topology
Faculty of Mechanics and Mathematics
Lomonosov Moscow State University
Leninskie Gory 1, 119991, Moscow, Russia
erochovetsn@hotmail.com

These lectures are devoted to a remarkable class of 3-dimensional polytopes, which are mathematical models of an important object in quantum physics, quantum chemistry and nanotechnology – fullerenes. The main goal is to show how results of toric topology help to build combinatorial invariants of fullerenes. Fundamental notions are introduced during the lectures. The lecture notes are addressed to a wide audience.

Contents

1. Introduction

These lecture notes are devoted to results on the crossroads of classical polytope theory, toric topology, and the mathematical theory of fullerenes. Toric topology is a new area of mathematics that emerged at the end of the 1990s on the border of equivariant topology, algebraic and symplectic geometry, combinatorics, and commutative algebra. The mathematical theory of fullerenes is a new area of mathematics focused on problems formulated on the basis of the outstanding achievements of quantum physics, quantum chemistry and nanotechnology.

This text is based on the lectures delivered by the first author at the Young Topologist Seminar during the program on Combinatorial and Toric Homotopy (1-31 August 2015) organized jointly by the Institute for Mathematical Sciences and the Department of Mathematics of the National University of Singapore.

The lectures are oriented to a wide audience. We give all necessary notions and constructions. For key results, including new results, we either give a full prove, or a sketch of a proof with an appropriate reference. These results are oriented for the applications to the combinatorial study and classification of fullerenes.

Lecture guide

- One of the main objects of toric topology is the moment-angle functor $P \to \mathcal{Z}_P$.
- It assigns to each simple n-polytope P, with m facets, an $(n+m)$-dimensional moment-angle complex $\mathcal{Z}_P$. This comes with an action of a compact torus T^m, whose orbit space $\mathcal{Z}_P/T^m$ can be identified with P.
- The space $\mathcal{Z}_P$ has the structure of a smooth manifold with a smooth action of T^m.
- A mathematical fullerene is a three dimensional convex simple polytope with all 2-faces being pentagons and hexagons.
- In this case the number p_5 of pentagons is 12.
- The number p_6 of hexagons can be arbitrary except for 1.
- Two combinatorially nonequivalent fullerenes with the same number p_6 are called combinatorial isomers. The number of combinatorial isomers of fullerenes grows fast as a function of p_6.
- The problem of the classification of fullerenes is well-known and is vital due to the applications in chemistry, physics, biology and nanotechnology.
- Our main goal is to apply methods of toric topology to build combinatorial invariants distinguishing isomers.
- Thanks to the toric topology, we can assign to each fullerene P its moment-angle manifold $\mathcal{Z}_P$.
- The cohomology ring $H^*(\mathcal{Z}_P)$ is a combinatorial invariant of the fullerene P.
- We shall focus upon results on the rings $H^*(\mathcal{Z}_P)$ and their applications based on a geometric interpretation of their cohomology classes and products.
- The multigrading in the ring $H^*(\mathcal{Z}_P)$, coming from the construction of $\mathcal{Z}_P$, and multigraded Poincare duality play an important role here.
- There exist 7 truncation operations on simple 3-polytopes such that any fullerene is combinatorially equivalent to a polytope obtained from the dodecahedron by a sequence of these operations.

2. Basic notions

2.1. *Convex polytopes*

Definition 2.1: A *convex polytope* P is a *bounded* set of the form

$$P = \{\boldsymbol{x} \in \mathbb{R}^n \colon \boldsymbol{a}_i\boldsymbol{x} + b_i \geqslant 0,\ i = 1, \dots, m\},$$

where $\boldsymbol{a}_i \in \mathbb{R}^n$, $b_i \in \mathbb{R}$, and $\boldsymbol{x}\boldsymbol{y} = x_1y_1 + \cdots + x_ny_n$ is the standard scalar product in $\mathbb{R}^n$. Let this representation be *irredundant*, that is a deletion of any inequality changes the set. Then each hyperplane $\mathcal{H}_i = \{\boldsymbol{x} \in \mathbb{R}^n \colon \boldsymbol{a}_i\boldsymbol{x} + b_i = 0\}$

defines a *facet* $F_i = P \cap \mathcal{H}_i$. Denote by $\mathcal{F}_P = \{F_1, \dots, F_m\}$ the ordered set of facets of P. For a subset $S \subset \mathcal{F}_P$ denote $|S| = \bigcup_{i \in S} F_i$. We have $|\mathcal{F}_P| = \partial P$ is the boundary of P.

A *face* is a subset of a polytope that is an intersection of facets. Two convex polytopes P and Q are *combinatorially equivalent* ($P \simeq Q$) if there is an inclusion-preserving bijection between their sets of faces. A *combinatorial polytope* is an equivalence class of combinatorially equivalent convex polytopes. In most cases we consider combinatorial polytopes and write $P = Q$ instead of $P \simeq Q$.

Example 2.2: An n-simplex Δ^n in $\mathbb{R}^n$ is the convex hull of $n+1$ affinely independent points. Let $\{\boldsymbol{e}_1, \dots, \boldsymbol{e}_n\}$ be the standard basis in $\mathbb{R}^n$. The n-simplex $\mathrm{conv}\{\boldsymbol{0}, \boldsymbol{e}_1, \dots, \boldsymbol{e}_n\}$ is called *standard*. It is defined in $\mathbb{R}^n$ by $n+1$ inequalities:

$$x_i \geqslant 0 \quad \text{for } i = 1, \dots, n, \quad \text{and} \quad -x_1 - \dots - x_n + 1 \geqslant 0.$$

The *standard n-cube* I^n is defined in $\mathbb{R}^n$ by $2n$ inequalities

$$x_i \geqslant 0, \quad -x_i + 1 \geqslant 0, \quad \text{for } i = 1, \dots, n.$$

Definition 2.3: An *orientation* of a combinatorial convex 3-polytope is a choice of the cyclic order of vertices of each facet such that for any two facets with a common edge the orders of vertices induced from the facets to this edge are the opposite. A combinatorial convex 3-polytope with a given orientation is called *oriented*.

Exercise:

- Any geometrical realization of a combinatorial 3-polytope P in $\mathbb{R}^3$ with the standard orientation induces an orientation of P.
- Any combinatorial 3-polytope has exactly two orientations.
- Define an oriented combinatorial convex n-polytope.

Definition 2.4: A polytope is called *combinatorially chiral* if any combinatorial equivalence to itself preserves the orientation.

The simplex Δ^3 and cube I^3 are not combinatorially chiral.
Exercise: Give an example of a combinatorially chiral 3-polytope.

There is the classical notion of a *(geometrically) chiral* polytope (connected with the right-hand and left-hand rules).

Definition 2.5: A convex 3-polytope $P \subset \mathbb{R}^3$ is called *(geometrically) chiral* if there is no orientation preserving isometry of $\mathbb{R}^3$ that maps P to its mirror image.

Proposition 2.6: *A combinatorially chiral polytope is geometrically chiral, whilst a geometrically chiral polytope is not necessarily combinatorially chiral.*

Proof: The orientation-preserving isometry of $\mathbb{R}^3$ that maps P to its mirror image defines the combinatorial equivalence that changes the orientation. On the other hand, the simplex Δ^3 realized such that all of the angles between the facets are distinct can not be mapped to itself by an isometry of $\mathbb{R}^3$ that is different from the identity. Hence it is chiral. An odd permutation of vertices defines the combinatorial equivalence that changes the orientation; hence Δ^3 is not combinatorially chiral. □

2.2. *Schlegel diagrams*

Schlegel diagrams were introduced by Victor Schlegel (1843-1905) in 1886.

Definition 2.7: A Schlegel diagram of a convex polytope P in $\mathbb{R}^3$ is a *projection* of P from $\mathbb{R}^3$ into $\mathbb{R}^2$ through a point beyond one of its facets.

The resulting entity is a subdivision of the projection of this facet that is a *combinatorial invariant* of the original polytope. It is clear that a Schlegel diagram depends on the choice of the facet.

Exercise: Describe the Schlegel diagram of the cube and the octahedron.

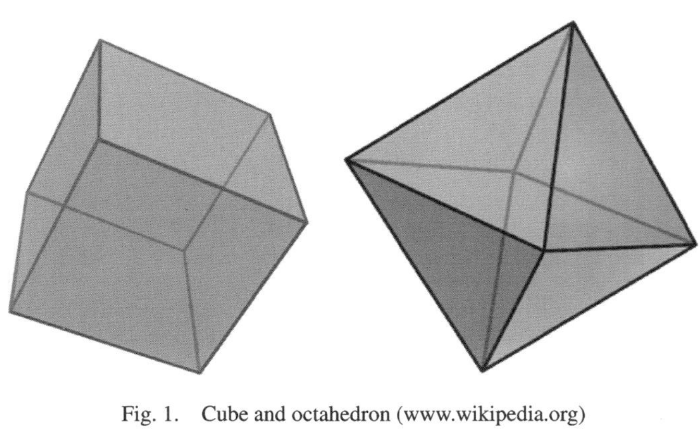

Fig. 1. Cube and octahedron (www.wikipedia.org)

Example 2.8:

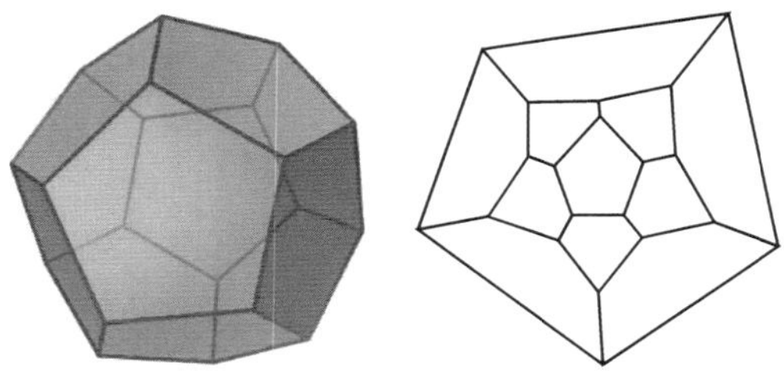

Fig. 2. Dodecahedron and its Schlegel digram (www.wikipedia.org)

2.3. *Euler's formula*

Let f_0, f_1, and f_2 be the numbers of vertices, edges, and 2-faces of a 3-polytope. Leonard Euler (1707-1783) proved the following fundamental relation:

$$f_0 - f_1 + f_2 = 2.$$

By a *fragment* we mean a subset $W \subset P$ that is a union of faces of P. Define the *Euler characteristic* of W by

$$\chi(W) = f_0(W) - f_1(W) + f_2(W).$$

If W_1 and W_2 are fragments, then $W_1 \cup W_2$ and $W_1 \cap W_2$ are fragments.
Exercise: Proof the *inclusion-exclusion formula*

$$\chi(W_1 \cup W_2) = \chi(W_1) + \chi(W_2) - \chi(W_1 \cap W_2).$$

2.4. *Platonic solids*

Definition 2.9: A *regular polytope (Platonic solid)* [18] is a convex 3-polytope with all facets being congruent regular polygons that are assembled in the same way around each vertex.

There are only 5 Platonic solids, see Fig. 3. All Platonic solids are vertex-, edge-, and facet-transitive. They are not combinatorially chiral.

2.5. *Archimedean solids*

Definition 2.10: An *Archimedean solid* [18] is a convex 3-polytope with all facets regular polygons of *two or more types*, such that for any pair of vertices there is a symmetry of the polytope that moves one vertex to another.

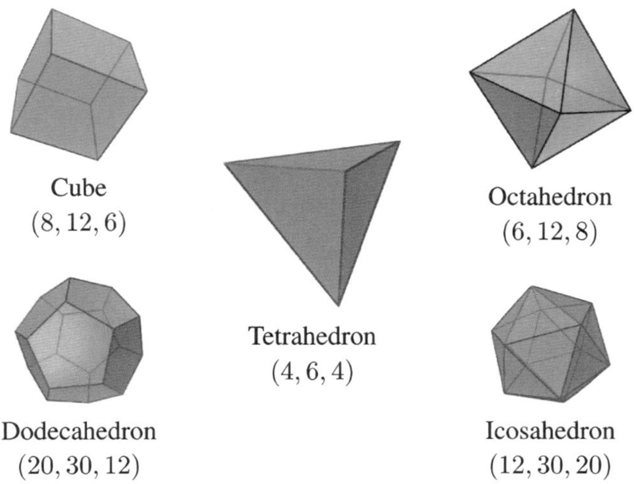

Fig. 3. Platonic solids with f-vectors (f_0, f_1, f_2) (www.wikipedia.org)

There are only 13 solids with these properties: 10 with facets of two types, and 3 with facets of three types. In the following figures we present the Archimedean solids. For any polytope we give vectors (f_0, f_1, f_2) and $(k_1, \ldots, k_p; q)$, where q is the valency of any vertex and a tuple $(k_1, \ldots, k_p)$ show which k-gons are present. The snub cube and snub dodecahedron are combinatorially chiral, whilst the other 11 Archimedean solids are not combinatorially chiral.

2.6. *Simple polytopes*

An n-polytope is *simple* if every vertex is contained in exactly n facets.

Example 2.11:

- 3 of 5 Platonic solids are simple.
- 7 of 13 Archimedean solids are simple.

Exercise: Show the following:

- A simple n-polytope with all 2-faces triangles is combinatorially equivalent to the n-simplex.
- A simple n-polytope with all 2-faces quadrangles is combinatorially equivalent to the n-cube.
- A simple 3-polytope with all 2-faces pentagons is combinatorially equivalent to the dodecahedron.

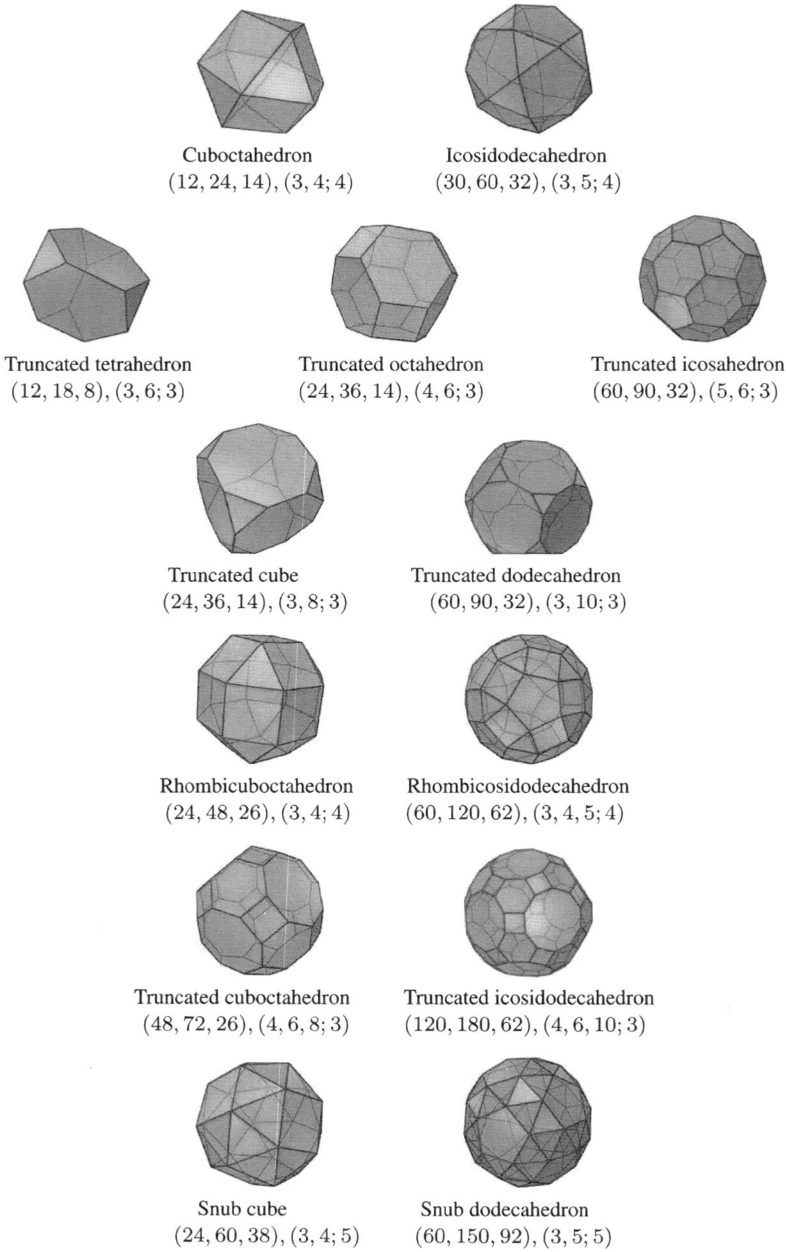

Fig. 4. Archimedean solids with f-vectors and facet-vertex types (www.wikipedia.org)

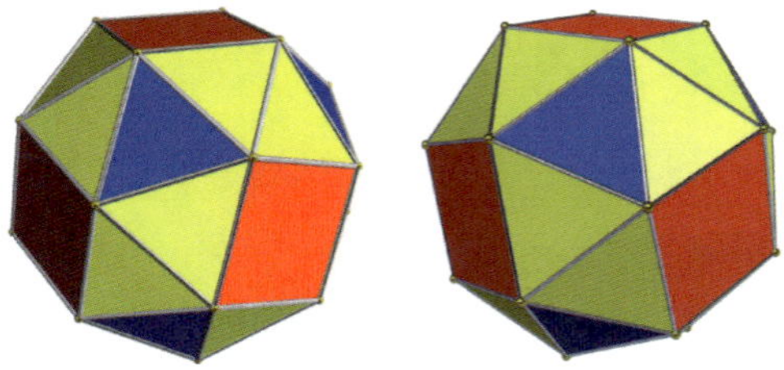

Fig. 5. Left and right snub cube. Fix the orientations induced from the ambient space. There is no combinatorial equivalence preserving this orientation. (www.wikipedia.org)

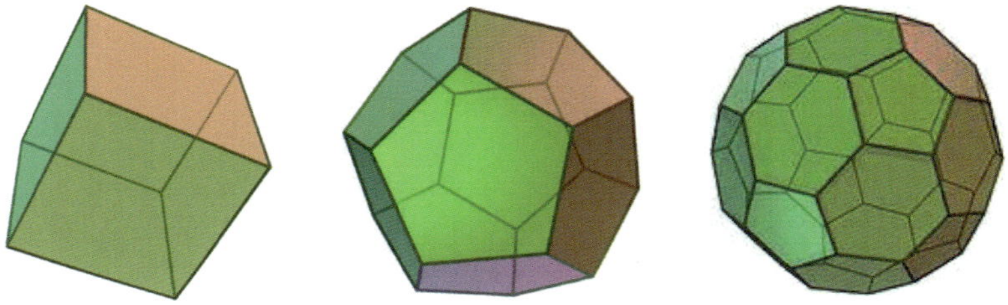

Fig. 6. Simple polytopes: cube, dodecahedron and truncated icosahedron (www.wikipedia.org)

2.7. *Realization of f-vector*

Theorem 2.12: [47] *(Ernst Steinitz (1871-1928)) An integer vector (f_0, f_1, f_2) is a face vector of a* ***three-dimensional*** *polytope if and only if*

$$f_0 - f_1 + f_2 = 2, \quad f_2 \leqslant 2f_0 - 4, \quad f_0 \leqslant 2f_2 - 4.$$

Corollary 2.13:

$$f_2 + 4 \leqslant 2f_0 \leqslant 4f_2 - 8.$$

The well-known g-theorem [46, 2] gives the criterion for when an integer vector $(f_0, \dots, f_{n-1})$ is an f-vector of a simple n-polytope (see also [14]).

For general polytopes there are only partial results about f-vectors.

Classical problem: For *four-dimensional* polytopes the conditions characterizing the face vector (f_0, f_1, f_2, f_3) are still *not known* [53].

2.8. *Dual polytopes*

For an n-polytope $P \subset \mathbb{R}^n$ with $\mathbf{0} \in \operatorname{int} P$ the *dual polytope* P^* is

$$P^* = \{\boldsymbol{y} \in (R^n)^* \colon \boldsymbol{y}\boldsymbol{x} + 1 \geqslant 0\}.$$

- i-faces of P^* are in an inclusion reversing bijection with $(n-i-1)$-faces of P.
- $(P^*)^* = P$.

An n-polytope is *simplicial* if all of its facets are simplicies.

Lemma 2.14: *A polytope dual to a simple polytope is simplicial and a polytope dual to a simplicial polytope is simple.*

Lemma 2.15: *Let a polytope P^n, $n \geqslant 2$, be simple and simplicial. Then either $n = 2$, or P^n is combinatorially equivalent to a simplex Δ^n, $n > 2$.*

Example 2.16: Amongst the 5 Platonic solids the tetrahedron is self-dual, the cube is dual to the octahedron, and the dodecahedron is dual to the icosahedron.

There are no simplicial polytopes amongst the Archimedean solids. Polytopes dual to Archimedean solids are called *Catalan solids*, since they where first described by E. C. Catalan (1814-1894). For example, the polytope dual to the truncated icosahedron is called the *pentakis dodecahedron*.

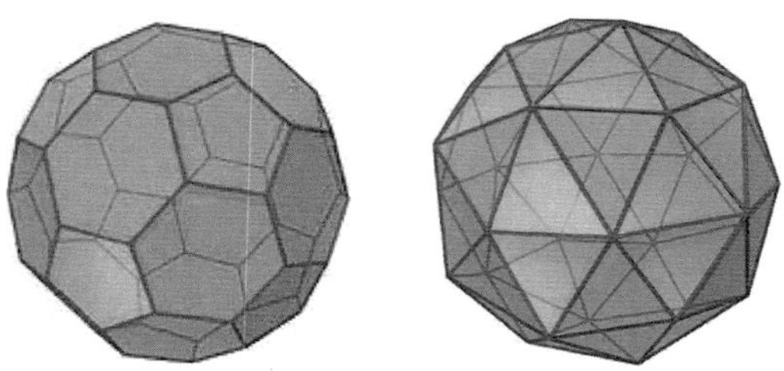

Fig. 7. Truncated icosahedron and pentakis dodecahedron (www.wikipedia.org)

In Figure 8 the point $(4, 4)$ corresponds to the tetrahedron. The bottom ray corresponds to simple polytopes, the upper ray to simplicial polytopes. For $k \geqslant 3$ self-dual pyramids over k-gons give points on the diagonal.

2.9. *k-belts*

Definition 2.17: Let P be a simple convex 3-polytope. A *thick path* is a sequence of facets $(F_{i_1}, \dots, F_{i_k})$ with $F_{i_j} \cap F_{i_{j+1}} \neq \varnothing$, for $j = 1, \dots, k-1$. A *k-loop* is a cyclic sequence $(F_{i_1}, \dots, F_{i_k})$ of facets, such that $F_{i_1} \cap F_{i_2}$, $\dots$, $F_{i_{k-1}} \cap F_{i_k}$, $F_{i_k} \cap F_{i_1}$ are edges. A k-loop is called *simple*, if the facets $(F_{i_1}, \dots, F_{i_k})$ are pairwise distinct.

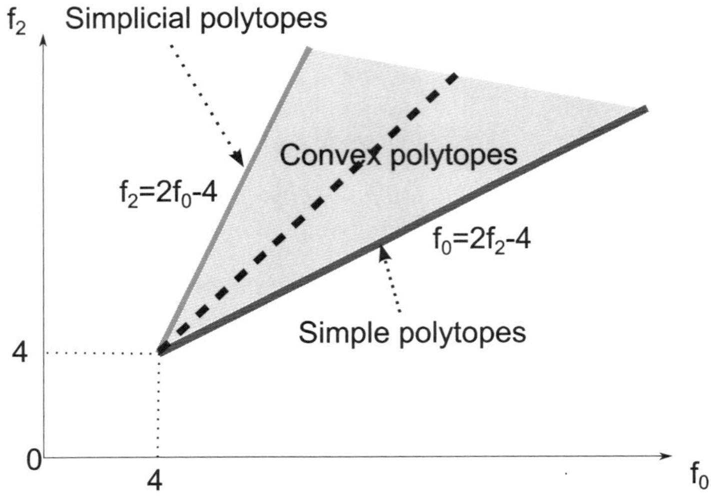

Fig. 8. By Steinitz's theorem and Euler's formula, integer points inside the cone are in the one-to-one correspondence with the f-vectors of convex 3-polytopes

Example 2.18: Any vertex of P is surrounded by a simple 3-loop. Any edge is surrounded by a simple 4-loop. Any k-gonal facet is surrounded by a simple k-loop.

Definition 2.19: A k-*belt* is a k-loop, such that $F_{i_1}\cap\cdots\cap F_{i_k} = \varnothing$ and $F_{i_p}\cap F_{i_q} \neq \varnothing$ if and only if $\{p, q\} \in \{\{1, 2\}, \ldots, \{k-1, k\}, \{k, 1\}\}$.

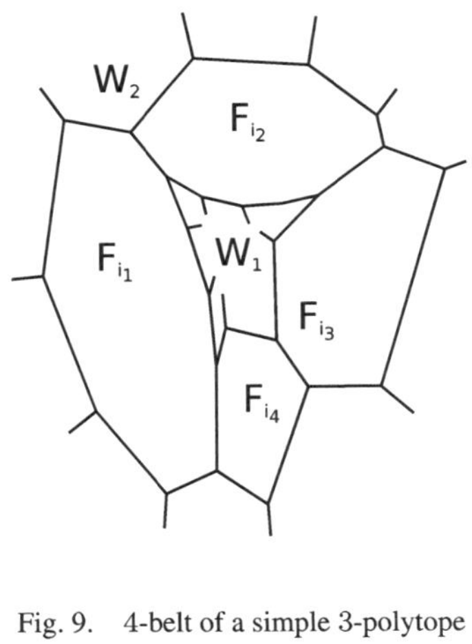

Fig. 9. 4-belt of a simple 3-polytope

2.10. *Simple paths and cycles*

By $G(P)$ we denote a vertex-edge graph of a simple 3-polytope P. We call it the *graph of a polytope*. Let G be a graph.

Definition 2.20:

- An *edge path* is a sequence of vertices $(v_1, \dots, v_k)$, $k > 1$, such that v_i and v_{i+1} are connected by some edge E_i for all $i < k$.
- An edge path is *simple* if it passes through any vertex of G at most once.
- A *cycle* is a simple edge path, such that $v_k = v_1$, where $k > 2$. We denote a cycle by $(v_1, \dots, v_{k-1})$.

A cycle $(v_1, \dots, v_k)$ in the graph of a simplicial 3-polytope P is *dual* to a k-belt in a simple 3-polytope P^* if all of its vertices do not lie in the same face, and v_i and v_j are connected by an edge if and only if $\{i, j\} \in \{\{1, 2\}, \dots, \{k-1, k\}, \{k, 1\}\}$.

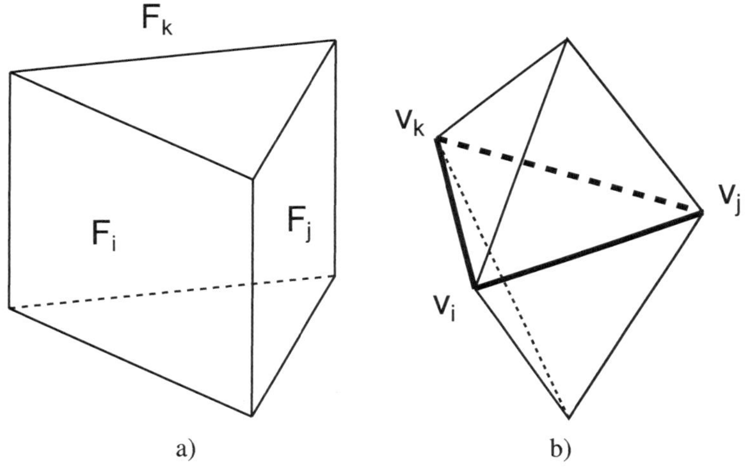

Fig. 10. a) (F_i, F_j, F_k) is a 3-belt; b) (v_i, v_j, v_k) is a cycle dual to the 3-belt

Definition 2.21: A *zigzag path* on a simple 3-polytope is an edge path with no 3 successive edges lying in the same facet.

Starting with one edge and choosing another sharing with it a common vertex, we obtain a unique way to construct a zigzag path.

Definition 2.22: A *zigzag cycle* on a simple 3-polytope is a cycle with no 3 successive edges lying in the same facet.

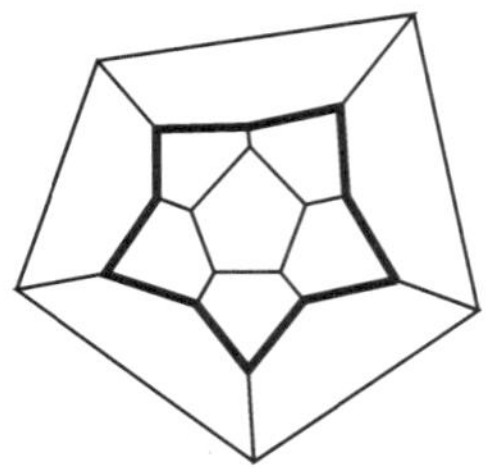

Fig. 11. A zigzag cycle on the Schlegel diagram of the dodecahedron

2.11. *The Steinitz theorem*

Definition 2.23: A graph G is called *simple* if it has no loops and multiple edges. A connected graph G is called 3-connected if it has at least 4 edges and the deletion of any one or two vertices along with all incident edges leaves G connected.

Theorem 2.24: (The Steinitz theorem, see [54]) *A graph G is a graph of a 3-polytope if and only if it is simple, planar and 3-connected.*

Remark 2.25: Moreover, the cycles in G corresponding to facets are exactly chordless simple edge cycles C with $G \setminus C$ disconnected; hence the combinatorics of the embedding $G \subset S^2$ is uniquely defined.

We will need the following version of the Jordan curve theorem. It can be proved rather directly similarly to the piecewise-linear version of this theorem on the plane.

Theorem 2.26: *Let γ be a simple piecewise-linear (in respect to some homeomorphism $S^2 \simeq \partial P$ for a 3-polytope P) closed curve on the sphere S^2. Then*

(1) $S^2 \setminus \gamma$ consists of two connected components $\mathcal{C}_1$ and $\mathcal{C}_2$.
(2) The closure $\overline{\mathcal{C}_\alpha}$ is homeomorphic to a disk, for each $\alpha = 1, 2$.

We will also need the following result.

Lemma 2.27: *Let $G \subset S^2$ be a finite connected simple graph with more than one vertex, realized piecewise-linearly in respect to some homeomorpism $S^2 \simeq \partial P$ for a 3-polytope P. Then G is 3-connected if and only if all connected components of $S^2 \setminus G$ are bounded by simple edge cycles and the closures of any two areas («facets») either do not intersect, or intersect in a single common vertex, or intersect in a single common edge.*

Proof: Let G satisfy the second condition of the lemma. We will prove that G is 3-connected. Let $v_1 \neq v_2$, $u_1, u_2 \notin \{v_1, v_2\}$, be vertices of G, perhaps $u_1 = u_2$. We need to prove that there is an edge-path from v_1 to v_2 in $G \setminus \{u_1, u_2\}$. Since G is connected, there is an edge-path γ connecting v_1 and v_2. Consider the vertex u_α, $\alpha \in \{1, 2\}$, and all facets $F_{i_1}, \ldots, F_{i_p}$ containing it. From the second condition of the lemma $p \geqslant 3$, otherwise either there is a face bounded by a non-simple cycle, or two faces intersect by more than one edge. Since the graph is embedded into the sphere, after relabelling we obtain a simple p-loop $(F_{i_1}, \ldots, F_{i_p})$. For $j \in \{1, \ldots, p\}$ denote by w_j the end of the edge $F_{i_j} \cap F_{i_{j+1}}$ different from u_α, where $F_{i_{p+1}} = F_{i_1}$. Let g_j be the simple edge-path connecting w_{j-1} and w_j in $F_{i_j} \setminus u_\alpha$ (see Fig. 12). Then $\eta_\alpha = (g_1, g_2, \ldots, g_p)$ is a simple edge-cycle. Indeed, if

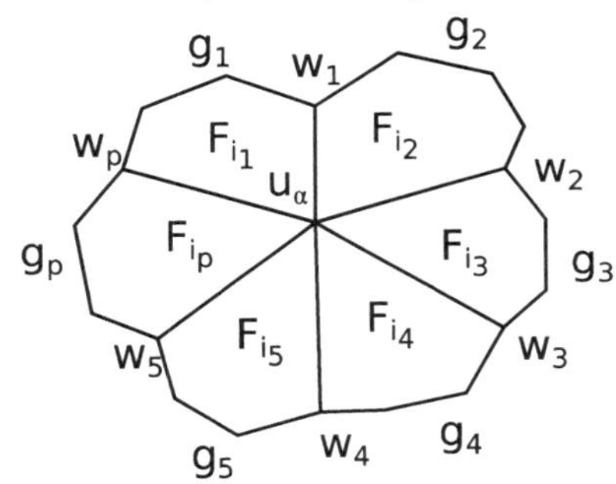

Fig. 12. Star of the vertex u_α

g_s and g_t have a common vertex, then this vertex belongs to $F_{i_s} \cap F_{i_t}$ together with u_α; hence it is connected to u_α by an edge; therefore $\{s, t\} = \{k, k+1 \mod p\}$ for some k, and the vertex is w_k. If u_1 and u_2 are different and are connected by an edge E, then $E = F_{i_s} \cap F_{i_t}$ for some $s, t \in \{1, \ldots, p\}$, $s - t = \pm 1 \mod p$, and we can change γ to γ_1 so as not to contain E by substituting the simple edge-path in $F_{i_s} \setminus E$ for E. Else set $\gamma_1 = \gamma$. Now for the path γ_1 consider all times it passes through u_α. We can remove all the fragments (w_i, u_α, w_j) and substitute the simple edge path in $\eta_\alpha \setminus u_\beta$ connecting w_i and w_j for each fragment (w_i, u_α, w_j). The same can be done for u_β, $\{\alpha, \beta\} = \{1, 2\}$. Thus we obtain the edge-path γ_2 connecting v_1 and v_2 in $G \setminus \{u_1, u_2\}$. G cannot contain less than 4 edges, since if it contains 0, 1, or 2 edges, its only face is not bounded by a simple edge-cycle, and if it contains 3 edges, then either its only face is not bounded by a simple edge-cycle, or its two faces have 3 common boundary edges. Thus, G is 3-connected.

Now let G be 3-connected. Consider the connected component D of $S^2 \setminus G$ and its boundary ∂D. If there is a hanging vertex $v \in \partial D$ of G, then deletion of the other end of the edge containing v makes the graph disconnected. Hence any vertex $v \in \partial D$ of G has valency at least 2, and D is surrounded by an edge-cycle η. If η is not simple, then there is a vertex $v \in \eta$ passed through several times. Then the area D appears several times when we walk around the vertex v. Since D is connected, there is a simple piecewise-linear (in respect to the homeomorphism $S^2 \simeq \partial P$) closed curve η in the closure $\overline{D}$ of D with the only point v on the boundary. Walking round v, we pass edges in both connected components of $S^2 \setminus \eta$; hence the deletion of v divides G into several connected components. Thus the cycle η is simple. Let facets $F_1 = \overline{D_1}$, $F_2 = \overline{D_2}$ have two common vertices v_1, v_2. Consider piecewise-linear simple curves $\eta_1 \subset F_1$, $\eta_2 \subset F_2$, with ends v_1 and v_2 and all other points lying in D_1 and D_2 respectively. Then $\eta_1 \cup \eta_2$ is a simple piecewise-linear closed curve; hence it separates the sphere S^2 into two connected components. If v_1 and v_2 are not adjacent in F_1 or F_2, then both connected components contain vertices of G; hence the deletion of v_1 and v_2 makes the graph disconnected. Thus any two common vertices are adjacent in both facets. Moreover, since there are no multiple edges, the corresponding edges belong to both facets. Then either both facets are surrounded by a common 3-cycle, and in this case G has only 3 edges, or any two facets either do not intersect, or intersect by a common vertex, or intersect by a common edge. This finishes the proof. □

Let $\mathcal{L}_k = (F_{i_1}, \dots, F_{i_k})$ be a simple k-loop for $k \geqslant 3$. Consider midpoints w_j of edges $F_{i_j} \cap F_{i_{j+1}}$, $F_{i_{k+1}} = F_{i_1}$ and segments E_j connecting w_j and w_{j+1} in F_{j+1}. Then $(E_1, \dots, E_k)$ is a simple piecewise-linear curve η on ∂P. It separates $\partial P \simeq S^2$ into two areas homeomorphic to discs D_1 and D_2 with $\partial D_1 = \partial D_2 = \eta$. Consider two graphs G_1 and G_2 obtained from the graph $G(P)$ of P by addition of vertices $\{w_j\}_{j=1}^k$ and edges $\{E_j\}_{j=1}^k$, and deletion of all vertices and edges with interior points inside D_1 or D_2 respectively.

Lemma 2.28: *There exist simple polytopes P_1 and P_2 with graphs $G_1 = G(P_1)$ and $G_2 = G(P_2)$.*

Proof: The proof is similar for both graphs; hence we consider the graph G_1. It is a simple connected piecewise-linear graph in the surface of the polytope P. It is sufficient to prove that the hypothesis of Lemma 2.27 is valid. For this we see each facet of G_1 is either a facet of P, or it is the part of a facet F_{i_j} for some j, or it is bounded by the cycle η. In particular, all facets are bounded by simple edge-cycles. If the facets F_i and F_j are both of the first two types they either do not intersect or intersect by common edge as it is in P. If F is the facet bounded by η, then it intersects only facets $(F_{i_1}, \dots, F_{i_k})$, and each intersection is an edge $F \cap F_{i_j}$, $j = 1, \dots, k$. □

Definition 2.29: We will call polytopes P_1 and P_2 *loop-cuts* (or, more precisely, $\mathcal{L}_k$*-cuts*) of P.

3. Combinatorics of simple polytopes

3.1. *Flag polytopes*

Definition 3.1: A simple polytope is called *flag* if any set of pairwise intersecting facets $F_{i_1}, \dots, F_{i_k}\colon F_{i_s} \cap F_{i_t} \neq \varnothing$, $s, t = 1, \dots, k$, has a nonempty intersection $F_{i_1} \cap \dots \cap F_{i_k} \neq \varnothing$.

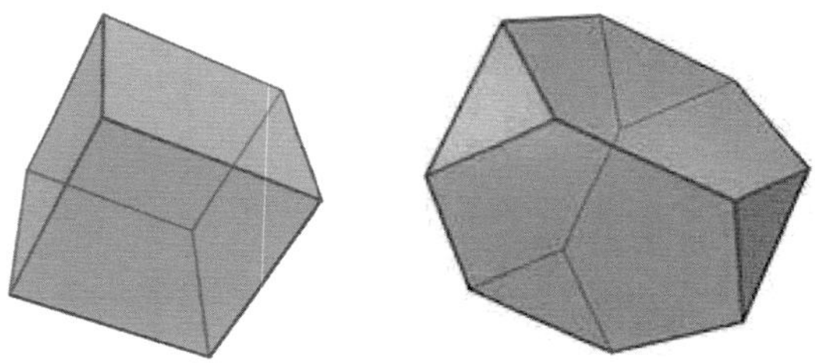

a) b)

Fig. 13. a) Flag polytope; b) non-flag polytope (www.wikipedia.org)

Example 3.2: The n-simplex Δ^n is not a flag polytope for $n \geqslant 2$.

Proposition 3.3: *A simple* 3*-polytope* P *is* ***not flag*** *if and only if either* $P = \Delta^3$ *or* P *contains a* 3*-belt.*

Corollary 3.4: *A simple* 3*-polytope* $P \neq \Delta^3$ *is flag if and only any* 3*-loop corresponds to a vertex.*

Proposition 3.5: *A simple* 3*-polytope* P *is flag if and only if any facet is surrounded by a* k*-belt, where* k *is the number of its edges.*

Proof: A simplex is not flag and has no 3-belts.

By Proposition 3.3 a simple 3-polytope $P \not\simeq \Delta^3$ is not flag if and only if it has a 3-belt. The facet $F \subset P$ is not surrounded by a belt if and only if it belongs to a 3-belt. □

Corollary 3.6: *For any flag simple* 3*-polytope* P *we have* $p_3 = 0$.

Later (see Section 10) we will need the following result.

Proposition 3.7: *A flag* 3*-polytope* P *has no* 4*-belts if and only if any pair of adjacent facets is surrounded by a belt.*

Proof: The pair (F_i, F_j) of adjacent facets is a 2-loop and is surrounded by a simple edge-cycle. Let $\mathcal{L} = (F_{i_1}, \dots, F_{i_k})$ be the k-loop that borders it. If $\mathcal{L}$ is not simple, then $F_{i_a} = F_{i_b}$ for $a \neq b$. Then F_{i_a} and F_{i_b} are not adjacent to the same facet F_i or F_j. Let F_{i_a} be adjacent to F_i, and F_{i_b} to F_j. Then (F_i, F_j, F_{i_a}) is a 3-belt. A contradiction. Hence $\mathcal{L}$ is a simple loop. If it is not a belt, then $F_{i_a} \cap F_{i_b} \neq \varnothing$ for non-successive facets F_{i_a} and F_{i_b}. From Proposition 3.5 we obtain that F_{i_a} and F_{i_b} are not adjacent to the same facet F_i or F_j. Let F_{i_a} be adjacent to F_i, and F_{i_b} to F_j. Then $(F_{i_a}, F_i, F_j, F_{i_b})$ is a 4-belt. On the other hand, if there is a 4-belt (F_i, F_j, F_k, F_l), then facets F_k and F_l belong to the loop surrounding the pair (F_i, F_j). Since $F_i \cap F_k = \varnothing = F_j \cap F_l$, they are not successive facets of this loop; hence the loop is not a belt. □

In the combinatorial study of fullerenes the following version of the Jordan curve theorem gives an important tool. It follows from Theorem 2.26.

Theorem 3.8: *Let γ be a simple edge-cycle on a simple 3-polytope P. Then*

(1) $\partial P \setminus \gamma$ consists of two connected components $\mathcal{C}_1$ and $\mathcal{C}_2$.
(2) Let $\mathcal{D}_\alpha = \{F_j \in \mathcal{F}_P \colon \operatorname{int} F_j \subset \mathcal{C}_\alpha\}$, $\alpha = 1, 2$. Then $\mathcal{D}_1 \sqcup \mathcal{D}_2 = \mathcal{F}_P$.
(3) The closure $\overline{\mathcal{C}_\alpha}$ is homeomorphic to a disk. We have $\overline{\mathcal{C}_\alpha} = |\mathcal{D}_\alpha|$.

Corollary 3.9: *If we remove the 3-belt from the surface of a simple 3-polytope, we obtain two parts W_1 and W_2, which are homeomorphic to disks.*

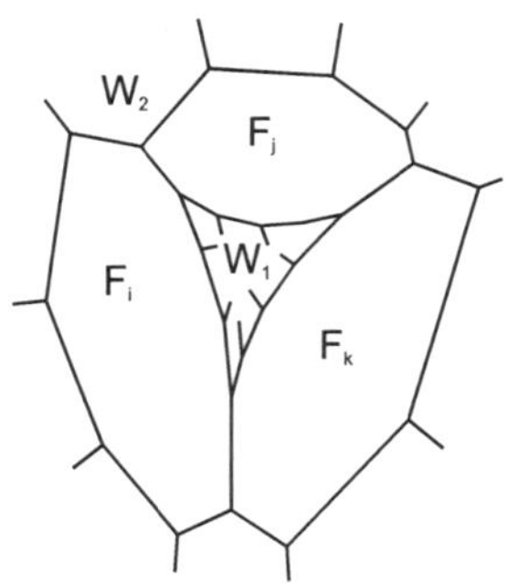

Fig. 14. 3-belt on the surface of a simple 3-polytope

Proposition 3.10: *Let P be a flag simple 3-polytope. Then $m \geqslant 6$, and $m = 6$ if and only if P is combinatorially equivalent to the cube I^3.*

Proof: Take a facet F_1. By Proposition 3.5 it is surrounded by a k-belt $\mathcal{B} = (F_{i_1}, \ldots, F_{i_k})$, $k \geqslant 4$. Since there is at least one facet in the connected component W_α of $\partial P \setminus \mathcal{B}$, $\operatorname{int} F_1 \notin W_\alpha$, we obtain $m \geqslant 2 + k \geqslant 6$. If $m = 6$, then $k = 4$, F_1 is a quadrangle, and $W_\alpha = \operatorname{int} F_j$ for some facet F_j Then $F_j \cap F_{i_1} \cap F_{i_2}$, $F_j \cap F_{i_2} \cap F_{i_3}$, $F_j \cap F_{i_3} \cap F_{i_4}$, $F_j \cap F_{i_4} \cap F_{i_1}$ are vertices, and P is combinatorially equivalent to I^3. □

Lemma 3.11: *Let P be a flag polytope, $\mathcal{L}_k$ be a simple k-loop, and P_1 and P_2 be $\mathcal{L}_k$-cuts of P. Then the following conditions are equivalent:*

(1) both polytopes P_1 and P_2 are flag;
(2) $\mathcal{L}_k$ is a k-belt.

Proof: Since P has no 3-belts, for $k = 3$ the loop $\mathcal{L}_k$ surrounds a vertex; hence one of the polytopes P_1 and P_2 is a simplex, and it is not flag. Let $k \geqslant 4$. Then P_1 and P_2 are not simplices. There are three types of facets in P: lying only in P_1, lying only in P_2, and lying in $\mathcal{L}_k$. Let $\mathcal{B}_3 = (F_i, F_j, F_k)$ be a 3-loop in P_α, $\alpha \in \{1, 2\}$. Let F_i, F_j, F_k correspond to facets of P. Since intersecting facets in P_α also intersect in P, (F_i, F_j, F_k) is also a 3-loop in P, and $F_i \cap F_j \cap F_k \in P$ is a vertex. Since $F_i \cap F_j \neq \varnothing$ in P_α, either the corresponding edge of P lies in P_α, or it intersects the new facet, and F_i and F_j are consequent facets of $\mathcal{L}_k$. Since $k \geqslant 4$, at least one edge of $F_i \cap F_j$, $F_j \cap F_k$, and $F_k \cap F_i$ of P lies in P_α; hence $F_i \cap F_j \cap F_k \in P_\alpha$, and $\mathcal{B}_3$ is not a 3-belt in P_α. If one of the facets, say F_i, is a new facet of P_α, then $F_j, F_k \in \mathcal{L}_k$, since $F_i \cap F_j, F_i \cap F_k \neq \varnothing$. Consider the edge $F_j \cap F_k$ of P. It intersects F_i in P_α if and only if F_j and F_k are consequent facets in $\mathcal{L}_k$. Thus if $\mathcal{B}_3$ is a 3-belt, then $\mathcal{L}_k$ is not a k-belt, and vice versa, if $\mathcal{L}_k$ is not a k-belt, then $F_j \cap F_k \neq \varnothing$ for some non-consequent facets of $\mathcal{L}_k$, and the corresponding 3-loop $\mathcal{B}_3$ is a 3-belt in the polytope P_1 or P_2 containing $F_j \cap F_k$. This finishes the proof. □

3.2. *Non-flag 3-polytopes as connected sums*

The existence of a 3-belt is equivalent to the fact that P is combinatorially equivalent to a *connected sum* $P = Q_1 \#_{v_1, v_2} Q_2$ of two simple 3-polytopes Q_1 and Q_2 along vertices v_1 and v_2. The part W_i appears if we remove from the surface of the polytope Q_i the facets containing the vertex v_i, $i = 1, 2$.

3.3. *Consequence of Euler's formula for simple 3-polytopes*

Let p_k be a number of k-gonal facets of a 3-polytope.

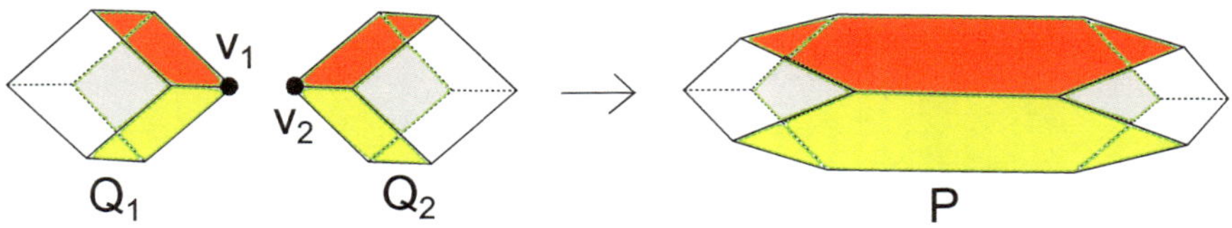

Fig. 15. Connected sum of two simple polytopes along vertices

Theorem 3.12: *(See [32]) For any **simple** 3-polytope P*

$$3p_3 + 2p_4 + p_5 = 12 + \sum_{k \geqslant 7} (k-6)p_k. \tag{3.1}$$

Proof: The number of pairs (edge, vertex of this edge) is equal, on the one hand, to $2f_1$ and, on the other hand (since the polytope is simple), to $3f_0$. Then $f_0 = \frac{2f_1}{3}$, and from the Euler formula we obtain $2f_1 = 6f_2 - 12$. Counting the pairs (facet, edge of this facet), we have

$$\sum_{k \geqslant 3} kp_k = 2f_1 = 6\left(\sum_{k \geqslant 3} p_k\right) - 12,$$

which implies formula (3.1). □

Corollary 3.13: *There is no simple polytope P with all facets hexagons. Moreover, if $p_k = 0$, for $k \neq 5, 6$, then $p_5 = 12$.*

Exercise: Show that the f-vector of a simple polytope is expressed in terms of the p-vector by the following formulas:

$$f_0 = 2\,(f_2 - 2)\,; \quad f_1 = 3\,(f_2 - 2)\,; \quad f_2 = \sum_k p_k.$$

3.4. *Realization theorems*

Definition 3.14: An integer sequence $(p_k | k \geqslant 3)$ is called 3-*realizable* is there is a simple 3-polytope P with $p_k(P) = p_k$.

Theorem 3.15: (Victor Eberhard [25], see [32]) *For a sequence $(p_k | 3 \leqslant k \neq 6)$ there exists p_6 such that the sequence $(p_k | k \geqslant 3)$ is 3-realizable if and only if it satisfies formula (3.1).*

There arises a natural question.

Problem: For a given sequence $(p_k|3 \leqslant k \neq 6)$ find all p_6 such that the sequence $(p_k|k \geqslant 3)$ is 3-realizable.

Notation: When we write a finite sequence $(p_3, p_4, \ldots, p_k)$ we mean that $p_l = 0$ for $l > k$.

Example 3.16: (see [32]) Sequences $(0, 6, 0, p_6)$ and $(0, 0, 12, p_6)$ are 3-realizable if and only if $p_6 \neq 1$. The sequence $(4, 0, 0, p_6)$ is 3-realizable if and only if p_6 is an even integer different from 2. The sequence $(3, 1, 1, p_6)$ is 3-realizable if and only if p_6 is an odd integer greater than 1.

Let us also mention the following results.

Theorem 3.17: *For a given sequence $(p_k|3 \leqslant k \neq 6)$ satisfying formula (3.1)*

- *there exists* $p_6 \leqslant 3 \left(\sum\limits_{k\neq 6} p_k \right)$ *such that the sequence $(p_k|k \geqslant 3)$ is 3-realizable* [30];
- *if $p_3 = p_4 = 0$ then any sequence $(p_k|k \geqslant 3, p_6 \geqslant 8)$ is 3-realizable* [31].

There are operations on simple 3-polytopes that do not effect p_k except for p_6. We call them *p_6-operations*. As we will see later they are important for applications.

Operation I: This operation affects all edges of the polytope P. We present a fragment on Fig. 16. On the right picture the initial polytope P is drawn by dotted

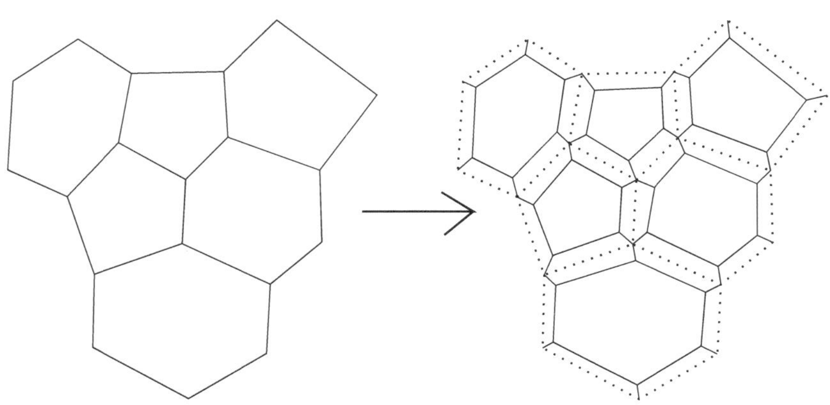

Fig. 16. Operation I

lines, whilst the resulting polytope is drawn by solid lines. We have

$$p_k(P') = \begin{cases} p_k(P), & k \neq 6; \\ p_6(P) + f_1(P), & k = 6. \end{cases}$$

Operation II: This operation affects all edges of the polytope P. We present a fragment in Figure 17. On the right picture the initial polytope P is drawn by

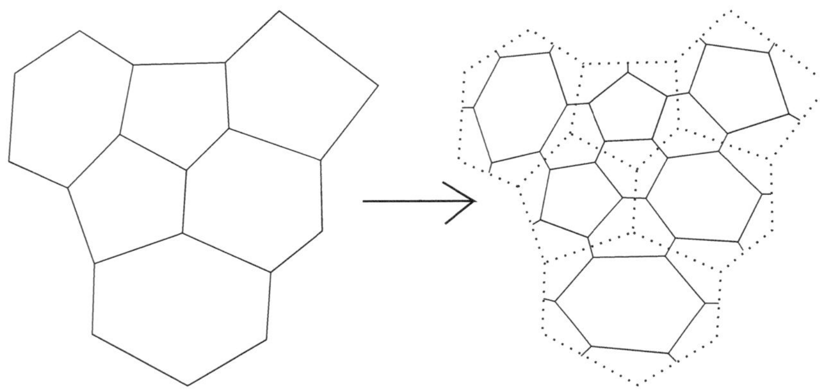

Fig. 17. Operation II

dotted lines, whilst the resulting polytope is drawn by solid lines. We have

$$p_k(P') = \begin{cases} p_k(P), & k \neq 6; \\ p_6(P) + f_0(P), & k = 6. \end{cases}$$

Operation I and Operation II are called iterative procedures (see [38]), since arbitrary compositions of them are well defined.

Exercise: Operation I and Operation II commute; therefore they define an action of the semigroup $\mathbb{Z}_{\geqslant 0} \times \mathbb{Z}_{\geqslant 0}$ on the set of all combinatorial simple 3-polytopes, where $\mathbb{Z}_{\geqslant 0}$ is the additive semigroup of nonnegative integers.

3.5. *Graph-truncation of simple 3-polytopes*

Consider a subgraph $\Gamma \subset G(P)$ without isolated vertices. For each edge

$$E_{i,j} = F_i \cap F_j = P \cap \{\boldsymbol{x} \in \mathbb{R}^3 \colon (\boldsymbol{a}_i + \boldsymbol{a}_j)\boldsymbol{x} + (b_i + b_j) = 0\}$$

consider the halfspace

$$\mathcal{H}^+_{ij,\varepsilon} = \{\boldsymbol{x} \in \mathbb{R}^3 \colon (\boldsymbol{a}_i + \boldsymbol{a}_j)\boldsymbol{x} + (b_i + b_j) \geqslant \varepsilon\}.$$

Set

$$P_{\Gamma,\varepsilon} = P \cap \bigcap_{E_{i,j} \in \Gamma} \mathcal{H}^+_{ij,\varepsilon}.$$

Exercise: Show that for small enough values of ε the combinatorial type of $P_{\Gamma,\varepsilon}$ does not depend on ε.

Definition 3.18: We will denote by P_Γ the combinatorial type of $P_{\Gamma,\varepsilon}$ for small enough values of ε and call it a Γ-*truncation* of P. When it is clear what Γ is we call P_Γ the *graph-truncation* of P.

Example 3.19: For $\Gamma = G(P)$ the polytope $P' = P_\Gamma$ is obtained from P by a p_6-operation I defined above.

Proposition 3.20: *Let P be a simple polytope with $p_3 = 0$. Then the polytope $P_{G(P)}$ is flag.*

We leave the proof as an exercise.

Corollary 3.21: *For a given sequence $(p_k | 3 \leqslant k \neq 6)$ satisfying formula (3.1) there are infinitely many values of p_6 such that the sequence $(p_k | k \geqslant 3)$ is 3-realizable.*

3.6. *Analog of Eberhard's theorem for flag polytopes*

Theorem 3.22: [7] *For every sequence $(p_k | 3 \leqslant k \neq 6, p_3 = 0)$ of nonnegative integers satisfying formula (3.1) there exists a value of p_6 such that there is a **flag simple** 3-polytope P^3 with $p_k = p_k(P^3)$, for all $k \geqslant 3$.*

Proof: For a given sequence $(p_k | 3 \leqslant k \neq 6, p_3 = 0)$ satisfying formula (3.1) by Eberhard's theorem there exists a simple polytope P with $p_k = p_k(P)$, $k \neq 6$. Then the polytope $P' = P_{G(P)}$ is flag by Proposition 3.20. We have $p_k(P') = p_k(P)$, $k \neq 6$, and $p_6(P') = p_6(P) + f_1(P)$. □

4. Combinatorial fullerenes

4.1. *Fullerenes*

A *fullerene* is a molecule of carbon that is topologically a sphere and any atom belongs to exactly three carbon rings, which are pentagons or hexagons.

The first fullerene C_{60} was generated by chemists-theorists Robert Curl, Harold Kroto, and Richard Smalley in 1985 (Nobel Prize in chemistry 1996, [19, 36, 45]). They called it the *Buckminsterfullerene*.

Buckminsterfullerene C_{60}
$(f_0, f_1, f_2) = (60, 90, 32)$
$(p_5, p_6) = (12, 20)$

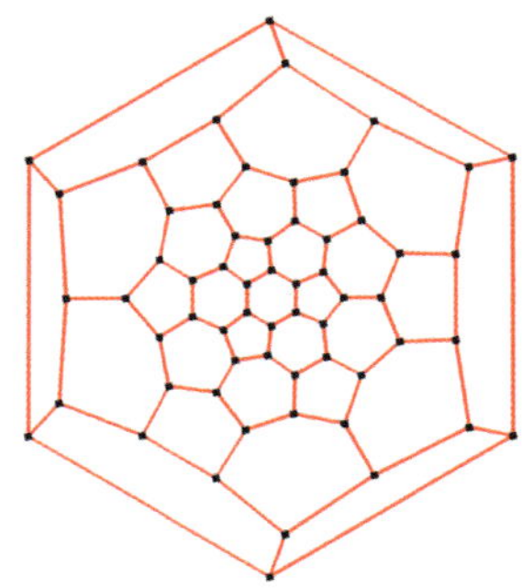

Schlegel diagram

Fig. 18. Buckminsterfullerene and its Schlegel diagram (www.wikipedia.org)

Fig. 19. Fuller's Biosphere, USA Pavillion on Expo-67 (Montreal, Canada) (www.wikipedia.org)

Fullerenes were named after Richard Buckminster Fuller (1895-1983) – a famous American architect, systems theorist, author, designer and inventor. In 1954 he patented an architectural construction in the form of polytopal spheres for roofing large areas.

They are also called *buckyballs*.

Definition 4.1: A *combinatorial fullerene* is a simple 3-polytope with all facets pentagons and hexagons.

To be short, by a fullerene below we mean a combinatorial fullerene. For any fullerene $p_5 = 12$, the expression of the f-vector in terms of the p-vector has the following form

$$f_0 = 2(10 + p_6), \quad f_1 = 3(10 + p_6), \quad f_2 = (10 + p_6) + 2.$$

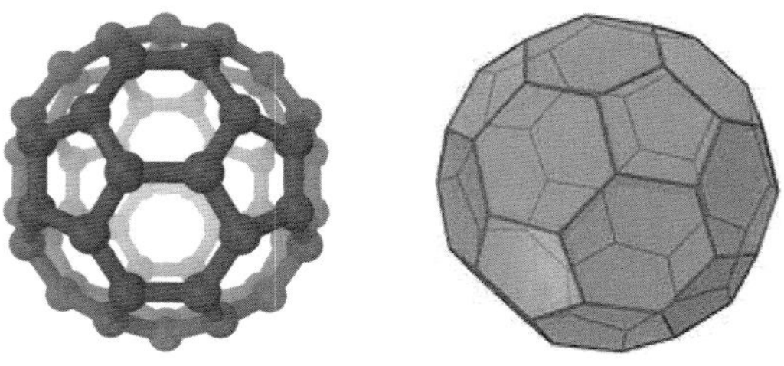

Fig. 20. Fullerene C_{60} and truncated icosahedron (www.wikipedia.org)

Remark 4.2: Since the combinatorially chiral polytope is geometrically chiral (see Proposition 2.6), the following problem is important for applications in the physical theory of fullerenes:

Problem: Find an algorithm to decide if a given fullerene is combinatorially chiral.

4.2. *Icosahedral fullerenes*

Operations I and II (see page 87) transform fullerenes into fullerenes. The first procedure increases f_0 by 4 times, the second by 3 times.

Applying Operation I to the dodecahedron we obtain the fullerene C_{80} with $p_6 = 30$. In total there are 31924 fullerenes with $p_6 = 30$. Applying Operation II to the dodecahedron we obtain the Buckminsterfullerene C_{60} with $p_6 = 20$. In total there are 1812 fullerenes with $p_6 = 20$.

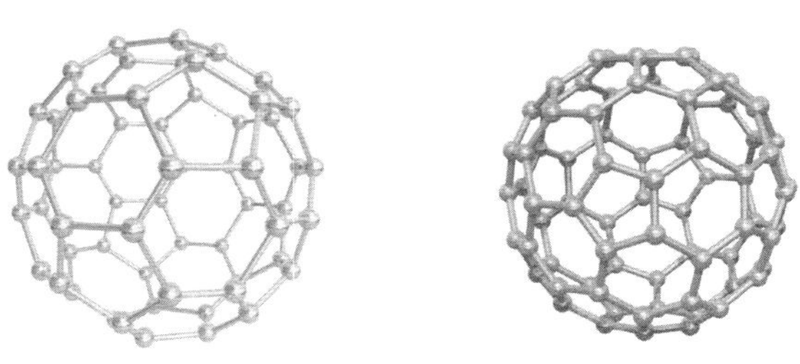

C_{60} C_{80}

Fig. 21. Icosahedral fullerenes C_{60} and C_{80} (http://previews.123rf.com)

Definition 4.3: A fullerene with a (combinatorial) group of symmetry of the icosahedron is called an *icosahedral fullerene*.

The construction implies that starting from the dodecahedron any combination of the first and the second iterative procedures gives an icosahedral fullerene.

Exercise: Prove that the opposite is also true.

Denote Operation I by T_1 and Operation II by T_2. Theses operations define the action of the semigroup $\mathbb{Z}^2_{\geqslant 0}$ on the set of combinatorial fullerenes.

Proposition 4.4: *The operations T_1 and T_2 change the number of hexagons of the fullerene P by the following rule:*

$$p_6(T_1P) = 30 + 4p_6(P); \quad p_6(T_2P) = 20 + 3p_6(P).$$

The proof we leave as an exercise.

Corollary 4.5: *The f-vector of a fullerene is changed by the following rule:*

$$T_1(f_0, f_1, f_2) = (4f_0, 4f_1, f_2 + f_1); \quad T_2(f_0, f_1, f_2) = (3f_0, 3f_1, f_2 + f_0).$$

4.3. *Cyclic k-edge cuts*

Definition 4.6: Let Γ be a graph. A *cyclic k-edge cut* is a set E of k edges of Γ, such that $\Gamma \setminus E$ consists of two connected components each containing a cycle, and for any subset $E' \subsetneq E$ the graph $\Gamma \setminus E'$ is connected.

For any k-belt $(F_1, \dots, F_k)$ of the simple 3-polytope P the set of edges $\{F_1 \cap F_2, \dots, F_{k-1} \cap F_k, F_k \cap F_1\}$ is a cyclic k-edge cut of the graph $G(P)$. For $k = 3$ any cyclic k-edge cut in $G(P)$ is obtained from a 3-belt in this way. For larger k not every cyclic k-edge cut is obtained from a k-belt.

In the paper [23] it was proved that for any fullerene P the graph $G(P)$ has no cyclic 3-edge cuts. In [24] it was proved that $G(P)$ has no cyclic 4-edge cuts. In [37] and [34] cyclic 5-edge cuts were classified. In [34] cyclic 6-edge cuts were classified. In [35] degenerated cyclic 7-edge cuts and fullerenes with non-degenerated cyclic 7-edge cuts were classified, where a cyclic k-edge cut is called degenerated, if one of the connected components has less than 6 pentagonal facets, otherwise it is called non-degenerated.

4.4. *Fullerenes as flag polytopes*

Let γ be a simple edge-cycle on a simple 3-polytope. We say that γ *borders* a k-loop $\mathcal{L}$ if $\mathcal{L}$ is a set of facets that appear when we walk along γ in one of the components $\mathcal{C}_\alpha$. We say that an l_1-loop $\mathcal{L}_1 = (F_{i_1}, \dots, F_{i_{l_1}})$ *borders* an l_2-loop $\mathcal{L}_2 = (F_{j_1}, \dots, F_{j_{l_2}})$

(along γ), if they border the same edge-cycle γ. If $l_2 = 1$, then we say that $\mathcal{L}_1$ *surrounds* F_{j_1}.

Let γ have a_p^1 successive edges corresponding to $F_{i_p} \in \mathcal{L}_1$, and a_q^2 successive edges corresponding to $F_{j_q} \in \mathcal{L}_2$.

Lemma 4.7: *Let a loop $\mathcal{L}_1$ border a loop $\mathcal{L}_2$ along γ. Then one of the following holds:*

(1) $\mathcal{L}_\alpha$ is a 1-loop and $\mathcal{L}_\beta$ is a a_1^α-loop, for $\{\alpha, \beta\} = \{1, 2\}$;
(2) For $l_1, l_2 \geqslant 2$, $l_1 + l_2 = l_\gamma = \sum_{r=1}^{l_1} a_r^1 = \sum_{r=1}^{l_2} a_r^2$.

Proof: If $l_2 = 1$, then γ is a boundary of the facet F_{j_1}, successive edges of γ belong to different facets in $\mathcal{L}_1$, and $l_1 = a_1^2$. A similar argument works for $l_1 = 1$.

Let $l_1, l_2 \geqslant 2$. Then any edge of γ is an intersection of a facet from $\mathcal{L}_1$ with a facet from $\mathcal{L}_2$. Successive edges of γ belong to the same facet in $\mathcal{L}_\alpha$ if and only if they belong to successive facets in $\mathcal{L}_\beta$, $\{\alpha, \beta\} = \{1, 2\}$; therefore $l_\alpha = \sum_{r=1}^{l_\beta}(a_r^\beta - 1) = \sum_{r=1}^{l_\beta} a_r^\beta - l_\beta$. We have $l_\gamma = \sum_{r=1}^{l_\beta} a_r^\beta = l_1 + l_2$. □

Lemma 4.8: *Let $\mathcal{B} = (F_{i_1}, \ldots, F_{i_k})$ be a k-belt. Then*

(1) $|\mathcal{B}| = F_{i_1} \cup \cdots \cup F_{i_k}$ is homeomorphic to a cylinder;
(2) $\partial|\mathcal{B}|$ consists of two simple edge-cycles γ_1 and γ_2;
(3) $\partial P \setminus |\mathcal{B}|$ consists of two connected components $\mathcal{P}_1$ and $\mathcal{P}_2$;
(4) Let $\mathcal{W}_\alpha = \{F_j \in \mathcal{F}_P \colon \operatorname{int} F_j \subset \mathcal{P}_\alpha\} \subset \mathcal{F}_P$, $\alpha = 1, 2$. Then $\mathcal{W}_1 \sqcup \mathcal{W}_2 \sqcup \mathcal{B} = \mathcal{F}_P$;
(5) $\overline{\mathcal{P}_\alpha} = |\mathcal{W}_\alpha|$ is homeomorphic to a disk, $\alpha = 1, 2$;
(6) $\partial\mathcal{P}_\alpha = \partial\overline{\mathcal{P}_\alpha} = \gamma_\alpha$, $\alpha = 1, 2$.

The proof is straightforward using Theorem 3.8.

Suppose a facet $F_{i_j} \in \mathcal{B}$ has α_j edges in γ_1 and β_j edges in γ_2. If F_{i_j} is an m_{i_j}-gon, then $\alpha_j + \beta_j = m_{i_j} - 2$.

Lemma 4.9: *Let P be a simple 3-polytope with $p_3 = 0$, $p_k = 0$, $k \geqslant 8$, $p_7 \leqslant 1$, and let $\mathcal{B}_k$ be a k-belt, $k \geqslant 3$, consisting of b_i i-gons, $4 \leqslant i \leqslant 7$. Then one of the following holds:*

(1) $\mathcal{B}_k$ surrounds two k-gonal facets $F_s : \{F_s\} = \mathcal{W}_1$, and $F_t : \{F_t\} = \mathcal{W}_2$, and all facets of $\mathcal{B}_k$ are quadrangles;
(2) $\mathcal{B}_k$ surrounds a k-gonal facet $F_s : \{F_s\} = \mathcal{W}_\alpha$, and borders an l_β-loop $\mathcal{L}_\beta \subset \mathcal{W}_\beta$, $\{\alpha, \beta\} = \{1, 2\}$, $l_\beta = b_5 + 2b_6 + 3b_7 \geqslant 2$;
(3) $\mathcal{B}_k$ borders an l_1-loop $\mathcal{L}_1 \subset \mathcal{W}_1$ and an l_2-loop $\mathcal{L}_2 \subset \mathcal{W}_2$, where

(a) $l_1 = \sum_{j=1}^{k}(\alpha_j - 1) \geqslant 2$, $l_2 = \sum_{j=1}^{k}(\beta_j - 1) \geqslant 2$;
(b) $l_1 + l_2 = 2k - 2b_4 - b_5 + b_7 \leqslant 2k + 1$;
(c) $\min\{l_1, l_2\} \leqslant k - b_4 - \lceil \frac{b_5 - b_7}{2} \rceil \leqslant k$;
(d) *If* $b_7 = 0$, $l_1, l_2 \geqslant k$, *then* $l_1 = l_2 = k$, $b_4 = b_5 = 0$, $b_6 = k$.

Proof: Walking round γ_α in $\mathcal{P}_\alpha$ we obtain an l_α-loop $\mathcal{L}_\alpha \subset \mathcal{W}_\alpha$.

If $\mathcal{B}_k$ surrounds two k-gons $F_s : \{F_s\} = \mathcal{W}_1$, and $F_t : \{F_t\} = \mathcal{W}_2$, then all facets in $\mathcal{B}_k$ are quadrangles.

If $\mathcal{B}_k$ surrounds a k-gon $F_s : \{F_s\} = \mathcal{W}_\alpha$ and borders an l_β-loop $\mathcal{L}_\beta \subset W_\beta$, $l_\beta \geqslant 2$, then from Lemma 4.7 we have

$$l_\beta = \sum_{j=1}^{k}(m_{i_j} - 3) - k = \sum_{j=1}^{k}(m_{i_j} - 3 - 1) = \sum_{j=4}^{7} jb_j - 4\sum_{j=4}^{7} b_j = b_5 + 2b_6 + 3b_7.$$

If $\mathcal{B}_k$ borders an l_1-loop $\mathcal{L}_1$ and an l_2-loop $\mathcal{L}_2$, $l_1, l_2 \geqslant 2$, then (a) follows from Lemma 4.7.

$$\begin{aligned} l_1 + l_2 &= \sum_{j=1}^{k}(\alpha_j + \beta_j - 2) = \sum_{i=1}^{k}(m_{i_j} - 4) \\ &= \sum_{j=4}^{7} jb_j - 4\sum_{j=4}^{7} b_j = b_5 + 2b_6 + 3b_7 = 2k - 2b_4 - b_5 + b_7. \end{aligned}$$

We have $\min\{l_1, l_2\} \leqslant \left[\frac{l_1 + l_2}{2}\right] = k - b_4 - \lceil \frac{b_5 - b_7}{2} \rceil \leqslant k$, since $b_7 \leqslant 1$.

If $b_7 = 0$ and $l_1, l_2 \geqslant k$, then from (3b) we have $l_1 = l_2 = k$, $b_4 = b_5 = 0$, $b_6 = k$. □

Lemma 4.10: *Let an* l_1*-loop* $\mathcal{L}_1 = (F_{i_1}, \ldots, F_{i_{l_1}})$ *border an* l_2*-loop* $\mathcal{L}_2$, $l_2 \geqslant 2$.

(1) If $l_1 = 2$, *then* $l_2 = m_{i_1} + m_{i_2} - 4$;
(2) If $l_1 = 3$ *and* $\mathcal{L}_1$ *is not a* 3*-belt, then* $F_{i_1} \cap F_{i_2} \cap F_{i_3}$ *is a vertex, and* $l_2 = m_{i_1} + m_{i_2} + m_{i_3} - 9$.

The proof is straightforward from Lemma 4.7.

Theorem 4.11: *Let* P *be simple* 3*-polytope with* $p_3 = 0$, $p_4 \leqslant 2$, $p_7 \leqslant 1$, *and* $p_k = 0$, $k \geqslant 8$. *Then it has no* 3*-belts. In particular, it is a flag polytope.*

Proof: Suppose P has a 3-belt $\mathcal{B}_3$. Since $p_3 = 0$, by Lemma 4.9 it borders an l_1-loop $\mathcal{L}_1$ and an l_2-loop $\mathcal{L}_2$, where $l_1, l_2 \geqslant 2$, $l_1 + l_2 \leqslant 7$. By Lemma 4.10 (1) we have $l_1, l_2 \geqslant 3$; hence $\min\{l_1, l_2\} = 3$. If $\mathcal{B}_3$ contains a heptagon, then $\mathcal{W}_1, \mathcal{W}_2$ contain no heptagons. If $\mathcal{B}_3$ contains no heptagons, then from Lemma 4.9

(3d) $l_1 = l_2 = 3$, and one of the sets $\mathcal{W}_1$ and $\mathcal{W}_2$, say $\mathcal{W}_\alpha$, contains no heptagons. In both cases we obtain a set $\mathcal{W}_\alpha$ without heptagons and a 3-loop $\mathcal{L}_\alpha \subset \mathcal{W}_\alpha$. Then $\mathcal{L}_\alpha$ is a 3-belt, else by Lemma 4.10 (2) the belt $\mathcal{B}_3$ should have at least $4+4+5-9=4$ facets. Considering the other boundary component of $\mathcal{L}_\alpha$ we obtain again a 3-belt there. Thus we obtain an infinite series of different 3-belts inside $|\mathcal{W}_\alpha|$. A contradiction. □

Corollary 4.12: *Any fullerene is a flag polytope.*

This result follows directly from the results of paper [23] about cyclic k-edge cuts of fullerenes. We present a different approach from [7, 8] based on the notion of a k-belt.

Corollary 4.13: *Let P be a fullerene. Then any 3-loop surrounds a vertex.*

In what follows we will implicitly use the fact that for any flag polytope, in particular satisfying the conditions of Theorem 4.11, if facets F_i, F_j, F_k pairwise intersect, then $F_i \cap F_j \cap F_k$ is a vertex.

4.5. *4-belts and 5-belts of fullerenes*

Lemma 4.14: *Let P be a flag 3-polytope, and let a 4-loop $\mathcal{L}_1 = (F_{i_1}, F_{i_2}, F_{i_3}, F_{i_4})$ border an l_2-loop $\mathcal{L}_2$, $l_2 \geqslant 2$, where the index j of i_j lies in $\mathbb{Z}_4 = \mathbb{Z}/(4)$. Then one of the following holds:*

*(1) $\mathcal{L}_1$ is a 4-**belt** (Figure 22 a);*
*(2) $\mathcal{L}_1$ is a **simple loop** consisting of facets surrounding an edge (Figure 22 b), and*
$l_2 = m_{i_1} + m_{i_2} + m_{i_3} + m_{i_4} - 14$;
*(3) $\mathcal{L}_1$ is **not a simple loop**: $F_{i_j} = F_{i_{j+2}}$ for some j, $F_{i_{j-1}} \cap F_{i_{j+1}} = \varnothing$ (Figure 22 c), and $l_2 = m_{i_{j-1}} + m_{i_j} + m_{i_{j+1}} - 8$.*

Proof: Suppose $\mathcal{L}_1$ is not a 4-belt. If $\mathcal{L}_1$ is simple, then $F_{i_j} \cap F_{i_{j+2}} \neq \varnothing$ for some j. Then $F_{i_j} \cap F_{j_{j+1}} \cap F_{i_{j+2}}$ and $F_{i_j} \cap F_{j_{j-1}} \cap F_{i_{j+2}}$ are vertices, $\mathcal{L}_1$ surrounds the edge $F_{i_j} \cap F_{i_{j+2}}$, and by Lemma 4.7 we have $l_2 = (m_{i_j} - 3) + (m_{i_{j+1}} - 2) + (m_{i_{j+2}} - 3) + (m_{i_{j-2}} - 2) - 4 = m_{i_1} + m_{i_2} + m_{i_3} + m_{i_4} - 14$. If $\mathcal{L}_1$ is not simple, then $F_{i_j} = F_{i_{j+2}}$ for some j. The successive facets of $\mathcal{L}_1$ are different by definition. Let $\mathcal{L}_1$ and $\mathcal{L}_2$ border the edge cycle γ and $\mathcal{L}_1 \subset \mathcal{D}_\alpha$ in notations of Theorem 3.8. Since F_{i_j} intersects γ by two paths, $\operatorname{int} F_{i_{j-1}}$ and $\operatorname{int} F_{i_{j+1}}$ lie in different connected components of $\mathcal{C}_\alpha \setminus \operatorname{int} F_{i_j}$; hence $F_{i_{j-1}} \cap F_{i_{j+1}} = \varnothing$.

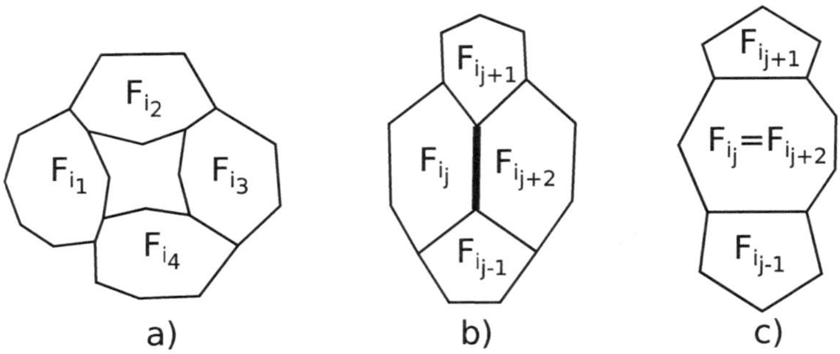

Fig. 22. Possibilities for a 4-loop $\mathcal{L}_1$

By Lemma 4.7 we have $l_2 = (m_{i_{j-1}} - 1) + (m_{i_j} - 2) + (m_{i_{j+1}} - 1) - 4 = m_{j_{j-1}} + m_{i_j} + m_{i_{j+1}} - 8$. □

Theorem 4.15: *Let P be a simple polytope with all facets pentagons and hexagons with at most one exceptional facet F being a quadrangle or a heptagon.*

(1) If P has no quadrangles, then P has no 4-belts.
(2) If P has a quadrangle F, then there is exactly one 4-belt. It surrounds F.

Proof: By Theorem 4.11 the polytope P is flag. By Lemma 3.5 a quadrangular facet is surrounded by a 4-belt. Let $\mathcal{B}_4$ be a 4-belt that does not surround a quadrangular facet. By Lemma 4.9 it borders an l_1-loop $\mathcal{L}_1$ and an l_2-loop $\mathcal{L}_2$, where $l_1, l_2 \geqslant 2$, and $l_1 + l_2 \leqslant 9$. We have $l_1, l_2 \geqslant 3$, since by Lemma 4.10 (1) a 2-loop borders a k-loop with $k \geqslant 4 + 5 - 4 = 5$. We have $l_1, l_2 \geqslant 4$ by Theorem 4.11 and Lemma 4.10 (2), since a 3-loop that is not a 3-belt borders a k-loop with $k \geqslant 4 + 5 + 5 - 9 = 5$. Also $\min\{l_1, l_2\} = 4$. If $\mathcal{B}_4$ contains a heptagon, then $\mathcal{W}_1, \mathcal{W}_2$ contain no heptagons. If $\mathcal{B}_4$ contains no heptagons, then $l_1 = l_2 = 4$ by Lemma 4.9 (3d), and one of the sets $\mathcal{W}_1$ and $\mathcal{W}_2$, say $\mathcal{W}_\alpha$, contains no heptagons. In both cases we obtain a set $\mathcal{W}_\alpha$ without heptagons and a 4-loop $\mathcal{L}_\alpha \subset \mathcal{W}_\alpha$. Then $\mathcal{L}_\alpha$ is a 4-belt, else by Lemma 4.14 the belt $\mathcal{B}_4$ should have at least $4 + 5 + 5 + 5 - 14 = 5$ or $4 + 5 + 5 - 8 = 6$ facets. Applying the same argument to $\mathcal{L}_\alpha$ instead of $\mathcal{B}_k$, we have that either $\mathcal{L}_\alpha$ surrounds on the opposite side a quadrangle, or it borders a 4-belt and consists of hexagons. In the first case by Lemma 4.9 (2) the 4-belt $\mathcal{L}_\alpha$ consists of pentagons. Thus we can move inside $\mathcal{W}_\alpha$ until we finish with a quadrangle. If P has no quadrangles, then we obtain a contradiction. If P has a quadrangle F, then it has no heptagons; therefore moving inside $\mathcal{W}_\beta$ we should meet some other quadrangle. A contradiction. □

Corollary 4.16: *Fullerenes have no* 4*-belts.*

This result follows directly from [24]. Above we prove more general Theorems 4.11 and 4.15, since we will need them in Section 10.

Corollary 4.17: *Let P be a fullerene. Then any simple* 4*-loop surrounds an edge.*

Remark 4.18: It follows from results by A.V. Pogorelov [42] and E.M. Andreev [1] that flag simple 3-polytopes without 4-belts are exactly 3-polytopes that can be realized in Lobachevsky (hyperbolic) 3-space as bounded polytopes with all dihedral angles right. Moreover, such realization is unique up to isometries. For more details see [50].

Now consider 5-belts of fullerenes. Describe a special family of fullerenes.

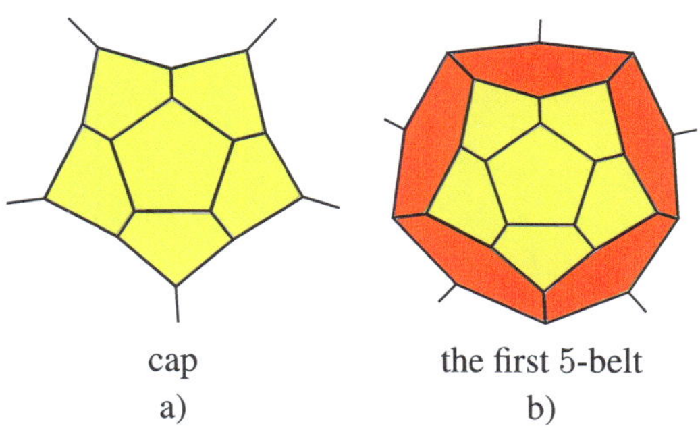

Fig. 23. Construction of fullerenes D_k

Construction (Series of polytopes D_k): Denote by D_0 the dodecahedron. If we cut its surface along the zigzag cycle (Figure 11), we obtain two caps on Figure 23a). Insert k successive 5-belts of hexagons with hexagons intersecting neighbours by opposite edges to obtain the combinatorial description of D_k. We have $p_6(D_k) = 5k$, $f_0(D_k) = 20 + 10k$, for $k \geqslant 0$.

The geometrical realization of the polytope D_k can be obtained from the geometrical realization of D_{k-1} by the following sequence of edge- and two-edge truncations, represented by Figure 24.

The polytopes D_k for $k \geqslant 1$ are exactly nanotubes of type $(5, 0)$ [37, 34, 35].

Lemma 4.19: *Let P be a flag 3-polytope without 4-belts, and let a 5-loop $\mathcal{L}_1 = (F_{i_1}, F_{i_2}, F_{i_3}, F_{i_4}, F_{i_5})$ border an l_2-loop $\mathcal{L}_2$, $l_2 \geqslant 2$, where the index j of i_j lies in $\mathbb{Z}_5 = \mathbb{Z}/(5)$. Then one of the following holds:*

*(1) $\mathcal{L}_1$ is a 5-**belt** (Figure 26a);*

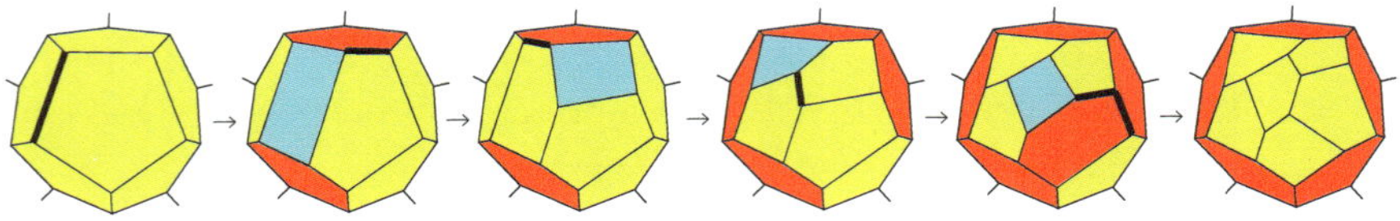

Fig. 24. Geometrical construction of a 5-belt of hexagons

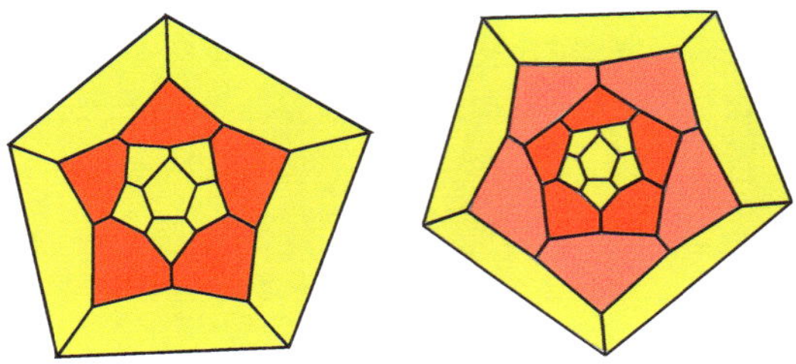

Fig. 25. Fullerenes D_1 and D_2

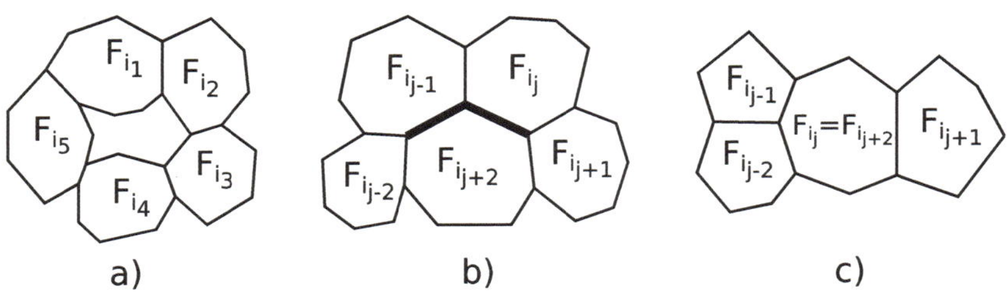

Fig. 26. Possibilities for a 5-loop $\mathcal{L}_1$

(2) *$\mathcal{L}_1$ is a **simple loop** consisting of facets surrounding two adjacent edges (Figure 26b), and $l_2 = m_{i_1} + m_{i_2} + m_{i_3} + m_{i_4} + m_{i_5} - 19 \geqslant 6$;*

(3) *$\mathcal{L}_1$ is **not a simple loop**: $F_{i_j} = F_{i_{j+2}}$ for some j, $F_{i_{j-2}} \cap F_{i_{j-1}} \cap F_{i_j}$ is a vertex, $F_{i_{j+1}}$ does not intersect $F_{i_{j-2}}$ and $F_{i_{j-1}}$ (Figure 26c), and $l_2 = m_{i_{j-2}} + m_{i_{j-1}} + m_{i_j} + m_{i_{j+1}} - 13 \geqslant 7$.*

Proof: Suppose $\mathcal{L}_1$ is not a 5-belt. If $\mathcal{L}_1$ is simple, then two non-successive facets F_{i_j} and $F_{i_{j+2}}$ intersect. Then $F_{i_j} \cap F_{i_{j+1}} \cap F_{i_{j+2}}$ is a vertex. By Theorem 4.15 the 4-loop $(F_{i_{j-2}}, F_{i_{j-1}}, F_{i_j}, F_{i_{j+2}})$ is not a 4-belt; hence either $F_{i_{j-2}} \cap F_{i_j} \neq \varnothing$, or $F_{i_{j-1}} \cap F_{i_{j+2}} \neq \varnothing$. Up to relabelling in the inverse order, we can assume that $F_{i_{j-1}} \cap F_{i_{j+2}} \neq \varnothing$. Then $F_{i_{j-1}} \cap F_{i_j} \cap F_{i_{j+2}}$ and $F_{i_{j-2}} \cap F_{i_{j-1}} \cap F_{i_{j+2}}$ are vertices. Thus $\mathcal{L}_1$ surrounds the adjacent edges $F_{i_{j-1}} \cap F_{i_{j+2}}$ and $F_{i_j} \cap F_{i_{j+2}}$. By

Lemma 4.7 we have $l_2 = (m_{i_{j-2}} - 2) + (m_{i_{j-1}} - 3) + (m_{i_j} - 3) + (m_{i_{j+1}} - 2) + (m_{i_{j+2}} - 4) - 5 = m_{i_1} + m_{i_2} + m_{i_3} + m_{i_4} + m_{i_5} - 19 \geqslant 6$. The last inequality holds since flag 3-polytopes without 4-belts have no triangles and quadrangles. If $\mathcal{L}_1$ is not simple, then $F_{i_j} = F_{i_{j+2}}$ for some j. The successive facets of $\mathcal{L}_1$ are different by definition. Let $\mathcal{L}_1$ and $\mathcal{L}_2$ border the edge cycle γ and $\mathcal{L}_1 \subset \mathcal{D}_\alpha$ in the notation of Theorem 3.8. Since F_{i_j} intersects γ by two paths, int $F_{i_{j-2}} \cup$ int $F_{i_{j-1}}$ and int $F_{i_{j+1}}$ lie in different connected components of $\mathcal{C}_\alpha \setminus \text{int } F_{i_j}$; hence $F_{i_{j-2}} \cap F_{i_{j+1}} = \varnothing = F_{i_{j-1}} \cap F_{i_{j+1}}$. Since P is flag, $F_{i_{j-2}} \cap F_{i_{j-1}} \cap F_{i_j}$ is a vertex, thus we obtain the configuration in Figure 26c. By Lemma 4.7 we have $l_2 = (m_{i_{j-2}} - 2) + (m_{i_{j-1}} - 2) + (m_{i_j} - 3) + (m_{i_{j+1}} - 1) - 5 = m_{i_{j-2}} + m_{j_{j-1}} + m_{i_j} + m_{i_{j+1}} - 13 \geqslant 7$. □

The next result follows directly from [34] or [37]. We develop the approach from [8] based on the notion of a k-belt.

Theorem 4.20: *Let P be a fullerene. Then the following statements hold:*
I. Any pentagonal facet is surrounded by a 5-belt. There are 12 belts of this type.
II. If there is a 5-belt not surrounding a pentagon, then

(1) it consists only of hexagons;
(2) the fullerene is combinatorially equivalent to the polytope D_k, $k \geqslant 1$.
(3) the number of 5-belts is $12 + k$.

Proof: I. Follows from Proposition 3.5 and Corollary 4.12.
II. Suppose the 5-belt $\mathcal{B}_5$ does not surround a pentagon. By Lemma 4.9 it borders an l_1-loop $\mathcal{L}_1 \subset \mathcal{W}_1$ and an l_2-loop $\mathcal{L}_2 \subset \mathcal{W}_2$, $l_1, l_2 \geqslant 2$, $l_1 + l_2 \leqslant 10$. By Lemma 4.10 (1) we have $l_1, l_2 \geqslant 3$. From Corollary 4.12 and Lemma 4.10 (2) we obtain $l_1, l_2 \geqslant 4$. From Corollary 4.16 and Lemma 4.14 we obtain $l_1, l_2 \geqslant 5$. Then $l_1 = l_2 = 5$ and all facets in $\mathcal{B}_5$ are hexagons by Lemma 4.9 (3d). From Lemma 4.19 we obtain that $\mathcal{L}_1$ and $\mathcal{L}_2$ are 5-belts. Moving inside $\mathcal{W}_1$ we obtain a series of hexagonal 5-belts, and this series can stop only if the last 5-belt $\mathcal{B}_l$ surrounds a pentagon. Since $\mathcal{B}_l$ borders a 5-belt, Lemma 4.9 (2) implies that $\mathcal{B}_l$ consists of pentagons, which have $(2, 2, 2, 2, 2)$ edges on the common boundary with a 5-belt. We obtain the fragment in Figure 23a). Moving from this fragment backward we obtain a series of hexagonal 5-belts including $\mathcal{B}_5$ with facets having $(2, 2, 2, 2, 2)$ edges on both boundaries. This series can finish only with fragment in Figure 23a) again. Thus any belt not surrounding a pentagon belongs to this series and the number of 5-belts is equal to $12 + k$. □

Theorem 4.21: *A fullerene P is combinatorially equivalent to a polytope D_k, for some $k \geqslant 0$, if and only if it contains the fragment in Figure 23a).*

Proof: By Proposition 3.5 the outer 5-loop of the fragment in Figure 23a) is a 5-belt. By the outer boundary component it borders a 5-loop $\mathcal{L}$. By Lemma 4.19 it is a 5-belt. If this belt surrounds a pentagon, then we obtain a combinatorial dodecahedron (case $k = 0$). If not, then P is combinatorially equivalent to D_k, $k \geqslant 1$, by Theorem 4.20. □

Corollary 4.22: *Any simple* 5*-loop of a fullerene*

(1) either surrounds a pentagon;
(2) or is a hexagonal 5*-belt of a fullerene* D_k, $k \geqslant 1$;
(3) or surrounds a pair of adjacent edges (Figure 26b).

Proof: Let $\mathcal{L} = (F_{i_1}, F_{i_2}, F_{i_3}, F_{i_4}, F_{i_5})$ be a simple 5-loop, where the index j of i_j lies in $\mathbb{Z}_5 = \mathbb{Z}/(5)$. If $\mathcal{L}$ is a 5-belt, then by Theorem 4.20, we obtain cases (1) or (2). Otherwise some non-successive facets intersect: $F_{i_j} \cap F_{i_{j+2}} \neq \varnothing$, for some j. Then $F_{i_j} \cap F_{i_{j+1}} \cap F_{i_{j+2}}$ is a vertex. Since a fullerene has no 4-belts in a simple 4-loop $(F_{i_{j-2}}, F_{i_{j-1}}, F_{i_j}, F_{i_{j+2}})$, either $F_{i_{j-2}} \cap F_{i_j} \neq \varnothing$ or $F_{i_{j-1}} \cap F_{i_{j+2}} \neq \varnothing$. Up to relabelling in the inverse order, we can assume that $F_{i_{j-1}} \cap F_{i_{j+2}} \neq \varnothing$. Then $F_{i_{j-1}} \cap F_{i_j} \cap F_{i_{j+2}}$ and $F_{i_{j-2}} \cap F_{i_{j-1}} \cap F_{i_{j+2}}$ are vertices. Thus $\mathcal{L}_1$ surrounds the adjacent edges $F_{i_{j-1}} \cap F_{i_{j+2}}$ and $F_{i_j} \cap F_{i_{j+2}}$. □

5. Moment-angle complexes and moment-angle manifolds

We discuss some fundamental notions, constructions and results of toric topology. Details can be found in the monograph [14], which we will follow.

5.1. *Toric topology*

Nowadays toric topology is a large research area. Below we discuss applications of toric topology to the mathematical theory of fullerenes based on the following correspondence.

Canonical correspondence

Simple polytope P number of facets $= m$ $\dim P = n$	$\longrightarrow$	moment-angle manifold $\mathcal{Z}_P$ canonical T^m-action on $\mathcal{Z}_P$ $\dim \mathcal{Z}_P = m + n$
Characteristic function $\{F_1, \dots, F_m\} \to \mathbb{Z}^n$	$\longrightarrow$	Quasitoric manifold $M^{2n} = \mathcal{Z}_P / T^{m-n}$

Algebraic-topological invariants of moment-angle manifolds $\mathcal{Z}_P$ give combinatorial invariants of polytopes P. As an application we obtain combinatorial invariants of mathematical fullerenes.

5.2. *Moment-angle complex of a simple polytope*

Set

$$D^2 = \{z \in \mathbb{C}; |z| \le 1\}, \qquad S^1 = \{z \in D^2, |z| = 1\}.$$

The multiplication of complex numbers gives the canonical action of the circle S^1 on the disk D^2 whose orbit space is the interval $\mathbb{I} = [0, 1]$.

We have the canonical projection

$$\pi : (D^2, S^1) \to (\mathbb{I}, 1) : z \to |z|^2.$$

By definition a *multigraded polydisk* is $\mathbb{D}^{2m} = D_1^2 \times \cdots \times D_m^2$.

Define the *standard torus* $\mathbb{T}^m = S_1^1 \times \cdots \times S_m^1$.

Proposition 5.1: *There is a canonical action of the torus $\mathbb{T}^m$ on the polydisk $\mathbb{D}^{2m}$ with the orbit space*

$$\mathbb{D}^{2m}/\mathbb{T}^m \simeq \mathbb{I}^m = \mathbb{I}_1^1 \times \cdots \times \mathbb{I}_m^1.$$

Consider a simple polytope P. Let $\{F_1, \dots, F_m\}$ be the set of facets and $\{v_1, \dots, v_{f_0}\}$ – the set of vertices. We have the face lattice $L(P)$ of P.

Construction (moment-angle complex of a simple polytope [12, 14]): For $P = \mathrm{pt}$, set $\mathcal{Z}_P = \mathrm{pt} = \{0\} = \mathbb{D}^0$. Let $\dim P > 0$. For any face $F \in L(P)$ set

$$\mathcal{Z}_{P,F} = \{(z_1, \dots, z_m) \in \mathbb{D}^{2m} : z_i \in D_i^2 \text{ if } F \subset F_i, z_i \in S_i^1 \text{ if } F \not\subset F_i\};$$
$$\mathbb{I}_{P,F} = \{(y_1, \dots, y_m) \in \mathbb{I}^m : y_i \in \mathbb{I}_i^1 \text{ if } F \subset F_i, y_i = 1 \text{ if } F \not\subset F_i\}.$$

Proposition 5.2:

(1) $\mathcal{Z}_{P,F} \simeq \mathbb{D}^{2k} \times \mathbb{T}^{m-k}$, $\mathbb{I}_{P,F} \simeq \mathbb{I}^k$, *where* $k = n - \dim F$.

(2) $\mathcal{Z}_{P,P} = \mathbb{T}^m$, $\mathcal{Z}_{P,\varnothing} = \mathbb{D}^{2m}$.

(3) If $G_1 \subset G_2$, *then* $\mathcal{Z}_{P,G_2} \subset \mathcal{Z}_{P,G_1}$, *and* $\mathbb{I}_{P,G_2} \subset \mathbb{I}_{P,G_1}$.

(4) $\mathcal{Z}_{P,F}$ *is invariant under the action of* $\mathbb{T}^m$, *and the mapping* $\pi^m : \mathbb{D}^{2m} \to \mathbb{I}^m$ *defines the homeomorphism* $\mathcal{Z}_{P,F}/\mathbb{T}^m \simeq \mathbb{I}_{P,F}$.

The *moment-angle complex* of a simple polytope P is a subset in $\mathbb{D}^{2m}$ of the form

$$\mathcal{Z}_P = \bigcup_{F \in L(P)\setminus\{\varnothing\}} \mathcal{Z}_{P,F} = \bigcup_{v-\text{ vertex}} \mathcal{Z}_{P,v}.$$

The cube $\mathbb{I}^m$ has the canonical structure of a *cubical complex*. It is a cellular complex with all cells being cubes with an appropriate boundary condition. The *cubical complex* of a simple polytope P is a cubical subcomplex in $\mathbb{I}^m$ of the form

$$\mathbb{I}_P = \bigcup_{F \in L(P)\setminus\{\varnothing\}} \mathbb{I}_{P,F} = \bigcup_{v-\text{ vertex}} \mathbb{I}_{P,v}.$$

From the construction of the space $\mathcal{Z}_P$ we obtain:

Proposition 5.3:

(1) The subset $\mathcal{Z}_P \subset \mathbb{D}^{2m}$ is $\mathbb{T}^m$-invariant; hence there is the canonical action of $\mathbb{T}^m$ on $\mathcal{Z}_P$.
(2) The mapping π^m defines the homeomorphism $\mathcal{Z}_P/\mathbb{T}^m \simeq \mathbb{I}_P$.
(3) We have $\mathcal{Z}_{P_1\times P_2} = \mathcal{Z}_{P_1} \times \mathcal{Z}_{P_2}$.

5.3. *Admissible mappings*

Definition 5.4: Let P_1, P_2 be two simple polytopes. We call a mapping of sets of facets $\varphi\colon \mathcal{F}_{P_1} \to \mathcal{F}_{P_2}$ *admissible* if $\varphi(F_{i_1})\cap\dots\cap\varphi(F_{i_k}) \neq \varnothing$ for any collection $F_{i_1},\dots,F_{i_k} \in \mathcal{F}_{P_1}$, with $F_{i_1}\cap\dots\cap F_{i_k} \neq \varnothing$.

Any admissible mapping $\varphi\colon \mathcal{F}_{P_1} \to \mathcal{F}_{P_2}$ induces the mapping $\varphi\colon L(P_1) \to L(P_2)$ by the rule: $\varphi(P_1) = P_2$, $\varphi(F_{i_1}\cap\dots\cap F_{i_k}) = \varphi(F_{i_1})\cap\dots\cap\varphi(F_{i_k})$. This mapping preserves the inclusion relation.

Proposition 5.5: *Any admissible mapping $\varphi\colon \mathcal{F}_{P_1} \to \mathcal{F}_{P_2}$ induces the mapping of triples $(\mathbb{D}^{2m_1}, \mathcal{Z}_{P_1}, \mathbb{T}^{m_1}) \to (\mathbb{D}^{2m_2}, \mathcal{Z}_{P_2}, \mathbb{T}^{m_2})$ and the mapping $\mathbb{I}_{P_1} \to \mathbb{I}_{P_2}$, which we will denote by the same letter $\widehat{\varphi}$:*

$$\widehat{\varphi}(x_1,\dots,x_{m_1}) = (y_1,\dots,y_{m_2}),\ y_j = \begin{cases} 1, & \text{if } \varphi^{-1}(j) = \varnothing, \\ \prod\limits_{i\in\varphi^{-1}(j)} x_i, & \text{else.} \end{cases}$$

In particular, we have the homomorphism of tori $\mathbb{T}^{m_1} \to \mathbb{T}^{m_2}$ such that the mapping $\mathcal{Z}_{P_1} \to \mathcal{Z}_{P_2}$ is equivariant.

We have the commutative diagram

$$\begin{array}{ccc} \mathcal{Z}_{P_1} & \xrightarrow{\widehat{\varphi}} & \mathcal{Z}_{P_2} \\ \downarrow{\scriptstyle \pi^m} & & \downarrow{\scriptstyle \pi^m} \\ \mathbb{I}_{P_1} & \xrightarrow{\widehat{\varphi}} & \mathbb{I}_{P_2} \end{array}$$

Example 5.6: Let $P_1 = \mathbb{I}^2$ and $P_2 = \mathbb{I}$. Then any admissible mapping $\mathcal{F}_{P_1} \to \mathcal{F}_{P_2}$ is a constant mapping. Indeed, there are two facets G_1 and G_2 in $\mathbb{I}$, which do not intersect. $\mathbb{I}^2$ has four facets F_1, F_2, F_3, F_4, such that $F_1\cap F_2$, $F_2\cap F_3$, $F_3\cap F_4$, and $F_4\cap F_1$ are vertices. Let $\varphi(F_1) = G_i$. Then $\varphi(F_2) = G_i$, since $\varphi(F_1)\cap\varphi(F_2) = \varnothing$. By the same reason we have $\varphi(F_3) = \varphi(F_4) = G_i$.

Without loss of generality let $i = 1$ and $G_1 = \{0\}$. Then the mapping of the moment-angle complexes

$$\begin{aligned}\mathcal{Z}_{\mathbb{I}^2} &= \{(z_1, z_2, z_3, z_4) \in \mathbb{D}^8 \colon |z_1| = 1 \text{ or } |z_3| = 1, \text{ and } |z_2| = 1 \text{ or } |z_4| = 1\}\\ &= (S_1^1 \times D_3^2 \cup D_1^2 \times S_3^1) \times (S_2^1 \times D_4^2 \cup D_2^2 \times S_4^1) \cong S^3 \times S^3,\\ \mathcal{Z}_{I^1} &= \{(w_1, w_2) \in \mathbb{D}^4 \colon |w_1| = 1 \text{ or } |w_2| = 1\}\\ &= (S_1^1 \times D_2^2) \cup (D_1^2 \times S_2^1) \cong S^3\end{aligned}$$

is

$$\widehat{\varphi} \colon \mathcal{Z}_{\mathbb{I}^2} \to \mathcal{Z}_{\mathbb{I}^1}, \qquad \widehat{\varphi}(z_1, z_2, z_3, z_4) = (z_1 z_2 z_3 z_4, 1).$$

Example 5.7: Let $P_1 = \mathbb{I}^2$, $P_2 = \Delta^2$. Then any mapping $\varphi \colon \mathcal{F}_{P_1} \to \mathcal{F}_{P_2}$ is admissible. Let $\mathcal{F}_{P_1} = \{F_1, F_2, F_3, F_3\}$ as in the previous example, and $\mathcal{F}_{P_2} = \{G_1, G_2, G_3\}$. The admissible mapping

$$\varphi(F_1) = G_1, \quad \varphi(F_2) = G_2, \quad \varphi(F_3) = \varphi(F_4) = G_3$$

induces the mapping of face lattices

$$\begin{gathered}\varphi(\mathbb{I}^2) = \Delta^2, \quad \varphi(\varnothing) = \varnothing,\\ \varphi(F_1 \cap F_2) = G_1 \cap G_2, \quad \varphi(F_2 \cap F_3) = G_2 \cap G_3,\\ \varphi(F_3 \cap F_4) = G_3, \quad \varphi(F_4 \cap F_1) = G_3 \cap G_1.\end{gathered}$$

The mapping of the moment-angle complexes

$$\begin{aligned}\mathcal{Z}_{\mathbb{I}^2} &= \{(z_1, z_2, z_3, z_4) \in \mathbb{D}^8 \colon |z_1| = 1 \text{ or } |z_3| = 1, \text{ and } |z_2| = 1 \text{ or } |z_4| = 1\}\\ &= (S_1^1 \times D_3^2 \cup D_1^2 \times S_3^1) \times (S_2^1 \times D_4^2 \cup D_2^2 \times S_4^1) \cong S^3 \times S^3,\\ \mathcal{Z}_{\Delta^2} &= \{(w_1, w_2, w_3) \in \mathbb{D}^6 \colon |w_1| = 1, \text{ or } |w_2| = 1, \text{ or } |w_3| = 1\}\\ &= (S_1^1 \times D_2^2 \times D_3^2) \cup (D_1^2 \times S_2^1 \times D_3^2) \cup (D_1^2 \times D_2^2 \times S_3^1) \cong S^5\end{aligned}$$

is

$$\widehat{\varphi} \colon \mathcal{Z}_{\mathbb{I}^2} \to \mathcal{Z}_{\Delta^2}, \qquad \varphi(z_1, z_2, z_3, z_4) = (z_1, z_2, z_3 \cdot z_4).$$

5.4. *Barycentric embedding and cubical subdivision of a simple polytope*

Construction (barycentric embedding of a simple polytope): Let P be a simple n-polytope with facets $F_1, \ldots, F_m$. For each face $G \subset P$ define a point $\boldsymbol{x}_G$ as a barycenter of its vertices. We have $\boldsymbol{x}_G \in \operatorname{relint} G$. The points $\boldsymbol{x}_G$, $G \in L(P) \backslash \{\varnothing\}$, define a barycentric simplicial subdivision $\Delta(P)$ of the polytope P. The simplices of $\Delta(P)$ correspond to flags of faces $F^{a_1} \subset F^{a_2} \subset \cdots \subset F^{a_k}$, $\dim F^i = i$:

$$\Delta_{F^{a_1} \subset F^{a_2} \subset \cdots \subset F^{a_k}} = \operatorname{conv}\{\boldsymbol{x}_{F^{a_1}}, \boldsymbol{x}_{F^{a_2}}, \ldots, \boldsymbol{x}_{F^{a_k}}\}.$$

The maximal simplices are $\Delta_{\boldsymbol{v}\subset F^1\subset F^2\subset\cdots\subset F^{n-1}\subset P}$, where $\boldsymbol{v}$ is a vertex. For any point $\boldsymbol{x}\in P$ the minimal simplex $\Delta(\boldsymbol{x})$ containing $\boldsymbol{x}$ can be found by the following rule. Let $G(\boldsymbol{x})=\bigcap\limits_{F_i\ni\boldsymbol{x}}F_i$. If $\boldsymbol{x}=\boldsymbol{x}_G$, then $\Delta(\boldsymbol{x})=\Delta_G$. Else take a ray starting in $\boldsymbol{x}_G$, passing through $\boldsymbol{x}$ and intersecting ∂G in $\boldsymbol{x}_1$. Iterating the argument we obtain either $\boldsymbol{x}_1=\boldsymbol{x}_{G_1}$ and $\Delta(\boldsymbol{x})=\Delta_{G_1\subset G}$, or a new point $\boldsymbol{x}_2$. In the end we will stop when $\boldsymbol{x}_l=\boldsymbol{x}_{G_l}$ and $\Delta(\boldsymbol{x})=\Delta_{G_l\subset\cdots\subset G_1\subset G}$.

Define a piecewise-linear mapping $b_P\colon P\to\mathbb{I}^m$ by the rule

$$\boldsymbol{x}_G\to\widehat{\varepsilon}(G)=(\varepsilon_1,\dots,\varepsilon_m)\in\mathbb{I}^m,\quad\text{where }\varepsilon_i=\begin{cases}0, & \text{if } G\subset F_i,\\ 1, & \text{if } G\not\subset F_i,\end{cases}$$

on the vertices of $\Delta(P)$, and for any simplex continue the mapping to the cube $\mathbb{I}^m$ via barycentric coordinates. In particular, $b_P(\boldsymbol{x}_P)=(1,1,\dots,1)$, and $b_P(\boldsymbol{x}_{\boldsymbol{v}})$ is a point with n zero coordinates.

Theorem 5.8: *The mapping b_P defines a homeomorphism $P\simeq\mathbb{I}_P\subset\mathbb{I}^m$.*

Proof: Let $\boldsymbol{x}\in P$, and $\Delta(\boldsymbol{x})=\Delta_{G_1\subset\cdots\subset G_r}$.

We have $\boldsymbol{x}=t_1\boldsymbol{x}_{G_1}+\cdots+t_r\boldsymbol{x}_{G_r}$, where $t_i>0$, and $t_1+\cdots+t_r=1$. The coordinates of the vector $b_P(\boldsymbol{x})=t_1\widehat{\varepsilon}(G_1)+\cdots+t_r\widehat{\varepsilon}(G_r)=(x_1,\dots,x_m)$ belong to the interval $[0,1]$. Arrange them in ascending order:

$$0=x_{i_1}=\cdots=x_{i_{p_1}}<x_{i_{p_1+1}}=\cdots=x_{i_{p_1+p_2}}$$
$$<\cdots<x_{i_{p_1+\cdots+p_r+1}}=\cdots=x_{i_m}=1.$$

Then

$$G_1=F_{i_1}\cap\cdots\cap F_{i_{p_1+\cdots+p_r}},G_2=F_{i_1}\cap\cdots\cap F_{i_{p_1+\cdots+p_{r-1}}},\dots,G_r=F_{i_1}\cap\cdots\cap F_{i_{p_1}},$$

and

$$t_1=1-x_{i_{p_1+\cdots+p_r}},t_2=x_{i_{p_1+\cdots+p_r}}-x_{i_{p_1+\cdots+p_{r-1}}},\dots,t_r=x_{i_{p_1+p_2}}.$$

Thus the mapping b_P is an embedding. Since P is compact and $\mathbb{I}^m$ is Hausdorff, we have the homeomorphism $P\simeq b_P(P)$. In the construction above we have $x_{i_j}\neq 1$ only if $F_{i_j}\supset G_1$; hence $b_P(\boldsymbol{x})\in\mathbb{I}_{P,G_1}$, and $b_P(P)\subset\mathbb{I}_P$. On the other hand, the above formulas imply that $\mathbb{I}_P\subset b_P(P)$. This finishes the proof. □

Corollary 5.9: *The homeomorphism $b_P\colon P\to\mathbb{I}_P\simeq\mathcal{Z}_P/\mathbb{T}^m$ defines a mapping $\pi_P\colon\mathcal{Z}_P\to P$ such that the following diagram is commutative:*

$$\begin{array}{ccc}\mathcal{Z}_P & \longrightarrow & \mathbb{D}^{2m}\\ \big\downarrow{\scriptstyle\pi_P} & & \big\downarrow\\ P & \xrightarrow{\ b_P\ } & \mathbb{I}^m\end{array}$$

Corollary 5.10: *Any admissible mapping $\varphi\colon \mathcal{F}_{P_1} \to \mathcal{F}_{P_2}$ induces the mapping of polytopes $\widehat{\varphi}\colon P_1 \to P_2$ such that the following diagram is commutative:*

$$\begin{array}{ccc} \mathcal{Z}_{P_1} & \xrightarrow{\widehat{\varphi}} & \mathcal{Z}_{P_2} \\ \big\downarrow{\scriptstyle \pi_{P_1}} & & \big\downarrow{\scriptstyle \pi_{P_2}} \\ P_1 & \xrightarrow{\widehat{\varphi}} & P_2 \end{array}$$

Construction (canonical section): The mapping

$$s : \mathbb{I} \to D^2 : s(y) = \sqrt{y}$$

induces the section $s^m\colon \mathbb{I}_P \to \mathcal{Z}_P$. Together with the homeomorphism $P \simeq \mathbb{I}_P$ this gives the *canonical section* $s_P = s^m \circ b_P : P \to \mathcal{Z}_P$, such that $\pi_P \circ s_P = \text{id}$.

Construction (cubical subdivision): The space $\mathbb{I}_P$ has the canonical partition into cubes $\mathbb{I}_{P,v}$, one for each vertex $v \in P^n$. The homeomorphism $\mathbb{I}_P = \operatorname{Im} b_P(P) \simeq P$ gives the *cubical subdivision* of the polytope P.

Example 5.11: For $P = I$ we have an embedding $I \subset I^2$.

Fig. 27. Barycentric embedding and cubical subdivision of the interval

Example 5.12: For $P = \Delta^2$ we have an embedding $\Delta^2 \subset I^3$.

Construction (product over space): Let $f\colon X \to Z$ and $g\colon Y \to Z$ be maps of topological spaces. The product $X \times_Z Y$ over space Z is described by the general pullback diagram:

$$\begin{array}{ccc} X \times_Z Y & \longrightarrow & X \\ \big\downarrow & & \big\downarrow{\scriptstyle f} \\ Y & \xrightarrow{g} & Z \end{array}$$

where $X \times_Z Y = \big\{(x, y) \in X \times Y \ : \ f(x) = g(y)\big\}$.

Proposition 5.13: *We have $\mathcal{Z}_P = \mathbb{D}^{2m} \times_{\mathbb{I}^m} P$.*

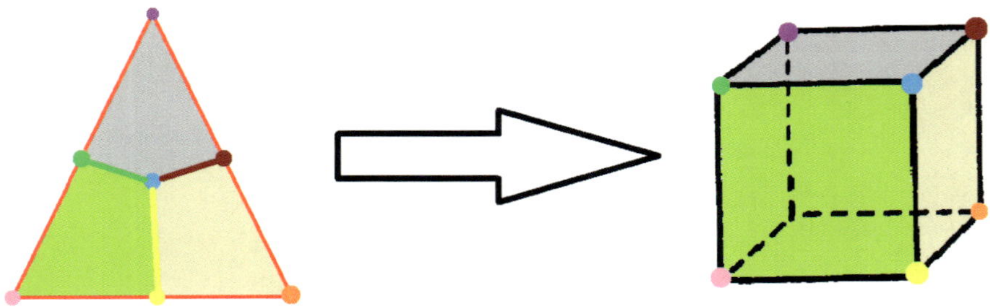

Fig. 28. Barycentric embedding and cubical subdivision of the triangle

5.5. *Pair of spaces in the power of a simple polytope*

Construction (raising to the power of a simple polytope): Let P be a simple polytope P with the face lattice $L(P)$ and the set of facets $\{F_1, \dots, F_m\}$. For m pairs of topological spaces $\{(X_i, W_i), i = 1, \dots, m\}$ set $(\boldsymbol{X}, \boldsymbol{W}) = \{(X_i, W_i), i = 1, \dots, m\}$. For a face $F \in L(P) \setminus \{\varnothing\}$ define

$$(\boldsymbol{X}, \boldsymbol{W})_F^P = \{(y_1, \dots, y_m) \in X_1 \times \dots \times X_m \colon y_i \in X_i \text{ if } F \in F_i, y_i \in W_i \text{ if } F \notin F_i\}.$$

The set of pairs $(\boldsymbol{X}, \boldsymbol{W})$ in degree of a simple polytope P is

$$(\boldsymbol{X}, \boldsymbol{W})^P = \bigcup_{F \in L(P) \setminus \{\varnothing\}} (\boldsymbol{X}, \boldsymbol{W})_F^P.$$

Example 5.14:

(1) Let $W_i = X_i$ for all i. Then $(\boldsymbol{X}, \boldsymbol{W})^P = X_1 \times \dots \times X_m$ for any P.
(2) Let $W_i = *_i$ be a fixed point in X_i, $i = 1, 2$, and $P = I$. Then $(\boldsymbol{X}, \boldsymbol{W})^I = X_1 \vee X_2$ is the wedge of the spaces X_1 and X_2.

Construction (pair of spaces in the power of a simple polytope): In the case $W_i = W$, $X_i = X$, $i = 1, \dots, m$, the space $(\boldsymbol{X}, \boldsymbol{W})^P$ is called a pair of spaces (X, W) to the power of a simple polytope P and is denoted by $(X, W)^P$.

Example 5.15: The space $(D^2, S^1)^P$ is the moment-angle complex $\mathcal{Z}_P$ of the polytope P (see Subsection 5.2).

Example 5.16: The space $(\mathbb{I}, 1)^P$ is the image $\mathbb{I}_P = b_P(P)$ of the barycentric embedding of the polytope P (see Subsections 5.2 and 5.4).

Exercise: Describe the space $(X, W)^P$, where P is a 5-gon.

Let us formulate properties of the construction. The proof we leave as an exercise.

Proposition 5.17:

(1) Let P_1 and P_2 be simple polytopes. Then

$$(X,W)^{P_1\times P_2} = (X,W)^{P_1} \times (X,W)^{P_2}.$$

(2) Let $\{v_1,\dots,v_{f_0}\}$ be the set of vertices of P. There is a homeomorphism

$$(X,W)^P \cong \bigcup_{k=1}^{f_0} (X,W)^P_{v_k}.$$

(3) Any mapping $f\colon (X_1,W_1)\to(X_2,W_2)$ gives the commutative diagram

$$\begin{array}{ccc} (X_1,W_1)^P & \xrightarrow{f^P} & (X_2,W_2)^P \\ \cap\big\downarrow & & \big\downarrow\cap \\ (X_1,X_1)^P = X_1^m & \xrightarrow{f^m} & X_2^m = (X_2,X_2)^P \end{array}$$

(4) We have $id^P = id$. For $f_1\colon (X_1,W_1)\to(X_2,W_2)$ and $f_2\colon (X_2,W_2)\to (X_1,W_1)$ we have

$$(f_2\circ f_1)^P = f_2^P\circ f_1^P.$$

5.6. *Davis-Januszkiewicz' construction*

See [20]. For $\boldsymbol{x}\in P$ we have the face $G(x) = \bigcap_{F_i\ni \boldsymbol{x}} F_i \in L(P)$. For a face $G\in L(P)$ define the subgroup $\mathbb{T}^G\subset\mathbb{T}^m$ as

$$\mathbb{T}^G = (S^1,1)^P_G = \{(t_1,\dots,t_m)\in\mathbb{T}^m\colon t_j=1,\ \text{if } F_j\not\supset G\}.$$

Set

$$\widetilde{\mathcal{Z}}_P = P\times\mathbb{T}^m/\sim,$$

where $(\boldsymbol{x}_1,\boldsymbol{t}_1)\sim(\boldsymbol{x}_2,\boldsymbol{t}_2)\Leftrightarrow \boldsymbol{x}_1=\boldsymbol{x}_2=\boldsymbol{x}$, and $\boldsymbol{t}_1\boldsymbol{t}_2^{-1}\in T^{G(\boldsymbol{x})}$.

There is a canonical action of $\mathbb{T}^m$ on $\widetilde{\mathcal{Z}}_P$ induced by the action of $\mathbb{T}^m$ on the second factor.

Theorem 5.18: *The canonical section $s_P\colon P\to\mathcal{Z}_P$ induces the $\mathbb{T}^m$-equivariant homeomorphism*

$$\widetilde{\mathcal{Z}}_P \longrightarrow \mathcal{Z}_P$$

defined by the formula $(\boldsymbol{x},\boldsymbol{t})\to \boldsymbol{t}s_P(\boldsymbol{x})$.

5.7. *Moment-angle manifold of a simple polytope*

Construction (moment-angle manifold of a simple polytope [16, 14]**):** Take a simple polytope

$$P = \{\boldsymbol{x} \in \mathbb{R}^n : \boldsymbol{a}_i\boldsymbol{x} + b_i \geqslant 0, i = 1, \ldots, m\}.$$

We have $\operatorname{rank} A = n$, where A is the $m \times n$-matrix with rows $\boldsymbol{a}_i$. Then there is an embedding

$$j_P : P \longrightarrow \mathbb{R}^m_{\geq} : j_P(\boldsymbol{x}) = (y_1, \ldots, y_m),$$

where $y_i = \boldsymbol{a}_i\boldsymbol{x} + b_i$, and we will consider P as the subset in $\mathbb{R}^m_{\geqslant}$.
A *moment-angle manifold* $\widehat{\mathcal{Z}_P}$ is the subset in $\mathbb{C}^m$ defined as $\rho^{-1} \circ j_P(P)$, where $\rho(z_1, \ldots, z_m) = \left(|z_1|^2, \ldots, |z_m|^2\right)$. The action of $\mathbb{T}^m$ on $\mathbb{C}^m$ induces the action of $\mathbb{T}^m$ on $\widehat{\mathcal{Z}_P}$.

For the embeddings $j_{\mathcal{Z}} : \widehat{\mathcal{Z}_P} \subset \mathbb{C}^m$ and $j_P : P \subset \mathbb{R}^m_{\geqslant}$ we have the commutative diagram:

$$\begin{array}{ccc} \widehat{\mathcal{Z}_P} & \xrightarrow{j_{\mathcal{Z}}} & \mathbb{C}^m \\ \rho_P \downarrow & & \downarrow \rho \\ P & \xrightarrow{j_P} & \mathbb{R}^m_{\geqslant} \end{array}$$

Proposition 5.19: *We have* $\widehat{\mathcal{Z}_P} \subset \mathbb{C}^m \setminus \{0\}$.

Proof: If $0 \in \widehat{\mathcal{Z}_P}$, then $0 = \rho(0) \in j_P(P)$. This corresponds to a point $\boldsymbol{x} \in P$ such that $\boldsymbol{a}_i\boldsymbol{x} + b_i = 0$, for all i. This is impossible, since any point of a simple n-polytope lies in at most n facets. □

Definition 5.20: For the set of vectors $(\boldsymbol{x}_1, \ldots, \boldsymbol{x}_m)$ spanning $\mathbb{R}^n$, the set of vectors $(\boldsymbol{y}_1, \ldots, \boldsymbol{y}_m)$ spanning $\mathbb{R}^{m-n}$ is called the *Gale dual*, if for the matrices X and Y with column vectors $\boldsymbol{x}_i$ and $\boldsymbol{y}_j$ we have $XY^T = 0$.

Take an $((m-n) \times m)$-matrix C such that $CA = 0$ and $\operatorname{rank} C = m - n$. Then the vectors $\boldsymbol{a}_i$ and the column vectors $\boldsymbol{c}_i$ of C are Gale dual to each other. Let $\boldsymbol{c}_i = (c_{1,i}, \ldots, c_{m-n,i})$.

Proposition 5.21: *We have*

$$\widehat{\mathcal{Z}_P} = \{\boldsymbol{z} \in \mathbb{C}^m : c_{i,1}|z_1|^2 + \cdots + c_{i,m}|z_m|^2 = c_i\},$$

where $c_i = c_{i,1}b_1 + \cdots + c_{i,m}b_m$.

Denote $\Phi_i(\boldsymbol{z}) = c_{i,1}|z_1|^2 + \cdots + c_{i,m}|z_m|^2 - c_i$. Consider the mapping

$$\Phi : \mathbb{C}^m \to \mathbb{R}^{m-n} : \ \Phi(\boldsymbol{z}) = (\Phi_1(\boldsymbol{z}), \ldots, \Phi_{m-n}(\boldsymbol{z})).$$

It is the $\mathbb{T}^m$-equivariant quadratic mapping with respect to the trivial action of $\mathbb{T}^m$ on $\mathbb{R}^{m-n}$.

Proposition 5.22:

(1) $\widehat{\mathcal{Z}_P}$ *is a* ***complete intersection*** *of* ***real quadratic hypersurfaces*** *in* $\mathbb{R}^{2m} \cong \mathbb{C}^m$*:*

$$\mathcal{F}_k = \{\boldsymbol{z} \in \mathbb{C}^m : \Phi_k(\boldsymbol{z}) = 0\}, \ k = 1, \ldots, m-n.$$

(2) There is a canonical ***trivialisation*** *of the normal bundle of the* $\mathbb{T}^m$*-equivariant embedding* $\widehat{\mathcal{Z}_P} \subset \mathbb{C}^m$*, that is* $\widehat{\mathcal{Z}_P}$ *has the canonical structure of a* ***framed*** *manifold.*

Proof: We have $\widehat{\mathcal{Z}_P} = \Phi^{-1}(0)$, where $\Phi\colon \mathbb{R}^{2m} \cong \mathbb{C}^m \to \mathbb{R}^{m-n}$. The next step is an exercise.
Exercise: The differential $d\Phi|_y\colon \mathbb{R}^{2m} \to \mathbb{R}^{m-n}$ is an epimorphism for any point of $y \in \Phi^{-1}(0)$. □

Corollary 5.23: *For an appropriate choice of* C

$$\widehat{\mathcal{Z}_P} = \bigcap_{k=1}^{m-n} \mathcal{F}_k$$

where any surface $\mathcal{F}_k \subset \mathbb{R}^{2m}$ *is a* $(2m-1)$*-dimensional smooth* $\mathbb{T}^m$*-manifold.*

Proof: We just need to find such a C such that the vector Cb has all coordinates nonzero. For any C as above, Cb has a nonzero coordinate since $0 \notin \widehat{\mathcal{Z}_P}$ by Proposition 5.19. Then we can obtain from it the matrix we need by elementary row transformations. □

Exercise: Describe the orbit space $\mathcal{F}_k/\mathbb{T}^m$.

Construction (canonical section): The projection ρ has the canonical section

$$s\colon \mathbb{R}^m_{\geqslant} \to \mathbb{C}^m, \qquad s(x_1, \ldots, x_m) = (\sqrt{x_1}, \ldots, \sqrt{x_m}),$$

which gives a *canonical section* $\widehat{s_P}\colon P \to \widehat{\mathcal{Z}_P}$ by the formula $\widehat{s_P} = s \circ j_P$.

Theorem 5.24: ***(Smooth structure on the moment-angle complex,*** [16]) *The section* $\widehat{s_P}\colon P \to \widehat{\mathcal{Z}_P}$ *induces the* $\mathbb{T}^m$*-equivariant homeomorphism*

$$\widetilde{\mathcal{Z}_P} \longrightarrow \widehat{\mathcal{Z}_P}$$

defined by the formula $(\boldsymbol{x}, \boldsymbol{t}) \to \boldsymbol{t}\widehat{s_P}(\boldsymbol{x})$.

Together with the $\mathbb{T}^m$*-equivariant homeomorphism* $\widetilde{\mathcal{Z}_P} \to \mathcal{Z}_P$ *this gives a smooth structure on the moment-angle complex* $\mathcal{Z}_P$.

Thus, in what follows we identify $\widehat{\mathcal{Z}_P}$ and $\mathcal{Z}_P$.

Exercise: Describe the manifold $\mathcal{Z}_P$ for $P = \{\boldsymbol{x} \in \mathbb{R}^2 : A\boldsymbol{x} + b \geqslant 0\}$, where

1. $A^\top = \begin{pmatrix} 1 & 0 & -1 & 1 \\ 0 & 1 & 0 & -1 \end{pmatrix}$, $\qquad b^\top = (0, 0, 1, 1)$.

2. $A^\top = \begin{pmatrix} 1 & 0 & -1 & 1 & -1 \\ 0 & 1 & 0 & -1 & -1 \end{pmatrix}$, $\qquad b^\top = (0, 0, 1, 1, 2)$.

Exercise: Let $G \subset P$ be a face of codimension k in a simple n-polytope P and let $\mathcal{Z}_P$ be the corresponding moment-angle manifold with the quotient projection $p\colon \mathcal{Z}_P \to P$. Show that $p^{-1}(G)$ is a smooth submanifold of $\mathcal{Z}_P$ of codimension $2k$. Furthermore, $p^{-1}(G)$ is diffeomorphic to $\mathcal{Z}_G \times T^\ell$, where $\mathcal{Z}_G$ is the moment-angle manifold corresponding to G and ℓ is the number of facets of P not intersecting G.

5.8. *Mappings of the moment-angle manifold into spheres*

For any set $\omega = \{j_1, \dots, j_k\} \subset \{1, \dots, m\}$ define

$$\mathbb{C}^{m-k}_\omega = \{(z_1, \dots, z_m) \in \mathbb{C}^m \colon z_j = 0, j \in \omega\};$$
$$S^{2m-2k-1}_\omega = \{(z_1, \dots, z_m) \in \mathbb{C}^m \colon z_j = 0, j \in \omega, \sum_{j \notin \omega} |z_j|^2 = 1\};$$
$$\mathbb{R}^{m-k}_\omega = \{(y_1, \dots, y_m) \in \mathbb{R}^m \colon y_j = 0, j \in \omega\}.$$

Exercise: Show that for $k \geqslant 1$ the sphere $S^{2k-1}_{[m]\setminus\omega}$ is a deformation retract of $S^{2m-1} \setminus S^{2m-2k-1}_\omega$.

Proposition 5.25:

(1) *The embedding* $\mathcal{Z}_P \subset \mathbb{C}^m$ *induces the embedding* $\mathcal{Z}_P \subset S^{2m-1}$ *via the projection* $\mathbb{C}^m \setminus \{0\} \to S^{2m-1}$.

(2) *For any set* ω, $|\omega| = k$, *such that* $\bigcap_{j\in\omega} F_j = \varnothing$ *the image of the embedding* $\mathcal{Z}_P \subset S^{2m-1}$ *lies in* $S^{2m-1} \setminus S_\omega^{2m-2k-1}$; *hence the embedding is homotopic to the mapping* $\varphi_\omega\colon \mathcal{Z}_P \to S^{2k-1}_{[m]\setminus\omega}$, *induced by the projection* $\mathbb{C}^m \to \mathbb{C}^k_{[m]\setminus\omega}$.

Proof:

(1) follows from Proposition 5.19.
(2) follows from the fact that if $\bigcap_{j\in\omega} F_j = \varnothing$, then there is no $\boldsymbol{x} \in P$ such that $\boldsymbol{a}_j\boldsymbol{x} + b_j = 0$ for all $j \in \omega$. □

We have the commutative diagram

$$\begin{array}{ccccc}
\mathcal{Z}_P & \longrightarrow & \mathbb{C}^m \setminus \mathbb{C}^{m-k}_\omega & \xrightarrow{\xi_\omega} & S^{2k-1}_{[m]\setminus\omega} \subset \mathbb{C}^k_{[m]\setminus\omega} \\
\downarrow & & \downarrow \rho & & \downarrow \\
P & \xrightarrow{A\boldsymbol{x}+\boldsymbol{b}} & \mathbb{R}^m_\geqslant \setminus \mathbb{R}^{m-k}_\omega & \xrightarrow{\pi_\omega} & \Delta^{k-1} \subset \mathbb{R}^k_\geqslant
\end{array}$$

where

$$\xi_\omega(z_1, \ldots, z_m) = \frac{\boldsymbol{z}_\omega}{|\boldsymbol{z}_\omega|},\ \boldsymbol{z}_\omega = (z_{j_1}, \ldots, z_{j_k}),\ |\boldsymbol{z}_\omega| = \sqrt{|z_{j_1}|^2 + \cdots + |z_{j_k}|^2}.$$

$$\pi_\omega(y_1, \ldots, y_m) = \frac{\boldsymbol{y}_\omega}{d_\omega},\ \boldsymbol{y}_\omega = (y_{j_1}, \ldots, y_{j_k}),\ d_\omega = |y_{j_1}| + \cdots + |y_{j_k}|.$$

Example 5.26: For any pair of facets F_i, F_j, such that $F_i \cap F_j = \varnothing$, there is a mapping $\mathcal{Z}_P \to S^3_{[m]\setminus\{i,j\}}$.

Definition 5.27: The class $a \in H^k(X, \mathbb{Z})$ is called *cospherical* if there is a mapping $\varphi\colon X \to S^k$ such that $\varphi^*\left([S^k]\right) = a$.

Corollary 5.28: *For each* $\omega \subset [m]$, $|\omega| = k$, *such that* $\bigcap_{i\in\omega} F_i = \varnothing$ *we have the cospherical class* $\varphi_\omega^*\left(\left[S^{2k-1}_{[m]\setminus\omega}\right]\right)$ *in* $H^{2k-1}(\mathcal{Z}_P)$.

5.9. *Projective moment-angle manifold*

Let S^1_Δ be the diagonal subgroup in $\mathbb{T}^m$. We have the *free action* of S^1_Δ on $\mathcal{Z}_P$ and therefore the *smooth* manifold

$$\mathcal{P}\mathcal{Z}_P = \mathcal{Z}_P / S^1_\Delta$$

is the projective version of the moment-angle manifold $\mathcal{Z}_P$.

Definition 5.29: For actions of the commutative group G on spaces X and Y define:

$$X \times_G Y = X \times Y / \{gx, gy) \sim (x, y) \, \forall x \in X, y \in Y, g \in G\}.$$

Corollary 5.30: *For any simple polytope P there exists the* ***smooth*** *manifold*

$$W = \mathcal{Z}_P \times_{S^1_\Delta} D^2$$

such that $\partial W = \mathcal{Z}_P$.

We have the fibration $W \longrightarrow \mathcal{P}\mathcal{Z}_P$ with the fibre D^2.

Exercise: $P = \Delta^n \Longleftrightarrow \mathcal{Z}_P = S^{2n+1} \Longrightarrow \mathcal{P}S^{2n+1} = \mathbb{C}P^n$.

The constructions of Subsection 5.8 respect the diagonal action of S^1; hence we obtain the following results:

For $k \geqslant 1$ the set $\mathbb{C}P^{k-1}_{[m]\setminus\omega}$ is a deformation retract of $\mathbb{C}P^{m-1} \setminus \mathbb{C}P^{m-k-1}_\omega$.

Proposition 5.31:

(1) *The embedding $\mathcal{Z}_P \subset \mathbb{C}^m$ induces the embedding $\mathcal{P}\mathcal{Z}_P \subset \mathbb{C}P^{m-1}$.*
(2) *For any set ω, $|\omega| = k$, such that $\bigcap_{j\in\omega} F_j = \varnothing$ the image of the embedding $\mathcal{P}\mathcal{Z}_P \subset \mathbb{C}P^m$ lies in $\mathbb{C}P^{m-1} \setminus \mathbb{C}P^{m-k-1}_\omega$; hence the embedding is homotopic to the mapping $\mathcal{P}\mathcal{Z}_P \to \mathbb{C}P^{k-1}_{[m]\setminus\omega}$, induced by the projection $\mathbb{C}^m \to \mathbb{C}^k_{[m]\setminus\omega}$.*

6. Cohomology of moment-angle manifolds

When we deal with homology and cohomology, if it is not specified, the notation $H^*(X)$ and $H_*(X)$ means that we consider integer coefficients.

6.1. *Cellular structure*

Define a cellular structure on D^2 consisting of 3 cells:

$$p = \{1\}, \quad U = S^1 \setminus \{1\}, \quad V = D^2 \setminus S^1.$$

Set on D^2 the standard orientation, with $(1, 0)$ and $(0, 1)$ being the positively oriented basis, and on S^1 the counterclockwise orientation induced from D^2. Then in the chain complex $C_*(D^2)$ we have

$$dp = 0, \quad dU = 0, \quad dV = U.$$

The coboundary operator $\partial \colon C^i(X) \to C^{i+1}(X)$ is defined by the rule $\langle \partial\varphi, a\rangle = \langle \varphi, da\rangle$. For a cell E let us denote by E^* the cochain such that $\langle E^*, E'\rangle =$

$\delta(E, E')$ for any cell E'. Let $p^* = 1$. Then the coboundary operator in $C^*(D^2)$ has the form

$$\partial 1 = 0, \quad \partial U^* = V^*, \quad \partial V^* = 0.$$

By definition the *multigraded polydisk* $\mathbb{D}^{2m}$ has the canonical multigraded cellular structure , which is a product of cellular structures of disks, with cells corresponding to pairs of sets σ, ω, with $\sigma \subset \omega \subset [m] = \{1, 2, \ldots, m\}$.

$$C_{\sigma,\omega} = \tau_1 \times \cdots \times \tau_m, \tau_j = \begin{cases} V_j, & j \in \sigma, \\ U_j, & j \in \omega \setminus \sigma, \\ p_j, & j \in [m] \setminus \omega, \end{cases} \quad m\deg C_{\sigma,\omega} = (-i, 2\omega),$$

where $i = |\omega \setminus \sigma|$. Then the cellular chain complex $C_*(\mathbb{D}^{2m})$ is the tensor product of m chain complexes $C_*(D_i^2)$, $i = 1, \ldots, m$. The boundary operator d of the chain complex respects the multigraded structure and can be considered as a multigraded operator of $m\deg\ d = (-1, 0)$. It can be calculated on the elements of the tensor product by the Leibnitz rule

$$d(a \times b) = (da) \times b + (-1)^{\dim a} a \times (db).$$

For cochains the $\times$-operation $C^i(X) \times C^j(Y) \to C^{i+j}(X \times Y)$ is defined by the rule $\langle \varphi \times \psi, a \times b \rangle = \langle \varphi, a \rangle \langle \psi, b \rangle$. Then

$$\langle \psi_1 \times \cdots \times \psi_m, a_1 \times \cdots \times a_m \rangle = \langle \psi_1, a_1 \rangle \ldots \langle \psi_m, a_m \rangle.$$

The basis in $C^*(\mathbb{D}^{2m})$ is formed by the cochains $C^*_{\sigma,\omega} = \tau_1^* \times \cdots \times \tau_m^*$, where $C_{\sigma,\omega} = \tau_1 \times \cdots \times \tau_m$.

The coboubdary operator ∂ is also multigraded. It has multidegree $m\deg\ \partial = (1, 0)$. It can be calculated on the elements of the tensor algebra $C^*(\mathbb{D}^{2m})$ by the rule $\partial(\varphi \times \psi) = (\partial\varphi) \times \psi + (-1)^{\dim \varphi} \varphi \times (\partial\psi)$.

Proposition 6.1: *The moment-angle complex $\mathcal{Z}_P$ has the canonical structure of a multigraded subcomplex in the multigraded cellular structure of $\mathbb{D}^{2m}$. The projection $\pi^m \colon \mathcal{Z}_P \to \mathbb{I}_P$ is cellular.*

Theorem 6.2: *There is a multigraded structure in the cohomology group*

$$H^n(\mathcal{Z}_P, \mathbb{Z}) \simeq \bigoplus_{2|\omega|=n+i} H^{-i,2\omega}(\mathcal{Z}_P, \mathbb{Z}),$$

where for $\omega = \{j_1, \ldots, j_k\}$, we have $|\omega| = k$.

Proof: The multigraded structure in cohomology is induced by the multigraded cellular structure described above. □

Example 6.3: Let $P = \Delta^n$, then $\mathcal{Z}_P = S^{2n+1}$. In the case $n = 1$ the simplex Δ^1 is an interval I, and we have the decomposition $\mathcal{Z}_I = S^3 = S^1 \times D^2 \cup D^2 \times S^1$. The space $\mathcal{Z}_I$ consists of 8 cells

$$\begin{gathered} p_1 \times p_2 \\ p_1 \times U_2, \quad U_1 \times p_2 \\ p_1 \times V_2, \quad U_1 \times U_2, \quad V_1 \times p_2, \\ U_1 \times V_2, \quad V_1 \times U_2 \end{gathered}$$

We have

$$H^*(S^3) = H^{0,2\varnothing}(S^3) \oplus H^{-1,2\{1,2\}}(S^3).$$

6.2. *Multiplication*

Now following [14] we will describe the cohomology ring of a moment-angle complex in terms of the cellular structure defined above. This result is non-trivial, since the problem of defining the multiplication in cohomology in terms of cellular cochains is unsolvable in general. The reason is that the diagonal mapping used in the definition of the cohomology product is not cellular, and a cellular approximation can not be made functorial with respect to arbitrary cellular mappings. We construct a canonical cellular diagonal approximation $\widetilde{\Delta}\colon \mathcal{Z}_P \to \mathcal{Z}_P \times \mathcal{Z}_P$, which is functorial with respect to mappings induced by admissible mappings of sets of facets of polytopes.

Recall that the product in the cohomology of a cell complex X is defined as follows: Consider the composite mapping of cellular cochain complexes

$$\mathcal{C}^*(X) \otimes \mathcal{C}^*(X) \xrightarrow{\times} \mathcal{C}^*(X \times X) \xrightarrow{\widetilde{\Delta}^*} \mathcal{C}^*(X). \tag{6.1}$$

Here the mapping $\times$ sends a cellular cochain $c_1 \otimes c_2 \in C^{q_1}(X) \otimes C^{q_2}(X)$ to the cochain $c_1 \times c_2 \in C^{q_1+q_2}(X \times X)$, whose value on a cell $e_1 \times e_2 \in C_*(X \times X)$ is $\langle c_1, e_1\rangle\langle c_2, e_2\rangle$. The mapping $\widetilde{\Delta}^*$ is induced by a cellular mapping $\widetilde{\Delta}$ (a cellular *diagonal approximation*) homotopic to the diagonal $\Delta\colon X \to X \times X$. In cohomology, the mapping (6.1) induces a multiplication $H^*(X) \otimes H^*(X) \to H^*(X)$ which does not depend on the choice of a cellular approximation and is functorial. However, the mapping (6.1) itself is not functorial because there is no choice of a cellular approximation compatible with arbitrary cellular mappings.

Define polar coordinates in D^2 by $z = \rho e^{i\varphi}$.

Proposition 6.4:

(1) *The mapping* $\Delta_t\colon \mathbb{I}\times D^2 \to D^2\times D^2$: $\rho e^{i\varphi}$

$$\to \begin{cases} \left((1-\rho)t+\rho e^{i(1+t)\varphi},(1-\rho)t+\rho e^{i(1-t)\varphi}\right), & \varphi\in[0,\pi],\\ \left((1-\rho)t+\rho e^{i(1-t)\varphi+2\pi it},(1-\rho)t+\rho e^{i(1+t)\varphi-2\pi it}\right), & \varphi\in[\pi,2\pi] \end{cases}$$

defines the homotopy of mappings of pairs $(D^2,S^1)\to(D^2\times D^2,S^1\times S^1)$.

(2) *The mapping* Δ_0 *is the diagonal mapping* $\Delta\colon D^2\to D^2\times D^2$.

(3) *The mapping* Δ_1 *is*

$$\rho e^{i\varphi}\to\begin{cases}((1-\rho)+\rho e^{2i\varphi},1), & \varphi\in[0,\pi],\\ (1,(1-\rho)+\rho e^{2i\varphi}), & \varphi\in[\pi,2\pi].\end{cases}$$

It is cellular and sends the pair (D^2,S^1) *to the pair of wedges* $(D^2\times 1\vee 1\times D^2,\ S^1\times 1\vee 1\times S^1)$ *in the point* $(1,1)$. *Hence it is a cellular approximation of* Δ.

(4) *We have*

$$(\Delta_1)_*p=p\times p,\ (\Delta_1)_*U=U\times p+p\times U,\ (\Delta_1)_*V=V\times p+p\times V;$$

hence

$$(U^*)^2=\langle U^*\times U^*,(\Delta_1)_*V\rangle V^*=\langle U^*\times U^*,V\times p+p\times V\rangle V^*=0,$$

and the multiplication of cochains in $\mathcal{C}^*(D^2)$ *induced by* Δ_1 *is trivial:*

$$1\cdot X=X=X\cdot 1,\quad (U^*)^2=U^*V^*=V^*U^*=(V^*)^2=0.$$

The proof we leave as an exercise.

Using the properties of the construction of the moment-angle complex we obtain the following result:

Corollary 6.5:

(1) *For any simple polytope* P *with* m *facets there is a homotopy*

$$\Delta_t^m\colon(\mathbb{D}^{2m},\mathcal{Z}_P)\to(\mathbb{D}^{2m}\times\mathbb{D}^{2m},\mathcal{Z}_P\times\mathcal{Z}_P),$$

where Δ_0^m *is the diagonal mapping and* Δ_1^m *is a cellular mapping.*

(2) *In the cellular cochain complex of* $\mathbb{D}^{2m}=D^2\times\cdots\times D^2$ *the multiplication defined by* Δ_1^m *is the tensor product of multiplications of the factors defined by the rule* $(\varphi_1\times\varphi_2)(\psi_1\times\psi_2)=(-1)^{\dim\varphi_2\dim\psi_1}\varphi_1\psi_1\times\varphi_2\psi_2$, *and*

$$(\varphi_1\times\cdots\times\varphi_m)(\psi_1\times\cdots\times\psi_m)=(-1)^{\sum\limits_{i>j}\dim\varphi_i\dim\psi_j}\varphi_1\psi_1\times\cdots\times\varphi_m\psi_m,$$

and respects the multigrading.

(3) *The multiplication in* $\mathcal{C}^*(\mathcal{Z}_P)$ *given by* Δ_1^m *is defined from the inclusion* $\mathcal{Z}_P\subset\mathbb{D}^{2m}$ *as a multigraded cellular subcomplex.*

6.3. *Description in terms of the Stanley-Reisner ring*

Definition 6.6: Let $\{F_1, \dots, F_m\}$ be the set of facets of a simple polytope P. Then a *Stanley-Reisner ring* of P over $\mathbb{Z}$ is defined as a monomial ring

$$\mathbb{Z}[P] = \mathbb{Z}[v_1, \dots, v_m]/J_{SR}(P),$$

where

$$J_{SR}(P) = (v_{i_1} \dots v_{i_k}, \text{ if } F_{i_1} \cap \dots \cap F_{i_k} = \varnothing)$$

is the Stanley-Reisner ideal.

Example 6.7: $\mathbb{Z}[\Delta^2] = \mathbb{Z}[v_1, v_2, v_3]/(v_1v_2v_3)$.

Theorem 6.8: *(see [5]) Two polytopes are combinatorially equivalent if and only if their Stanley-Reisner rings are isomorphic.*

Corollary 6.9: *Fullerenes P_1 and P_2 are combinatorially equivalent if and only if there is an isomorphism $\mathbb{Z}[P_1] \cong \mathbb{Z}[P_2]$.*

Theorem 6.10: *The Stanley-Reisner ring of a flag polytope is a monomial quadratic ring:*

$$J_{SR}(P) = \{v_iv_j : F_i \cap F_j = \varnothing\}.$$

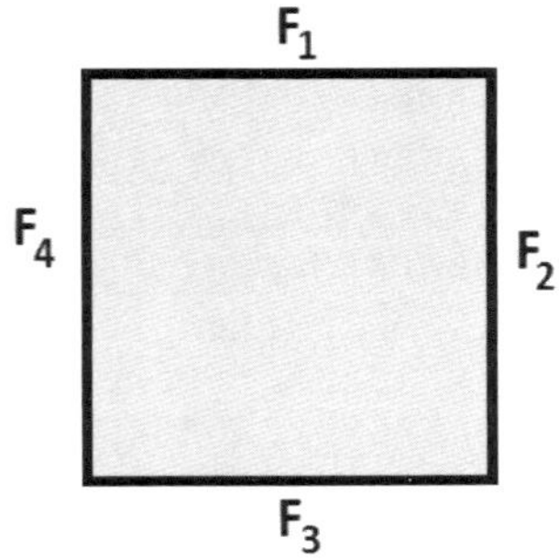

Fig. 29. Cube I^2. We have $J_{SR}(I^2) = \{v_1v_3, v_2v_4\}$

Each fullerene is a simple flag polytope (see Corollary 4.12).

Corollary 6.11: *The Stanley-Reisner ring of a fullerene is monomial quadratic.*

Construction (multigraded complex): For a set $\sigma \subset [m]$ define $G(\sigma) = \bigcap\limits_{i\in\sigma} F_i$. Conversely, for a face G define $\sigma(G) = \{i\colon G \subset F_i\} \subset [m]$. Then $\sigma(G(\sigma)) = \sigma$, and $G(\sigma(G)) = G$. Let

$$R^*(P) = \Lambda[u_1, \dots, u_m] \otimes \mathbb{Z}[P]/(u_i v_i, v_i^2),$$
$$\text{mdeg } u_i = (-1, 2\{i\}), \text{mdeg } v_i = (0, 2\{i\}), du_i = v_i, dv_i = 0$$

be a multigraded differential algebra. It is additively generated by monomials $v_\sigma u_{\omega\setminus\sigma}$, where $v_\sigma = \prod\limits_{i\in\sigma} v_i$, $G(\sigma) \neq \varnothing$, and $u_{\omega\setminus\sigma} = u_{j_1} \wedge \cdots \wedge u_{j_l}$ for $\omega \setminus \sigma = \{j_1, \dots, j_l\}$.

Theorem 6.12: [14] *We have a mutigraded ring isomorphism*

$$H[R^*(P), d] \simeq H^*(\mathcal{Z}_P, \mathbb{Z}).$$

Proof: Define the mapping $\zeta\colon R^*(P) \to C^*(\mathcal{Z}_P)$ by the rule $\zeta(v_\sigma u_{\omega\setminus\sigma}) = C^*_{\sigma,\omega}$. It is a graded ring isomorphism from Proposition 6.4(4), and Corollary 6.5. The formula $\zeta(dv_\sigma u_{\omega\setminus\sigma}) = \partial C^*_{\sigma,\omega}$ follows from the Leibnitz rule. □

Exercise: Prove that for the cospherical class $\varphi^*_\omega\left(\left[S^{2k-1}_{[m]\setminus\omega}\right]\right)$, $\omega = \{i_1, \dots, i_k\}$, (see Corollary 5.28) we have $\varphi^*_\omega\left(\left[S^{2k-1}_{[m]\setminus\omega}\right]\right) = \pm[u_{i_1} v_{i_2} \dots v_{i_k}] \in H[R^*(P), d]$.

6.4. *Description in terms of unions of facets*

Let $P_\omega = \bigcup\limits_{i\in\omega} F_i$ for a subset $\omega \subset [m]$. By definition $P_\varnothing = \varnothing$, and $P_{[m]} = \partial P$.

Definition 6.13: For two sets $\sigma, \tau \subset [m]$ define $l(\sigma, \tau)$ to be the number of pairs $\{(i, j)\colon i \in \sigma, j \in \tau, i > j\}$. We write $l(i, \tau)$ and $l(\sigma, j)$ for $\sigma = \{i\}$ and $\tau = \{j\}$ respectively.

Comment: The number $(-1)^{l(\sigma,\omega)}$ is used in the definition of the multiplication of cubical chain complexes (see [44]). In the discrete mathematics the number $l(\sigma, \tau)$ is a characteristic of two subsets σ, τ of an ordered set.

Proposition 6.14: *We have:*

(1) $l(\sigma, \tau) = \sum\limits_{i\in\sigma} l(i, \tau) = \sum\limits_{j\in\tau} l(\sigma, j) = \sum\limits_{i\in\sigma, j\in\tau} l(i, j)$,
(2) $l(\sigma, \tau_1 \sqcup \tau_2) = l(\sigma, \tau_1) + l(\sigma, \tau_2)$, $l(\sigma_1 \sqcup \sigma_2, \tau) = l(\sigma_1, \tau) + l(\sigma_2, \tau)$,
(3) $l(\sigma, \tau) + l(\tau, \sigma) = |\sigma||\tau| - |\sigma \cap \tau|$.
In particular, if $\sigma \cap \tau = \varnothing$, *then* $l(\sigma, \tau) + l(\tau, \sigma) = |\tau||\sigma|$.

Definition 6.15: Set

$$\mathbb{I}_{P,\omega} = \bigcup_{G\neq\varnothing:\,\sigma(G)\subset\omega} \mathbb{I}_{P,G} = \{(x_1,\dots,x_m)\in\mathbb{I}_P : x_i = 1, i\notin\omega\}.$$

Theorem 6.16: [14] *For any $\omega\subset[m]$ there is an isomorphism:*

$$H^{-i,2\omega}(\mathcal{Z}_P,\mathbb{Z})\cong H^{|\omega|-i}(P,P_\omega,\mathbb{Z}).$$

Proof: For subsets $A\subset\mathbb{I}^m$ and $\omega\subset[m]$ define

$$A^\omega = \{(y_1,\dots,y_m)\in A : y_i = 0 \text{ for some } i\in\omega\},\quad A^0 = A^{[m]}.$$

We have $A^\varnothing = A$. There is a homeomorphism of pairs $(P,P_\omega)\simeq(\mathbb{I}_P,\mathbb{I}_P^\omega)$.

The homotopy $r_t^\omega\colon\mathbb{I}^m\to\mathbb{I}^m$:

$$r_t^\omega(y_1,\dots,y_m) = (y_1',\dots,y_m'), y_j' = \begin{cases}(1-t)y_j+t, & j\notin\omega;\\ y_j, & j\in\omega,\end{cases}$$

gives a deformation retraction $r^\omega = r_1^\omega\colon(\mathbb{I}_P,\mathbb{I}_P^\omega)\to(\mathbb{I}_{P,\omega},\mathbb{I}_{P,\omega}^0)$.

There is a natural multigraded cell structure on the cube $\mathbb{I}^m$, induced by the cell structure on $\mathbb{I}$ consisting of 3 cells: $\mathbf{0}=\{0\}$, $\mathbf{1}=\{1\}$ and $J=(0,1)$. All the sets $\mathbb{I}_P$, $\mathbb{I}_{P,G}$, $\mathbb{I}_P^\omega$, $\mathbb{I}_{P,\omega}$, $\mathbb{I}_{P,\omega}^0$ are cellular subcomplexes. There is a natural orientation in J such that $\mathbf{0}$ is the beginning, and $\mathbf{1}$ is the end. We have

$$d\mathbf{0} = d\mathbf{1} = 0,\quad dJ = \mathbf{1}-\mathbf{0};$$
$$\partial\mathbf{1}^* = -\partial\mathbf{0}^* = J^*,\quad \partial J^* = 0.$$

The cells in $\mathbb{I}^m$ have the form $\eta_1\times\cdots\times\eta_m$, $\eta_i\in\{\mathbf{0}_i,\mathbf{1}_i,J_i\}$. There is natural cellular approximation for the diagonal mapping $\Delta\colon\mathbb{I}\to\mathbb{I}\times\mathbb{I}$ by the mapping Δ_1:

$$\Delta_1(x) = \begin{cases}(2x,1), & x\in[0,\frac{1}{2}];\\ (1,2x-1), & x\in[\frac{1}{2},1],\end{cases}$$

connected with Δ by the homotopy $\Delta_t = (1-t)\Delta + t\Delta_1$. Then

$$(\Delta_1)_*\mathbf{0} = \mathbf{0}\times\mathbf{0},\quad (\Delta_1)_*\mathbf{1} = \mathbf{1}\times\mathbf{1},\quad (\Delta_1)_*J = J\times\mathbf{0}+\mathbf{1}\times J,$$

and for the induced multiplication we have

$$(\mathbf{0}^*)^2 = \mathbf{0}^*,\quad (\mathbf{1}^*)^2 = \mathbf{1}^*,\quad \mathbf{0}^*\mathbf{1}^* = \mathbf{1}^*\mathbf{0}^* = 0,\quad \mathbf{0}^*J^* = J^*\mathbf{1}^* = 0,$$
$$J^*\mathbf{0}^* = \mathbf{1}^*J^* = J^*,\quad (J^*)^2 = 0.$$

The cells in $\mathbb{I}_{P,\omega} \setminus \mathbb{I}^0_{P,\omega}$ have the form

$$E_\sigma = \eta_1 \times \cdots \times \eta_m, \qquad \eta_j = \begin{cases} J_j, & j \in \sigma; \\ 1, & j \notin \sigma, \end{cases}$$

where $\sigma \subset \omega$, and $G(\sigma) \neq \varnothing$. Then $E^*_\sigma = \eta^*_1 \times \cdots \times \eta^*_m$.

Now define the mapping $\xi_\omega \colon R^{-i,2\omega} \to C^{|\omega|-i}(\mathbb{I}_{P,\omega}, \mathbb{I}^0_{P,\omega})$ by the rule

$$\xi_\omega(u_{\omega\setminus\sigma} v_\sigma) = (-1)^{l(\sigma,\omega)} E^*_\sigma.$$

By construction ξ_ω is an additive isomorphism. For $\sigma \subset \omega$ we have

$$\partial\xi_\omega(v_\sigma u_{\omega\setminus\sigma}) = \partial\left((-1)^{l(\sigma,\omega)} E^*_\sigma\right) = (-1)^{l(\sigma,\omega)} \sum_{j\in\omega\setminus\sigma, G(\sigma\sqcup\{j\})\neq\varnothing} (-1)^{l(j,\sigma)} E^*_{\sigma\sqcup\{j\}}.$$

On the other hand,

$$\begin{aligned} \xi_\omega(dv_\sigma u_{\omega\setminus\sigma}) &= \xi_\omega\left(\sum_{j\in\omega\setminus\sigma, G(\sigma\sqcup\{j\})\neq\varnothing} (-1)^{l(j,\omega\setminus\sigma)} v_{\sigma\sqcup\{j\}} u_{\omega\setminus(\sigma\sqcup\{j\})}\right) \\ &= \sum_{j\in\omega\setminus\sigma, G(\sigma\sqcup\{j\})\neq\varnothing} (-1)^{l(\sigma\sqcup\{j\},\omega)} (-1)^{l(j,\omega\setminus\sigma)} E^*_{\sigma\sqcup\{j\}}. \end{aligned}$$

Now the proof follows from the formula

$$\begin{aligned} l(\sigma \sqcup \{j\}, \omega) + l(j, \omega \setminus \sigma) &= l(\sigma, \omega) + l(j, \omega) + l(j, \omega \setminus \sigma) \\ &= l(\sigma, \omega) + l(j, \sigma) + 2l(j, \omega \setminus \sigma). \end{aligned}$$ □

Corollary 6.17: [14] *For any $\omega \subset [m]$ there is an isomorphism*

$$H^{-i,2\omega}(\mathcal{Z}_P, \mathbb{Z}) \cong \widetilde{H}^{|\omega|-i-1}(P_\omega, \mathbb{Z}),$$

where by definition $\widetilde{H}^{-1}(\varnothing, \mathbb{Z}) = \mathbb{Z}$.

The proof follows from the long exact sequence in the reduced cohomology of the pair (P, P_ω), since P is contractible.

6.5. *Multigraded Betti numbers and the Poincare duality*

Definition 6.18: Define *multigraded Betti numbers* $\beta^{-i,2\omega} = \operatorname{rank} H^{-i,2\omega}(\mathcal{Z}_P)$. We have

$$\beta^{-i,2\omega} = \operatorname{rank} H^{|\omega|-i}(P, P_\omega) = \operatorname{rank} \widetilde{H}^{|\omega|-i-1}(P_\omega, \mathbb{Z}).$$

From Proposition 5.22 the manifold $\widehat{\mathcal{Z}_P}$ is oriented.

Proposition 6.19: *We have*

$$\beta^{-i,2\omega} = \beta^{-(m-n-i),2([m]\backslash\omega)}.$$

Proof: From the Poincare duality theorem the bilinear form $H^*(\mathcal{Z}_P) \otimes H^*(\mathcal{Z}_P) \to \mathbb{Z}$ defined by

$$\langle \varphi, \psi \rangle = \langle \varphi\psi, [\mathcal{Z}_P] \rangle,$$

where $[\mathcal{Z}_P]$ is a fundamental cycle, is non-degenerate if we factor out the torsion. This means that there is a basis for which the matrix of the bilinear form has determinant ± 1. For a mutligraded ring this means that the matrix consists of blocks corresponding to the forms

$$H^{-i,2\omega}(\mathcal{Z}_P) \otimes H^{-(m-n-i),2([m]\backslash\omega)}(\mathcal{Z}_P) \to \mathbb{Z}.$$

Hence all the blocks are squares and have determinant ± 1, which finishes the proof. □

Let the polytope P be given in the irredundant form $\{\boldsymbol{x} \in \mathbb{R}^n : A\boldsymbol{x} + b \geqslant 0\}$. For the vertex $v = F_{i_1} \cap \dots \cap F_{i_n}$ define the submatrix A_v in A corresponding to the rows $i_1, \dots, i_n$.

Proposition 6.20: *The fundamental cycle $[\mathcal{Z}_P]$ can be represented by the following element in $C_{-(m-n),[m]}(\mathcal{Z}_P)$:*

$$Z = \sum_{v - \text{vertex}} (-1)^{l(\sigma(v),[m])} \operatorname{sign}(\det A_v) C_{\sigma(v),[m]}.$$

Then the form

$$C^{-i,\omega}(\mathcal{Z}_P) \otimes C^{-(m-n-i),[m]\backslash\omega}(\mathcal{Z}_P) \to \mathbb{Z}$$

is defined by the property

$$\langle u_{[m]\backslash\sigma(v)} v_{\sigma(v)}, Z \rangle = (-1)^{l(\sigma(v),m)} \operatorname{sign}(\det A_v).$$

The idea of the proof is to use the Davis–Januszkiewicz' construction. The space $P^n \times \mathbb{T}^m$ has the orientation defined by orientations of P^n and S^1. Then the mapping

$$P^n \times \mathbb{T}^m \to \mathcal{Z}_P : (\boldsymbol{x}, \boldsymbol{t}) \to \boldsymbol{t} s_P(\boldsymbol{x})$$

defines the orientation of the cells $C_{\sigma(v),[m]}$.

6.6. *Multiplication in terms of unions of facets*

For pairs of spaces define the direct product as

$$(X,A)\times(Y,B)=(X\times Y, A\times Y\cup X\times B).$$

There is a canonical multiplication in the cohomology of cellular pairs

$$H^k(X,A)\otimes H^l(X,B)\to H^{k+l}(X,A\cup B)$$

defined in the cellular cohomology by the rule

$$H^k(X,A)\otimes H^l(X,B)\xrightarrow{\times} H^{k+l}((X,A)\times(X,B))\xrightarrow{\widetilde{\Delta}^*} H^{k+l}(X,A\cup B),$$

where $\widetilde{\Delta}$ is a cellular approximation of the diagonal mapping

$$\Delta\colon (X,A\cup B)\to(X,A)\times(X,B).$$

Thus for any simple polytope P and subsets $\omega_1,\omega_2\subset[m]$, we have the canonical multiplication

$$H^k(P,P_{\omega_1})\otimes H^l(P,P_{\omega_2})\to H^{k+l}(P,P_{\omega_1\cup\omega_2}).$$

Theorem 6.21: *There is the ring isomorphism*

$$H^*(\mathcal{Z}_P)\simeq\bigoplus_{\omega\subset[m]}H^*(P,P_\omega)$$

where the multiplication on the right hand side

$$H^{|\omega_1|-k}(P,P_{\omega_1})\otimes H^{|\omega_2|-l}(P,P_{\omega_2})\to H^{|\omega_1|+|\omega_2|-k-l}(P,P_{\omega_1\cup\omega_2})$$

is trivial if $\omega_1\cap\omega_2\neq\varnothing$, and for the case $\omega_1\cap\omega_2=\varnothing$ is given by the rule

$$a\otimes b\to(-1)^{l(\omega_2,\omega_1)+|\omega_1|l}ab,$$

where $a\otimes b\to ab$ is the canonical multiplication.

Comment: The statement of the theorem presented in [14] as Exercise 3.2.14 does not contain the specialization of the sign.

Proof: We will identify (P,P_ω) with $(\mathbb{I}_P,\mathbb{I}_P^\omega)$ and $H^*(\mathcal{Z}_P)$ with $H[R^*(P),d]$. If $\omega_1\cap\omega_2\neq\varnothing$, then the multiplication

$$H^{-k,2\omega_1}(\mathcal{Z}_P)\otimes H^{-l,2\omega_2}(\mathcal{Z}_P)\to H^{-(k+l),2(\omega_1\cup\omega_2)}(\mathcal{Z}_P)$$

is trivial by Theorem 6.12. Let $\omega_1\cap\omega_2=\varnothing$. We have the commutative diagram of mappings

$$\begin{array}{ccc}
(\mathbb{I}_{P,\omega_1\sqcup\omega_2},\mathbb{I}^0_{P,\omega_1\sqcup\omega_2}) & \xrightarrow{i_{\omega_1,\omega_2}} & (\mathbb{I}_{P,\omega_1},\mathbb{I}^0_{P,\omega_1})\times(\mathbb{I}_{P,\omega_2},\mathbb{I}^0_{P,\omega_2})\\
{\scriptstyle r^{\omega_1\sqcup\omega_2}}\uparrow & & \uparrow{\scriptstyle r^{\omega_1}\times r^{\omega_2}}\\
(\mathbb{I}_P,\mathbb{I}_P^{\omega_1\sqcup\omega_2}) & \xrightarrow{\Delta} & (\mathbb{I}_P,\mathbb{I}_P^{\omega_1})\times(\mathbb{I}_P,\mathbb{I}_P^{\omega_2})
\end{array}$$

which gives the commutative diagram

$$\begin{array}{ccc}
H^*\left((\mathbb{I}_{P,\omega_1},\mathbb{I}^0_{P,\omega_1})\times(\mathbb{I}_{P,\omega_2},\mathbb{I}^0_{P,\omega_2})\right) & \xrightarrow{i^*_{\omega_1,\omega_2}} & H^*\left(\mathbb{I}_{P,\omega_1\sqcup\omega_2},\mathbb{I}^0_{P,\omega_1\sqcup\omega_2}\right)\\
{\scriptstyle (r^{\omega_1}\times r^{\omega_2})^*}\downarrow & & \downarrow{\scriptstyle (r^{\omega_1\sqcup\omega_2})^*}\\
H^*\left((\mathbb{I}_P,\mathbb{I}_P^{\omega_1})\times(\mathbb{I}_P,\mathbb{I}_P^{\omega_2})\right) & \xrightarrow{\Delta^*} & H^*\left(\mathbb{I}_P,\mathbb{I}_P^{\omega_1\sqcup\omega_2}\right)
\end{array}$$

where the vertical mappings are isomorphisms. Together with the functoriality of the $\times$-product in cohomology this proves the theorem provided the commutativity of the diagram

$$\begin{array}{ccc}
C^{-k,2\omega_1}(\mathcal{Z}_P)\otimes C^{-l,2\omega_2}(\mathcal{Z}_P) & \xrightarrow{\tilde{\Delta}\circ\times} & C^{-(k+l),2(\omega_1\sqcup\omega_2)}(\mathcal{Z}_P)\\
{\scriptstyle \xi_{\omega_1}\otimes\xi_{\omega_2}}\downarrow & & \downarrow{\scriptstyle \xi_{\omega_1\sqcup\omega_2}}\\
C^{|\omega_1|-k}\left(\mathbb{I}_{P,\omega_1},\mathbb{I}^0_{P,\omega_1}\right)\otimes C^{|\omega_2|-l}\left(\mathbb{I}_{P,\omega_2},\mathbb{I}^0_{P,\omega_2}\right) & \xrightarrow{i^*_{\omega_1,\omega_2}\circ\times} & C^{|\omega_1|+|\omega_2|-k-l}\left(\mathbb{I}_{P,\omega_1\sqcup\omega_2},\mathbb{I}^0_{P,\omega_1\sqcup\omega_2}\right)
\end{array}$$

where the lower arrow is the composition of two mappings:

$$C^{|\omega_1|-k}\left(\mathbb{I}_{P,\omega_1},\mathbb{I}^0_{P,\omega_1}\right)\otimes C^{|\omega_2|-l}\left(\mathbb{I}_{P,\omega_2},\mathbb{I}^0_{P,\omega_2}\right)\xrightarrow{\times} C^{|\omega_1|+|\omega_2|-k-l}\left((\mathbb{I}_{P,\omega_1},\mathbb{I}^0_{P,\omega_1})\times(\mathbb{I}_{P,\omega_2},\mathbb{I}^0_{P,\omega_2})\right)$$

$$C^{|\omega_1|+|\omega_2|-k-l}\left((\mathbb{I}_{P,\omega_1},\mathbb{I}^0_{P,\omega_1})\times(\mathbb{I}_{P,\omega_2},\mathbb{I}^0_{P,\omega_2})\right)\xrightarrow{i^*_{\omega_1,\omega_2}} C^{|\omega_1|+|\omega_2|-k-l}\left(\mathbb{I}_{P,\omega_1\sqcup\omega_2},\mathbb{I}^0_{P,\omega_1\sqcup\omega_2}\right).$$

For this we have

$$\begin{aligned}
&\xi_{\omega_1\sqcup\omega_2}\big((u_{\omega_1\backslash\sigma_1}v_{\sigma_1})(u_{\omega_2\backslash\sigma_2}v_{\sigma_2})\big)\\
&\qquad=(-1)^{l(\omega_1\backslash\sigma_1,\omega_2\backslash\sigma_2)}\xi_{\omega_1\sqcup\omega_2}\big(u_{(\omega_1\sqcup\omega_2)\backslash(\sigma_1\sqcup\sigma_2)}v_{\sigma_1\sqcup\sigma_2}\big)\\
&\qquad\qquad=(-1)^{l(\omega_1\backslash\sigma_1,\omega_2\backslash\sigma_2)}(-1)^{l(\sigma_1\sqcup\sigma_2,\omega_1\sqcup\omega_2)}E^*_{\sigma_1\sqcup\sigma_2}.
\end{aligned}$$

On the other hand

$$\begin{aligned}
&i^*_{\omega_1,\omega_2}\left(\xi_{\omega_1}(u_{\omega_1\backslash\sigma_1}v_{\sigma_1})\times\xi_{\omega_2}(u_{\omega_2\backslash\sigma_2}v_{\sigma_2})\right)\\
&\qquad=(-1)^{l(\sigma_1,\omega_1)}(-1)^{l(\sigma_2,\omega_2)}i^*_{\omega_1,\omega_2}(E^*_{\sigma_1}\times E^*_{\sigma_2})\\
&\qquad\qquad=(-1)^{l(\sigma_1,\omega_1)}(-1)^{l(\sigma_2,\omega_2)}(-1)^{l(\sigma_1,\sigma_2)}E^*_{\sigma_1\sqcup\sigma_2},
\end{aligned}$$

where the last equality follows from the following calculation:

$$(i_{\omega_1,\omega_2})_*(E_{\sigma_1\sqcup\sigma_2}) = (-1)^{l(\sigma_1,\sigma_2)}E_{\sigma_1} \times E_{\sigma_2}.$$

Now let us calculate the difference of signs:

$$\begin{aligned}
&(l(\omega_1 \setminus \sigma_1, \omega_2 \setminus \sigma_2) + l(\sigma_1 \sqcup \sigma_2, \omega_1 \sqcup \omega_2)) \\
&\qquad - (l(\sigma_1,\omega_1) + l(\sigma_2,\omega_2) + l(\sigma_1,\sigma_2)) \mod 2 \\
&= l(\omega_1 \setminus \sigma_1, \omega_2 \setminus \sigma_2) + l(\sigma_1,\omega_2) + l(\sigma_2,\omega_1) + l(\sigma_1,\sigma_2) \mod 2 \\
&= l(\omega_1 \setminus \sigma_1, \omega_2 \setminus \sigma_2) + l(\sigma_1,\omega_2 \setminus \sigma_2) + l(\sigma_2,\omega_1) \mod 2 \\
&= l(\omega_1, \omega_2 \setminus \sigma_2) + l(\sigma_2,\omega_1) \mod 2 \\
&= l(\omega_1,\omega_2 \setminus \sigma_2) + l(\omega_1,\sigma_2) + |\sigma_2||\omega_1| \mod 2 = l(\omega_1,\omega_2) + |\sigma_2||\omega_1| \mod 2 \\
&= l(\omega_2,\omega_1) + |\omega_1||\omega_2| + |\omega_1|(|\omega_2| - l) \mod 2 = l(\omega_2,\omega_1) + |\omega_1|l \mod 2. \square
\end{aligned}$$

6.7. *Description in terms of related simplicial complexes*

Definition 6.22: An *(abstract) simplicial complex* K on the vertex set $[m] = \{1,\dots,m\}$ is the set of subsets $K \subset 2^{[m]}$ such that

(1) $\varnothing \in K$;
(2) $\{i\} \in K$, for $i = 1,\dots,m$;
(3) If $\sigma \subset \tau$ and $\tau \in K$, then $\sigma \in K$.

The sets $\sigma \in K$ are called *simplices* . For an abstract simplicial complex K there is a *geometric realization* $|K|$ as a subcomplex in the simplex Δ^{m-1} with the vertex set $[m]$.

For a simple polytope P define an abstract simplicial complex K on the vertex set $[m]$ by the rule

$$\sigma \in K \text{ if and only if } \sigma = \sigma(G) = \{i \colon G \subset F_i\} \text{ for some } G \in L(P) \setminus \{\varnothing\}.$$

We have the combinatorial equivalence $K \simeq \partial P^*$. For any subset $\omega \subset [m]$ define the full subcomplex $K_\omega = \{\sigma \in K \colon \sigma \subset \omega\}$.

Definition 6.23: For two simplicial complexes K_1 and K_2 on the vertex sets $\mathrm{vert}(K_1)$ and $\mathrm{vert}(K_2)$ the *join* $K_1 * K_2$ is the simplicial complex on the vertex set $\mathrm{vert}(K_1) \sqcup \mathrm{vert}(K_2)$ with simplices $\sigma_1 \sqcup \sigma_2$, $\sigma_1 \in K_1$, $\sigma_2 \in K_2$.

A cone CK_ω is by definition $\{0\} * K_\omega$, where $\{0\}$ is the simplicial complex with one vertex $\{0\}$.

Proposition 6.24: *For any $\varnothing \neq \omega \subset [m]$ we have a homeomorphism of pairs*

$$(\mathbb{I}_{P,\omega}, \mathbb{I}^0_{P,\omega}) \simeq (C|K_\omega|, |K_\omega|).$$

Proof: For any simplex $\sigma \in K$ consider its barycenter $\boldsymbol{y}_\sigma \in |K|$. Then we have a barycentric subdivision of K consisting of simplices

$$\Delta_{\sigma_1 \subset \cdots \subset \sigma_k} = \operatorname{conv}\{\boldsymbol{y}_{\sigma_1}, \ldots, \boldsymbol{y}_{\sigma_k}\},\ k \geqslant 1.$$

Define the mapping $c_K\colon K \to \mathbb{I}^m$ as

$$c_K(\boldsymbol{y}_{\sigma_k}) = (y_1, \ldots, y_m), y_i = \begin{cases} 0, & i \in \sigma, \\ 1, & i \notin \sigma, \end{cases}$$

on the vertices of the barycentric subdivision, $c_K(\{0\}) = (1, \ldots, 1)$, and on the simplices and cones on simplices by linearity. This defines the piecewise linear homeomorphisms of pairs

$$(C|K|, |K|) \to (\mathbb{I}_P, \mathbb{I}^0_P), \text{ and } (C|K_\omega|, |K_\omega|) \to (\mathbb{I}_{P,\omega}, \mathbb{I}^0_{P,\omega}). \qquad \square$$

Corollary 6.25: *We have the homotopy equivalence $P_\omega \sim |K_\omega|$.*

For the simplicial complex K_ω we have the simplicial chain complex with the free abelian groups of chains $C_i(K_\omega)$, $i \geqslant -1$, generated by simplices $\sigma \in K_\omega$, $|\sigma| = i + 1$, (including the empty simplex $\varnothing$, $|\varnothing| = 0$), and the boundary homomorphism

$$d\colon C_i(K_\omega) \to C_{i-1}(K_\omega), \quad d\sigma = \sum_{i \in \sigma} (-1)^{l(i,\sigma)} (\sigma \setminus \{i\}).$$

There is the cochain complex of groups $C^i(K_\omega) = \operatorname{Hom}(C_i(K_\omega), \mathbb{Z})$. Define the cochain σ^* by the rule $\langle \sigma^*, \tau \rangle = \delta(\sigma, \tau)$. The coboundary homomorphism $\partial = d^*$ can be calculated by the rule

$$\partial \sigma^* = \sum_{j \in \omega \setminus \sigma, \sigma \sqcup \{j\} \in K_\omega} (-1)^{l(j,\sigma)} (\sigma \sqcup \{j\})^*.$$

The homology groups of the chain and cochain complexes are $\widetilde{H}_i(K_\omega)$ and $\widetilde{H}^i(K_\omega)$ respectively. The following result is proved similarly to Theorem 6.16 and Theorem 6.21.

Theorem 6.26: *For any $\omega \subset [m]$ the mapping*

$$\widehat{\xi}_\omega\colon R^{-i,2\omega} \to C^{|\omega|-i-1}(K_\omega), \quad \widehat{\xi}_\omega(u_{\omega \setminus \sigma} v_\sigma) = (-1)^{l(\sigma,\omega)} \sigma^*$$

is the isomorphism of cochain complexes

$$\{C^{-i,2\omega}(\mathcal{Z}_P)\}_{i \geqslant 0} \quad \text{and} \quad \{C^{|\omega|-i-1}(K_\omega)\}_{i \geqslant 0}.$$

It induces the isomorphism $H^{-i,2\omega}(\mathcal{Z}_P) \simeq \widetilde{H}^{|\omega|-i-1}(K_\omega)$ *and the isomorphism of rings*

$$H^*(\mathcal{Z}_P) \simeq \bigoplus_{\omega\subset[m]} \widetilde{H}^*(K_\omega),$$

where the multiplication on the right hand side

$$\widetilde{H}^p(K_{\omega_1}) \otimes \widetilde{H}^q(K_{\omega_2}) \to \widetilde{H}^{p+q+1}(K_{\omega_1\cup\omega_2})$$

is trivial if $\omega_1 \cap \omega_2 \neq \varnothing$, *and for the case* $\omega_1 \cap \omega_2 = \varnothing$ *is given by the mapping of cochains defined by the rule*

$$\sigma_1^* \otimes \sigma_2^* \mapsto (-1)^{l(\omega_1,\omega_2)+l(\sigma_1,\sigma_2)+|\omega_1||\sigma_2|}(\sigma_1 \sqcup \sigma_2)^*.$$

Remark 6.27: The calculation of the sign in the multiplication appears in [14] in a different way. In [29] there is another calculation that gives the result we present.

6.8. *Description in terms of unions of facets modulo boundary*

The embeddings $b_P\colon P \to \mathbb{I}_P$ and $c_K\colon K \to \mathbb{I}_P^0$ define the simplicial isomorphism of barycentric subdivisions of ∂P and K: the vertex $\boldsymbol{y}_\sigma$, $\sigma \neq \varnothing$, is mapped to the vertex $\boldsymbol{x}_{G(\sigma)}$ and on simplices we have the linear isomorphism. Then K_ω is embedded into P_ω.

For the set P_ω, considered in the space ∂P, the boundary ∂P_ω consists of all points $\boldsymbol{x} \in P_\omega$ such that $\boldsymbol{x} \in F_j$ for some $j \notin \omega$. Hence ∂P_ω consists of all faces $G \subset P$ such that $\sigma(G) \cap \omega \neq \varnothing$ and $\sigma(G) \not\subset \omega$.

Define on P the orientation induced from $\mathbb{R}^n$, and on ∂P the orientation induced from P by the rule: a basis $(\boldsymbol{e}_1, \dots, \boldsymbol{e}_{n-1})$ in ∂P is positively oriented if and only if the basis $(\boldsymbol{n}, \boldsymbol{e}_1, \dots, \boldsymbol{e}_{n-1})$ is positively oriented, where $\boldsymbol{n}$ is the outward pointing normal vector.

We have the orientation of simplices in K_ω defined by the canonical order of the vertices of the set $\omega \subset [m]$. We have the cellular structure on P_ω defined by the faces of P. Fix some orientation of faces in P such that for facets the orientation coincides with ∂P. For a cell E with fixed orientation in some cellular or simplicial structure it is convenient to consider the chain $-E$ as a cell with an opposite orientation. Then the boundary operator just sends the cell to the sum of cells on the boundary with induced orientations.

Lemma 6.28: *The orientation of the simplex* $\sigma = \{i_1, \dots, i_l\} \in |K_\omega|$ *coincides with the orientation of the simplex*

$$\operatorname{conv}\{\boldsymbol{y}_\sigma, \boldsymbol{y}_{\sigma\setminus\{i_1\}}, \boldsymbol{y}_{\sigma\setminus\{i_1,i_2\}}, \dots, \boldsymbol{y}_{\{i_l\}}\}.$$

The proof we leave as an exercise.

Now we establish the Poincare duality between the groups $\widetilde{H}^i(K_\omega)$ and $H_{n-i-1}(P_\omega, \partial P_\omega)$.

Definition 6.29: For a face $G \subset P_\omega$, $G \not\subset \partial P_\omega$, with a positively oriented basis $(\boldsymbol{e}_1, \dots, \boldsymbol{e}_k)$ and a simplex $\sigma \in K_\omega$ define the *intersection index*

$$C_*(P_\omega, \partial P_\omega) \otimes C_*(K_\omega) \to \mathbb{Z}$$

by the rule

$$\langle G, \sigma\rangle = \begin{cases} 0, & \text{if } G(\sigma) \neq G; \\ 1, & \text{if } G(\sigma) = G, \text{and the basis } (\boldsymbol{e}_1, \dots, \boldsymbol{e}_k, \boldsymbol{h}_1, \dots, \boldsymbol{h}_l) \text{ is positive;} \\ -1, & \text{if } G(\sigma) = G, \text{and the basis } (\boldsymbol{e}_1, \dots, \boldsymbol{e}_k, \boldsymbol{h}_1, \dots, \boldsymbol{h}_l) \text{ is negative,} \end{cases}$$

where $l = n - k - 1$, and $(\boldsymbol{h}_1, \dots, \boldsymbol{h}_l)$ is any basis defining the orientation of any maximal simplex in the barycentric subdivision of $\sigma \subset P_\omega$ consistent with the orientation of σ, for example

$$(\boldsymbol{h}_1, \dots, \boldsymbol{h}_l) = (\boldsymbol{y}_{\sigma\setminus\{i_1\}} - \boldsymbol{y}_\sigma, \boldsymbol{y}_{\sigma\setminus\{i_1,i_2\}} - \boldsymbol{y}_\sigma, \dots, \boldsymbol{y}_{\{i_l\}} - \boldsymbol{y}_\sigma).$$

Proposition 6.30: *We have* $\langle dG, \tau\rangle = (-1)^{\dim G}\langle G, d\tau\rangle$.

Proof: Both left and right sides are equal to zero, if $\tau \neq \sigma(G) \sqcup \{j\}$ for some $j \in \omega\setminus\sigma$. Let $\tau = \sigma(G) \sqcup \{j\}$. Then $\tau = \sigma(G_j)$ for $G_j = G \cap F_j$. Let $\sigma = \sigma(G)$. The vector corresponding to $\boldsymbol{u}_j = \boldsymbol{y}_{\sigma\sqcup\{j\}} - \boldsymbol{y}_\sigma$ and the outward normal vector to the facet σ of the simplex $\sigma \sqcup \{j\}$ look to opposite sides of affσ in aff$(\sigma \sqcup \{j\})$ in the geometric realization of K; hence the orientation of the basis $(\boldsymbol{u}_j, \boldsymbol{h}_1, \dots, \boldsymbol{h}_l)$ is negative in $\sigma\sqcup\{j\}$. On the other hand, $\boldsymbol{u}_j = \boldsymbol{x}_{G(\sigma\sqcup\{j\})} - \boldsymbol{x}_{G(\sigma)}$; hence this vector looks to the same side of aff(G_j) in aff(G) with the outer normal vector to G_j, the orientation of the basis $(\boldsymbol{u}_j, \boldsymbol{g}_1, \dots, \boldsymbol{g}_{k-1})$ is positive for the basis $(\boldsymbol{g}_1, \dots, \boldsymbol{g}_{k-1})$ defining the induced orientation of G_j. Hence for the induced orientations of G_j and σ we have

- $\langle G \cap F_j, \sigma \sqcup \{j\}\rangle$ is opposite to the sign of the orientation of $(\boldsymbol{g}_1, \dots, \boldsymbol{g}_{k-1}, \boldsymbol{u}_j, \boldsymbol{h}_1, \dots, \boldsymbol{h}_l)$;
- $\langle G, \sigma \sqcup \{j\}\rangle$ coinsides with the sign of the orientation of $(\boldsymbol{u}_j, \boldsymbol{g}_1, \dots, \boldsymbol{g}_{k-1}, \boldsymbol{h}_1, \dots, \boldsymbol{h}_l)$.

Hence these numbers differ by the sign $(-1)^k$. □

Definition 6.31: Set

$$\widehat{H}_i(P_\omega, \partial P_\omega) = \begin{cases} H_i(P_\omega, \partial P_\omega), & 0 \leqslant i \leqslant n-2; \\ H_{n-1}(P_\omega, \partial P_\omega)/([\sum\limits_{i\in\omega} F_i]), & i = n-1. \end{cases}$$

Theorem 6.32: *The mapping $G \to \langle G, \sigma(G)\rangle \sigma(G)^*$ induces the isomorphism*

$$\widehat{H}_{n-i-1}(P_\omega, \partial P_\omega) \simeq \widetilde{H}^i(K_\omega),\ 0 \leqslant i \leqslant n-1,\ \omega \neq \varnothing.$$

Moreover, for $\omega_1 \cap \omega_2 = \varnothing$ the multiplication

$$\widehat{H}_{n-p-1}(P_{\omega_1}, \partial P_{\omega_1}) \otimes \widehat{H}_{n-q-1}(P_{\omega_2}, \partial P_{\omega_2}) \to \widehat{H}_{n-(p+q)-2}(P_{\omega_1 \sqcup \omega_2}, \partial P_{\omega_1 \sqcup \omega_2})$$

induced by the isomorphism is defined by the rule

$$G_1 \otimes G_2 \longmapsto$$

$$\frac{\langle G_1, \sigma(G_1)\rangle \langle G_2, \sigma(G_2)\rangle}{\langle G_1 \cap G_2, \sigma(G_1 \cap G_2)\rangle} (-1)^{l(\omega_1,\omega_2)+|\omega_1|(n-\dim G_2)+l(\sigma(G_1),\sigma(G_2))} G_1 \cap G_2.$$

The proof follows directly from Proposition 6.30.

6.9. *Geometrical interpretation of the cohomological groups*

Let P be a simple polytope. From Corollary 6.25 we obtain the following results:

Proposition 6.33:

(1) If $\omega = \varnothing$, then $P_\omega = \varnothing$; hence

$$H^{-i,2\varnothing}(\mathcal{Z}_P) = \widetilde{H}^{-i-1}(P_\omega) = \begin{cases} \mathbb{Z}, & i = 0; \\ 0, & \textit{otherwise}. \end{cases}$$

(2) If $G(\omega) \neq \varnothing$, then P_ω is contractible; hence

$$H^{-i,2\omega}(\mathcal{Z}_P) = \widetilde{H}^{|\omega|-i-1}(P_\omega) = 0, \quad \textit{for all } i.$$

In particular, this is the case for $|\omega| = 1$.

(3) If $\omega = \{p, q\}$, then either P_ω is contractible, if $F_p \cap F_q \neq \varnothing$, or $P_\omega = F_p \sqcup F_q$, where both F_p and F_q are contractible, if $F_p \cap F_q = \varnothing$. Hence

$$H^{-i,2\{p,q\}}(\mathcal{Z}_P) = \widetilde{H}^{1-i}(P_\omega) \begin{cases} \mathbb{Z}, & i = 1, F_p \cap F_q \neq \varnothing; \\ 0, & \textit{otherwise}. \end{cases}$$

(4) If $G(\omega) = \varnothing$ and $\omega \neq \varnothing$, then $\dim K_\omega \leqslant \min\{n-1, |\omega|-2\}$; hence

$$H^{-i,2\omega}(\mathcal{Z}_P) = \widetilde{H}^{|\omega|-i-1}(P_\omega) = 0, \textit{ for } |\omega| - i - 1 > \min\{n-1, |\omega|-2\}.$$

(5) If $\omega = [m]$, *then* $P_{[m]} = \partial P \simeq S^{n-1}$; *hence*

$$H^{-i,2[m]}(\mathcal{Z}_P) = \widetilde{H}^{m-i-1}(P_\omega) = \begin{cases} \mathbb{Z}, & i = m-n; \\ 0, & \textit{otherwise.} \end{cases}$$

(6) P_ω *is a subcomplex in* $\partial P \simeq S^{n-1}$; *hence*

$$\widetilde{H}^{n-1}(P_\omega) = \begin{cases} \mathbb{Z}, & \omega = [m]; \\ 0, & \textit{otherwise.} \end{cases}$$

(7) $H^{0,2\omega}(\mathcal{Z}_P) = \widetilde{H}^{|\omega|-1}(K_\omega) = \begin{cases} \mathbb{Z}, & \omega = \varnothing; \\ 0, & \textit{otherwise.} \end{cases}$

Corollary 6.34: *For* $k \geqslant 0$ *we have*

$$H^k(\mathcal{Z}_P) = \bigoplus_\omega \widetilde{H}^{k-1-|\omega|}(P_\omega).$$

More precisely,

$$H^0(\mathcal{Z}_P) = \widetilde{H}^{-1}(\varnothing) = \mathbb{Z} = \widetilde{H}^{n-1}(P_{[m]}) = H^{m+n}(\mathcal{Z}_P),$$

and for $0 < k < m+n$ *we have*

$$H^k(\mathcal{Z}_P) = \bigoplus_{\max\{\lceil \frac{k+1}{2}\rceil, k-n+1\} \leqslant |\omega| \leqslant \min\{k-1,m-1\}, G(\omega)=\varnothing} \widetilde{H}^{k-1-|\omega|}(P_\omega).$$

In particular,

$$H^1(\mathcal{Z}_P) = H^2(\mathcal{Z}_P) = 0 = H^{m+n-2}(\mathcal{Z}_P) = H^{m+n-1}(\mathcal{Z}_P);$$

$$H^3(\mathcal{Z}_P) \simeq \bigoplus_{|\omega|=2} \widetilde{H}^0(P_\omega) = \bigoplus_{F_i \cap F_j = \varnothing} \mathbb{Z} \simeq H^{m+n-3}(\mathcal{Z}_P);$$

$$H^4(\mathcal{Z}_P) \simeq \bigoplus_{|\omega|=3} \widetilde{H}^0(P_\omega) \simeq H^{m+n-4}(\mathcal{Z}_P);$$

$$H^5(\mathcal{Z}_P) \simeq \bigoplus_{|\omega|=3} \widetilde{H}^1(P_\omega) + \bigoplus_{|\omega|=4} \widetilde{H}^0(P_\omega) \simeq H^{m+n-5}(\mathcal{Z}_P);$$

$$H^6(\mathcal{Z}_P) \simeq \bigoplus_{|\omega|=4} \widetilde{H}^1(P_\omega) + \bigoplus_{|\omega|=5} \widetilde{H}^0(P_\omega);$$

$$H^7(\mathcal{Z}_P) \simeq \bigoplus_{|\omega|=4} \widetilde{H}^2(P_\omega) + \bigoplus_{|\omega|=5} \widetilde{H}^1(P_\omega) + \bigoplus_{|\omega|=6} \widetilde{H}^0(P_\omega).$$

Proof: From Corollary 6.17 we obtain

$$H^k(\mathcal{Z}_P) = \bigoplus_{2|\omega|-i=k} H^{-i,2\omega}(\mathcal{Z}_P) \simeq \bigoplus_{2|\omega|-i=k} \widetilde{H}^{|\omega|-i-1}(P_\omega) = \bigoplus_{|\omega| \leqslant k} \widetilde{H}^{k-|\omega|-1}(P_\omega).$$

If $|\omega| = 0$, then $\widetilde{H}^{k-|\omega|-1}(P_\omega) = \widetilde{H}^{k-1}(\varnothing) = \begin{cases} \mathbb{Z}, & k = 0; \\ 0, & \text{otherwise.} \end{cases}$

If $|\omega| = k$, then $\widetilde{H}^{k-|\omega|-1}(P_\omega) = \widetilde{H}^{-1}(P_\omega) = \begin{cases} \mathbb{Z}, & k = 0; \\ 0, & \text{otherwise.} \end{cases}$

Thus we have $H^0(\mathcal{Z}_P) = \widetilde{H}^{-1}(\varnothing) = \mathbb{Z}$, and for $k > 0$ nontrivial summands appear only for $0 < |\omega| < k$, and $k - 1 - |\omega| \leqslant \dim K_\omega \leqslant \min\{n-1, |\omega| - 2\}$. Hence $|\omega| \geqslant \max\{k-n, \lceil \frac{k+1}{2} \rceil\}$.

If $|\omega| = k-n$, then $\widetilde{H}^{k-|\omega|-1}(P_\omega) = \widetilde{H}^{n-1}(P_\omega) = \begin{cases} \mathbb{Z}, & |\omega| = m, k = m+n; \\ 0, & \text{otherwise.} \end{cases}$

If $k = m + n$, then $|\omega| \geqslant m$; hence $|\omega| = m$, $H^{m+n}(\mathcal{Z}_P) = \widetilde{H}^{n-1}(\partial P) = \mathbb{Z}$.

If $|\omega| = m$, then $\widetilde{H}^{k-|\omega|-1}(P_\omega) = \widetilde{H}^{k-m-1}(\partial P) = \begin{cases} \mathbb{Z}, & k = m+n; \\ 0, & \text{otherwise.} \end{cases}$

Thus, for $0 < k < m + n$ nontrivial summands appear only for

$$\max\left\{k - n + 1, \left\lceil \frac{k+1}{2} \right\rceil\right\} \leqslant |\omega| \leqslant \min\{k-1, m-1\}.$$

If $|\omega| = 1$, then $\widetilde{H}^{k-|\omega|-1}(P_\omega) = 0$ for all k.

If $|\omega| = 2$, then $\widetilde{H}^{k-|\omega|-1}(P_\omega) = \begin{cases} \mathbb{Z}, & k = 3 \text{ and } G(\omega) = \varnothing; \\ 0, & \text{otherwise.} \end{cases}$

Thus, for $k = 3, 4, 5, 6, 7$ we have the left parts of the formulas above; in particular the corresponding cohomology groups have no torsion. From the universal coefficient formula the homology groups $H_k(\mathcal{Z}_P)$, $k \leqslant 5$, have no torsion. Then the right parts follow from Poincare duality. □

Corollary 6.35: *If the group* $H^k(\mathcal{Z}_P)$ *has torsion, then* $7 \leqslant k \leqslant m + n - 6$.

7. Moment-angle manifolds of 3-polytopes

7.1. *Corollaries of general results*

From Corollary 6.34 for a 3-polytope P we have

Proposition 7.1:

$$H^0(\mathcal{Z}_P) = \widetilde{H}^{-1}(\varnothing) = \mathbb{Z} = \widetilde{H}^2(P_{[m]}) = H^{m+3}(\mathcal{Z}_P);$$
$$H^1(\mathcal{Z}_P) = H^2(\mathcal{Z}_P) = 0 = H^{m+1}(\mathcal{Z}_P) = H^{m+2}(\mathcal{Z}_P);$$
$$H^3(\mathcal{Z}_P) \simeq \bigoplus_{|\omega|=2} \widetilde{H}^0(P_\omega) = \bigoplus_{F_i \cap F_j = \varnothing} \mathbb{Z} \simeq H^m(\mathcal{Z}_P);$$
$$H^4(\mathcal{Z}_P) \simeq \bigoplus_{|\omega|=3, G(\omega)=\varnothing} \widetilde{H}^0(P_\omega) \simeq H^{m-1}(\mathcal{Z}_P);$$
$$H^k(\mathcal{Z}_P) \simeq \oplus \bigoplus_{|\omega|=k-2} \widetilde{H}^1(P_\omega) \oplus \bigoplus_{|\omega|=k-1} \widetilde{H}^0(P_\omega), \text{ for } 5 \leqslant k \leqslant m-2.$$

In particular, $H^*(\mathcal{Z}_P)$ has no torsion, and so $H^k(\mathcal{Z}_P) \simeq H^{m+3-k}(\mathcal{Z}_P)$.

Proposition 7.2: *For a 3-polytope P, nonzero Betti numbers could be*

$$\operatorname{rank} \widetilde{H}^{-1}(\varnothing) = \beta^{0,2\varnothing} = 1 = \beta^{-(m-3),2[m]} = \operatorname{rank} \widetilde{H}^2(\partial P);$$
$$\operatorname{rank} \widetilde{H}^0(P_\omega) = \beta^{-i,2\omega} = \beta^{-(m-3-i),2([m]\setminus\omega)} = \operatorname{rank} \widetilde{H}^1(P_{[m]\setminus\omega}),$$
$$|\omega| = i+1, i = 1, \dots, m-4.$$

The proof we leave as an exercise.

For $|\omega| = i+1$ the number $\beta^{-i,2\omega} + 1$ is equal to the number of connected components of the set $P_\omega \subset P$.

Definition 7.3: *Bigraded Betti numbers* are defined as

$$\beta^{-i,2j} = \operatorname{rank} H^{-i,2j}(\mathcal{Z}_P) = \sum_{|\omega|=j} \beta^{-i,2\omega}.$$

Exercise: $\beta^{-1,4} = \frac{m(m-1)}{2} - f_1 = \frac{(m-3)(m-4)}{2}$.

Proposition 7.4: *Let $\omega \subset [m]$ and P_ω be connected. Then topologically P_ω is a sphere with k holes bounded by connected components η_i of ∂P_ω, which are simple edge cycles.*

Proof: It is easy to prove that P_ω is an orientable 2-manifold with boundary, which proves the statement. □

Let the 3-polytope P have the standard orientation induced from $\mathbb{R}^3$, and the boundary ∂P have the orientation induced from P by the rule: the basis $(\boldsymbol{e}_1, \boldsymbol{e}_2)$ in ∂P is positively oriented if and only if the basis $(n, \boldsymbol{e}_1, \boldsymbol{e}_2)$ is positively oriented in P, where n is the outward pointing normal vector. Then any set P_ω is an oriented surface with the boundary ∂P_ω consisting of simple edge cycles. We now describe the Poincare duality given by Theorem 6.32. We have the orientation of simplices in K_ω defined by the canonical order of the vertices induced from the set $\omega \subset [m]$. We have the cellular structure on P_ω defined by vertices, edges and facets of P. Orient the faces of P by the following rule:

- facets F_i orient similarly to ∂P;
- for $i < j$ orient the edge $F_i \cap F_j$ in such a way that the pair of vectors $(F_i \cap F_j, \boldsymbol{y}_{\{j\}} - \boldsymbol{y}_{\{i,j\}})$ has a positive orientation in F_j;
- for $i < j < k$ assign $+$ to the vertex $F_i \cap F_j \cap F_k$, if the pair of vectors $(\boldsymbol{y}_{\{j,k\}} - \boldsymbol{y}_{\{i,j,k\}}, \boldsymbol{y}_{\{k\}} - \boldsymbol{y}_{\{i,j,k\}})$ has positive orientation in F_k, and – otherwise.

Corollary 7.5: *The mapping*

$$C^i(K_\omega) \to C_{2-i}(P_\omega, \partial P_\omega), \quad \sigma^* \mapsto G(\sigma)$$

defines an isomorphism

$$\widetilde{H}^i(K_\omega) \simeq \widehat{H}_{2-i}(P_\omega, \partial P_\omega).$$

We have the following computations.

Proposition 7.6: *For the set ω, let $P_\omega = P_{\omega^1} \sqcup \cdots \sqcup P_{\omega^s}$ be the decomposition into connected components. Then*

(1) $H_0(P_\omega, \partial P_\omega) = 0$ for $\omega \neq [m]$, and $H_0(\partial P, \varnothing) = \mathbb{Z}$ for $\omega = [m]$ with the basis $[v]$, where $v \in P$ is any vertex with the orientation $+$.

(2) $H_1(P_\omega, \partial P_\omega) = \bigoplus\limits_{i=1}^{s} H_1(P_{\omega^i}, \partial P_{\omega^i})$, and $H_1(P_{\omega^i}, \partial P_{\omega^i}) \simeq \mathbb{Z}^{q_i - 1}$, where q_i is the number of cycles in ∂P_{ω^i}. The basis is given by any set of edge paths in P_{ω^i} connecting one fixed boundary cycle with other boundary cycles.

(3) $H_2(P_\omega, \partial P_\omega)/(\sum\limits_{i \in \omega}[F_i]) \simeq \mathbb{Z}^s/(1, 1, \ldots, 1)$, where $\mathbb{Z}^s$ has the basis $e_{\omega^j} = [\sum\limits_{i \in \omega^j} F_i]$.

The nontrivial multiplication is defined by the following rule: each set P_{ω^j} is a sphere with holes. If $\omega_1 \cap \omega_2 = \varnothing$, then $P_{\omega_1^i} \cap P_{\omega_2^j}$ is the intersection of a boundary cycle in $\partial P_{\omega_1^i}$ with a boundary cycle in $\partial P_{\omega_2^j}$, which is the union $\gamma_1 \sqcup \cdots \sqcup \gamma_l$ of edge-paths.

Proposition 7.7: *We have $e_{\omega_1^i} \cdot e_{\omega_2^j} = 0$, if $P_{\omega_1^i} \cap P_{\omega_2^j} = \varnothing$. Else up to the sign $(-1)^{l(\omega_1, \omega_2) + |\omega_1|}$ it is the sum of the elements $[\gamma_i]$ given by the paths with the orientations such that an edge on the path and the transversal edge lying in one facet and oriented from $P_{\omega_1^i}$ to $P_{\omega_2^j}$ form a positively oriented pair of vectors.*

Proof: For the facets $F_i \in P_{\omega_1}$ and $F_j \in P_{\omega_2}$ we have

$$F_i \otimes F_j \mapsto (-1)^{l(\omega_1,\omega_2)+|\omega_1|}(-1)^{l(i,j)} F_i \cap F_j,$$

where the pair of vectors $\left((-1)^{l(i,j)} F_i \cap F_j, \boldsymbol{y}_j - \boldsymbol{y}_{\{i,j\}}\right)$ is positively oriented in F_j. □

Proposition 7.8: *Let $\omega_1 \sqcup \omega_2 = [m]$, and let the element $[\gamma]$ correspond to the oriented edge path γ, connecting two boundary cycles of $P_{\omega_2^j}$. Then $e_{\omega_1^i} \cdot [\gamma] = 0$, if $P_{\omega_1^i} \cap \gamma = \varnothing$, and up to the sign $(-1)^{l(\omega_1,\omega_2)}$ it is $+1$, if γ starts at $P_{\omega_1^i}$, and -1, if γ ends at $P_{\omega_1^i}$.*

Proof:

$$F_i \otimes (F_j \cap F_k) \mapsto (-1)^{l(\omega_1,\omega_2)}(-1)^{l(i,\{j,k\})} F_i \cap F_j \cap F_k,$$

where $(-1)^{l(i,\{j,k\})} F_i \cap F_j \cap F_k$ is the vertex $F_i \cap F_j \cap F_k$ with the sign $+$, if $F_j \cap F_k$ starts at F_i, and $-$, if $F_j \cap F_k$ ends at F_i. □

7.2. *k-belts and Betti numbers*

Definition 7.9: For any k-belt $\mathcal{B}_k = \{F_{i_1}, \dots, F_{i_k}\}$ define $\omega(\mathcal{B}_k) = \{i_1, \dots, i_k\}$, and $\widetilde{\mathcal{B}_k}$ to be the generator in the group

$$\mathbb{Z} \simeq H^{-(k-2),2\omega}(\mathcal{Z}_P) \simeq H^1(P_\omega) \simeq H^1(K_\omega) \simeq H_1(P_\omega, \partial P_\omega),$$

where $\omega = \omega(\mathcal{B}_k)$.

Remark 7.10: It is easy to prove that $\mathcal{B}_k$ is a k-belt if and only if $K_{\omega(\mathcal{B}_k)}$ is combinatorially equivalent to the boundary of a k-gon.

Let P be a simple 3-polytope with m facets.

Proposition 7.11: *Let $\omega = \{i, j, k\} \subset [m]$. Then*

$$H^{-1,2\omega}(\mathcal{Z}_P) = \begin{cases} \mathbb{Z}, & (F_i, F_j, F_k) \text{ is a 3-belt;} \\ 0, & \text{otherwise .} \end{cases}$$

In particular, $\beta^{-1,6}$ is equal to the number of 3-belts, and the set of elements $\{\widetilde{\mathcal{B}_3}\}$ is a basis in $H^{-1,6}(\mathcal{Z}_P)$.

Proof: We have $H^{-1,2\omega}(\mathcal{Z}_P) \simeq \widetilde{H}^1(K_\omega)$. Consider all possibilities for the simplicial complex K_ω on 3 vertices. If $\{i, j, k\} \in K_\omega$, then K_ω is a 3-simplex, and it is contractible. Else K_ω is a graph. If K_ω has no cycles, then each connected component is a tree, else K_ω is a cycle with 3 vertices. □

Proposition 7.12: *Let P be a simple 3-polytope without 3-belts, and $\omega \subset [m]$, $|\omega| = 4$. Then*

$$H^{-2,2\omega}(\mathcal{Z}_P) = \begin{cases} \mathbb{Z}, & \omega = \omega(\mathcal{B}) \text{ for some 4-belt } \mathcal{B}; \\ 0, & \text{otherwise}, \end{cases}$$

where the belt $\mathcal{B}$ is defined in a unique way (we will denote it $\mathcal{B}(\omega)$). In particular, $\beta^{-2,8}$ is equal to the number of 4-belts, and the set of elements $\{\widetilde{\mathcal{B}_4}\}$ is a basis in $H^{-2,8}(\mathcal{Z}_P)$.

Proof: We have $H^{-2,2\omega}(\mathcal{Z}_P) \simeq \widetilde{H}^1(K_\omega)$. Consider the 1-skeleton K_ω^1. If it has no cycles, then $K_\omega = K_\omega^1$ is a disjoint union of trees. If K_ω^1 has a 3-cycle on vertices $\{i,j,k\}$, then $\{i,j,k\} \in K_\omega$. Let $l = \omega \setminus \{i,j,k\}$. Then l is either disconnected from $\{i,j,k\}$, or connected to it by one edge, or connected to it by two edges, say $\{i,l\}$ and $\{j,l\}$, with $\{i,j,l\} \in K_\omega$, or connected to it by three edges with $K_\omega \simeq \partial\Delta^3$. In all these cases $\widetilde{H}^1(K_\omega) = 0$. If K_ω^1 has no 3-cycles, but has a 4-cycle $\{i,j\},\{j,k\},\{k,l\},\{l,i\}$, then K_ω coincides with this cycle and (F_i,F_j,F_k,F_l) is a 4-belt. This proves the statement. □

Theorem 7.13: *Let P be a simple 3-polytope without 3-belts and 4-belts, and $\omega \subset [m]$, $|\omega| = 5$. Then*

$$H^{-3,2\omega}(\mathcal{Z}_P) = \begin{cases} \mathbb{Z}, & \omega = \omega(\mathcal{B}) \text{ for some 5-belt } \mathcal{B}; \\ 0, & \text{otherwise}, \end{cases}$$

where the belt $\mathcal{B}$ is defined in a unique way (we will denote it $\mathcal{B}(\omega)$). In particular, $\beta^{-3,10}$ is equal to the number of 5-belts, and the set of elements $\{\widetilde{\mathcal{B}_5}\}$ is a basis in $H^{-3,10}(\mathcal{Z}_P)$.

Proof: We have $H^{-3,2\omega}(\mathcal{Z}_P) \simeq \widetilde{H}^1(K_\omega)$. Since $\widetilde{H}^1(K_\omega) = 0$ for $|\omega| \leqslant 2$, from Propositions 7.11 and 7.12 we have $\widetilde{H}^1(K_\omega) = 0$, if K_ω is disconnected. Let it be connected. Consider the sphere with holes P_ω. If $H^1(P_\omega) \neq 0$, then there are at least two holes. Consider a simple edge cycle γ bounding one of the holes. Walking round γ we obtain a k-loop $\mathcal{L}_k = (F_{i_1},\dots,F_{i_k})$, $k \geqslant 3$ in P_ω. If $k = 3$, then the absence of 3-belts implies that $F_{i_1} \cap F_{i_2} \cap F_{i_3}$ is a vertex; hence $P_\omega = \{F_{i_1},F_{i_2},F_{i_3}\}$, which is a contradiction. If $k = 4$, then the absence of 4-belts implies that $F_{i_1} \cap F_{i_3} \neq \varnothing$, or $F_{i_2} \cap F_{i_4} \neq \varnothing$. Without loss of generality let $F_{i_1} \cap F_{i_3} \neq \varnothing$. Then $F_{i_1} \cap F_{i_2} \cap F_{i_3}$ and $F_{i_3} \cap F_{i_4} \cap F_{i_1}$ are vertices; hence

$P_\omega = \{F_{i_1}, F_{i_2}, F_{i_3}, F_{i_4}\}$, which is a contradiction. Let $k = 5$. If $\mathcal{L}_5$ is not a 5-belt, then some two nonsuccessive facets intersect. They are adjacent to some facet of $\mathcal{L}_5$. Without loss of generality let it be F_{i_2}, and $F_{i_1} \cap F_{i_3} \neq \varnothing$. Then $F_{i_1} \cap F_{i_2} \cap F_{i_3}$ is a vertex. The absence of 4-belts implies that $F_{i_3} \cap F_{i_5} \neq \varnothing$, or $F_{i_4} \cap F_{i_1} \neq \varnothing$. Without loss of generality let $F_{i_3} \cap F_{i_5} \neq \varnothing$. Then $F_{i_3} \cap F_{i_4} \cap F_{i_5}$ and $F_{i_1} \cap F_{i_3} \cap F_{i_5}$ are vertices, and P_ω is a disc bounded by γ. A contradiction. Thus $\mathcal{L}_5$ is a 5-belt, and $H^1(P_\omega) \simeq \mathbb{Z}$ generated by $\widetilde{\mathcal{L}_5}$. □

Proposition 7.14: *Any simple 3-polytope $P \neq \Delta^3$ has either a 3-belt, or a 4-belt, or a 5-belt.*

Proof: If $P \neq \Delta^3$ has no 3-belts, then it is a flag polytope and any facet of P is surrounded by a belt. Theorem 3.12 implies that any flag simple 3-polytope has a quadrangular or pentagonal facet. This finishes the proof. □

Corollary 7.15: *For a fullerene P*

- $\beta^{-1,6} = 0$ – *the number of 3-belts;*
- $\beta^{-2,8} = 0$ – *the number of 4-belts;*
- $\beta^{-3,10} = 12 + k$, $k \geqslant 0$, – *the number of 5-belts. If $k > 0$, then $p_6 = 5k$;*
- *the product mapping $H^3(\mathcal{Z}_P) \otimes H^3(\mathcal{Z}_P) \to H^6(\mathcal{Z}_P)$ is trivial.*

7.3. *Relations between Betti numbers*

Theorem 7.16: *(Theorem 4.6.2, [14]) For any simple polytope P with m facets*

$$(1 - t^2)^{m-n}(h_0 + h_1 t^2 + \cdots + h_n t^{2n}) = \sum_{-i,2j} (-1)^i \beta^{-i,2j} t^{2j},$$

where $h_0 + h_1 t + \cdots + h_n t^n = (t-1)^n + f_{n-1}(t-1)^{n-1} + \cdots + f_0$.

Corollary 7.17: *Set $h = m - 3$. For a simple 3-polytope $P \neq \Delta^3$ with m facets*

$$\begin{aligned}(1 - t^2)^h&(1 + ht^2 + ht^4 + t^6)\\ &= 1 - \beta^{-1,4}t^4 + \sum_{j=3}^{h}(-1)^{j-1}(\beta^{-(j-1),2j} - \beta^{-(j-2),2j})t^{2j}\\ &\qquad + (-1)^{h-1}\beta^{-(h-1),2(h+1)}t^{2(h+1)} + (-1)^h t^{2(h+3)}.\end{aligned}$$

Exercise: For any simple 3-polytope P we have:

- $\beta^{-1,4}$ – the number of pairs (F_i, F_j), $F_i \cap F_j = \varnothing$;

- $\beta^{-1,6}$ – the number of 3-belts;
- $\beta^{-2,6} = \sum\limits_{i<j<k} s_{i,j,k}$, where $s_{i,j,k} + 1$ is equal to the number of connected components of the set $F_i \cup F_j \cup F_k$;
- $\beta^{-3,8} = \sum\limits_{i<j<k<r} s_{i,j,k,r}$, where $s_{i,j,k,r} + 1$ is equal to the number of connected components of the set $F_i \cup F_j \cup F_k \cup F_r$.

Corollary 7.18: *For any simple* 3*-polytope* P

- $\beta^{-1,4} = \frac{h(h-1)}{2}$;
- $\beta^{-2,6} - \beta^{-1,6} = \frac{(h^2-1)(h-3)}{3}$;
- $\beta^{-3,8} - \beta^{-2,8} = \frac{(h+1)h(h-2)(h-5)}{8}$.

Corollary 7.19: *For any fullerene*

- $\beta^{-1,4} = \frac{(8+p_6)(9+p_6)}{2}$;
- $\beta^{-2,6} = \frac{(6+p_6)(8+p_6)(10+p_6)}{3}$;
- $\beta^{-3,8} = \frac{(4+p_6)(7+p_6)(9+p_6)(10+p_6)}{8}$.

8. Rigidity for 3-polytopes

8.1. *Notions of cohomological rigidity*

There is a fundamental problem in algebraic topology: to understand whether the isomorphims of cohomology rings or the homotopy equivalence of manifolds implies the diffeomorphism. In this chapter we describe some results in the framework of toric topology. For more results see also [28, 11, 15].

Definition 8.1: Let $\mathfrak{P}$ be some set of polytopes.

We call a property of simple polytopes *rigid* in $\mathfrak{P}$ if for any polytope $P \in \mathfrak{P}$ with this property the isomorphism of graded rings $H^*(\mathcal{Z}_P) \simeq H^*(\mathcal{Z}_Q)$, $Q \in \mathfrak{P}$, implies that Q also has this property.

We call sets $\mathfrak{S}_P \subset H^*(\mathcal{Z}_P)$, defined for any polytope $P \in \mathfrak{P}$, *rigid* in $\mathfrak{P}$ if for any isomorphism φ of graded rings $H^*(\mathcal{Z}_P) \simeq H^*(\mathcal{Z}_Q)$, $P, Q \in \mathfrak{P}$, we have $\varphi(\mathfrak{S}_P) = \mathfrak{S}_Q$.

We call a polytope *rigid* (or *B-rigid*) in $\mathfrak{P}$, if for any isomorphism of graded rings $H^*(\mathcal{Z}_P) \simeq H^*(\mathcal{Z}_Q)$, $Q \in \mathfrak{P}$, implies that Q is combinatorially equivalent to P.

In this section we mainly follow the works [29] and [28]. We give a full proof of all facts, sometimes different from the proof in [28, 29].

8.2. *Straightening along an edge*

For any edge $F_p \cap F_q$ of a simple 3-polytope P there is an operation of *straightening along the edge* (see Fig. 30). In this subsection we discuss its properties we will need below.

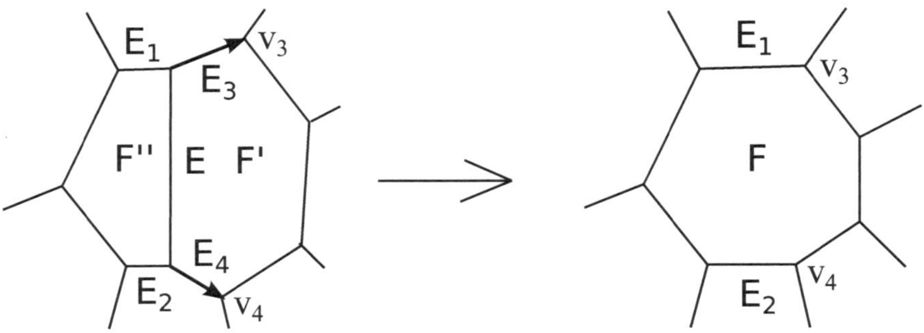

Fig. 30. Straightening along the edge

Lemma 8.2: [7] *For $P \simeq \Delta^3$ no straightening operations are defined. Let $P \not\simeq \Delta^3$ be a simple polytope. The operation of straightening along $E = F_i \cap F_j$ is not defined if and only if there is a 3-belt (F_i, F_j, F_k) for some F_k.*

Proof: For $P = \Delta^3$ straightening along any edge transforms a triangle into double edge; hence it is not allowed.

Let $P \not\simeq \Delta^3$ be a simple polytope. We will use Lemma 2.27 to establish whether a graph G' arising from the graph $G(P)$ under straightening along the edge corresponds to a polytope (which will be simple by construction). Let $F_i \cap F_j$ be an edge we want to straighten along. Let $F_i \cap F_j \cap F_p$ and $F_i \cap F_j \cap F_q$ be its vertices (see Figure 31). Then, after straightening, the number of edges of F_p and F_q decreases; hence F_p and F_q should have at least 4 edges. From the construction any facet of G', except for F, is surrounded by a simple edge-cycle. Since F_i and F_j intersect by a single edge, F also is surrounded by a simple edge cycle. Let F'_a and F'_b be facets of G'. If $F'_a, F'_b \neq F$, then we have $F'_a \cap F'_b = F_a \cap F_b$, where F_a, F_b are the corresponding facets of P. Hence this is either an empty set, or an edge. If $F'_a = F$, then we have that $F'_a \cap F'_b$ consists of more than one edge if and only if the corresponding facet F_b of P intersects both F_i and F_j, and $F_b \neq F_p, F_q$. This is equivalent to the fact that (F_i, F_j, F_b) is a 3-belt. □

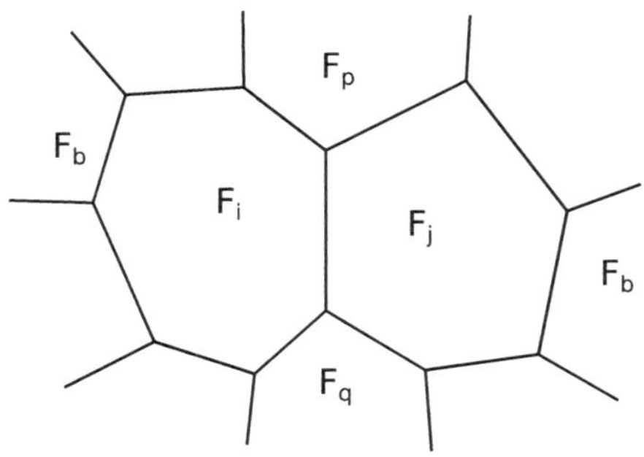

Fig. 31. Neighborhood of the edge $F_i \cap F_j$

Lemma 8.3: [7] *The polytope Q obtained by straightening a flag polytope P along the edge $F_i \cap F_j$ is not flag if and only if there is a 4-belt (F_i, F_j, F_k, F_l).*

Proof: Since P is flag, it has no triangles. Hence $Q \not\simeq \Delta^3$. Then Q is not flag if and only if it contains a 3-belt (F'_a, F'_b, F'_c). If $F \notin \{F'_a, F'_b, F'_c\}$, then (F_a, F_b, F_c) is a 3-loop of P; hence $F_a \cap F_b \cap F_c$ is a vertex. Then $F'_a \cap F'_b \cap F'_c$ is also a vertex. A contradiction. Let $F'_a = F$. Then both F_b, F_c intersect $F_i \cup F_j$, and $F_b \cap F_c \neq \varnothing$ in P. If (F_α, F_b, F_c) is a 3-loop in P for $\alpha \in \{i, j\}$, then $F_\alpha \cap F_b \cap F_c$ is a vertex different from the ends of $F_i \cap F_j$. This vertex is also a vertex in G' and is the intersection $F'_a \cap F'_b \cap F'_c$. A contradiction. Thus each of the facets F_i and F_j intersects only one facet among F_b and F_c, and these facets are different. This is possible if and only if (F_i, F_j, F_b, F_c) or (F_i, F_j, F_c, F_b) is a 4-belt. □

Remark 8.4: Straightenings along opposite edges of a quadrangle give the same combinatorial polytope.

Lemma 8.5: [7] *Let P be a flag simple 3-polytope and F be its quadrangular facet. If $P \not\simeq I^3$, then one of the two combinatorial polytopes obtained by straightening along edges of F is flag.*

Proof: Let F be surrounded by a 4-belt $(F_{i_1}, F_{i_2}, F_{i_3}, F_{i_4})$. If straightening along any edge of F gives non-flag polytopes, then by Lemma 8.3 there are some 4-belts $\mathcal{B}_1 = (F_{i_1}, F, F_{i_3}, F_{i_5})$ and $\mathcal{B}_2 = (F_{i_2}, F, F_{i_4}, F_{i_6})$. Since $\partial P \setminus \mathcal{B}_1$ consists of two connected components, one of them containing F_{i_2} and the other containing F_{i_4}, the belt $\mathcal{B}_2$ should intersect the belt $\mathcal{B}_1$ by one more facet except for F. Hence $F_{i_5} = F_{i_6}$. Since P is flag, we obtain vertices $F_{i_1} \cap F_{i_2} \cap F_{i_5}$, $F_{i_2} \cap F_{i_3} \cap F_{i_5}$, $F_{i_3} \cap F_{i_4} \cap F_{i_5}$, and $F_{i_4} \cap F_{i_1} \cap F_{i_5}$; hence all facets are quadrangles, and $P \simeq I^3$. □

8.3. *Rigidity of the property of being a flag polytope*

Proposition 8.6: (Lemma 5.2, [29]) *Let P be a flag simple 3-polytope. Then for any three different facets $\{F_i, F_j, F_k\}$, with $F_i \cap F_j = \varnothing$, there exist $l \geqslant 4$ and an l-belt $\mathcal{B}_l$ such that $F_i, F_j \in \mathcal{B}_l$ and $F_k \notin \mathcal{B}_l$.*

Proof: We will prove this statement by induction on the number of facets of P. By Proposition 3.10 for flag 3-polytopes we have $m \geqslant 6$, and $m = 6$ if and only if $P \simeq I^3$. In this case $F_i \cap F_j = \varnothing$ means that F_i and F_j are opposite facets. Then one of the two 4-belts passing through F_i and F_j does not pass F_k.

Now let the statement be true for all flag 3-polytopes with less than m facets and let P be a flag simple polytope with m facets. If $F_k \cap F_i \neq \varnothing$ and $F_k \cap F_j \neq \varnothing$, then we can take $\mathcal{B}_l$ to be the belt surrounding F_k. Now let at least one of the facets F_i or F_j do not intersect F_k, say F_i.

Consider facets $F_p \notin \{F_i, F_j\}$ adjacent to F_k. If for some p the facets (F_p, F_k) do not belong to a 4-belt, then straighten along $F_p \cap F_k$ to obtain a flag polytope Q with $m - 1$ facets. Then there is an l-belt $\mathcal{B}_l$, $F_i, F_j \in \mathcal{B}_l \not\ni F_k$ in Q. This belt corresponds to the belt on P we need. Suppose that for any $p \notin \{i, j\}$ there is a 4-belt containing (F_p, F_k). Since $F_k \not\simeq \Delta^2$, there are at least 3 values of p; hence there are at least two 4-belts. If any of these belts surrounds a quadrangle, then each quadrangle is adjacent to F_k. Consider the quadrangle F_q different from F_i and F_j (it exists, since $F_k \cap F_i = \varnothing$ or $F_k \cap F_j = \varnothing$ by assumption). By the inductive hypothesis $m > 6$, $P \not\simeq I^3$; hence by Lemma 8.5 straightening along $F_q \cap F_k$, or any of the edges adjacent to this edge in F_q gives a flag polytope. By assumption the first case is not allowed. Consider the edge $F_p \cap F_q$ adjacent to $F_q \cap F_k$ in F_q with $F_p \neq F_i, F_j$. After straightening along $F_p \cap F_q$ we obtain a flag polytope Q, which has a k-belt containing F_i, F_j and not containing F_k. This belt corresponds either to a k-belt, or to a $(k + 1)$-belt on P with the same properties.

At last consider the case when there is a 4-belt $\mathcal{B} = (F_p, F_k, F_q, F_r)$, $p \notin \{i, j\}$, not surrounding a quadrangle. By Lemma 3.11 $\mathcal{B}$-cuts P_1 and P_2 are flag polytopes. Since $\mathcal{B}$ is not surrounding a quadrangle, they have less facets than P. If F_i and F_j belong to one of the polytopes, say P_1, then by the induction hypothesis there is an l-belt $\mathcal{B}_l$ with $F_i, F_j \in \mathcal{B}_l \not\ni F_k$. Consider the new facet F of P_1. If $F \notin \mathcal{B}_l$, then $\mathcal{B}_l$ is an l-belt on P and the lemma is proved. Else, since $F_k \notin \mathcal{B}_l$, the l-belt $\mathcal{B}_l$ contains the fragment (F_p, F, F_q) and does not contain F_r. By the induction hypothesis there is an l'-belt $\mathcal{B}'_{l'}$ on P_2, $F_p, F_q, \in \mathcal{B}'_{l'} \not\ni F_k$. Then the segment of $\mathcal{B}'_{l'}$ with ends F_p and F_q does not contain the new facet F' and F_r, and the segment $\mathcal{B}_l \setminus \{F\}$ together form a belt we need. If any of the polytopes P_1 and P_2 contain exactly one of the facets F_i and F_j, say $F_i \in P_1$, $F_j \in P_2$, then consider an l_1-loop $\mathcal{B}_1 \subset P_1$, $F, F_i \in \mathcal{B}_1 \not\ni F_k$, and an l_2-loop $\mathcal{B}_2 \subset P_2$, $F', F_j \in \mathcal{B}_2 \not\ni F_k$. Then $F_r \notin \mathcal{B}_1, \mathcal{B}_2$; hence $(\mathcal{B}_1 \setminus (F_p, F, F_q)) \cup (\mathcal{B}_2 \setminus \{F'\})$ is a belt we need. This finishes the proof. □

Corollary 8.7: (Proposition 5.4, [29]) *Let P be a flag simple 3-polytope. Then for any $\omega \subset [m]$, $\omega \neq \varnothing$, the mapping*

$$\bigoplus_{\omega_1 \sqcup \omega_2 = \omega} \widehat{H}_2(P_{\omega_1}, \partial P_{\omega_1}) \otimes \widehat{H}_2(P_{\omega_2}, \partial P_{\omega_2}) \longrightarrow H_1(P_\omega, \partial P_\omega)$$

is an epimorphism.

Proof: We will use the notation of Lemma 4.8. Let $P_\omega = P_{\omega^1} \sqcup \cdots \sqcup P_{\omega^s}$ be the decomposition into connected components. Consider P_{ω^r}, $r \in \{1, \dots, s\}$. Let $\partial P_{\omega^r} = \eta_1 \sqcup \cdots \sqcup \eta_t$ be the decomposition into boundary components. If $t = 1$, then P_{ω^r} is a disk and is contractible. Let $t \geqslant 2$. Take $a \neq b$, and facets F_{i_1} and F_{i_2} in $\partial P \setminus P_{\omega^r}$ intersecting η_a and η_b respectively. By Proposition 8.6 there is an l-belt $\mathcal{B}_l$ of the form $(F_{j_1}, \dots, F_{j_l})$ with $F_{j_1} = F_{i_1}$, and $F_{j_p} = F_{i_2}$ for some p, $3 \leqslant p \leqslant l-1$. Set $\Pi_1 = (F_{j_1}, \dots, F_{j_p})$. Take

$$\begin{aligned} \omega_1 &= \{j \colon F_j \in \mathcal{B}_l \cap P_{\omega^r}\},\ \omega_2 = \omega \setminus \omega_1, \\ A &= [\sum_{F_j \in P_{\omega^r} \cap \Pi_1} F_j] \in \widehat{H}_2(P_{\omega_1}, \partial P_{\omega_1}), \\ B &= [\sum_{F_k \in P_{\omega^r} \cap \mathcal{W}_1} F_k] \in \widehat{H}_2(P_{\omega_2}, \partial P_{\omega_2}). \end{aligned}$$

Then $A \cdot B = [\gamma_1] + \cdots + [\gamma_q]$, where γ_i is an edge path in P_{ω^r} that starts at $\eta_{\alpha_{i-1}}$ and ends at η_{α_i}, $\alpha_j \in [s]$, $j = 0, \dots, q$, $i = 1, \dots, q$, and $\{\alpha_0, \alpha_q\} = \{a, b\}$. This element corresponds to a path connecting η_a and η_b in P_{ω^r}. Thus we can realize any element from the basis given by Proposition 7.6. □

The following simple result is well-known:

Lemma 8.8: *The simplex Δ^3 is rigid in the class of all simple 3-polytopes.*

Proof: This is equivalent to the fact that any two facets intersect, that is $H^3(\mathcal{Z}_P) = 0$. □

Theorem 8.9: *The polytope $P \neq \Delta^3$ is flag if and only if*

$$H^{m-2}(\mathcal{Z}_P) \subset (\widetilde{H}^*(\mathcal{Z}_P))^2.$$

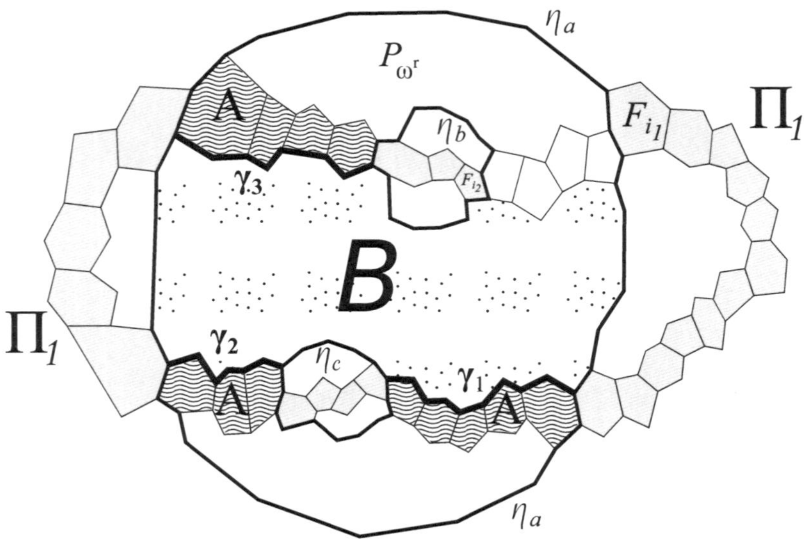

Fig. 32. The belt $\mathcal{B}_l$ intersecting P_{ω^r}

Proof: The polytope $P \neq \Delta^3$ is not flag if and only if it has a 3-belt. This corresponds to an element of a basis in $H^{-1,2\omega}(\mathcal{Z}_P) \simeq H_1(P_\omega, \partial P_\omega)$, $|\omega| = 3$. By the Poincare duality this element corresponds to an element of a basis in $H^{-(m-4),2([m]\setminus\omega)}(\mathcal{Z}_P) \simeq \widehat{H}_2(P_\omega, \partial P_\omega)$. The latter element belongs to $H^{m-2}(\mathcal{Z}_P)$ but does not belong to $(\widetilde{H}^*(\mathcal{Z}_P))^2$.

If the polytope is flag, then it has no 3-belts, and by Proposition 7.11

$$H^5(\mathcal{Z}_P) = \bigoplus_{|\omega|=4} H^{-3,2\omega}(\mathcal{Z}_P) = \bigoplus_{|\omega|=4} \widehat{H}_2(P_\omega, \partial P_\omega).$$

Hence by Poincare duality

$$H^{m-2}(\mathcal{Z}_P) = \bigoplus_{|\omega|=m-4} H_1(P_\omega, \partial P_\omega).$$

By Corollary 8.7 we have $H^{m-2}(\mathcal{Z}_P) \subset (\widetilde{H}^*(\mathcal{Z}_P))^2$.

By Lemma 8.8 the simplex is a rigid polytope. This finishes the proof. □

Corollary 8.10: *The property of being a flag polytope is rigid in the class of simple 3-polytopes.*

Remark 8.11: Corollary 8.10 originally was proved in [29] as a corollary of Theorems 4.2 and 6.6 in [29]. Our Theorem 8.9 gives more direct prove of this fact.

8.4. *Rigidity of the property of having a 4-belt*

Recall that for any set $\omega = \{i, j\} \subset [m]$ we have

$$H^{-1,2\omega}(\mathcal{Z}_P) = \widehat{H}_2(P_\omega, \partial P_\omega) = \begin{cases} \mathbb{Z} \text{ with generator } [F_i] = -[F_j], & F_i \cap F_j = \varnothing; \\ 0, & F_i \cap F_j \neq \varnothing, \end{cases}$$

and

$$H^3(\mathcal{Z}_P) = \bigoplus_{\{i,j\}:\ F_i \cap F_j = \varnothing} \mathbb{Z}.$$

Definition 8.12: The set $\{F_{i_1}, \dots, F_{i_k}\}$ with $F_{i_1} \cap \dots \cap F_{i_k} = \varnothing$ is called a *nonface* of P, and the corresponding set $\{i_1, \dots, i_k\}$ – a *nonface* of K_P. A nonface, minimal by inclusion, is called a *minimal* nonface, or a *missing face*. Define $N(K)$ to be the set of all minimal nonfaces of the simplicial complex K.

For any nonface $\omega = \{i, j\}$ choose a generator $\widetilde{\omega} \in H^{-1,2\omega}(\mathcal{Z}_P)$.

Proposition 8.13: *The multiplication $H^3(\mathcal{Z}_P) \otimes H^3(\mathcal{Z}_P) \to H^6(\mathcal{Z}_P)$ is trivial if and only if P has no 4-belts.*

Proof: For $\omega_1 = \{i, j\}, \omega_2 = \{p, q\} \in N(K_P)$, $\omega_1 \cap \omega_2 = \varnothing$, the simplicial complex $K_{\omega_1 \sqcup \omega_2}$ has no 2-simplices; hence it is at most 1-dimensional and can be considered as a graph. Moreover, this graph has no 3-cycles. If it has a 4-cycle, then $K_{\omega_1 \sqcup \omega_2}$ is a boundary of a 4-gon, (F_i, F_p, F_j, F_q) is a 4-belt, and $\widetilde{\omega_1} \cdot \widetilde{\omega_2}$ is a generator of $H_1(P_{\omega_1 \sqcup \omega_2}, \partial P_{\omega_1 \sqcup \omega_2})$. If $K_{\omega_1 \sqcup \omega_2}$ has no 4-cycles, then it has no cycles at all, $H_1(P_{\omega_1 \sqcup \omega_2}, \partial P_{\omega_1 \sqcup \omega_2}) \simeq \widetilde{H}^1(K_{\omega_1 \sqcup \omega_2}) = 0$, and $\widetilde{\omega_1} \cdot \widetilde{\omega_2} = 0$. This proves the statement. □

Corollary 8.14: *The property of having a 4-belt is rigid in the class of all simple 3-polytopes.*

Remark 8.15: Corollary 8.14 originally follows from Corollary 7.3 in version 1 of [29]. Our Proposition 8.13 gives another proof here.

8.5. *Rigidity of flag 3-polytopes without 4-belts*

As it was mentioned in Section 4.5 flag 3-polytopes without 4-belts correspond to right-angled polytopes in Lobachevsky (hyperbolic) 3-space (Pogorelov polytopes). In this Section we give a proof of B-rigidity of such polytopes among all simple 3-polytopes following [28]. For more recent results on Pogorelov polytopes see [11, 9, 50].

First we prove the following technical result, which we will need below.

Proposition 8.16: (Lemma 3.2, [28]) *Let P be a flag 3-polytope without 4-belts. Then for any three different facets $\{F_i, F_j, F_k\}$, with $F_i \cap F_j = \varnothing$, there exist $l \geqslant 5$ and an l-belt $\mathcal{B}_l$ such that $F_i, F_j \in \mathcal{B}_l$, $F_k \notin \mathcal{B}_l$, and F_k does not intersect at least one of the two connected components of $\mathcal{B}_l \setminus \{F_i, F_j\}$.*

Remark 8.17: The proof of Lemma 3.2 in [28] contains several additional assumptions. We give a full proof following the same idea.

Proof: From Proposition 8.6 there is an s-belt $\mathcal{B}_1$, with $F_i, F_j \in \mathcal{B}_1 \not\ni F_k$. We have $\mathcal{B}_1 = (F_i, F_{i_1}, \ldots, F_{i_p}, F_j, F_{j_1}, \ldots, F_{j_q})$, $s = p + q + 2$, $p, q \geqslant 1$. According to Lemma 4.8 the belt $\mathcal{B}_1$ divides the surface $\partial P \setminus \mathcal{B}_1$ into two connected components $\mathcal{P}_1$ and $\mathcal{P}_2$, both homeomorphic to disks. Consider the component $\mathcal{P}_\alpha$ containing $\operatorname{int} F_k$. Set $\beta = 3 - \alpha$. Then either $\partial\mathcal{P}_\alpha = \partial F_k$ or $\partial\mathcal{P}_\alpha \cap \partial F_k$ consists of a finite set of disjoint edge-segments $\gamma_1, \ldots, \gamma_d$.

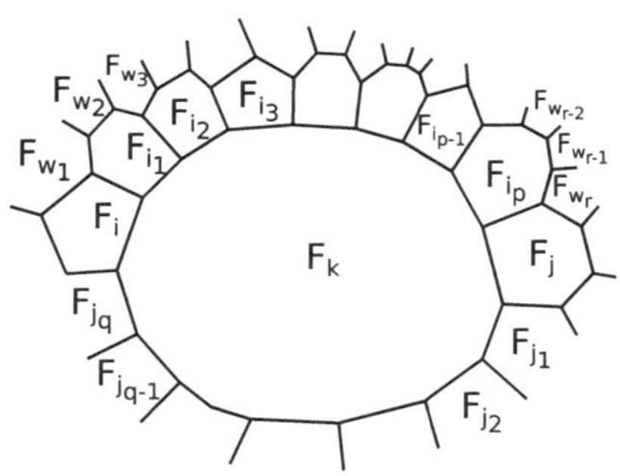

Fig. 33. Case 1

Consider the first case. Then $\mathcal{B}_1$ surrounds F_k, and F_i and F_j are adjacent to F_k. Consider all facets $\{F_{w_1}, \ldots, F_{w_r}\}$ in $\mathcal{W}_\beta$ (in the notation of Lemma 4.8), adjacent to facets in $\{F_{i_1}, \ldots, F_{i_p}\}$ (see Figure 33), in the order we meet them while walking round $\partial\mathcal{B}_1$ from F_i to F_j. Then $F_{w_a} \cap F_{j_b} = \varnothing$ for any a, b, else $(F_k, F_{j_b}, F_{w_a}, F_{i_c})$ is a 4-belt for any i_c with $F_{i_c} \cap F_{w_a} \neq \varnothing$, since $F_k \cap F_{w_a} = \varnothing$ (because $\operatorname{int} F_{w_a} \subset \mathcal{P}_\beta$) and $F_{j_b} \cap F_{i_c} = \varnothing$. We have a thick path $(F_i, F_{w_1}, \ldots, F_{w_r}, F_j)$. Consider the shortest thick path of the form $(F_i, F_{w_{s_1}}, \ldots, F_{w_{s_t}}, F_j)$. If two facets of this path intersect, then they are successive, else there is a shorter thick path. Thus we have a belt $(F_i, F_{w_{s_1}}, \ldots, F_{w_{s_t}}, F_j, F_{j_1}, \ldots, F_{j_q})$ containing F_i, F_j, not containing F_k, and the segment $(F_{w_{s_1}}, \ldots, F_{w_{s_t}})$ does not intersect F_k.

Now consider the second case. We can assume that $F_i \cap F_k = \varnothing$ or $F_j \cap F_k = \varnothing$, say $F_i \cap F_k \neq \varnothing$, else consider the belt $\mathcal{B}_1$ surrounding F_k and apply the arguments of the first case. Let $\gamma_a = (F_k \cap F_{u_{a,1}}, \ldots, F_k \cap F_{u_{a,l_a}})$. Set $\mathcal{U}_a = (F_{u_{a,1}}, \ldots, F_{u_{a,l_a}})$. Denote the segment $(F_{s_{a,1}}, \ldots, F_{s_{a,t_a}})$ of $\mathcal{B}_1$ between $\mathcal{U}_a$ and $\mathcal{U}_{a+1}$ by $\mathcal{S}_a$. Then $\mathcal{B}_1 = (\mathcal{U}_1, \mathcal{S}_1, \mathcal{U}_2, \ldots, \mathcal{U}_d, \mathcal{S}_d)$ for some d.

Consider the thick path $W_a = (F_{w_{a,1}}, \ldots, F_{w_{a,r_a}}) \subset \mathcal{W}_\beta$ (see the notation in Lemma 4.8) arising whilst walking around the facets in $\mathcal{W}_\beta$ intersecting facets in $\mathcal{U}_a$ (see Figure 35). Then $W_a \cap W_b = \varnothing$ for $a \neq b$, else $(F_w, F_{u_{a,j_1}}, F_k, F_{u_{b,j_2}})$ is a 4-belt for any $F_w \in W_a \cap W_b$ such that $F_w \cap F_{u_{a,j_1}} \neq \varnothing$, $F_w \cap F_{u_{b,j_2}} \neq \varnothing$. Also $F_{w_{a,j_1}} \neq F_{w_{a,j_2}}$ for $j_1 \neq j_2$. This is true for facets adjacent to the same facet $F_{u_{a,i}}$. Let $F_{w_{a,j_1}} = F_{w_{a,j_2}}$. If the facets are adjacent to the successive facets $F_{u_{a,i}}$ and $F_{u_{a,i+1}}$, then the flagness condition implies that $j_1 = j_2$ and $F_{w_{a,j_1}}$ is the facet in $\mathcal{W}_\beta$ intersecting $F_{u_{a,i}} \cap F_{u_{a,i+1}}$. If the facets are adjacent to non-successive facets $F_{u_{a,i}}$ and $F_{u_{a,j}}$, then $(F_{w_{a,j_1}}, F_{u_{a,i}}, F_k, F_{u_{a,j}})$ is a 4-belt, which is a contradiction.

Now consider the thick path $\mathcal{V}_b = (F_{v_{b,1}}, \ldots, F_{v_{b,c_b}})$ arising whilst walking around the facets in $\mathcal{W}_\alpha$ intersecting facets in $\mathcal{S}_b$ (see Figure 35). Then $\mathcal{V}_a \cap \mathcal{V}_b = \varnothing$ for $a \neq b$, and $W_a \cap \mathcal{V}_b = \varnothing$ for any a, b, since interiors of the corresponding facets lie in different connected components of $\partial P \setminus (\mathcal{B}_1 \cup F_k)$, moreover by the same reason we have $F_{v_{a,j}} \cap F_{v_{b,j}} = \varnothing$ for any i, j, and $a \neq b$.

Now we will deform the segments $\mathcal{I} = (F_{i_1}, \ldots, F_{i_p})$ and $\mathcal{J} = (F_{j_1}, \ldots, F_{j_q})$ of the belt $\mathcal{B}_1$ to obtain a new belt $(F_i, \mathcal{I}', F_j, \mathcal{J}')$ with $\mathcal{I}'$ not intersecting F_k. First substitute the thick path W_a for each segment $\mathcal{U}_a \subset \mathcal{I}$ and the thick path $\mathcal{V}_b$ for each segment $\mathcal{S}_b \subset \mathcal{J}$. Since $F_{s_{a,t_a}} \cap F_{w_{a+1,1}} \neq \varnothing$, $F_{w_{a,r_a}} \cap F_{s_{a,1}} \neq \varnothing$, $F_{v_{a,c_a}} \cap F_{u_{a+1,1}} \neq \varnothing$, and $F_{u_{a,l_a}} \cap F_{s_{a,1}} \neq \varnothing$ for any a and $a+1$ considered $\mod d$, we obtain a loop $\mathcal{L}_1 = (F_i, \mathcal{I}_1, F_j, \mathcal{J}_1)$ instead of $\mathcal{B}_1$.

Since $F_i \cap F_k = \varnothing$, we have $F_i = F_{s_{a_i,f_i}}$ for some a_i, f_i. If $F_j = F_{s_{a_j,f_j}}$ for some a_j, f_j, then we can assume that $a_i \neq a_j$, else the facets in $\mathcal{I}$ or $\mathcal{J}$ already do not intersect F_k, and $\mathcal{B}_1$ is the belt we need. If $F_j = F_{u_{a_j,f_j}}$ for some a_j and some $f_j > 1$, then substitute the thick path $(F_{w_{a_j,1}}, \ldots, F_{w_{a_j,g_j}})$, where g_j – the first integer with $F_{w_{a_j,g_j}} \cap F_j \neq \varnothing$ (then $F_j \cap F_{u_{a_j,f_j-1}} \cap F_{w_{a_j,g_j}}$ is a vertex), for the segment $(F_{u_{a_j,1}}, \ldots, F_{u_{a_j,f_j-1}})$ to obtain a loop $\mathcal{L}_2 = (F_i, \mathcal{I}_2, F_j, \mathcal{J}_1)$ (else set $\mathcal{L}_2 = \mathcal{L}_1$) with facets in $\mathcal{I}_2$ not intersecting F_k. If $f_j < l_{a_j}$, then $F_{w_{a_j,g_j}} \cap F_{u_{a_j,f_j+1}} = \varnothing$, else $(F_k, F_{u_{a_j,f_j-1}}, F_{w_{a_j,g_j}}, F_{u_{a_j,f_j+1}})$ is a 4-belt. Then $F_{w_{a,l}} \cap F_{u_{a_j,r}} = \varnothing$ for any $r \in \{f_j+1, \ldots, l_{a_j}\}$ and a, l, such that either $a \neq a_j$, or $a = a_j$, and $l \in \{1, \ldots, g_j\}$. Hence facets of the segment $(F_{u_{a_j,f_j+1}}, \ldots, F_{u_{a_j,l_{a_j}}})$ do not intersect facets in $\mathcal{I}_2$.

Now a facet $F_{i'_a}$ of $\mathcal{I}_2$ can intersect a facet $F_{j'_b}$ of $\mathcal{J}_1$ only if $F_{i'_a} = F_{w_{c,h}}$ for some c, h, and $F_{j'_b} = F_{s_{a_i,l}}$ for $l < f_i$, or $F_{j'_b} = F_{s_{a_j,l}}$ for $F_j = F_{s_{a_j,f_j}}$

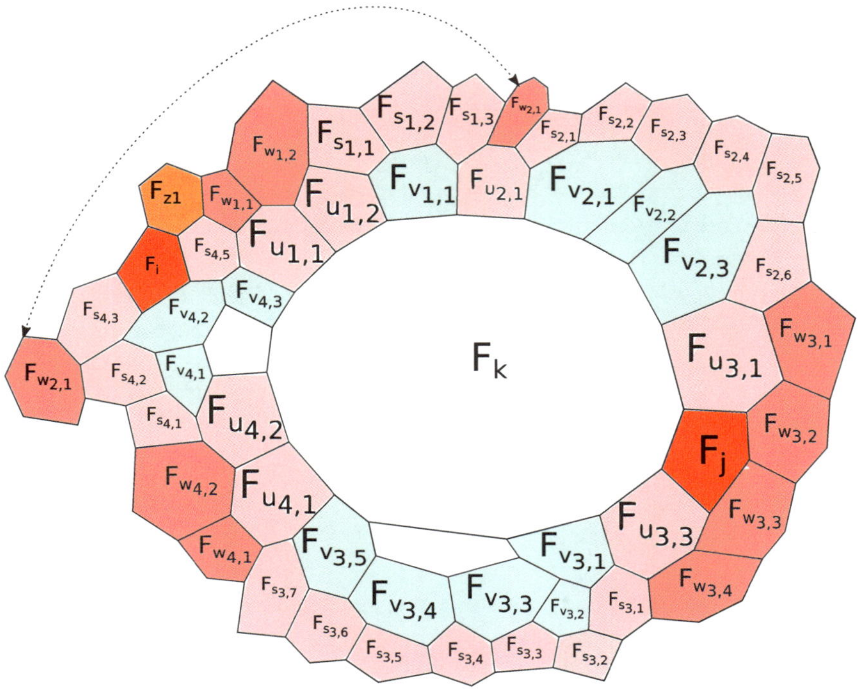

Fig. 34. Case 2

and $l > f_j$. In the first case take the smallest l for all c, h, and the corresponding facet $F_{w_{c,h}}$. Consider the facet $F_{u_{b,g}} = F_{i_e} \in \mathcal{I}$ with $F_{u_{b,g}} \cap F_{w_{c,h}} \neq \varnothing$. Then $\mathcal{L}' = (F_{s_{a_i,l}}, F_{s_{a_i,l+1}}, \dots, F_i, F_{i_1}, \dots, F_{i_e}, F_{w_{c,h}})$ is a simple loop. If $f_i < t_{a_i}$, then consider the thick path $\mathcal{Z}_1 = (F_{z_{1,1}}, \dots, F_{z_{1,y_1}})$ arising whilst walking along the boundary of $\mathcal{B}_1$ in $\mathcal{W}_\beta$ from the facet $F_{z_{1,1}}$ intersecting $F_i \cap F_{i_1}$ by the vertex, to the facet $F_{z_{1,y_1}}$ preceding $F_{w_{a_i+1,1}}$. Consider the thick path $\mathcal{X}_1 = (F_{v_{a_i,1}}, \dots, F_{v_{a_i,x_1}})$ with x_1 being the first integer with $F_{v_{a_i,x_1}} \cap F_i \neq \varnothing$. Consider the simple curve $\eta \subset \partial P$ consisting of segments connecting the midpoints of the successive edges of the intersection of the successive facets of $\mathcal{L}'$. It divides ∂P into two connected components $\mathcal{E}_1$ and $\mathcal{E}_2$ with $\mathcal{J}_1 \setminus (F_{s_{a_i,l}}, \dots, F_{s_{a_i,f_i-1}})$ lying in one connected component $\mathcal{E}_\alpha$, and $\mathcal{Z}_1$ – in $\mathcal{E}_\beta \cup F_{w_{c,h}}$, $\beta = 3 - \alpha$. Now substitute $\mathcal{X}_1$ for the segment $(F_{s_{a_i,1}}, \dots, F_{s_{a_i,f_i-1}})$ of $\mathcal{J}_1$. If $f_i < t_{a_i}$ substitute $\mathcal{Z}_1$ for the segment $(F_{s_{a_i,f_i+1}}, \dots, F_{s_{a_i,t_{a_i}}})$ of $\mathcal{I}_2$ to obtain a new loop $(F_i, \mathcal{I}_3, F_j, \mathcal{J}_2)$ with facets in $\mathcal{I}_3$ not intersecting F_k. A facet F''_{i_a} in $\mathcal{I}_3$ can intersect a facet F''_{j_b} in $\mathcal{J}_2$ only if $F_{i''_a} = F_{w_{c',h'}}$ for some c', h', $F_j = F_{s_{a_j,f_j}}$, and

$F_{j''_b} = F_{s_{a_j},l}$ for $l > f_j$. The thick path $\mathcal{Z}_1$ lies in $\mathcal{E}_\beta \cup F_{w_{c,h}}$ and the segment $(F_j = F_{s_{a_j},f_j}, \dots, F_{s_{a_j},t_{a_j}})$ lies in $\mathcal{E}_\alpha$; hence intersections of facets in $\mathcal{I}_3$ with facets in $\mathcal{J}_2$ are also intersections of the same facets in $\mathcal{I}_2$ and $\mathcal{J}_1$, and $F_{w_{c',h'}}$ is either $F_{w_{c,h}}$, or lies in $\mathcal{E}_\alpha$. We can apply the same argument for $\mathcal{S}_{a_j}$ as for $\mathcal{S}_{a_i}$ to obtain a new loop $\mathcal{L}_4 = (F_i, \mathcal{I}_4, F_j, \mathcal{J}_3)$ with facets in $\mathcal{I}_4$ not intersecting F_k and facets in $\mathcal{J}_3$. Then take the shortest thick path from F_i to F_j in $F_i \cup \mathcal{I}_4 \cup F_j$ and the shortest thick path from F_j to F_i in $F_j \cup \mathcal{J}_3 \cup F_i$ to obtain the belt we need.

□

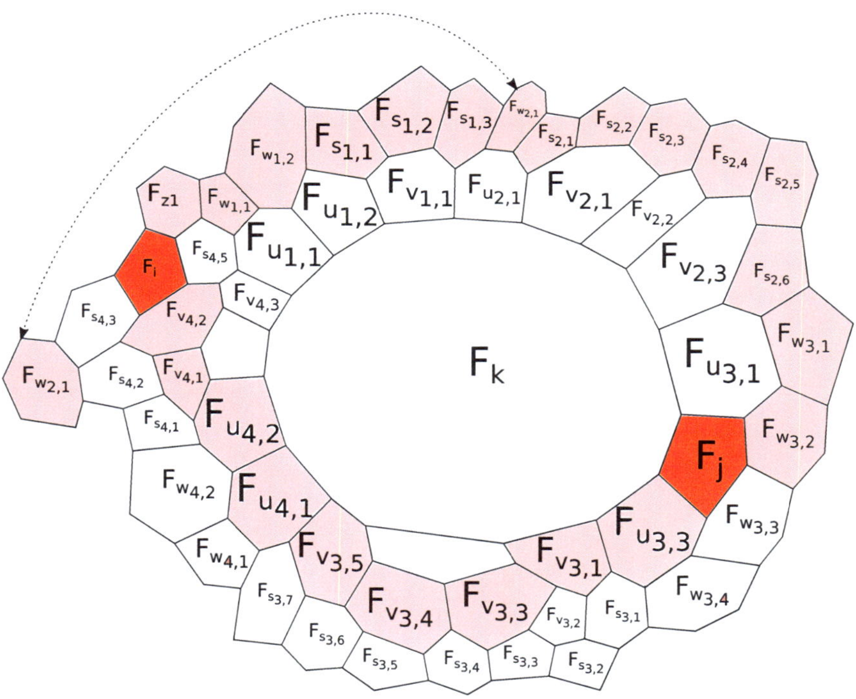

Fig. 35. Modified belt

Definition 8.18: An *annihilator* of an element r in a ring R is defined as

$$\mathrm{Ann}_R(r) = \{s \in R \colon rs = 0\}.$$

Proposition 8.19: (Corollary 3.4, [28]) *The set of elements in* $H^3(\mathcal{Z}_P)$ *corresponding to*

$$\bigcup_{\{i,j\}\colon F_i\cap F_j=\varnothing} \{[F_i],[F_j]\in \widehat{H}_2(P_{\{i,j\}},\partial P_{\{i,j\}})\}$$

is rigid in the class of all simple flag 3*-polytopes without* 4*-belts.*

Proof: Since the group $H^*(\mathcal{Z}_P)$ has no torsion, we have the isomorphism $H^*(\mathcal{Z}_P,\mathbb{Q})\simeq H^*(\mathcal{Z}_P)\otimes\mathbb{Q}$ and the embedding $H^*(\mathcal{Z}_P)\subset H^*(\mathcal{Z}_P)\otimes\mathbb{Q}$. For polytopes P and Q the isomorphism $H^*(\mathcal{Z}_P)\simeq H^*(\mathcal{Z}_Q)$ implies the isomorphism over $\mathbb{Q}$. For the cohomology over $\mathbb{Q}$ all theorems about the structure of $H^*(\mathcal{Z}_P,\mathbb{Q})$ are still valid. In what follows we consider cohomology over $\mathbb{Q}$. Set $H=H^*(\mathcal{Z}_P,\mathbb{Q})$. We will need the following result:

Lemma 8.20: (Lemma 3.3, [28]) *For an element*

$$\alpha=\sum_{\omega\in N(K_P),|\omega|=2} r_\omega\widetilde{\omega}\quad \textit{with } |\{\omega\colon r_\omega\neq 0\}|\geqslant 2$$

we have

$$\dim \operatorname{Ann}_H(\alpha)<\dim\operatorname{Ann}_H(\widetilde{\omega}),\quad \textit{if } r_\omega\neq 0.$$

Proof: Choose a complementary subspace C_ω to $\operatorname{Ann}_H(\widetilde{\omega})$ in H as a direct sum of complements $C_{\omega,\tau}$ to $\operatorname{Ann}_H(\widetilde{\omega})\cap\widehat{H}_*(P_\tau,\partial P_\tau)$ in $\widehat{H}_*(P_\tau,\partial P_\tau)$ for all $\tau\subset[m]\setminus\omega$. Then for any $\beta\in C_\omega\setminus\{0\}$ we have $\beta\widetilde{\omega}\neq 0$, which is equivalent to the fact that $\beta=\sum\beta_\tau$, $\beta_\tau\in C_{\omega,\tau}$, $\tau\subset[m]\setminus\omega$, with $\beta_{\tau_\beta}\widetilde{\omega}\neq 0$ for some $\tau_\beta\subset[m]\setminus\omega$. Moreover for any $\omega'\neq\omega$ with $r_{\omega'}\neq 0$ and $\tau\subset[m]\setminus\omega$, $\tau\neq\tau_\beta$, we have $\tau_\beta\sqcup\omega\notin\{\tau\cup\omega',\tau_\beta\cup\omega',\tau\sqcup\omega\}$; hence $(\beta\cdot\alpha)_{\tau_\beta\sqcup\omega}=r_\omega\beta_{\tau_\beta}\cdot\widetilde{\omega}\neq 0$, and $\beta\alpha\neq 0$. Then C_ω forms a direct sum with $\operatorname{Ann}(\alpha)$. Now consider some $\omega'\neq\omega$, $|\omega'|=2$, $r_{\omega'}\neq 0$. Let $\omega=\{p,q\}$, $\omega'=\{s,t\}$, $q\notin\omega'$. By Proposition 8.16 there is an l-belt $\mathcal{B}_l$ such that $F_s,F_t\in\mathcal{B}_l$, $F_q\notin\mathcal{B}_l$, and F_q does not intersect one of the two connected components B_1 and B_2 of $\mathcal{B}_l\setminus\{F_s,F_t\}$, say B_1. Take $\xi=[\sum_{i\colon F_i\subset B_1}F_i]\in\widehat{H}_2(P_\tau,\partial P_\tau)$, $\tau=\{i\colon F_i\in\mathcal{B}_l\setminus\{F_s,F_t\}\}$, and $[F_s]\in\widehat{H}_2(P_{\omega'},\partial P_{\omega'})$. Then $\xi\cdot[F_s]$ is a generator in $H_1(\mathcal{B}_l,\partial\mathcal{B}_l)\simeq\mathbb{Z}$. On the other hand, take $[F_q]\in\widehat{H}_2(P_\omega,\partial P_\omega)$. Then either $F_p\in\mathcal{B}_l\setminus\{F_s,F_t\}$, and $\xi\cdot\widetilde{\omega}=0$, since $\tau\cap\omega\neq\varnothing$, or $F_p\notin\mathcal{B}_l\setminus\{F_s,F_t\}$, and $\pm\xi\cdot\widetilde{\omega}=\xi\cdot[F_q]=0$, since F_q does not intersect B_1. In both cases $\xi\in\operatorname{Ann}(\widetilde{\omega})$ and $\xi\cdot\widetilde{\omega'}\neq 0$. Then $\xi\cdot\alpha\neq 0$, since $\tau\sqcup\omega'\neq\tau\sqcup\omega_1$ for $\omega_1\neq\omega'$. Consider any $\beta=\sum_{\tau\subset[m]\setminus\omega}\beta_\tau\in C_\omega\setminus\{0\}$. We have $(\beta\cdot\alpha)_{\tau_\beta\sqcup\omega}\neq 0$. If $(\xi\cdot\alpha)_{\tau_\beta\sqcup\omega}\neq 0$, then since ξ is a homogeneous element, $(\xi\cdot\alpha)_{\tau_\beta\sqcup\omega}=r_{\omega_1}\xi\cdot\widetilde{\omega_1}$ for $\omega_1=(\tau_\beta\sqcup\omega)\setminus\tau=\{q,r\}$, $r\in[m]$. We have $\xi\cdot\widetilde{\omega_1}=\pm\xi\cdot[F_q]=0$, since F_q does not intersect B_1. A contradiction. Thus, $((\xi+\beta)\cdot\alpha)_{\tau_\beta\sqcup\omega}=(\beta\cdot\alpha)_{\tau_\beta\sqcup\omega}\neq 0$; hence $(\xi+\beta)\cdot\alpha\neq 0$, and the space $\langle\xi\rangle\oplus C_\omega$ forms a direct sum with $\mathrm{Ann}_H(\alpha)$. This finishes the proof. □

Now let us prove Proposition 8.19. Let $\varphi\colon H^*(\mathcal{Z}_P,\mathbb{Z})\to H^*(\mathcal{Z}_Q,\mathbb{Z})$ be an isomorphism of graded rings for flag simple 3-polytopes P and Q without 4-belts. Let $\omega\in N(K_P)$, $|\omega|=2$, and

$$\varphi(\widetilde{\omega})=\alpha=\sum_{\omega'\in N(K_Q),|\omega'|=2} r_{\omega'}\widetilde{\omega'} \text{ with } |\{\omega'\colon r_{\omega'}\neq 0\}|\geqslant 2.$$

Then there is some ω' such that $r_{\omega'}\neq 0$ and

$$\varphi^{-1}(\widetilde{\omega'})=\alpha'=\sum_{\omega''\in N(K_P),|\omega''|=2} r'_{\omega''}\widetilde{\omega''}$$

with $r'_\omega\neq 0$. Now consider all the mappings in cohomology over $\mathbb{Q}$. Since the dimension of the annihilator of an element is invariant under isomorphisms, Lemma 8.20 gives a contradiction:

$$\dim\mathrm{Ann}(\widetilde{\omega})=\dim\mathrm{Ann}(\alpha)<\dim\mathrm{Ann}(\widetilde{\omega'})=\dim\mathrm{Ann}(\alpha')<\dim\mathrm{Ann}(\widetilde{\omega}).$$

Thus $\varphi(\widetilde{\omega})=r_{\omega'}\widetilde{\omega'}$ for some ω'. Since the isomorphism is over $\mathbb{Z}$, we have $r_{\omega'}=\pm 1$. This finishes the proof.

Definition 8.21: Following [29] and [28] for a graded algebra $A=\bigoplus_{i\geqslant 0}A^i$ over the field k, and a nonzero element $\alpha\in A$ define a *p-factorspace* V to be a vector subspace in A^p such that for any $v\in V\setminus\{0\}$ there exists $u_v\in A$ with $vu_v=\alpha$. A *p-factorindex* $\mathrm{ind}_p(\alpha)$ is defined to be the maximal dimension of p-factorspaces of α.

Definition 8.22: Define $\mathbf{B}_k=\bigoplus_{\mathcal{B}_k-k\text{-belt}}H_1(\mathcal{B}_k,\partial\mathcal{B}_k)$ to be the subgroup in $H^{k+2}(\mathcal{Z}_P)$ generated by all elements $\widetilde{\mathcal{B}_k}$ corresponding to k-belts.

Definition 8.23: For the rest of the Section let $\{\omega_i\}_{i=1}^{N_P}$ be the set of all *missing edges* of the complex K_P of the polytope P.

The following result is our interpretation of the argument in the proof of Corollary 3.4 in [28].

Proposition 8.24: *Let P be a simple 3-polytope. Then*

(1) for any element $\alpha \in H^{k+2}(\mathcal{Z}_P, \mathbb{Q})$, $4 \leqslant k \leqslant m-2$, we have $\mathrm{ind}_3(\alpha) \leqslant \frac{k(k-3)}{2}$, and the equality $\mathrm{ind}_3(\alpha) = \frac{k(k-3)}{2}$ implies $\alpha \in (\mathbf{B}_k \otimes \mathbb{Q}) \setminus \{0\}$;
(2) for any k-belt $\mathcal{B}_k$, $4 \leqslant k \leqslant m-2$, we have $\mathrm{ind}_3(\widetilde{\mathcal{B}_k}) = \frac{k(k-3)}{2}$.

In particular, the group $\mathbf{B}_k \subset H^{k+2}(\mathcal{Z}_P, \mathbb{Z})$, $4 \leqslant k \leqslant m-2$, is B-rigid in the class of all simple 3-polytopes.

Proof:

(1) We have

$$\alpha = \sum_{\omega} \alpha_\omega \in \bigoplus_{|\omega|=k} H_1(P_\omega, \partial P_\omega, \mathbb{Q}) \oplus \bigoplus_{|\omega|=k+1} \widehat{H}_2(P_\omega, \partial P_\omega, \mathbb{Q}).$$

Let $0 \neq \beta = \sum\limits_{i=1}^{N(P)} \lambda_i \widetilde{\omega}_i$ be the divisor of α. Then there exists

$$\gamma = \sum_{\eta} \gamma_\eta \in \bigoplus_{|\eta|=k-3} H_1(P_\eta, \partial P_\eta, \mathbb{Q}) \oplus \bigoplus_{|\eta|=k-2} \widehat{H}_2(P_\eta, \partial P_\eta, \mathbb{Q}),$$

with $\beta \cdot \gamma = \alpha$. Then $\alpha_\omega = 0$, for all ω with $|\omega| = k+1$, $\gamma_\eta = 0$ for all η with $|\eta| = k-3$, and $\alpha_\omega = \sum\limits_{\omega_i \subset \omega} \lambda_i \widetilde{\omega}_i \cdot \gamma_{\omega \setminus \omega_i} = \left(\sum\limits_{\omega_i \subset \omega} \lambda_i \widetilde{\omega}_i\right) \cdot \left(\sum\limits_{\eta \subset \omega, |\eta|=k-2} \gamma_\eta\right)$.

Thus for any 3-factorspace V of α and any ω with $\alpha_\omega \neq 0$ the linear mapping

$$\varphi_\omega \colon V \longrightarrow H^3(\mathcal{Z}_P, \mathbb{Q}) \colon \beta \longmapsto \beta_\omega = \sum_{\omega_i \subset \omega} \lambda_i \widetilde{\omega}_i$$

is a monomorphism; hence it is a linear isomorphism of V to the factorspace $\varphi_\omega(V)$ of α_ω. Let $P_\omega = P_{\omega^1} \sqcup \dots \sqcup P_{\omega^s}$ be the decomposition into the connected components. Then $H_1(P_\omega, \partial P_\omega) = \bigoplus\limits_{l=1}^{s} H_1(P_{\omega^l}, \partial P_{\omega^l})$, and $\alpha_\omega = \sum\limits_{l=1}^{s} \alpha_{\omega^l}$. Let $\omega_i = \{p, q\}$, with $p \in \omega^a$, $q \in \omega^b$. If $a \neq b$, then $\widetilde{\omega}_i \cdot \gamma_{\omega \setminus \omega_i} = 0$, since $\widetilde{\omega}_i = \pm[F_p] = \mp[F_q]$, and the cohomology class $\widetilde{\omega}_i \cdot \gamma_{\omega \setminus \omega_i}$ should lie in $H_1(P_{\omega^a}, \partial P_{\omega^a}) \cap H_1(P_{\omega^b}, \partial P_{\omega^b}) = 0$. Consider $\omega_i = \{p, q\} \subset \omega^a$. Each connected component of $P_{\omega \setminus \omega_i}$ lies in some P_{ω^l}. We have $\gamma_{\omega \setminus \omega_i} = \sum\limits_{l=1}^{s} \gamma_{\omega^l \setminus \omega_i}$, where each summand corresponds to the connected components lying in $\omega^l \setminus \omega_i$. Since $\widetilde{\omega_i} \cdot \gamma_{\omega^l\setminus\omega_i} = 0$ for $l \neq a$, we have

$$\sum_{l=1}^{s} \alpha_{\omega^l} = \alpha_\omega = \left(\sum_{l=1}^{s} \sum_{\omega_i \subset \omega^l} \lambda_i\widetilde{\omega_i} + \sum_{\omega_i \not\subset \omega^l \forall l} \lambda_i\widetilde{\omega_i}\right) \cdot \left(\sum_{\omega_i \subset \omega} \gamma_{\omega\setminus\omega_i}\right)$$

$$= \left(\sum_{l=1}^{s} \sum_{\omega_i \subset \omega^l} \lambda_i\widetilde{\omega_i}\right) \cdot \left(\sum_{l=1}^{s} \sum_{\omega_j \subset \omega^l} \gamma_{\omega^l\setminus\omega_j}\right)$$

$$= \sum_{l=1}^{s} \left(\sum_{\omega_i \subset \omega^l} \lambda_i\widetilde{\omega_i}\right) \cdot \left(\sum_{\omega_j \subset \omega^l} \gamma_{\omega^l\setminus\omega_j}\right);$$

hence for any $\alpha_{\omega^l} \neq 0$ the projection $\psi_l\colon \sum\limits_{\omega_i \subset \omega} \lambda_i\widetilde{\omega_i} \to \sum\limits_{\omega_i \subset \omega^l} \lambda_i\widetilde{\omega_i}$ sends the space $\varphi_\omega(V)$ isomorphically to the 3-factorspace $\psi_l\varphi_\omega(V)$ of α_{ω^l}. Now consider the connected space P_{ω^l}.

Let the graph $K^1_{\omega^l}$ have a hanging vertex a. Then the facet F_a intersects only one facet among $\{F_t\}_{t\in\omega^l\setminus\{a\}}$, say F_b. Then for any $\omega_i = \{a, r\} \subset \omega^l$ we have $\widetilde{\omega_i} \cdot \gamma_{\omega^l\setminus\omega_i} = \pm[F_a] \cdot \gamma_{\omega^l\setminus\omega_i}$ is equal up to a scalar to the class in $H_1(P_{\omega^l}, \partial P_{\omega^l}, \mathbb{Q})$ of the single edge $F_a \cap F_b$ connecting two points on the same boundary cycle of P_{ω^l}. Hence $\widetilde{\omega_i} \cap \gamma_{\omega^l\setminus\omega_i} = 0$. Thus we have

$$\alpha_{\omega^l} = \left(\sum_{\omega_i \subset \omega^l} \lambda_i\widetilde{\omega_i}\right)\left(\sum_{\omega_j \subset \omega^l} \gamma_{\omega^l\setminus\omega_j}\right)$$

$$= \left(\sum_{\omega_i \subset \omega^l\setminus\{a\}} \lambda_i\widetilde{\omega_i}\right)\left(\sum_{\omega_j \subset \omega^l\setminus\{a\}} \gamma_{\omega^l\setminus\omega_j}\right).$$

Hence the mapping $\xi_a\colon \sum\limits_{\omega_i \subset \omega^l} \lambda_i\widetilde{\omega_i} \to \sum\limits_{\omega_i \subset \omega^l\setminus\{a\}} \lambda_i\widetilde{\omega_i}$ sends any nonzero vector in $\psi_l\varphi_\omega(V)$ to a nozero vector; therefore the 3-factorspace $\psi_l\varphi_\omega(V)$ of α_{ω^l} is mapped isomorphically to the 3-factorspace

$$\xi_a\psi_l\varphi_\omega(V) \subset \bigoplus_{\omega_i \subset \omega^l\setminus\{a\}} \widehat{H}_2(P_{\omega_i}, \partial P_{\omega_i})$$

of α_{ω^l}. This space has the dimension of at most the number of missing edges in $K^1_{\omega^l\setminus\{a\}}$. Let $r = |\omega^l \setminus \{a\}|$. Since $\alpha_{\omega^l} \neq 0$, $r \geqslant 3$. Since $P_{\omega^l\setminus\{a\}}$ is connected, the graph $K^1_{\omega^l\setminus\{a\}}$ has at least $r-1$ edges. Then the number of missing edges is at most $\frac{r(r-1)}{2} - (r-1) = \frac{(r-1)(r-2)}{2}$. Thus we have

$\dim V = \dim \xi_a\psi_l\varphi_\omega(V) \leqslant \frac{(r-1)(r-2)}{2} \leqslant \frac{(k-2)(k-3)}{2} < \frac{k(k-3)}{2}$, since $r \leqslant k-1$.

Now let the graph $K^1_{\omega^l}$ have no hanging vertices. Set l to be the number of its edges and $r = |\omega^l|$. We have $r \leqslant k$. Then $\dim V \leqslant \frac{r(r-1)}{2} - l$. Since the graph is connected and has no hanging vertices, $r \geqslant 3$ and $l \geqslant r$. Therefore $\dim V \leqslant \frac{r(r-1)}{2} - r = \frac{r(r-3)}{2} \leqslant \frac{k(k-3)}{2}$. If the equality holds, then $r = k = l$, and $\varphi_\omega(V) = \mathbb{Q}\langle\widetilde{\omega}_i \colon \omega_i \subset \omega\rangle$. Then K^1_ω is connected, has no hanging vertices and $l = k = |\omega|$ edges. We have that $2k$ is the sum of k vertex degrees of K^1_ω, each degree being at least 2. Then each degree is exactly 2; therefore K_ω is a chordless cycle; hence P_ω is a k-belt. This holds for any ω with $\alpha_\omega \neq 0$; hence $\alpha \in (\mathbf{B}_k \otimes \mathbb{Q}) \setminus \{0\}$.

(2) For a k-belt $\mathcal{B}_j$, $k \geqslant 4$, the space $\mathbb{Q}\langle\widetilde{\omega}_i \colon \omega_i \subset \omega(\mathcal{B}_j)\rangle$ is a $\frac{k(k-3)}{2}$-dimensional 3-factorspace of $\widetilde{\mathcal{B}_j}$. Indeed, for any $\omega_i \subset \omega(\mathcal{B}_j)$ take $\gamma_{i,j}$ to be the fundamental cycle in $\widehat{H}_2(P_{\omega(\mathcal{B}_j)\setminus\omega_i}, \partial P_{\omega(\mathcal{B}_j)\setminus\omega_i}, \mathbb{Q})$. Then $\widetilde{\omega}_p \cdot \gamma_{q,j} = \pm\delta_{p,q}\widetilde{\mathcal{B}_j}$ for any $\omega_p, \omega_q \subset \omega(\mathcal{B}_j)$, and for a combination $\tau = \sum_{\omega_i \subset \omega(\mathcal{B}_j)} \lambda_i\widetilde{\omega}_i$ with $\lambda_p \neq 0$ we have $\tau \cdot (\pm\frac{1}{\lambda_p}\gamma_{p,j}) = \widetilde{\mathcal{B}_j}$.

Now for any graded isomorphism $\varphi \colon H^*(\mathcal{Z}_P, \mathbb{Z}) \to H^*(\mathcal{Z}_Q, \mathbb{Z})$ we have the graded isomorphism $\widehat{\varphi} \colon H^*(\mathcal{Z}_P, \mathbb{Q}) \to H^*(\mathcal{Z}_Q, \mathbb{Q})$ with the embeddings $H^*(\mathcal{Z}_P, \mathbb{Z}) \subset H^*(\mathcal{Z}_P, \mathbb{Q})$, and $H^*(\mathcal{Z}_P, \mathbb{Z}) \subset H^*(\mathcal{Z}_P, \mathbb{Q})$. For any $\alpha \in H^{k+2}(\mathcal{Z}_P, \mathbb{Q})$ the isomorphism $\widehat{\varphi}$ induces the bijection between the 3-factorspaces of α and $\widehat{\varphi}(\alpha)$; hence $\mathrm{ind}_3(\alpha) = \mathrm{ind}_3(\widehat{\varphi}(\alpha))$. In particular, for any k-belt $\mathcal{B}_k$, $4 \leqslant k \leqslant m-2$, we have $\frac{k(k-3)}{2} = \mathrm{ind}_3(\widetilde{\mathcal{B}_k}) = \mathrm{ind}_3(\widehat{\varphi}(\widetilde{\mathcal{B}_k}))$; hence (1) implies that $\widehat{\varphi}(\widetilde{\mathcal{B}_k}) = \sum_j \mu_j\widetilde{\mathcal{B}'_{k,j}}$ for k-belts $\mathcal{B}'_{k,j}$ of Q. Since $\widehat{\varphi}(\widetilde{\mathcal{B}_k}) = \varphi(\widetilde{\mathcal{B}_k})$, we have $\mu_j \in \mathbb{Z}$, $\varphi(\widetilde{\mathcal{B}_k}) \in \mathbf{B}_k(Q)$; hence $\varphi(\mathbf{B}_k(P)) \subset \mathbf{B}_k(Q)$. The same argument for the inverse isomorphism implies that $\varphi(\mathbf{B}_k(P)) = \mathbf{B}_k(Q)$. □

Proposition 8.25: (Corollary 3.4, [28]) *For any k, $5 \leqslant k \leqslant m-2$, the set*

$$\{\pm\widetilde{\mathcal{B}_k} \colon \mathcal{B}_k \text{ is a } k\text{-belt}\} \subset H^{k+2}(\mathcal{Z}_P)$$

is B-rigid in the class of flag simple 3-polytopes without 4-belts.

Proof: Let P and Q be flag 3-polytopes without 4-belts, and $\varphi \colon H^*(\mathcal{Z}_P, \mathbb{Z}) \to H^*(\mathcal{Z}_Q, \mathbb{Z})$ be a graded isomorphism. From Proposition 8.24 we have $\varphi(\widetilde{\mathcal{B}_k}) = \sum_j \mu_j\widetilde{\mathcal{B}'_{k,j}}$ for k-belts $\mathcal{B}'_{k,j}$ of Q. Then for any $\omega_i \subset \omega(\mathcal{B}_k)$ we have $\widetilde{\omega}_i\gamma_{\omega(\mathcal{B}_k)\setminus\omega_i} = \widetilde{\mathcal{B}_k}$ for some $\gamma_{\omega(\mathcal{B}_k)\setminus\omega_i}$. Then $\varphi(\widetilde{\omega}_i)\varphi(\gamma_{\omega(\mathcal{B}_k)\setminus\omega_i}) = \sum_j \mu_j\widetilde{\mathcal{B}'_{k,j}}$.

Lemma 8.26: *Let* $\alpha \in H^{k+2}(\mathcal{Z}_P, \mathbb{Z})$, $4 \leqslant k \leqslant m-2$,

$$\alpha = \sum_{\omega} \alpha_\omega \in \bigoplus_{|\omega|=k} H_1(P_\omega, \partial P_\omega, \mathbb{Z}) \oplus \bigoplus_{|\omega|=k+1} \widehat{H}_2(P_\omega, \partial P_\omega, \mathbb{Z}).$$

If $\beta \in \widehat{H}_2(P_\tau, \partial P_\tau, \mathbb{Z})$, $\tau \neq \varnothing$, *divides* α, *then condition* $\alpha_\omega \neq 0$ *implies that* $|\omega| = k$, $\tau \subset \omega$, *and* β *divides* α_ω.

Proof: Let $\beta\gamma = \alpha$, where $\gamma = \sum_\eta \gamma_\eta$. Then from the multiplication rule we have $\alpha_\omega = 0$ for $|\omega| = k+1$, and $\beta\gamma_{\omega\setminus\tau} = \alpha_\omega$ for each nonzero α_ω. □

Proposition 8.19 implies that $\varphi(\widetilde{\omega_i}) = \pm\widetilde{\omega'_j}$; therefore by Lemma 8.26 the element $\widetilde{\omega'_j}$ is a divisor of any $\widetilde{\mathcal{B}'_{k,j}}$ with $\mu_j \neq 0$. But for a k-belt $\mathcal{B}'_{k,j}$ the element $\widetilde{\omega'_j}$ is a divisor if and only if $\omega'_j \subset \omega(\mathcal{B}'_{k,j})$. We see that the isomorphism φ maps the set $\{\pm\widetilde{\omega_i} \colon \omega_i \subset \omega(\mathcal{B}_k)\}$ bijectively to the corresponding set of any $\mathcal{B}'_{k,j}$ with $\mu_j \neq 0$. But such a set uniquely defines the k-belt; hence we have only one nonzero μ_j, which should be equal to ± 1. This finishes the proof. □

Proposition 8.27: (Lemma 3.5, [28]) *For any* k, $5 \leqslant k \leqslant m-2$ *the set*

$$\{\pm\widetilde{\mathcal{B}_k} \colon \mathcal{B}_k \textit{ is a } k\textit{-belt surrounding a facet}\} \subset H^{k+2}(\mathcal{Z}_P)$$

is B*-rigid in the class of flag simple* 3*-polytopes without* 4*-belts.*

Proof: Let the k-belt $\mathcal{B}_k = (F_{i_1}, \ldots, F_{i_k})$ surround a facet F_j of a flag simple 3-polytope P without 4-belts. Consider any facet F_l, $l \notin \{i_1, \ldots, i_k, j\}$. If $F_l \cap F_{i_p} \neq \varnothing$, and $F_l \cap F_{i_q} \neq \varnothing$, then $F_{i_p} \cap F_{i_q} \neq \varnothing$, else $(F_j, F_{i_p}, F_l, F_{i_q})$ is a 4-belt. Then $F_{i_p} \cap F_{i_q} \cap F_l$ is a vertex, since P is flag. Then $p - q = \pm 1 \mod k$, and $F_l \cap F_{i_r} = \varnothing$ for any $r \neq \{p, q\}$. Thus either F_l does not intersect facets in $\mathcal{B}_k$, or it intersects exactly one facet in $\mathcal{B}_k$, or it intersects two successive facets in $\mathcal{B}_k$ by their common vertex. Consider all elements $\beta \in H^{k+3}(\mathcal{Z}_P, \mathbb{Z})$ such that β is divided by any $\widetilde{\omega_i}$ with $\omega_i \subset \omega(\mathcal{B}_k)$. By Lemma 8.26 we have $\beta = \sum_{|\omega|=k+1} \beta_\omega$. Moreover, since any $\omega_i \subset \omega(\mathcal{B}_k)$ lies in ω, we have $\omega(\mathcal{B}_k) \subset \omega$; hence $\omega = \omega(\mathcal{B}_k) \sqcup \{s\}$ for some s. Since $P_{j \sqcup \omega(\mathcal{B}_k)}$ is contractible, we have $s \notin j \sqcup \omega(\mathcal{B}_k)$.

Lemma 8.28: *If* F_l *either does not intersect facets in the* k*-belt* $\mathcal{B}_k$, *or intersects exactly one facet in* $\mathcal{B}_k$, *or intersects exactly two successive facets in* $\mathcal{B}_k$ *by their common vertex, then the generator of the group* $H_1(P_{\omega(\mathcal{B}_k)\sqcup\{l\}}, \partial P_{\omega(\mathcal{B}_k)\sqcup\{l\}}, \mathbb{Z}) \simeq \mathbb{Z}$ *is divisible by* $\widetilde{\omega_i}$, *for any* $\omega_i \subset \omega(\mathcal{B}_k)$.

Proof: Let $\omega_i = \{i_p, i_q\}$. Since the facets F_{i_p} and F_{i_q} are not successive in $\mathcal{B}_k$, one of the facets F_{i_p} and F_{i_q} does not intersect F_l, say F_{i_p}. The facet F_l can not intersect both connected components of $P_{\omega(\mathcal{B}_k)\setminus\{i_p,i_q\}}$; hence $P_{\omega(\mathcal{B}_k)\sqcup\{l\}\setminus\{i_p,i_q\}}$ is disconnected. Let γ be the fundamental cycle of the connected component intersecting F_{i_p}. Then $\widetilde{\omega_i}\cdot\gamma = \pm[F_{i_p}]\gamma$ is a single-edge path connecting two boundary components of $P_{\omega(\mathcal{B}_k)\sqcup\{l\}}$; hence it is a generator of $H_1(P_{\omega(\mathcal{B}_k)\sqcup\{l\}}, \partial P_{\omega(\mathcal{B}_k)\sqcup\{l\}}, \mathbb{Z})$. This finishes the proof. □

From Lemma 8.28 we obtain that the are exactly $m-k-1$ linearly independent elements in $H^{k+3}(\mathcal{Z}_P,\mathbb{Z})$ divisible by all $\widetilde{\omega_i}$, $\omega_i \subset \omega(\mathcal{B}_k)$.

Now let $\varphi\colon H^*(\mathcal{Z}_P,\mathbb{Z}) \to H^*(\mathcal{Z}_Q,\mathbb{Z})$ be the isomorphims of graded rings for a flag 3-polytope Q without 4-belts, and let $\varphi(\widetilde{\mathcal{B}_k}) = \pm\widetilde{\mathcal{B}_k}'$ for $\mathcal{B}_k' = (F_{j_1}',\dots,F_{j_k}')$. Assume that $\mathcal{B}_k'$ does not surround any facet. If there is a facet $F_l', l\notin\omega(\mathcal{B}_k')$ such that $F_l'\cap F_{j_p}'\neq\varnothing$, $F_l'\cap F_{j_q}'\neq\varnothing$, and $F_{j_p}'\cap F_{j_q}'=\varnothing$ for some p,q, then without loss of generality assume that $p<q$, and $F_l'\cap F_{j_t}'=\varnothing$ for all $t\in\{p+1,\dots,q-1\}$. Then $\mathcal{B}_r' = (F_l', F_{j_p}', F_{j_{p+1}}',\dots,F_{j_q}')$ is an r-belt for $r=q-p+2\leqslant k$, and there are $\frac{r(r-3)}{2}-(r-3)=\frac{(r-2)(r-3)}{2}$ common divisors of $\widetilde{\mathcal{B}_r'}$ and $\widetilde{\mathcal{B}_k'}$ of the form $\widetilde{\omega_i'}$. We have $\varphi^{-1}(\widetilde{\mathcal{B}_r'}) = \pm\widetilde{\mathcal{B}_r}$ for some r-belt $\mathcal{B}_r$ with $\widetilde{\mathcal{B}_r}$ having $\frac{(r-2)(r-3)}{2}$ common divisors of the form $\widetilde{\omega_i}$ with $\widetilde{\mathcal{B}_k}$. Since $\mathcal{B}_k\neq\mathcal{B}_r$, there is $F_u\in\mathcal{B}_r\setminus\mathcal{B}_k$; hence $\widetilde{\mathcal{B}_k}$ and $\widetilde{\mathcal{B}_r}$ have at most $\frac{(r-2)(r-3)}{2}$ common divisors of the form $\widetilde{\omega_i}$, and the equality holds if and only if $\mathcal{B}_r\setminus\{F_u\}\subset\mathcal{B}_k$. Then $F_u\neq F_j$. Suppose F_u follows $F_v=F_{i_s}$ and is followed by $F_w=F_{i_t}$ in $\mathcal{B}_r$. Then $F_u\cap F_{i_s}\neq\varnothing$, $F_u\cap F_{i_t}\neq\varnothing$, and $F_{i_s}\cap F_{i_t}=\varnothing$. We have the 4-belt $(F_j,F_{i_s},F_u,F_{i_t})$. A contradiction. Hence any facet F_l', $l\in[m]\setminus\omega(\mathcal{B}_k')$ does not intersect two non-successive facets of $\mathcal{B}_k'$; hence either it does not intersect $\mathcal{B}_k'$, or intersects in exactly one facet, or intersects exactly two successive facets by their common vertex. Then by Lemma 8.28 we obtain $m-k$ linearly independent elements in $H^{k+3}(\mathcal{Z}_Q,\mathbb{Z})$ divisible by all $\widetilde{\omega_i'}$, $\omega_i'\subset\omega(\mathcal{B}_k')$. A contradiction. This proves that $\mathcal{B}_k'$ surrounds a facet. □

Proposition 8.29: (Lemma 3.6, [28]) *Let $\varphi\colon H^*(\mathcal{Z}_P,\mathbb{Z})\to H^*(\mathcal{Z}_Q,\mathbb{Z})$ be an isomorphism of graded rings, where P and Q are flag simple 3-polytopes without 4-belts. If $\mathcal{B}_1$ and $\mathcal{B}_2$ are belts surrounding adjacent facets, and $\varphi(\widetilde{\mathcal{B}_i}) = \pm\widetilde{\mathcal{B}_i'}$, $i=1,2$, then the belts $\mathcal{B}_1'$ and $\mathcal{B}_2'$ also surround adjacent facets.*

Proof: The proof follows directly from the following result:

Lemma 8.30: *Let P be a flag simple 3-polytope without 4-belts. Let a belt $\mathcal{B}_1$ surround a facet F_p, and a belt $\mathcal{B}_2$ surround a facet $F_q\neq F_p$. Then $F_p\cap F_q\neq\varnothing$ if and only if $\widetilde{\mathcal{B}_1}$ and $\widetilde{\mathcal{B}_2}$ have exactly one common divisor among $\widetilde{\omega_i}$.*

Proof: If $F_p \cap F_q \neq \varnothing$, then, since P is flag, $\mathcal{B}_1 \cap \mathcal{B}_2$ consists of two facets which do not intersect. On the other hand, let $F_p \cap F_q = \varnothing$, and $\{u, v\} \subset \omega(\mathcal{B}_1) \cap \omega(\mathcal{B}_2)$ with $F_u \cap F_v = \varnothing$. Then (F_u, F_p, F_v, F_q) is a 4-belt, which is a contradiction. □

Now let us prove the main theorem.

Theorem 8.31: (Theorem 3.1, [28]) *Let P be a flag simple 3-polytope without 4-belts, and Q be a simple 3-polytope. Then the isomorphism of graded rings $\varphi\colon H^*(\mathcal{Z}_P, \mathbb{Z}) \simeq H^*(\mathcal{Z}_Q, \mathbb{Z})$ implies the combinatorial equivalence $P \simeq Q$. In other words, any flag simple 3-polytope without 4-belts is B-rigid in the class of all simple 3-polytopes.*

Proof: By Corollaries 8.10 and 8.14 the polytope Q is also flag and has no 4-belts. Since P is flag, any of its facet is surrounded by a belt. By Proposition 8.27, for any belt $\mathcal{B}_k$ surrounding a facet, $\varphi(\widetilde{\mathcal{B}_k}) = \pm\widetilde{\mathcal{B}'_k}$ for a belt $\mathcal{B}'_k$ surrounding a facet.

Lemma 8.32: *Any belt $\mathcal{B}_k$ surrounds at most one facet of a flag simple 3-polytope without 4-belts.*

Proof: If a belt $\mathcal{B}_k = (F_{i_1}, \dots, F_{i_k})$ surrounds facets F_p and F_q on both sides, then $(F_{i_1}, F_p, F_{i_3}, F_q)$ is a 4-belt, which is a contradiction. □

From this lemma we obtain that the correspondence $\mathcal{B}_k \to \mathcal{B}'_k$ induces a bijection between the facets of P and the facets of Q. Then Proposition 8.29 implies that this bijection is a combinatorial equivalence. □

Remark 8.33: In fact Theorem 3.1 in [28] proves more general result involving arbitrary 2-dimensional simplicial complexes instead of simple 3-polytopes.

9. Quasitoric manifolds

9.1. *Finely ordered polytope*

Every face of codimension k may be written uniquely as

$$G(\omega) = F_{i_1} \cap \dots \cap F_{i_k}$$

for some subset $\omega = \{i_1, \dots, i_k\} \subset [m]$. Then faces $G(\omega)$ may be ordered lexicographically for each $1 \leqslant k \leqslant n$.

By permuting the facets of P if necessary, we may assume that the intersection $F_1 \cap \dots \cap F_n$ is a vertex v. In this case we describe P as *finely ordered*, and refer to v as the *initial vertex*, since it is the first vertex of P with respect to the lexicographical ordering.

Up to an affine transformation we can assume that $\boldsymbol{a}_1 = \boldsymbol{e}_1, \dots, \boldsymbol{a}_n = \boldsymbol{e}_n$.

9.2. *Canonical orientation*

We consider $\mathbb{R}^n$ as the standard real n-dimensional Euclidean space with the *standard basis* consisting of vectors $\boldsymbol{e}_j = (0, \dots, 1, \dots, 0)$ with 1 on the j-th place, for $1 \leqslant j \leqslant n$; and similarly for $\mathbb{Z}^n$ and $\mathbb{C}^n$. The standard basis gives rise to the *canonical orientation* of $\mathbb{R}^n$.

We identify $\mathbb{C}^n$ with $\mathbb{R}^{2n}$, sending $\boldsymbol{e}_j$ to $\boldsymbol{e}_{2j-1}$ and $i\boldsymbol{e}_j$ to $\boldsymbol{e}_{2j}$ for $1 \leqslant j \leqslant n$. This provides the *canonical orientation* for $\mathbb{C}^n$.

Since $\mathbb{C}$-linear maps from $\mathbb{C}^n$ to $\mathbb{C}^n$ preserve the canonical orientation, we may also regard an arbitrary complex vector space as canonically oriented.

We consider $\mathbb{T}^n$ as the standard n-dimensional torus $\mathbb{R}^n/\mathbb{Z}^n$ which we identify with the product of n unit circles in $\mathbb{C}^n$:

$$\mathbb{T}^n = \{(e^{2\pi i\varphi_1}, \dots, e^{2\pi i\varphi_n}) \in \mathbb{C}^n\},$$

where $(\varphi_1, \dots, \varphi_n) \in \mathbb{R}^n$. The torus $\mathbb{T}^n$ is also *canonically oriented.*

9.3. *Freely acting subgroups*

Let $H \subset \mathbb{T}^m$ be a subgroup of dimension $r \leqslant m - n$. Choosing a basis, we can write it in the form

$$H = \left\{(e^{2\pi i(s_{11}\varphi_1+\dots+s_{1r}\varphi_r)}, \dots, e^{2\pi i(s_{m1}\varphi_1+\dots+s_{mr}\varphi_r)}) \in \mathbb{T}^m\right\},$$

where $\varphi_i \in \mathbb{R},\ i = 1, \dots, r$ and $S = (s_{ij})$ is an integer $(m \times r)$-matrix which defines a monomorphism $\mathbb{Z}^r \to \mathbb{Z}^m$ onto a direct summand. For any subset $\omega = \{i_1, \dots, i_n\} \subset [m]$ denote by $S_{\widehat{\omega}}$ the $((m-n) \times r)$-submatrix of S obtained by deleting the rows $i_1, \dots, i_n$.

Write each vertex $v \in P^n$ as v_ω if $v = F_{i_1} \cap \dots \cap F_{i_n}$.

Exercise: The subgroup H acts freely on $\mathcal{Z}_P$ if and only if for every vertex v_ω the submatrix $S_{\widehat{\omega}}$ defines a monomorphism $\mathbb{Z}^r \hookrightarrow \mathbb{Z}^{m-n}$ onto a direct summand.

Corollary 9.1: *The subgroup H of rank $r = m - n$ acts freely on $\mathcal{Z}_P$ if and only if for any vertex $v_\omega \in P$ we have:*

$$\det S_{\widehat{\omega}} = \pm 1.$$

9.4. *Characteristic mapping*

Definition 9.2: An $(n \times m)$-matrix Λ gives a *characteristic mapping*

$$\ell\colon \{F_1, \dots, F_m\} \longrightarrow \mathbb{Z}^n$$

for a given simple polytope P^n with facets $\{F_1, \dots, F_m\}$ if the columns $\ell(F_{j_1}) = \lambda_{j_1}, \dots, \ell(F_{j_n}) = \lambda_{j_n}$ of Λ corresponding to any vertex v_ω form a basis for $\mathbb{Z}^n$.

Example: For a pentagon P_5^2 we have a matrix $\Lambda = \begin{pmatrix} 1\,0\,1\,0\,1 \\ 0\,1\,0\,1\,1 \end{pmatrix}$

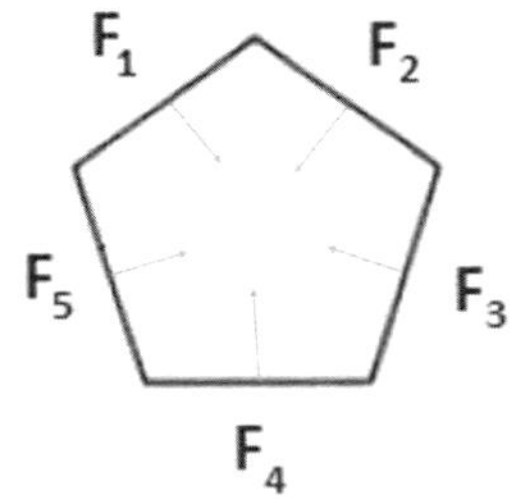

Fig. 36. Pentagon with normal vectors

Problem: For any simple n-polytope P find all integral $(n \times m)$-matrices

$$\Lambda = \begin{pmatrix} 1 & 0 & \dots & 0 & \lambda_{1,n+1} & \dots & \lambda_{1,m} \\ 0 & 1 & \dots & 0 & \lambda_{2,n+1} & \dots & \lambda_{2,m} \\ \vdots & \vdots & \ddots & \vdots & \vdots & \ddots & \vdots \\ 0 & 0 & \dots & 1 & \lambda_{n,n+1} & \dots & \lambda_{n,m} \end{pmatrix},$$

in which the column $\lambda_j = (\lambda_{1,j}, \dots, \lambda_{n,j})$ corresponds to the facet F_j, $j = 1, \dots, m$, and the columns $\lambda_{j_1}, \dots, \lambda_{j_n}$ corresponding to any vertex $v_\omega = F_{j_1} \cap \dots \cap F_{j_n}$ form a basis for $\mathbb{Z}^n$.

Note that there are simple n-polytopes, $n \geqslant 4$, admitting no characteristic functions.

Exercise: Let $C^n(m)$ be a combinatorial type of a *cyclic polytope* built as follows: take real numbers $t_1 < \dots < t_m$ and

$$C^n(t_1, \dots, t_m) = \operatorname{conv}\{(t_i, t_i^2, \dots, t_i^n), i = 1, \dots, m\}.$$

Prove that

(1) the combinatorial type of $C^n(t_1, \dots, t_m)$ does not depend on $t_1 < \dots < t_m$;

(2) the polytope $C^n(m)$ is simplicial;
(3) for $n \geqslant 4$ any two vertices of $C^n(m)$ are connected by an edge.

Conclude that for large m the dual simple polytope $C^n(m)^*$ admits no characteristic functions.

9.5. *Combinatorial data*

Definition 9.3: The *combinatorial quasitoric data* (P, Λ) consists of an *oriented* combinatorial simple polytope P and an integer $(n \times m)$-matrix Λ with the properties above.

The matrix Λ defines an epimorphism

$$\ell\colon \mathbb{T}^m \to \mathbb{T}^n.$$

The kernel of ℓ (which we denote $K(\Lambda)$) is isomorphic to $\mathbb{T}^{m-n}$.
Exercise: The action of $K(\Lambda)$ on $\mathcal{Z}_P$ is *free* due to the condition on the minors of Λ.

9.6. *Quasitoric manifold with the* (A, Λ)*-structure*

Construction: The quotient $M = \mathcal{Z}_P/K(\Lambda)$ is a $2n$-dimensional *smooth* manifold with an action of the n-dimensional torus $T^n = \mathbb{T}^m/K(\Lambda)$. We denote this action by α. It satisfies the *Davis–Januszkiewicz conditions*:

(1) α is locally isomorphic to the standard coordinatewise representation of $\mathbb{T}^n$ in $\mathbb{C}^n$;
(2) there is a projection $\pi\colon M \to P$ whose fibres are orbits of α.

We refer to $M = M(P, \Lambda)$ as the *quasitoric manifold associated with the combinatorial data* (P, Λ).

Let

$$P = \{\boldsymbol{x} \in \mathbb{R}^n \colon A\boldsymbol{x} + b \geqslant 0\}.$$

Definition 9.4: The manifold $M = M(P, \Lambda)$ is called the quasitoric manifold with (A, Λ)*-structure*.

Exercise: Suppose the $(n \times m)$-matrix $\Lambda = (I_n, \Lambda_*)$, where I_n is the unit matrix, gives a characteristic mapping

$$\ell\colon \{F_1, \dots, F_m\} \longrightarrow \mathbb{Z}^n.$$

Then the matrix $S = \begin{pmatrix} -\Lambda_* \\ I_{m-n} \end{pmatrix}$ gives the $(m-n)$-dimensional subgroup

$$H = \left\{(e^{2\pi i\psi_1}, \ldots, e^{2\pi i\psi_m}) \in \mathbb{T}^m\right\},$$

where

$$\psi_k = -\sum_{j=n+1}^{m} \lambda_{k,j}\varphi_{j-n},\ k = 1, \ldots, n;\ \psi_k = \varphi_{k-n},\ k = n+1, \ldots, m,$$

acting freely on $\mathcal{Z}_P$.

Example 9.5: Take $P = \Delta^2$. Let us describe the matrices A and Λ:

$$A = \begin{pmatrix} 1 & 0 \\ 0 & 1 \\ a_{31} & a_{32} \end{pmatrix}, \qquad \Lambda = \begin{pmatrix} 1 & 0 & \lambda_{13} \\ 0 & 1 & \lambda_{23} \end{pmatrix}, \qquad a_{31}, a_{32}, \lambda_{13}, \lambda_{23} \in \mathbb{Z}.$$

Since the normal vectors are oriented inside the polytope, $a_{31} < 0,\ a_{32} < 0$. Thus, up to combinatorial equivalence, one can take $a_{31} = a_{32} = -1$. The conditions on the characteristic mapping give

$$\begin{vmatrix} 0 & \lambda_{13} \\ 1 & \lambda_{23} \end{vmatrix} = \pm 1, \quad \begin{vmatrix} 1 & \lambda_{13} \\ 0 & \lambda_{23} \end{vmatrix} = \pm 1, \quad \Rightarrow \quad \lambda_{13} = \pm 1, \quad \lambda_{23} = \pm 1.$$

Therefore we have 4 structures (A, Λ).

Exercise: Let $P = \Delta^2$ and $\mathbb{C}P^2$ be the complex projective space with the canonical action of torus $\mathbb{T}^3$: $(t_1, t_2, t_3)(z_1 : z_2 : z_3) = (t_1 z_1 : t_2 z_2 : t_3 z_3)$.

(1) describe $\mathbb{C}P^2$ as $(S^5 \times_{\mathbb{T}^3} \mathbb{T}^2)$;
(2) describe the structure (A, Λ) such that $M(A, \Lambda)$ is $\mathbb{C}P^2$.

9.7. *A partition of a quasitoric manifold*

We have the homeomorphism

$$\mathcal{Z}_P \simeq \bigcup_{v_\omega - \text{vertex}} \mathcal{Z}_{P, v_\omega},$$

where

$$\mathcal{Z}_{P, v_\omega} = \prod_{j \in \omega} D_j^2 \times \prod_{j \in [m] \setminus \omega} S_j^1 \subset \mathbb{D}^{2m}.$$

Exercise: $\mathcal{Z}_{P, v_\omega} / K(\Lambda) \simeq \mathbb{D}_\omega^{2n}$.

Corollary 9.6: *We have the partition:*

$$M(P, \Lambda) = \bigcup_{v_\omega - \textit{vertex}} \mathbb{D}_\omega^{2n}.$$

9.8. *Stably complex structure and characteristic classes*

Denote by $\mathbb{C}_i$ the space of the 1-dimensional complex representation of the torus $\mathbb{T}^m$ induced from the standard representation in $\mathbb{C}^m$ by the projection $\mathbb{C}^m \to \mathbb{C}_i$ onto the ith coordinate. Let $\mathcal{Z}_P \times \mathbb{C}_i \to \mathcal{Z}_P$ be the trivial complex line bundle; we view it as an equivariant $\mathbb{T}^m$-bundle with the diagonal action of $\mathbb{T}^m$. By taking the quotient with respect to the diagonal action of $K = K(\Lambda)$ we obtain a T^n-equivariant complex line bundle

$$\rho_i\colon \mathcal{Z}_P \times_K \mathbb{C}_i \to \mathcal{Z}_P/K = M(P,\Lambda) \tag{9.1}$$

over the quasitoric manifold $M = M(P,\Lambda)$. Here

$$\mathcal{Z}_P \times_K \mathbb{C}_i = \mathcal{Z}_P \times \mathbb{C}_i \,/\, (\boldsymbol{tz}, \boldsymbol{t}w) \sim (\boldsymbol{z}, w) \text{ for any } \boldsymbol{t} \in K, \boldsymbol{z} \in \mathcal{Z}_P, w \in \mathbb{C}_i.$$

Theorem 9.7: (Theorem 6.6, [20]) *There is an isomorphism of real T^n-bundles over $M = M(P,\Lambda)$:*

$$\mathcal{T}M \oplus \underline{\mathbb{R}}^{2(m-n)} \cong \rho_1 \oplus \cdots \oplus \rho_m; \tag{9.2}$$

here $\underline{\mathbb{R}}^{2(m-n)}$ denotes the trivial real $2(m-n)$-dimensional T^n-bundle over M.

For the proof see (Theorem 7.3.15, [14]) .

Corollary 9.8: *Let $v_i = c_1(\rho_i) \in H^2(M(P,\Lambda),\mathbb{Z})$. Then for the total Chern class we have*

$$C(M(P,\Lambda)) = 1 + c_1 + \cdots + c_n = (1+v_1)\dots(1+v_m),$$

and for the total Pontryagin class we have

$$P(M(P,\Lambda)) = 1 + p_1 + \cdots + p_{[\frac{n}{2}]} = (1+v_1^2)\dots(1+v_m^2).$$

9.9. *Cohomology ring of the quasitoric manifold*

Theorem 9.9: [20] *We have*

$$H^*(M(P,\Lambda)) = \mathbb{Z}[v_1,\dots,v_m]/(J_{SR}(P) + I_{P,\Lambda}),$$

where $v_i = c_1(\rho_i)$, $J_{SR}(P)$ is the Stanley-Reisner ideal generated by monomials $\{v_{i_1}\dots v_{i_k}\colon F_{i_1} \cap \dots \cap F_{i_k} = \varnothing\}$, and $I_{P,\Lambda}$ is the ideal generated by the linear forms $\lambda_{i,1}v_1 + \cdots + \lambda_{i,m}v_m$ arising from the equality

$$\ell(F_1)v_1 + \cdots + \ell(F_m)v_m = 0.$$

For the proof see (Theorem 7.3.28, [14]).

Corollary 9.10: *If* $\Lambda = (I_n, \Lambda_*)$, *then*

$$H^2(M(P, \Lambda)) = \mathbb{Z}^{m-n}$$

with the generators $v_{n+1}, \ldots, v_m$.

Corollary 9.11:

(1) The group $H^k(M(P, \Lambda))$ *is nontrivial only for* k *even;*
(2) $M(P, \Lambda)$ *is even dimensional and orientable, hence the group* $H_k(M(P, \Lambda))$ *is nontrivial only for* k *even;*
(3) from the universal coefficient formula the abelian groups $H^*(M(P, \Lambda))$ *and* $H_*(M(P, \Lambda))$ *have no torsion.*

Corollary 9.12: *Let* P *be a flag polytope and* ℓ *be its characteristic function. Then*

$$H^*(M(P, \Lambda)) = \mathbb{Z}[v_1, \ldots, v_m]/(J_{SR} + I_{P,\Lambda}),$$

where the ideal J_{SR} *is generated by monomials* $v_i v_j$, *where* $F_i \cap F_j = \varnothing$, *and* $I_{P,\Lambda}$ *is generated by linear forms* $\lambda_{i,1} v_1 + \cdots + \lambda_{i,m} v_m$.

Corollary 9.13: *For any* $l = 1, \ldots, n$, *the cohomology group* $H^{2l}(M(P, \Lambda), \mathbb{Z})$ *is generated by monomials* $v_{i_1} \ldots v_{i_l}$, $i_1 < \cdots < i_l$, *corresponding to* $(n-l)$*-faces* $F_{i_1} \cap \cdots \cap F_{i_l}$.

Proof: We will prove this by induction on characteristic $\delta = \sum_{p_i > 1} p_i$ of a monomial $v_{i_1}^{p_1} \ldots v_{i_s}^{p_s}$ with $i_1 < \cdots < i_s$. Due to the relations from the ideal J_{SR} nonzero monomials correspond to faces $F_{i_1} \cap \cdots \cap F_{i_s} \neq \varnothing$. If $\delta = 0$, then we have the monomial we need. If $\delta > 0$, then take a vertex v in $F_{i_1} \cap \cdots \cap F_{i_s} \neq \varnothing$. Let Λ_v be the submatrix of Λ corresponding to the columns $\{j \colon v \in F_j\}$. Then by the definition of a characteristic function $\det \Lambda_v = \pm 1$. By integer elementary transformations of rows of the matrix Λ (hence of linear relations in the ideal $I_{P,\Lambda}$) we can make $\Lambda_v = E$. Let $p_k > 1$. The variable v_{i_k} can be expressed as a linear combination $v_{i_k} = \sum\limits_{j \notin \{i_1, \ldots, i_s\}} a_j v_j$. Then

$$v_{i_1}^{p_1} \ldots v_{i_s}^{p_s} = \sum_{j \notin \{i_1, \ldots, i_s\}} a_j v_{i_1}^{p_1} \ldots v_{i_k}^{p_k - 1} \ldots v_{i_s}^{p_s} v_j,$$

where on the right side we have the sum of monomials with value less than δ. This finishes the proof. □

For any $\xi = (i_1, \ldots, i_{n-1}) \subset [m]$ set $\xi_i = (\xi, i),\ i \notin \xi$.

Exercise:

1. Show that for any $\xi = (i_1, \ldots, i_{n-1}) \subset [m]$ there are the relations

$$\sum_{j=1}^{m} \varepsilon(\xi_j) v_j = 0 \tag{9.3}$$

where $\varepsilon(\xi_j) = \det |\ell(F_{i_1}), \ldots, \ell(F_{i_{n-1}}), \ell(F_j)|$.

2. Show that there is a graded ring isomorphism

$$H^*(M(P, \Lambda)) = \mathbb{Z}[P]/J$$

where J is the ideal generated by the relations (9.3).

Exercise: Show that for any vertex $v_\omega = F_{i_1} \bigcap \cdots \bigcap F_{i_n},\ \omega = (i_1, \ldots, i_n)$, there are the relations

$$v_{i_n} = -\varepsilon(\xi_{i_n}) \sum_j \varepsilon(\xi_j)\, v_j$$

where $\xi = (i_1, \ldots, i_{n-1}),\ j \in [m \backslash \xi_{i_n}]$.

Exercise: Show that for any vertex $v_\omega = F_{i_1} \bigcap \cdots \bigcap F_{i_n},\ \omega = (i_1, \ldots, i_n)$, there are the relations

$$v_{i_n}^2 = -\varepsilon(\xi_{i_n}) \sum_j \varepsilon(\xi_j)\, v_{i_n} v_j$$

where $j \in [m \backslash \xi_{i_n}]$, but $F_{i_n} \bigcap F_j \neq \emptyset$.

9.10. *Geometrical realization of cycles of quasitoric manifolds*

The fundamental notions of algebraic topology were introduced in the classical work by Poincaré [43]. Among them there were notions of cycles and homology. Quasitoric manifolds give nice examples of manifolds such that original notions by Poincaré obtain explicit geometric realization.

Let M^k be a smooth oriented manifold such that the groups $H_*(M^k, \mathbb{Z})$ have no torsion. There is the classical Poincaré duality $H_i(M^k, \mathbb{Z}) \simeq H^{k-i}(M^k, \mathbb{Z})$. Moreover, according to the Milnor-Novikov theorem [39, 40, 41] for any cycle $a \in H_l(M^k, \mathbb{Z})$ there is a smooth oriented manifold N^l and a continuous mapping $f\colon N^l \to M^k$ such that $f_*[N^l] = a$. For the homology groups of any quasitoric manifold there is the following remarkable geometrical interpretation of this result. Note that the odd homology groups of any quasitoric manifold are trivial.

Theorem 9.14:

(1) The homology group $H_{2n-2}(M(P,\Lambda),\mathbb{Z})$ of the quasitoric manifold $M^{2n}(P,\Lambda)$ is generated by embedded quasitoric manifolds $M_i^{2n-2}(P,\Lambda)$, $i=1,\ldots,m$, of facets of P. The embedding of the manifold $M_i^{2n-2}(P,\Lambda)\subset M(P,\Lambda)$ gives the geometric realization of the cycle Poincaré dual to the cohomology class $v_i\in H^2(M(P,\Lambda),\mathbb{Z})$ defined above.

(2) For any i the homology group $H_{2i}(M(P,\Lambda),\mathbb{Z})$ is generated by embedded quasitoric manifolds corresponding to all i-faces $F_{j_1}\cap\cdots\cap F_{j_{n-i}}$ of the polytope P. These manifolds can be described as complete intersections of manifolds $M_{j_1}^{2n-2}(P,\Lambda)$, $\ldots$, $M_{j_{n-i}}^{2n-2}(P,\Lambda)$.

The proof of the theorem follows directly from the above results on the cohomology of quasitoric manifolds and geometric interpretation of the Poincare duality in terms of Thom spaces [48].

9.11. *Four-color problem*

Classical formulation: Given any partition of a plane into *contiguous regions*, producing a figure called a map, two regions are called *adjacent* if they share a common boundary that is not a corner, where corners are the points shared by three or more regions.

Problem: *No more than four colors are required to color the regions of the map so that no two adjacent regions have the same color.*

The problem was first proposed on October 23, 1852, when Francis Guthrie, while trying to color the map of counties of England, noticed that only four different colors were needed.

The four-color problem became well-known in 1878 as a hard problem when Arthur Cayley suggested it for discussion during the meeting of the London mathematical society.

The four-color problem was solved in 1976 by Kenneth Appel and Wolfgang Haken. It became the first major problem solved using a computer. For the details and the history of the problem see [52]. One of the central topics of this monograph is "how the problem was solved".

Example 9.15: Platonic solids.
The octahedron can be colored in 2 colors.
The cube and the icosahedron can be colored into 3 colors.
The tetrahedron and the dodecahedron can be colored into 4 colors.

Exercise: Color all the Archimedean solids.

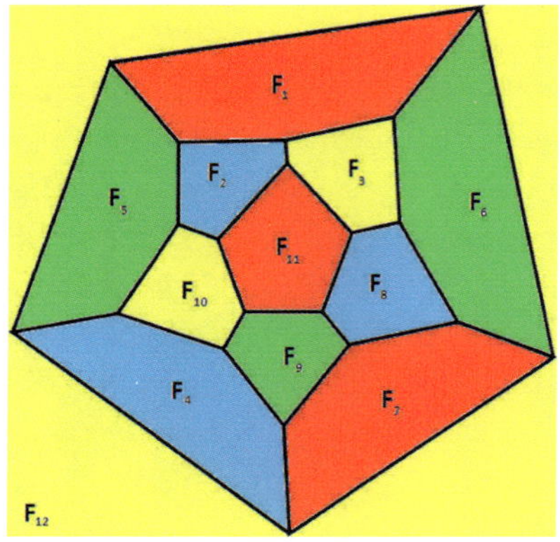

Fig. 37. Coloring of the dodecahedron

9.12. *Quasitoric manifolds of 3-dimensional polytopes*

Let P be a simple 3-polytope. Then ∂P is homeomorphic to the sphere S^2 partitioned into polygons $F_1, \dots, F_m$. By the four color theorem there is a coloring $\varphi\colon \{F_1, \dots, F_m\} \to \{1, 2, 3, 4\}$ such that adjacent facets have different colors.

Let $\boldsymbol{e}_1, \boldsymbol{e}_2, \boldsymbol{e}_3$ be the standard basis for $\mathbb{Z}^3$, and $\boldsymbol{e}_4 = \boldsymbol{e}_1 + \boldsymbol{e}_2 + \boldsymbol{e}_3$.

Proposition 9.16: *The mapping* $\ell\colon \{F_1, \dots, F_m\} \to \mathbb{Z}^3\colon \ell(F_i) = \boldsymbol{e}_{\varphi(F_i)}$ *is a characteristic function.*

Corollary 9.17:

- *Any simple 3-polytope P has combinatorial data (P, Λ) and the quasitoric manifold $M(P, \Lambda)$;*
- *Any fullerene has a quasitoric manifold.*

Since a fullerene is a flag polytope, the cohomology ring of any its quasitoric manifold is described by Corollary 9.12.
Exercise: Find a 4-coloring of the barrel (Figure 38).

9.13. *Cohomological rigidity of Pogorelov polytopes*

As it was mentioned in Section 8 one of the initial problems of algebraic topology is to find a wide class of manifolds such that in this class an isomorphism of cohomology rings implies a diffeomorphism of manifolds. Toric topology produces a large family of such manifolds consisting of 6-dimensional quasitoric manifolds over Pogorelov polytopes.

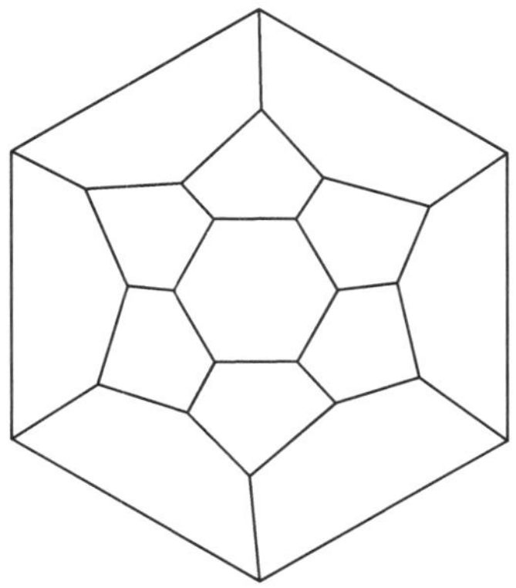

Fig. 38. Schlegel diagram of the barrel

Theorem 9.18: ([11]) *Let (P, Λ_P) and (Q, Λ_Q) be two flag simple 3-polytopes without 4-belts equipped with characteristic mappings. Then the following conditions are equivalent:*

(1) There is an isomorphism of graded rings $H^*(M(P, \Lambda_P), \mathbb{Z}) \simeq H^*(M(Q, \Lambda_Q), \mathbb{Z})$;
(2) There is a diffeomorphism $M(P, \Lambda_P) \simeq M(Q, \Lambda_Q)$;
(3) There is a combinatorial equivalence $P \simeq Q$ such that $A\Lambda_P B = \Lambda_Q$, where $A \in Gl_3(\mathbb{Z})$, and B is a diagonal $m \times m$ matrix consisting of ± 1.

10. Construction of fullerenes

10.1. *Number of combinatorial types of fullerenes*

Definition 10.1: Two combinatorially nonequivalent fullerenes with the same number p_6 are called *combinatorial isomers*.

Let $F(p_6)$ be the number of combinatorial isomers with given p_6.

From the results by W. Thurston [49] it follows that $F(p_6)$ grows like p_6^9.

There is an effective algorithm of combinatorial enumeration of fullerenes using supercomputer (Brinkmann-Dress [3], 1997). It gives:

p_6	0	1	2	3	4	5	6	7	8	...	75
$F(p_6)$	1	0	1	1	2	3	6	6	15	...	46.088.157

We see that for large values of p_6 the number of combinatorial isomers is very large. Hence there is an important problem to study different structures on the set of fullerenes.

10.2. *Growth operations*

The well-known problem [4, 33] is to find a simple set of operations sufficient to construct arbitrary fullerenes from the dodecahedron.

Definition 10.2: A *patch* is a disk bounded by a simple edge-cycle on the boundary of a simple 3-polytope.

Definition 10.3: A *growth operation* is a combinatorial operation that gives a new 3-polytope Q from a simple 3-polytope P by substituting a new patch with the same boundary and more facets for the patch on the boundary of P.

The Endo-Kroto operation [26] (Figure 39) is the simplest example of a growth operation that changes a fullerene into a fullerene. It was proved in [4] that there is

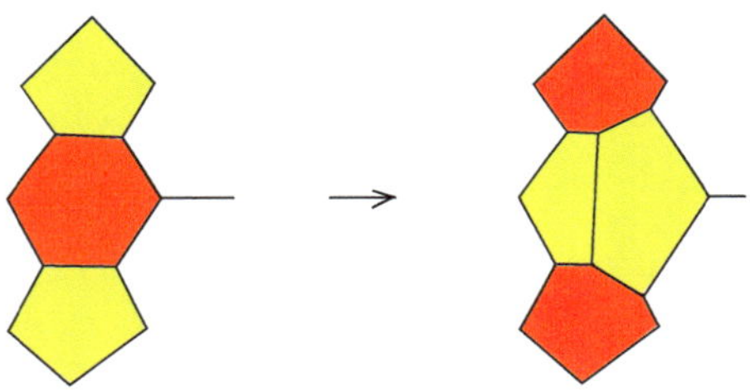

Fig. 39. Endo-Kroto operation

no finite set of growth operations transforming fullerenes into fullerenes sufficient to construct arbitrary fullerenes from the dodecahedron. In [33] the example of an infinite set was found. Our main result is the following (see [8]): if we allow at intermediate steps polytopes with at most one singular face (a quadrangle or a heptagon), then only 9 growth operations (induced by 7 truncations) are sufficient.
Exercise: Starting from the Barrel fullerene (see Figure 38) using the Endo-Kroto operation construct a fullerene with arbitrary $p_6 \geqslant 2$.

10.3. (s, k)*-truncations*

First we mention a well-known result about construction of simple 3-polytopes.

Theorem 10.4: *(Eberhard (1891), Brückner (1900)) A 3-polytope is simple if and only if it is combinatorially equivalent to a polytope obtained from the tetrahedron by a sequence of* ***vertex, edge*** *and* $(2, k)$*-truncations.*

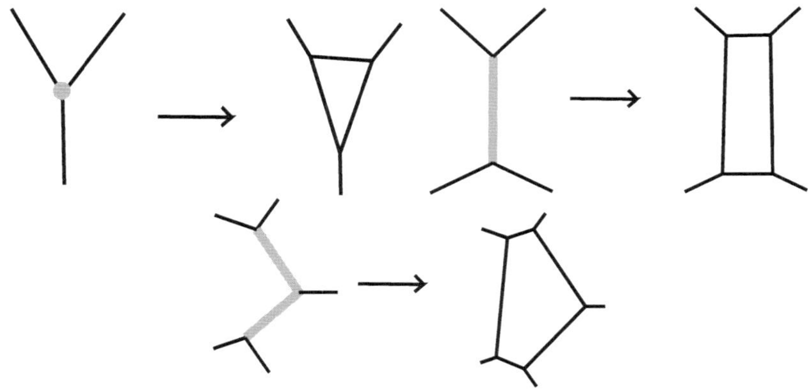

Fig. 40. Vertex-, edge- and $(2, k)$-truncations

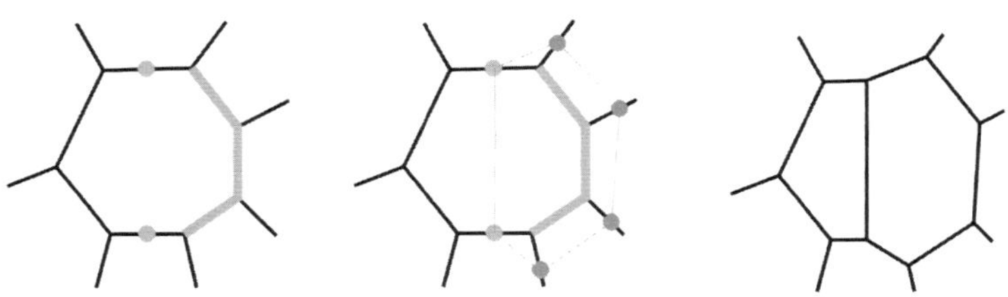

Fig. 41. $(3, 7)$-truncation

Construction (**(s, k)-truncation**): Let F_i be a *k-gonal face* of a simple 3-polytope P.

- choose *s consequent edges* of F_i;
- rotate the supporting hyperplane of F_i around the axis passing through the midpoints of adjacent two edges (one on each side);
- take the corresponding hyperplane truncation.

We call it an *(s, k)-truncation.*

Example 10.5:

(1) Vertex truncation is a $(0, k)$-truncation.
(2) Edge truncation is a $(1, k)$-truncation.
(3) The Endo-Kroto operation is a $(2, 6)$-truncation.

The next result follows from the definitions.

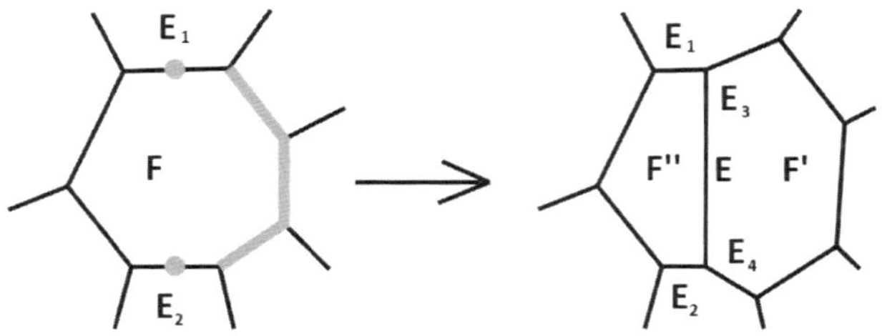

Fig. 42. (s, k)-truncation

Proposition 10.6:

- *Under the (s, k)-truncation of the polytope P its facets that do not contain the edges E_1 and E_2 (see Figure 42) preserve the number of sides.*
- *The facet F is split into two facets: an $(s+3)$-gonal facet F' and a $(k-s+1)$-gonal facet F'', $F' \cap F'' = E$.*
- *The number of sides of each of the two facets adjacent to F along the edges E_1 and E_2 increases by one.*

Remark 10.7: We see that an (s, k)-truncation is a combinatorial operation and is always defined. It is easy to show that the straightening along the edge E on the right side is a combinatorially inverse operation. It is not always defined.

Definition 10.8: If the facets intersecting F by E_1 and E_2 (see Figure 42) are m_1- and m_2-gons respectively, then we also call the corresponding operation an *$(s, k; m_1, m_2)$-truncation.*

For $s = 1$, combinatorially the $(1, k; m_1, m_2)$-truncation is the same operation as the $(1, t; m_1, m_2)$-truncation of the same edge of the other facet containing it. We call this operation simply a *$(1; m_1, m_2)$-truncation.*

Remark 10.9: Let P be a flag simple polytope. Then any (s, k)-truncation is a growth operation. Indeed, for $s = 0$ and $s = k - 2$ we have the vertex truncation, which can be considered as the substitution of the corresponding fragment for the three facets containing the vertex. For $0 < s < k - 2$, since P is flag, the facets F_{i_1} and $F_{i_{s+2}}$ intersecting F by edges adjacent to truncated edges do not intersect; hence the union $F_{i_1} \cup F \cup F_{i_{s+2}}$ is bounded by a simple edge-cycle (see Figure 43 on the left). After the (s, k)-truncation the union of facets $F' \cup F'' \cup F_{i_1} \cup F_{i_{s+2}}$ is bounded by combinatorially the same simple edge-cycle. We describe this operation by the scheme on Figure 43 on the right.

For $s = 1$ as mentioned above the edge-truncation can be considered as a $(1, k)$-truncation and a $(1, t)$-truncation for two facets containing the truncated edge: an

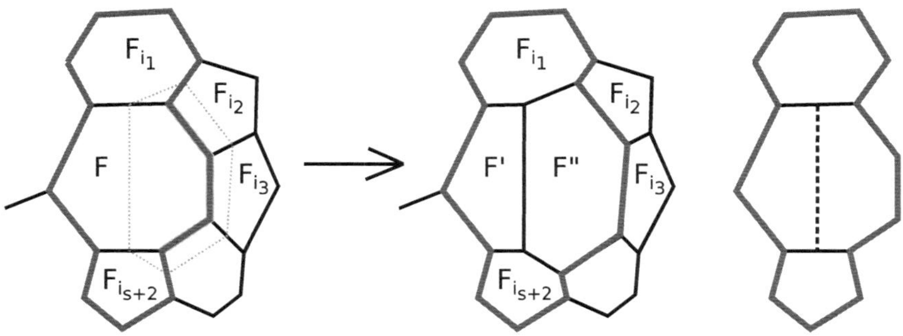

Fig. 43. (s, k)-truncation as a growth operation

s-gon and a t-gon. This gives two different patches, which differ by one facet.

Exercise: Consider the set of $k - s - 2$ edges of the face F that are not adjacent to the s edges defining the (s, k)-truncation. The polytope Q' obtained by the $(k - s - 2, k)$-truncation along these edges is combinatorially equivalent to the polytope Q. In particular

- The $(k - 3, k)$-truncation is combinatorially equivalent to the edge truncation;
- The $(k - 2, k)$-truncation is combinatorially equivalent to the vertex truncation.

Exercise: Let P be a flag 3-polytope. Then the polytope obtained from P by an (s, k)-truncation is flag if and only if $0 < s < k - 2$.

In [7] the analog of Theorem 10.4 for flag polytopes was proved.

Theorem 10.10: *A simple 3-polytope is flag if and only if it is combinatorially equivalent to a polytope obtained from the cube by a sequence of* ***edge truncations*** *and* $(2, k)$***-truncations,*** $k \geqslant 6$.

10.4. *Construction of fullerenes by truncations*

Definition 10.11: Let $\mathfrak{F}_{-1}$ be the set of combinatorial simple polytopes with all facets pentagons and hexagons except for one singular quadrangle facet.

Let $\mathfrak{F}$ be the set of all fullerenes.

Let $\mathfrak{F}_1$ be the set of simple polytopes with one singular heptagon facet adjacent to a pentagon such that either there are two pentagons with the common edge intersecting the heptagon and a hexagon (we will denote this fragment F_{5567}, see Figure 44), or for any two adjacent pentagons exactly one of them is adjacent to the heptagon. Set $\mathfrak{F}_s = \mathfrak{F}_{-1} \sqcup \mathfrak{F} \sqcup \mathfrak{F}_1$ to be the set of *singular fullerenes*.

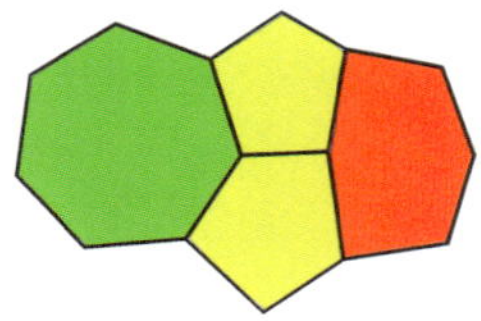

Fig. 44. Fragment F_{5567}

Theorem 10.12: *Any polytope in $\mathfrak{F}_s$ can be obtained from the dodecahedron by a sequence of p_6 truncations:* $(1;4,5)$-, $(1;5,5)$-, $(2,6;4,5)$-, $(2,6;5,5)$-, $(2,6;5,6)$-, $(2,7;5,5)$-, *and* $(2,7;5,6)$-, *in such a way that intermediate polytopes belong to $\mathfrak{F}_s$.*

More precisely:

(1) any polytope in $\mathfrak{F}_{-1}$ can be obtained by a $(1;5,5)$- *or* $(1;4,5)$-*truncation from a fullerene or a polytope in $\mathfrak{F}_{-1}$ respectively;*
(2) any polytope in $\mathfrak{F}_1$ can be obtained by a $(2,6;5,6)$- *or* $(2,7;5,6)$-*truncation from a fullerene or a polytope in $\mathfrak{F}_1$ respectively;*
(3) any fullerene can be obtained by a $(2,6;5,5)$-, $(2,6;4,5)$-, *or* $(2,7;5,5)$-*truncation from a fullerene or a polytope from $\mathfrak{F}_{-1}$ or $\mathfrak{F}_1$ respectively.*

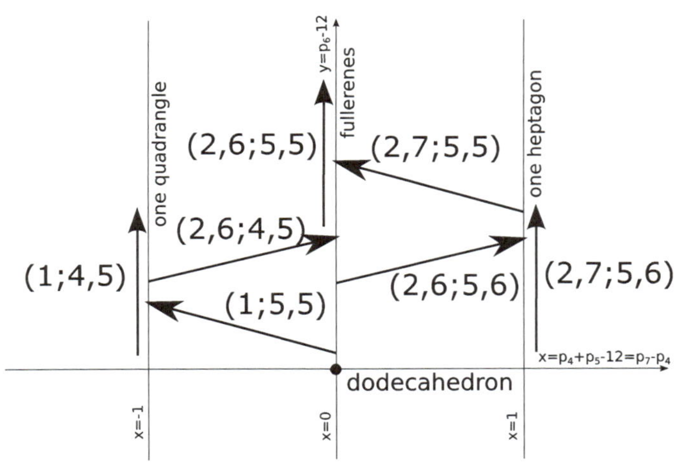

Fig. 45. Scheme of the truncation operations

Proof: By Theorems 4.11 and 4.15 any polytope in $\mathfrak{F}_s$ has no 3-belts and the only possible 4-belt surrounds a quadrangular facet. Hence for any edge the operation of straightening is well-defined.

For (1) we need the following result.

Lemma 10.13: *There are no polytopes in $\mathfrak{F}_{-1}$ with the quadrangle surrounded by pentagons.*

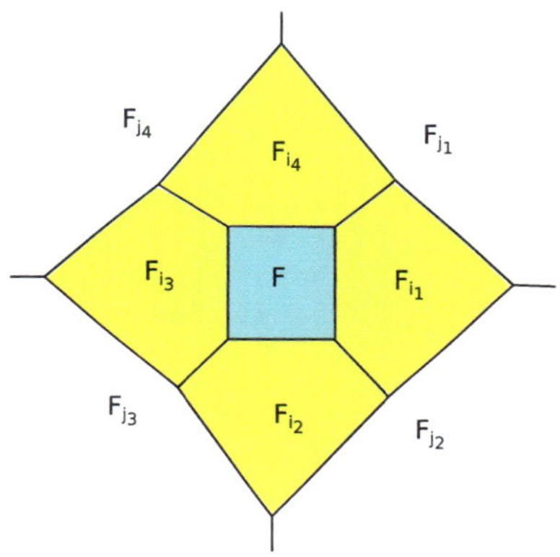

Fig. 46. Quadrangle surrounded by pentagons

Proof: Let the quadrangle F be surrounded by pentagons F_{i_1}, F_{i_2},F_{i_3}, and F_{i_4} as drawn in Figure 46. By Theorem 4.15 we have the 4-belt $\mathcal{B} = (F_{i_1}, F_{i_2}, F_{i_3}, F_{i_4})$ surrounding F, and there are no other 4-belts. Let $\mathcal{L} = (F_{j_1}, F_{j_2}, F_{j_3}, F_{j_4})$ be a 4-loop that borders $\mathcal{B}$ along its boundary component different from ∂F. Its consequent facets are different. If $F_{j_1} = F_{j_3}$, then we obtain a 4-belt $(F, F_{i_1}, F_{j_1}, F_{i_3})$, which is a contradiction. Similarly $F_{j_2} \neq F_{j_4}$. Hence $\mathcal{L}$ is a simple 4-loop. Since it is not a 4-belt its two opposite facets intersect, say $F_{j_1} \cap F_{j_3} \neq \varnothing$. Then $F_{j_1} \cap F_{j_2} \cap F_{j_3}$ is a vertex and F_{j_2} is a quadrangle. A contradiction. This proves the lemma. □

Thus, for any polytope P in $\mathfrak{F}_{-1}$ its quadrangle F is adjacent to some hexagon F_i by some edge E. Now straighten the polytope P along the edge of F adjacent to E to obtain a new polytope Q with a pentagon instead of F_i and a pentagon or a quadrangle instead of the facet F_j adjacent to F by the edge of F opposite to E. In the first case Q is a fullerene and P is obtained from Q by a $(1; 5, 5)$-truncation.

In the second case $Q \in \mathfrak{F}_{-1}$ and P is obtained from Q by a $(1; 4, 5)$-truncation. This proves (1).

To prove (2) note that if $P \in \mathfrak{F}_1$ contains the fragment F_{5567}, then straightening along the common edge of pentagons gives a fullerene Q such that P is obtained from Q by a $(2, 6; 5, 6)$-truncation. □

Lemma 10.14: *If $P \in \mathfrak{F}_1$ does not contain the fragment F_{5567}, then*

(1) P does not contain fragments in Figure 47;
(2) for any pair of adjacent pentagons they do not intersect any other pentagons.

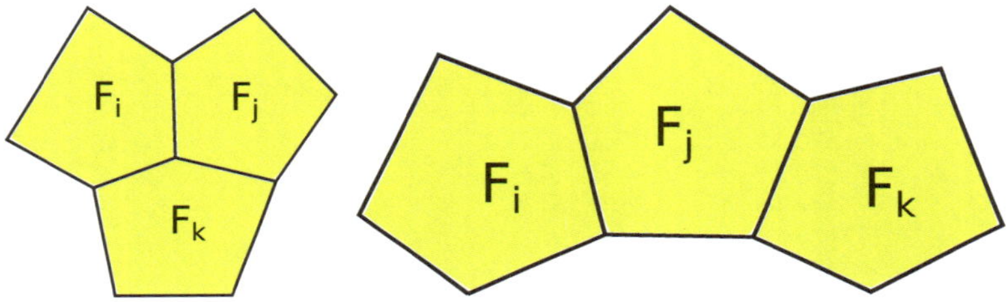

Fig. 47. Fragments that can not be present on the polytope in $\mathfrak{F}_1$ without the fragment F_{5567}

Proof: Let F_i, F_j, F_k be pentagons with a common vertex. Then for the pair (F_i, F_j) exactly one pentagon intersects the heptagon F, say F_i. Also for the pair (F_j, F_k) exactly one pentagon intersects F. This should be F_k. For the pair (F_i, F_k) this is a contradiction.

Suppose the pentagon F_j intersects pentagons F_i and F_k by non-adjacent edges as shown in Figure 47 on the right. The heptagon F should intersect exactly one pentagon of each pair (F_i, F_j) and (F_j, F_k). Then either it intersects F_i and F_k and does not intersect F_j, or it intersects F_j and does not intersect F_i and F_k. By Theorem 4.11 P has no 3-belts; hence $F_i \cap F_k = \varnothing$. In the first case we obtain the 4-belt (F, F_i, F_j, F_k), which contradicts Theorem 4.15. In the second case F intersects F_j by one of the three edges different from $F_i \cap F_j$ and $F_j \cap F_k$. But any of these edges intersects either F_i, or F_k, which is a contradiction.

Thus we have proved part (1) of the lemma. Let some pentagon of the pair of adjacent pentagons (F_i, F_j), say F_j, intersects some other pentagon F_k. If the edges of intersection are adjacent in F_j, then we obtain the fragment in Figure 47 on the left. Else we obtain the fragment in Figure 47 on the right. A contradiction. This proves part (2) of the lemma. □

Now assume that P does not contain the fragment F_{5567}.

Let (F_i, F_j) be a pair of two adjacent pentagons with F_i intersecting the heptagon F. Then by Lemma 10.14 we obtain the fragment in Figure 48 a). Since by Proposition 3.7 the pair of adjacent facets is surrounded by a belt, the adjacent pentagons do not intersect other pentagons and exactly one of them intersects the heptagon. The straightening along the edge $F_i \cap F_p$ gives a polytope Q such that P is obtained from Q by a $(2,7;5,6)$-truncation. Q has all facets pentagons and hexagons except for one heptagon adjacent to a pentagon. Q contains the fragment F_{5567}; hence it belongs to $\mathfrak{F}_1$.

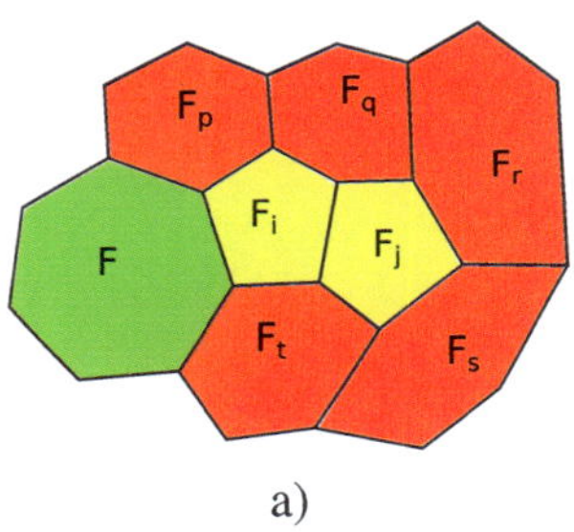

Fig. 48. a) Facets surrounding the pair of adjacent pentagons; b) the same fragment after the straightening

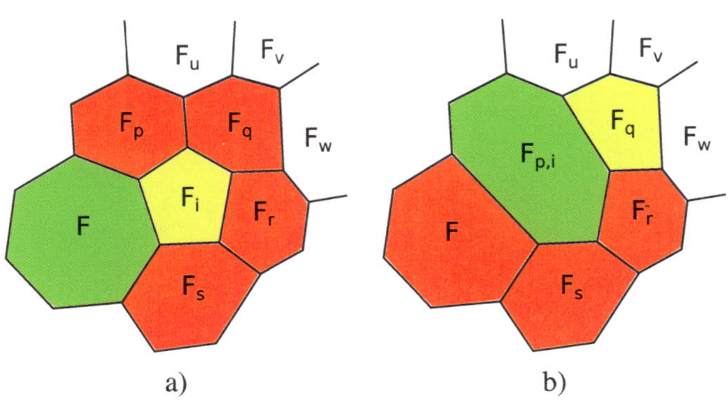

Fig. 49. a) Facets surrounding a pentagon adjacent to the heptagon; b) the same fragment after the straightening

Now let P have no adjacent pentagons. Consider a pentagon adjacent to the heptagon F. Then it is surrounded by a 5-belt $\mathcal{B}$ consisting of the heptagon and four hexagons (Figure 49 a). The straightening along the edge $F_p \cap F_i$ gives a simple polytope Q with the fragment on Figure 49 b) instead the fragment on Figure 49 a). The polytope Q has all facets pentagons and hexagons except for one heptagon $F_{p,i}$ adjacent to the pentagon F_q. Then P is obtained from Q by a $(2,7;5,6)$-truncation. We claim that $Q \in \mathfrak{F}_1$. Indeed, if Q has the fragment F_{5567}, it is true. If Q has no such fragments consider two adjacent pentagons of Q. The polytopes P and Q have the same structure outside the fragments in consideration; hence Q has the same pentagons as P except for F_q, which appeared instead of F_i. Also P has all pentagons isolated; hence one of the adjacent pentagons is F_q. The second pentagon F_t should be adjacent to the hexagon F_q in P; hence it should be one of the facets F_u, F_v, or F_w in Figure 49 a). Each of these facets is different from F, since they lie outside the 5-belt $\mathcal{B}$ containing F. And in each case the pentagon F_t is isolated in P by assumption.

If $F_t = F_u$, then F_v is a hexagon, since $F_v \neq F$ and F_v is not a pentagon. Then Q contains the fragment F_{5567}, which is a contradiction. Thus F_u is a hexagon.

If F_t is one of the facets F_v and F_w, then the other facet is a hexagon and there are no pairs of adjacent pentagons in Q other than (F_q, F_t). Each of the facets F_v, F_w in Q belongs to the 5-belt surrounding F_q together with $F_{p,i}$ and is not successive with it; hence F_v and F_w do not intersect $F_{p,i}$ in Q. Thus $F_t \cap F_{p,i} = \varnothing$ and $Q \in \mathfrak{F}_1$. This proves (2).

To prove (3) consider a fullerene P. If it contains the fragment in Figure 50 a) then the straightening along the edge $F_i \cap F_j$ gives a fullerene Q such that P is obtained from Q by a $(2,6;5,5)$-truncation (the Endo-Kroto operation). Let P contain no such fragments.

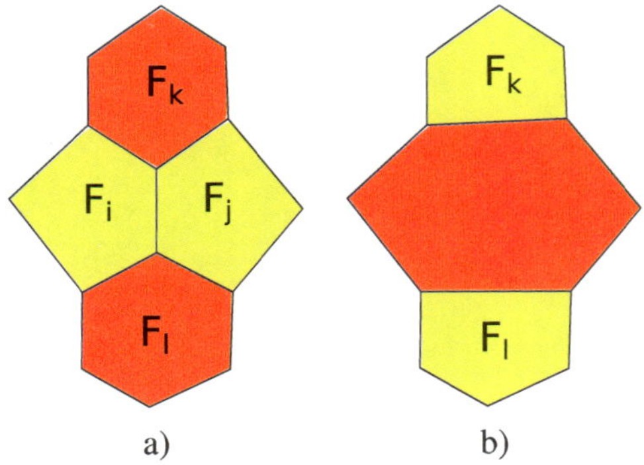

Fig. 50. a) Two adjacent pentagons with two hexagons; b) the same fragment after the straightening

If P has two adjacent pentagons, then one of the connected components of the union of pentagons has more than two pentagons. If P is not combinatorially equivalent to the dodecahedron, then each component is a sphere with holes. Consider the connected component with more than one pentagon and a vertex v on its boundary lying in two pentagons F_i and F_j. Then the third face containing v is a hexagon. Since P contains no fragments in Figure 50 a), the other facet intersecting the edge $F_i \cap F_j$ by the vertex is a pentagon and we obtain the fragment in Figure 51 a). Then the straightening along the edge $F_i \cap F_j$ gives the polytope $Q \in \mathfrak{F}_{-1}$ such that P is obtained from Q by a $(2,6;4,5)$-truncation.

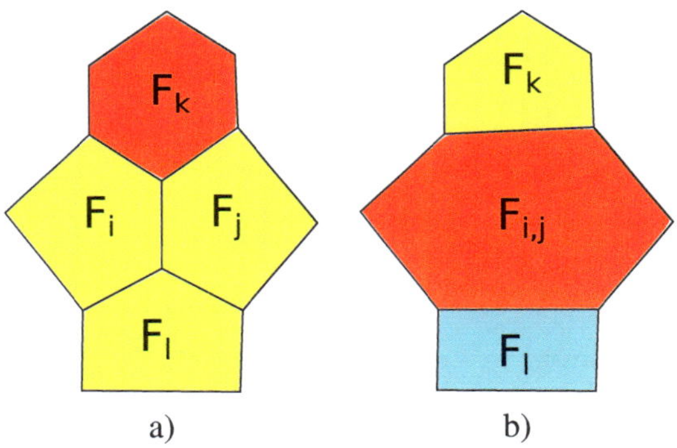

Fig. 51. a) Three adjacent pentagons and a hexagon; b) the same fragment after the straightening

If P has no adjacent pentagons, then consider the pentagon F_i adjacent to a hexagon F_j. The straightening along the edge $F_i \cap F_j$ gives the polytope Q with all facets pentagons and hexagons except for one heptagon $F_{i,j}$ adjacent to a pentagon. P is obtained from Q by a $(2,7;5,5)$-truncation. We claim that $Q \in \mathfrak{F}_1$. Indeed, if Q contains the fragment F_{5567}, then it is true. Else consider two adjacent pentagons in Q. The polytopes P and Q have the same structure outside the fragments in Figure 52; hence Q has the same pentagons as P except for pentagons F_k and F_l, which appeared instead of F_i. Also P has all pentagons isolated; hence one of the adjacent pentagons is F_k or F_l. We have $F_k \cap F_l = \varnothing$, else $(F_k, F_l, F_{i,j})$ is a 3-belt. Hence the other adjacent pentagon F_t does not belong to $\{F_k, F_l\}$. If F_t is adjacent to the heptagon $F_{i,j}$, then in P it is adjacent to F_i or F_j. Since F_i is an isolated pentagon, this is impossible. Hence F_t should be adjacent to F_j. Then F_t is one of the facets F_u, F_v, F_w in Figure 52. Let $F_t = F_u$. Since F_u is an isolated pentagon in P, the facet F_p is a hexagon on P and on Q, since $F_p \neq F_l$ because $F_k \cap F_l = \varnothing$. Then we obtain the fragment F_{5567}, which is a contradiction. The same argument works for F_w instead of F_u. If $F_t = F_v$, then

$F_v \cap F_k \neq \varnothing$, or $F_v \cap F_l \neq \varnothing$, which is impossible, since this gives the 3-belts (F_k, F_j, F_v), or (F_l, F_j, F_v). Thus, F_t does not intersect the heptagon $F_{i,j}$, and $Q \in \mathfrak{F}_1$. This finishes the proof of (3) and of the theorem. □

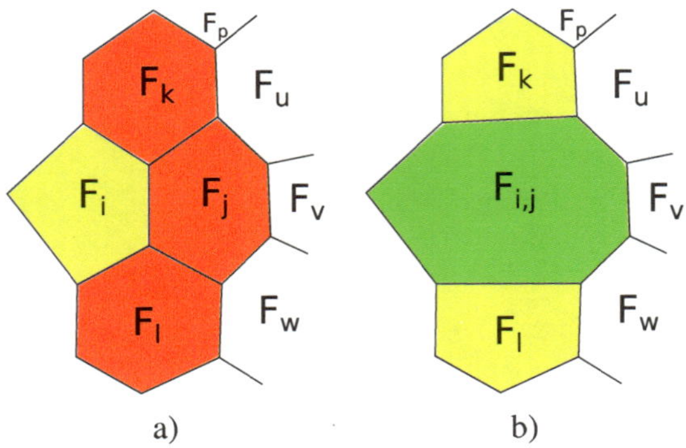

Fig. 52. a) Pentagon adjacent to three hexagons; b) the same fragment after the straightening

Remark 10.15: According to Remark 10.9 the seven truncations from Theorem 10.12 give rise to nine different growth operations (see Figure 53):

- Each $(1; m_1, m_2)$-truncation gives rise to two growth operations:
  - (a) if the truncated edge belongs to a pentagon, then we have the patch consisting of the pentagon adjacent to an m_1-gon and an m_2-gon by non-adjacent edges;
  - (b) if the truncated edge belongs to two hexagons, then we have the patch consisting of the hexagon adjacent to an m_1-gon and an m_2-gon by two edges that are not adjacent and not opposite;
- Each of the truncations $(2, 6; 4, 5)$-, $(2, 6; 5, 5)$-, $(2, 6; 5, 6)$-, $(2, 7; 5, 5)$-, and $(2, 7; 5, 6)$- gives rise to one growth operation.

If we take care of the orientation, then three of the operations have left and right versions.

Remark 10.16: For recent results on construction of fullerenes and Pogorelov polytopes see [9, 10].

Acknowledgments

The content of this lecture notes is based on lectures given by the first author at IMS of National University of Singapore in August 2015 during the Program

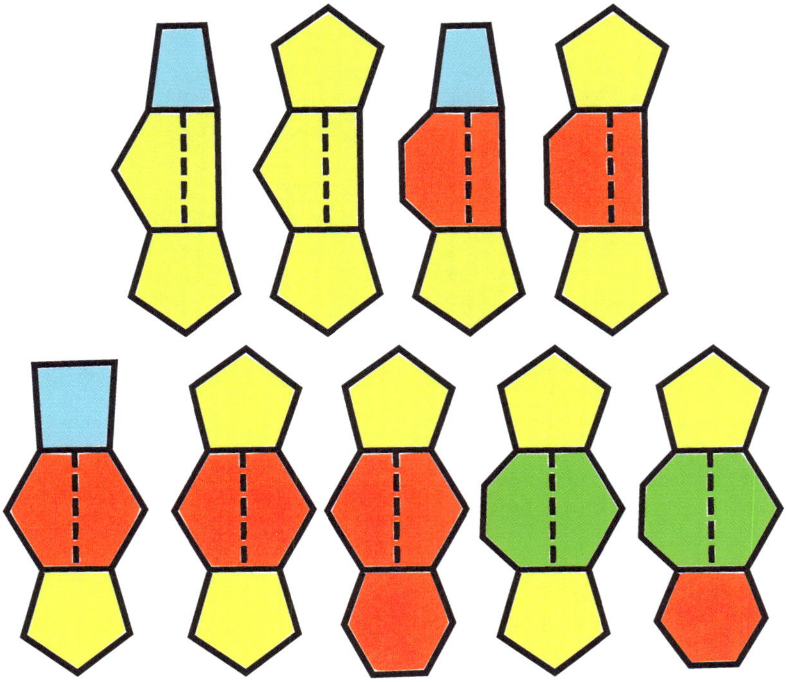

Fig. 53. 9 growth operations induced by 7 truncations

on Combinatorial and Toric Homotopy, and the work originated from the second authors participation in this Program. The authors thank Professor Jelena Grbić (University of Southampton), Professor Jie Wu (National University of Singapore), and IMS for organizing the Program and providing such a nice opportunity.

This work was partially supported by the RFBR grants 14-01-00537, 16-51-55017, and 17-01-00671, and the Young Russian Mathematics award.

References

1. E.M. Andreev, "Convex polyhedra in Lobačevskii spaces", *Mat. Sbornik*, 81 (123):3 (1970), 445–478 (Russian); *Math. USSR Sbornik*, 10:3 (1970), 413–440 (English translation).
2. L.J. Billera, C.W. Lee, "A proof of sufficiency of McMullen's conditions for f-vectors of simplicial polytopes", *J. Combin. Theory Ser. A*, 31:3 (1981), pp. 237–255.
3. G. Brinkmann, A.W.M. Dress, "A constructive enumeration of fullerenes," *J. Algorithms* 23 (2), (1997), pp. 345–358.
4. G. Brinkmann, J.E. Graver, C. Justus, "Numbers of faces in disordered patches", *J. Math. Chem.* 45:2 (2009), pp. 263–278.

5. W. Bruns, J. Gubeladze, "Polytopes, Rings, and K-Theory", Springer, 2009.
6. V.M. Buchstaber, "Toric Topology of Stasheff Polytopes," *MIMS EPrint*: 2007.232.
7. V.M. Buchstaber, N. Erokhovets, "Graph-truncations of simple polytopes", *Proc. of Steklov Math Inst*, MAIK, Moscow, vol. 289, 2015, pp. 104–133.
8. V.M. Buchstaber, N.Yu. Erokhovets, "Construction of fullerenes", arXiv 1510.02948, 2015.
9. V.M. Buchstaber, N.Yu. Erokhovets, "Construction of families of three-dimensional polytopes, characteristic patches of fullerenes, and Pogorelov polytopes", *Izvestiya: Mathematics*, 81:5 (2017).
10. V.M. Buchstaber, N.Yu. Erokhovets, "Finite sets of operations sufficient to construct any fullerene from C_{20}", *Structural Chemistry*, 28:1 (2017), pp. 225–234.
11. V.M. Buchstaber, N.Yu. Erokhovets, M. Masuda, T.E. Panov, S. Park, "Cohomological rigidity of manifolds defined by 3-dimensional polytopes", *Russian Math. Surveys*, 72:2 (2017), pp. 199–256.
12. V.M. Buchstaber, T.E. Panov, "Algebraic topology of manifolds defined by simple polytopes", *Russian Mathematical Surveys*, 1998, 53:3, pp. 623–625.
13. V.M. Buchstaber and T.E. Panov, "Torus actions and their applications in topology and combinatorics," *AMS University Lectures Series 24*, American Mathematical Society, Providence, RI, 2002.
14. V.M. Buchstaber, T.E. Panov, "Toric Topology," AMS Math. Surveys and Monographs, vol. 204, 2015. 518 pp.
15. V.M. Buchstaber and T.E. Panov, "On manifolds defined by 4-colourings of simple 3-polytopes", *Russian Math. Surveys*, 71:6 (2016), 1137–1139.
16. V.M. Buchstaber, T.E. Panov, N. Ray, "Spaces of polytopes and cobordism of quasitoric manifolds", *Mosc. Math. J.*, 7:2 (2007), 219–242 , arXiv 0609346.
17. V.M. Buchstaber, N. Ray "An invitation to toric topology: vertex four of a remarkable tetrahedron", In *Toric Topology*, M. Harada *et al.*, eds. Contemp. Math., 460, Amer. Math. Soc., Providence, RI, 2008, pp. 1–27.
18. H.S.M. Coxeter, "Regular Polytopes", (3rd edition, 1973), Dover edition, ISBN 0-486-61480-8.
19. R.F. Curl, "Dawn of the Fullerenes: Experiment and Conjecture", Nobel Lecture, December 7, 1996.
20. M. Davis, T. Januszkiewicz, "Convex polytopes, Coxeter orbifolds and torus actions", Duke Math. J., 1991. V.62, N2, pp. 417–451.
21. M. Deza, M. Dutour Sikirič, M.I. Shtogrin, "Fullerenes and disk-fullerenes," *Russian Math. Surveys*, 68:4 (2013), pp. 665–720.
22. M.-M. Deza, M. Dutour Sikirič, M.I. Shtogrin, "Geometric Structure of Chemistry-Relevant Graphs. Zigzags and Central Circuits", *Forum for Interdisciplinary Mathematics*, 1, eds. P.V. Subrahmanyam, B.D. Sharma, J. Matkowski, M. Dutour Sikirič, T. Parthasarathy, Y.P. Chaubey, Springer India, New Dehli, 2015, ISBN: 978-81-322-2448-8, 211 pp.
23. T. Došlić, "On lower bounds of number of perfect matchings in fullerene graphs", *Journal of Mathematical Chemistry*, 24 (1998), pp. 359–364.
24. T. Došlić, "Cyclical edge-connectivity of fullerene graphs and $(k, 6)$-cages", *Journal of Mathematical Chemistry*, 33:2 (2003), pp. 103–112.
25. V. Eberhard, "Zur Morphologie der Polyheder", Leipzig, 1891.

26. M. Endo, H.W. Kroto, "Formation of carbon nanofibers", *J. Phys. Chem.*, 96 (1992), pp. 6941–6944.
27. N.Yu. Erokhovets, "k-belts and edge cycles of simple 3-polytopes with at most hexagonal facets" (in Russian), *Far Eastern Mathematical Journal*, 15:2 (2015), pp. 197–213.
28. F. Fan, J. Ma, X. Wang, "B-Rigidity of flag 2-spheres without 4-belt", arXiv:1511.03624.
29. F. Fan, X. Wang, "On the cohomology of moment-angle complexes associated to Gorenstein* complexes", arXiv:1508.00159v3.
30. J.C. Fisher, "An existence theorem for simple convex polyhedra", *Discrete Math.*, 7 (1974), pp. 75–97.
31. B. Grünbaum, "Some analogs of Eberhard's theorem on convex polytopes", *Isr. J. Math.*, 6, 1968, pp. 398–411.
32. B. Grünbaum, "Convex polytopes" (2nd Edition), *Graduate Texts in Mathematics* 221, Springer-Verlag, New York, 2003.
33. M. Hasheminezhad, H. Fleischner, B.D. McKay, "A universal set of growth operations for fullerenes", *Chem. Phys. Letters*, 464 (2008), 118–121.
34. F. Kardoš, R. Skrekovski, "Cyclic edge-cuts in fullerene graphs", *J. Math. Chem*, 22 (2008), pp. 121–132.
35. F. Kardoš, M. Krnc, B. Lužar, R. Skrekovski, "Cyclic 7-edge-cuts in fullerene graphs", *Journal of Mathematical Chemistry*, Springer Verlag (Germany). 47:2 (2010), pp. 771–789.
36. H. Kroto, "Symmetry, Space, Stars and C_{60}", Nobel Lecture, December 7, 1996.
37. K. Kutnar, D. Marušič, "On cyclic edge-connectivity of fullerenes", *Discr. Appl. Math.* 156 (2008), pp. 1661–1669.
38. E.A. Lord, A.L. Mackay, S. Ranganathan, "New Geometries for New Materials", Cambridge University Press, 2006.
39. J. Milnor, "On the cobordism ring Ω^* and a complex analogue. I", *Amer. J. Math.* **82** (1960), pp. 505–521.
40. S.P. Novikov, "Some problems in the topology of manifolds connected with the theory of Thom spaces", *Dokl. Akad. Nauk SSSR* **132** (1960), pp. 1031–1034 (Russian); Soviet Math. Dokl. **1** (1960), pp. 717–720 (English translation).
41. S.P. Novikov, "Homotopy properties of Thom complexes", *Mat. Sbornik* **57** (1962), no. 4, pp. 407–442 (Russian); English translation at `http://www.mi.ras.ru/~snovikov/6.pdf`.
42. A.V. Pogorelov, "Regular decomposition of the Lobačevskii space", *Mat. Zametki*, 1:1 (1967), 3–8 (Russian).
43. H. Poincaré, "Analysis situs", *Journal de l'École Polytechnique.* (2) 1: pp. 1–123, (1895).
44. J.P. Serre, "Homologie singuliere des espaces fibres. Applications" (These), *Paris et Ann. of Math.* 54 (1951), pp. 425–505.
45. R.E. Smalley, "Discovering the Fullerenes", Nobel Lecture, December 7, 1996.
46. R.P. Stanley, "The number of faces of simplicial convex polytope", *Adv. in Math.*, 35:3 (1980), pp. 236–238.
47. E. Steinitz, "Über die Eulerschen Polyederrelationen", *Archiv für Mathematik und Physik* 11 (1906), pp. 86–88.

48. R. Thom, "Quelques propriétés globales des variétés différentiables", *Comment. Math. Helv.* **28** (1954), pp. 17–86 (French).
49. W.P. Thurston, "Shapes of polyhedra and triangulations of the sphere", *Geometry and Topology Monographs*, Volume 1 (1998), pp. 511–549.
50. A.Yu. Vesnin, "Right-angled polytopes and hyperbolic 3-manifolds", *Russian Math. Surveys*, 72:2 (2017), DOI:10.1070/RM9762.
51. V.D. Volodin, "Combinatorics of flag simplicial 3-polytopes", *Russian Math. Surveys*, 70:1 (2015), pp. 168–170; arXiv: 1212.4696.
52. R.J. Wilson, "Four colors suffice: how the map problem was solved", Princeton Univ. Press, Princeton 2014, 199 pp.
53. G.M. Ziegler, "Face numbers of 4-polytopes and 3-spheres", *Proceedings of the International Congress of Mathematicians (Beijing, China, 2002)*, III, Higher Ed. Press, Beijing, 2002, pp. 625–634; arXiv: math/0208073.
54. G.M. Ziegler, "Lectures on polytopes" (7th Printing), *Graduate texts in Mathematics* 152, Springer 2007.

AROUND BRAIDS

Vladimir Vershinin
Département des Sciences Mathématiques, Université de Montpellier
Place Eugène Bataillon, 34095 Montpellier cedex 5, France
vladimir.verchinine@univ-montp2.fr
Sobolev Institute of Mathematics, Novosibirsk 630090, Russia
versh@math.nsc.ru

These notes are based on the lectures given by the author at the Young Topologist Seminar, IMS, Singapore (11–21 August 2015).

The author was asked to speak on non-classical material concerning braids. The choice of this material is based on the author's interests.

The definition of braid groups and their initial properties can be found in books [13, 31, 40] or for example in the surveys by the author [46, 48]. Here in Section 2 we give only two standard presentations: that of Artin of the classical braid group [2, 3] and that of Burau of the pure braid group [18].

In Section 3 we give a survey of the Brunnian braids and in Section 4 of the Cohen braids. In Section 5 we study connections between braids and Lie algebras and in Section 6 we discuss some topics of virtual braids. This subject has become big now, so we touch only several selected items. In the last section some relations between braids and Thompson's groups are given.

Homological properties of braids can be found, for example, in the survey [45].

Contents

1. Standard facts about braids

1.1. *Artin presentation for the braid group*

Usually braids are depicted by projections on the plane. They are supposed to be in general position so that there is only finite number of double points of intersection which are lying on pairwise different levels and the intersections are transversal. The simplest braid σ_i (Fig. 1) corresponds to the transposition $(i, i+1)$.

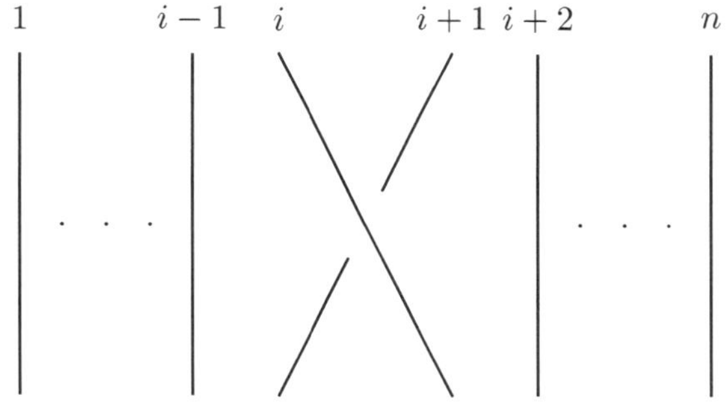

Fig. 1. Braid σ_i

Artin presentation of the braid group B_n [2, 3] has generators σ_i, $i = 1, ..., n-1$ and relations:

$$\begin{cases} \sigma_i \sigma_j & = \sigma_j \, \sigma_i, \quad \text{if} \;\; |i-j| > 1, \\ \sigma_i \sigma_{i+1} \sigma_i & = \sigma_{i+1} \sigma_i \sigma_{i+1}. \end{cases}$$

1.2. *Presentation of the pure braid group of a disc*

Define the elements $a_{i,j}$, $1 \leq i < j \leq n$, of B_n by:

$$a_{i,j} = \sigma_{j-1} ... \sigma_{i+1} \sigma_i^2 \sigma_{i+1}^{-1} ... \sigma_{j-1}^{-1}.$$

Geometrically generator of this type is depicted at Figure 2.

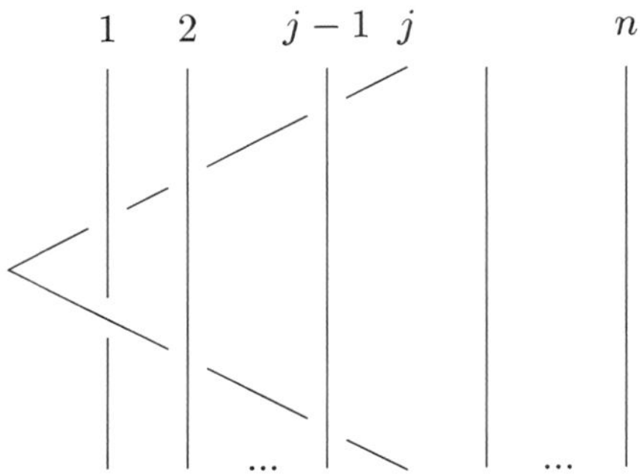

Fig. 2. Generator $a_{1,j}$

These elements satisfy the Burau relations:

$$\begin{aligned} a_{i,j}a_{k,l} = a_{k,l}a_{i,j} & \text{ for } i < j < k < l \text{ and } i < k < l < j, \\ a_{i,j}a_{i,k}a_{j,k} = a_{i,k}a_{j,k}a_{i,j} & \text{ for } i < j < k, \\ a_{i,k}a_{j,k}a_{i,j} = a_{j,k}a_{i,j}a_{i,k} & \text{ for } i < j < k, \\ a_{i,k}a_{j,k}a_{j,l}a_{j,k}^{-1} = a_{j,k}a_{j,l}a_{j,k}^{-1}a_{i,k} & \text{ for } i < j < k < l. \end{aligned}$$

W. Burau proved that this gives a presentation of the pure braid group P_n [18].

2. Brunnian braids

2.1. *Definition*

Let M be a compact connected surface, possibly with boundary, and let $B_n(M)$ denote the n-strand braid group on a surface M [40, 48]. From the point of view of braids, compactness of a surface is not essential: braids stay the same if you replace a boundary component by a puncture. However the number of punctures must be finite, so that the fundamental group and the braid groups will be finitely generated.

A *Brunnian braid* means a braid that becomes trivial after removing any one of its strands. A typical example of a 3-strand Brunnian braid on a disk is the braid given by the expression $(\sigma_1^{-1}\sigma_2)^3$, where σ_1 and σ_2 are the standard braid generators. If we make a closure of this braid (definition of the closure see for example in [48]) we get the Boromean rings, depicted in Figure 3.

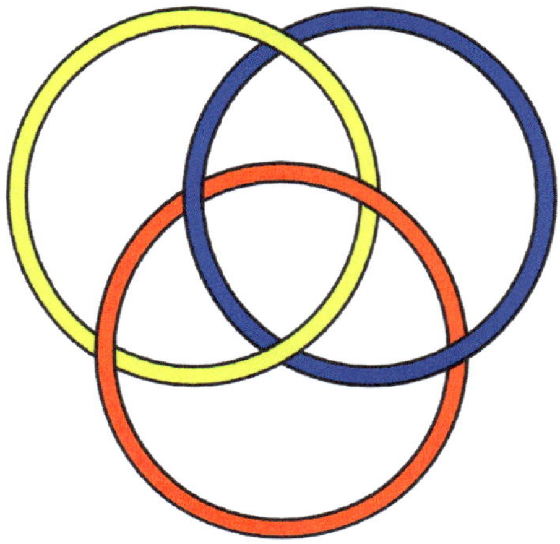

Fig. 3. Boromean rings

In picture, $\sigma_1^{-1}\sigma_2$ corresponds to the braid operation on three strands which consists of taking the crossing of the first two strands with the first strand above the second one followed by the crossing of the last two strands with the last strand above the second one. If the braid operation $\sigma_1^{-1}\sigma_2$ is repeated k times, where k is a multiple of 3, then one obtains a Brunnian braid.

Let $\mathrm{Brun}_n(M)$ denote the set of the n-strand Brunnian braids. Then $\mathrm{Brun}_n(M)$ forms a normal subgroup of $B_n(M)$. A classical question proposed by G. S. Makanin [37] in 1980 is to determine a set of generators for Brunnian braids over the disk. Brunnian braids were called *smooth braids* by Makanin. The question of Makaninin was answered by D. L. Johnson [27] and G. G. Gurzo [24]. A different approach to this question can be found in [35, 36]. In the 1970s, H. W. Levinson [32, 33] defined a notion of k-*decomposable* braid. It means a braid which becomes trivial after removal of any arbitrary k strings. In his terminology a *decomposable* braid means 1-*decomposable* and therefore, Brunnian.

A connection between Brunnian braids and the homotopy groups of spheres was given in [12]. In particular, the following exact sequence

$$1 \to \mathrm{Brun}_{n+1}(S^2) \to \mathrm{Brun}_n(D^2) \to \mathrm{Brun}_n(S^2) \to \pi_{n-1}(S^2) \to 1$$

was proved for $n > 4$.

We consider operations

$$d_i \colon B_n(M) \to B_{n-1}(M)$$

which are obtained (roughly speaking) by forgetting the i-th strand, $1 \leq i \leq n$. We can interpret a Brunnian braid $\beta \in B_n(M)$ as a solution of system of n equations

$$\begin{cases} d_1(\beta) = 1, \\ \dots \\ d_n(\beta) = 1. \end{cases} \tag{2.1}$$

2.2. *Symmetric commutator product*

Recall the notion of the symmetric commutator product. Given a group G, and a set of normal subgroups $R_1, \dots, R_n$, $(n \geq 2)$, the symmetric commutator product of these subgroups is defined as

$$[R_1, \dots, R_n]_S := \prod_{\sigma \in \Sigma_n} [[R_{\sigma(1)}, R_{\sigma(2)}], \dots, R_{\sigma(n)}],$$

where Σ_n is the symmetric group of degree n.

2.3. *Brunnian braids on surfaces*

Let $P_n(M)$ be the n-strand pure braid group on M. Let D^2 be a small disk in M. Then the inclusion $f\colon D^2 \hookrightarrow M$ induces a group homomorphism

$$f_*\colon P_n(D^2) \longrightarrow P_n(M).$$

Let $a_{i,j}[M] = f_*(a_{i,j})$ and let $\langle\langle a_{i,j}[M]\rangle\rangle^P$ be the normal closure of $a_{i,j}[M]$ in $P_n(M)$. Note that a set of generators for $\langle\langle a_{i,j}[M]\rangle\rangle^P$ is given by $\beta a_{i,j}[M]\beta^{-1}$ for $\beta \in P_n(M)$. Thus a set of generators for the iterated subgroup

$$[\langle\langle a_{1,n}[M]\rangle\rangle^P, \langle\langle a_{2,n}[M]\rangle\rangle^P, \dots, \langle\langle a_{n-1,n}[M]\rangle\rangle^P]_S$$

can be given.

The following results is a joint work with Bardakov, Mikhailov and Jie Wu [6].

Theorem 2.1: *Let M be a connected 2-manifold and let $n \geq 2$. Let*

$$R_n(M) = [\langle\langle a_{1,n}[M]\rangle\rangle^P, \langle\langle a_{2,n}[M]\rangle\rangle^P, \dots, \langle\langle a_{n-1,n}[M]\rangle\rangle^P]_S$$

(1) If $M \neq S^2$ or $\mathbb{RP}^2$, then

$$\mathrm{Brun}_n(M) = R_n(M).$$

(2) *If* $M = S^2$ *and* $n \geq 5$, *then there is a short exact sequence*

$$R_n(S^2) \hookrightarrow \mathrm{Brun}_n(S^2) \twoheadrightarrow \pi_{n-1}(S^2).$$

(3) *If* $M = \mathbb{R}\mathrm{P}^2$ *and* $n \geq 4$ *then there is a short exact sequence*

$$R_n(\mathbb{R}\mathrm{P}^2) \hookrightarrow \mathrm{Brun}_n(\mathbb{R}\mathrm{P}^2) \twoheadrightarrow \pi_{n-1}(S^2).$$

Assertion 3 gives a new connection between Brunnian braids and homotopy groups. The first case in assertion 3 ($n = 4$) is that the Hopf map $S^3 \to S^2$ lifts to a 4-strand Brunnian braid on $\mathbb{R}\mathrm{P}^2$.

$\mathrm{Brun}_n(M)$ is a normal subgroup of $B_n(M)$ for $n \geq 3$. As an abstract group, $\mathrm{Brun}_n(M)$ is a free group of infinite rank for $n \geq 3$ with $M \neq S^2$ or $\mathbb{R}\mathrm{P}^2$, for $n \geq 5$ with $M = S^2$ and for $n \geq 4$ with $M = \mathbb{R}\mathrm{P}^2$. A natural question is whether the quotient group $B_n(M)/\mathrm{Brun}_n(M)$ is finitely presented.

Theorem 2.2: [6] *Let* M *be a connected compact* 2*-manifold. Then the quotient groups* $P_n(M)/\mathrm{Brun}_n(M)$ *and* $B_n(M)/\mathrm{Brun}_n(M)$ *are finitely presented for each* $n \geq 3$.

3. Cohen braids

3.1. *Equations in braid groups*

All results of this section are from the work [8].

Let us replace the unit element of the group in (2.1) by an arbitrary braid $\alpha \in B_{n-1}(M)$ and we ask the following question: does there exist a braid $\beta \in B_n(M)$ such that it is a solution of the following system of equations

$$\begin{cases} d_1(\beta) = \alpha, \\ \cdots \\ d_n(\beta) = \alpha. \end{cases} \tag{3.1}$$

Apart from Brunnian braids the following example can be given. Let α be the Garside element $\Delta_{n-1} \in B_{n-1}(M)$ [21]. Then $\Delta_n \in B_n(M)$ is a solution of system (3.1), this is the half-twist braid

$$\Delta_n = (\sigma_1\sigma_2 \cdots \sigma_{n-1})(\sigma_1\sigma_2 \cdots \sigma_{n-2}) \cdots (\sigma_1\sigma_2)\sigma_1.$$

Theorem 3.1: *Let* M *be any connected* 2*-manifold such that* $M \neq S^2$ *or* $\mathbb{R}\mathrm{P}^2$ *and let* $\alpha \in B_{n-1}(M)$. *Then the equation (3.1) for* n*-strand braids* β *has a solution if and only if* α *satisfies the condition that*

$$d_1\alpha = \cdots = d_{n-1}\alpha.$$

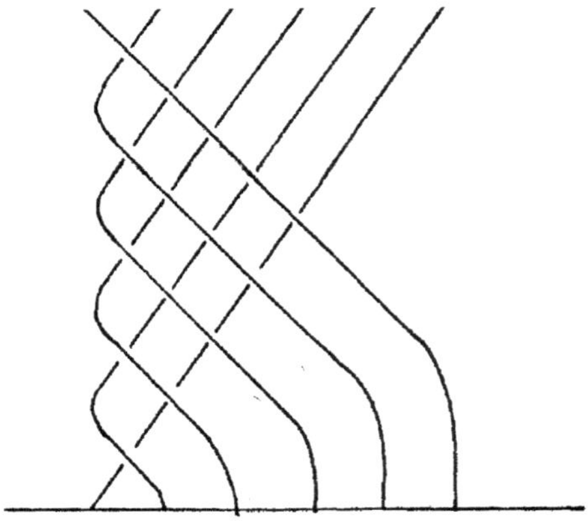

Fig. 4. Garside element Δ_6

3.2. Bi-Δ-structure for pure braids

The technique of the proof is based on the bi-Δ-group structure (see Appendix at the end of the text) on the pure braid groups over connected 2-manifolds with nonempty boundaries [49] as well as the determination of Brunnian braids on general connected 2-manifolds [6].

Let M be a connected manifold with $\partial M \neq \emptyset$. Let a be a point in a collar of ∂M

$$\partial M \times [0,1) \subseteq M.$$

Then the map

$$F(M,n) \simeq F(M \setminus \partial M \times [0,1), n) \to F(M, n+1),$$
$$(z_1, \ldots, z_n) \mapsto (z_1, \ldots, z_{i-1}, a, z_{i+1}, \ldots, z_n)$$

induces a group homomorphism

$$d^i \colon B_n(M) = \pi_1(F(M,n)/\Sigma_n) \to B_{n+1}(M) = \pi_1(F(M,n+1)/\Sigma_{n+1})$$

for $1 \leq i \leq n+1$.

Intuitively, d^i is given by adding a trivial strand in position i. The sequence of groups $\{B_{n+1}(M)\}_{n\geq 0}$ with faces relabeled as $\{d_0, d_1, \ldots\}$ and cofaces relabeled as $\{d^0, d^1, \ldots\}$ forms a bi-Δ-set structure.

So, the following identities hold:

(1) $d_j d_i = d_i d_{j+1}$ for $j \geq i$;
(2) $d^j d^i = d^{i+1} d^j$ for $j \leq i$;

$$(3)\ d_j d_i = \begin{cases} d^{i-1} d_j & \text{if } j < i, \\ \text{id} & \text{if } j = i, \\ d^i d_{j-1} & \text{if } j > i. \end{cases}$$

Let $\mathbb{P}_n = P_{n+1}$. The sequence of groups $\mathbb{P} = \{\mathbb{P}_n\}_{n\geq 0}$ with faces relabeled as $\{\mathbb{d}_0, \mathbb{d}_1, \ldots\}$ and co-faces relabeled as $\{\mathbb{d}^0, \mathbb{d}^1, \ldots\}$ forms a bi-Δ-group structure. The face operation $\mathbb{d}_i : \mathbb{P}_n \to \mathbb{P}_{n-1} = d_{i+1} : P_{n+1} \to P_n$ is obtained by deleting the $i+1$st string, the co-face operation $\mathbb{d}^i : \mathbb{P}_n \to \mathbb{P}_{n+1}$ is obtained by adding a trivial $i+1$st string in front of the other strings $(i = 0, 1, 2, \ldots, n)$.

Proposition 3.2: *Let M be a connected 2-manifold with nonempty boundary. Then $P_n(M)$ is the (iterated) semi-direct product of the subgroups*

$$d^{i_k} d^{i_{k-1}} \cdots d^{i_1}(\mathrm{Brun}_{n-k}(M)),$$

$1 \leq i_1 < i_2 < \cdots < i_k \leq n$, $0 \leq k \leq n-1$, *with lexicographical order from the right.*

3.3. *Cohen braids*

We return to the case when M is an arbitrary connected 2-manifold.

Define a set

$$\mathfrak{H}^B_n(M) = \{\beta \in B_n(M) \mid d_1\beta = d_2\beta = \cdots = d_n\beta\}.$$

Namely $\mathfrak{H}^B_n(M)$ consists of n-strand braids such that it stays the same braid after removing any one of its strands. We call this *Cohen braids.* We denote Cohen braids for the disc simply by $\mathfrak{H}^B_n$ as well as Brunnian braids of the disc by Brun_n.

Proposition 3.3: *Let M be any connected 2-manifold. Then the set $\mathfrak{H}^B_n(M)$ is subgroup of $B_n(M)$. Moreover $d_i(\mathfrak{H}^B_n(M)) \subseteq \mathfrak{H}^B_{n-1}(M)$ and the function*

$$d_1 = d_2 = \cdots = d_n \colon \mathfrak{H}^B_n(M) \to \mathfrak{H}^B_{n-1}(M)$$

is a group homomorphism.

Proposition 3.4: *Let M be any connected 2-manifold. Let $n \geq 2$. Then $\mathfrak{H}^B_n(M) \cap P_n(M)$ is a subgroup of $\mathfrak{H}^B_n(M)$ of index 2.*

Let $\mathfrak{H}_n(M) = \mathfrak{H}^B_n(M) \cap P_n(M)$. Then $d_1(\mathfrak{H}_n(M)) \subseteq \mathfrak{H}_{n-1}(M)$. This gives a tower of groups

$$\cdots \to^{d_1} \mathfrak{H}_n(M) \to^{d_1} \mathfrak{H}_{n-1}(M) \to^{d_1} \cdots.$$

Let $\mathfrak{H}(M) = \lim\limits_{n} \mathfrak{H}_n(M)$ be the inverse limit of this tower of groups.

Proposition 3.5: *Let M be any connected 2-manifold such that $M \neq S^2$ or $\mathbb{R}\mathrm{P}^2$. Then*

$$d_1 : \mathfrak{H}_n(M) \to \mathfrak{H}_{n-1}(M)$$

is an epimorphism for each $n \geq 2$.

Corollary 3.6: *Let M be any connected 2-manifold such that $M \neq S^2$ or $\mathbb{R}\mathrm{P}^2$. Then there exists the following short exact sequence*

$$1 \to \mathrm{Brun}_n(M) \to \mathfrak{H}_n(M) \xrightarrow{d_1} \mathfrak{H}_{n-1}(M) \to 1 \tag{3.2}$$

which connects the nth $(n-1)$th Cohen braid groups and Brunnian braids on n-strands.

In small dimensions we have $\mathfrak{H}_2 = \mathrm{Brun}_2 = P_2 = \mathbb{Z}$ and the exact sequence (3.2) gives

$$1 \to \mathrm{Brun}_3 \to \mathfrak{H}_3 \xrightarrow{d_1} \mathbb{Z} \to 1.$$

In $\mathfrak{H}_3$ it is the central element Δ_3^2 which is mapped to the generator of $\mathfrak{H}_2$, so $\mathfrak{H}_3 \cong \mathbb{Z} \times \mathrm{Brun}_3$ and for the next step we have

$$1 \to \mathrm{Brun}_4 \to \mathfrak{H}_4 \longrightarrow \mathbb{Z} \times \mathrm{Brun}_3 \to 1.$$

Question: What can one say about Cohen braids on the sphere and projective plane?

In dimension 3 we have $\mathfrak{H}_3(S^2) = \mathrm{Brun}_3(S^2) = P_3(S^2) = \mathbb{Z}/2$, the map $\mathfrak{H}_4(S^2) \xrightarrow{d_1} \mathfrak{H}_3(S^2)$ is obviously onto and Δ_3^2 is central in $\mathfrak{H}_4(S^2)$, so we have

$$\mathfrak{H}_4(S^2) \cong \mathbb{Z}/2 \times \mathrm{Brun}_4(S^2).$$

Calculations show that

$$\mathfrak{H}_2(\mathbb{R}\mathrm{P}^2) \cong \mathbb{Z}/4.$$

4. Braids and Lie algebras

4.1. *Lie algebras from a descending central series of groups*

Let K be a commutative ring with unit.

Definition 4.1: An algebra L over K is called a Lie algebra over K if its multiplication (denoted by $(x, y) \mapsto [x, y]$) satisfies:

(1) $[x, x] = 0$
(2) $[x, [y, z]] + [y, [z, x]] + [z, [x, y]] = 0$

for all x, y, z in L.

For a group G the descending central series

$$G = \Gamma_1 \geq \Gamma_2 \geq \cdots \geq \Gamma_i \geq \Gamma_{i+1} \geq \cdots \tag{4.1}$$

is defined by the formulae

$$\Gamma_1 = G, \quad \Gamma_{i+1} = [\Gamma_i, G]. \tag{4.2}$$

The descending central series of a discrete group G gives rise to the associated graded Lie algebra (over $\mathbb{Z}$) $L(G)$

$$L_i(G) = \Gamma_i(G)/\Gamma_{i+1}(G).$$

The details of this standard material can be found, for example, in Chapters I and II of the book of J.-P. Serre [44].

4.2. *Presentation of the Lie algebra* $L(P_n)$

This presentation was given by Toshitake Kohno [30]. It is the quotient of the free Lie algebra $L[A_{i,j} \,|\, 1 \leq i < j \leq n]$ generated by elements $A_{i,j}$ with $1 \leq i < j \leq n$ modulo the "infinitesimal braid relations" or "horizontal $4T$ relations" given by the following three relations:

$$\begin{cases} [A_{i,j}, A_{s,t}] = 0, \text{ if } \{i,j\} \cap \{s,t\} = \phi, \\ [A_{i,j}, A_{i,k} + A_{j,k}] = 0, \text{ if } i < j < k, \\ [A_{i,k}, A_{i,j} + A_{j,k}] = 0, \text{ if } i < j < k. \end{cases} \tag{4.3}$$

Where $A_{i,j}$ is the projection of the $a_{i,j}$ to $L(P_n)$.

4.3. *Definition and properties of* $\mathbf{L^P(Brun_n)}$

The matter of this subsection and till the end of the section is from the work [34].

We consider the restriction $\{\Gamma_q(P_n) \cap \mathrm{Brun}_n\}$ of the lower central series of P_n to Brun_n. This gives a relative Lie algebra

$$L^P(\mathrm{Brun}_n) = \bigoplus_{q=1}^{\infty} (\Gamma_q(P_n) \cap \mathrm{Brun}_n)/(\Gamma_{q+1}(P_n) \cap \mathrm{Brun}_n).$$

The purpose is to study the Lie algebra $L^P(\mathrm{Brun}_n)$.

This is a Lie subalgebra of $L(P_n)$, we call it the *relative Lie algebra associated with Brunnian subgroup* of the pure braid group.

Proposition 4.2: $\mathrm{L^P(Brun_n)}$ *is a two-sided Lie ideal in* $L(P_n)$.

The removing-strand operation on braids induces an operation

$$d_k \colon L(P_n) \longrightarrow L(P_{n-1})$$

formulated by

$$d_k(A_{i,j}) = \begin{cases} A_{i,j}, & \text{if } i < j < k, \\ 0, & \text{if } k = j, \\ A_{i,j-1}, & \text{if } i < k < j, \\ 0, & \text{if } k = i, \\ A_{i-1,j-1}, & \text{if } k < i < j. \end{cases} \tag{4.4}$$

Proposition 4.3: *The relative Lie algebra* $\mathrm{L^P(Brun_n)}$ *is the Lie subalgebra* $\bigcap_{i=1}^{n} \ker(d_i : L(P_n) \to L(P_{n-1}))$.

4.4. *Generators for the Lie algebra* $\mathbf{L^P(Brun_n)}$

The following fact is a Lie algebra analogue of the theorem proved by A. A. Markov for the pure braid group [38].

Proposition 4.4: *The kernel of the homomorphism* $d_n : L(P_n) \to L(P_{n-1})$ *is a free Lie algebra, generated by the free generators* $A_{i,n}$, *for* $1 \le i \le n-1$.

$$\mathrm{Ker}(d_n : L(P_n) \to L(P_{n-1})) = L[A_{1,n}, \dots, A_{n-1,n}].$$

For a set Z, let $L[Z]$ denote the free Lie algebra freely generated by Z. Let X and Y be non-empty sets with $X \cap Y = \emptyset$, $X \cup Y = Z$. Let π be the Lie homomorphism

$$\pi \colon L[Z] \longrightarrow L[Y]$$

such that $\pi(x) = 0$ for $x \in X$ and $\pi(y) = y$ for $y \in Y$.

Proposition 4.5: *The kernel of* π *is a free Lie algebra, generated by the following family of free generators:*

$$x, [\cdots [x, y_1], \dots, y_t] \tag{4.5}$$

for $x \in X, y_i \in Y$ *for* $1 \le i \le t$.

Proposition 4.6: *The intersection of the kernels of the homomorphisms d_n and d_k, $k \neq n$, is a free Lie algebra, generated by the following infinite family of free generators:*

$$A_{k,n}, [\cdots [A_{k,n}, A_{j_1,n}], \ldots, A_{j_m,n}] \tag{4.6}$$

for $j_i \neq k, n;\ j_i \leq n-1;\ i \leq m;\ m \geq 1.$

Another set of free generators of $\mathrm{Ker}(d_n) \cap \mathrm{Ker}(d_k)$ can be obtained using Hall bases [14], [25].

We suppose that all Lie monomials on $B_1, \ldots, B_k$ are ordered lexicographically.

Lie monomials $B_1, \ldots, B_k$ are the *standard* monomials of degree 1. If we have defined standard monomials of degrees $1, \ldots, n-1$, then $[u, v]$ is a *standard* monomial if both of the following conditions hold:

(1) u and v are standard monomials and $u > v$.
(2) If $u = [x, y]$ is the form of the standard monomial u, then $v \geq y$.

Standard monomials form the *Hall basis* of a free Lie algebra (also over $\mathbb{Z}$).

Examples of standard monomials are the products of the type:

$$[\cdots [B_{j_1}, B_{j_2}], B_{j_3}], \ldots, B_{j_t}],\ j_1 > j_2 \leq j_3 \leq \cdots \leq j_t. \tag{4.7}$$

Proposition 4.7: *The intersection* $\mathrm{Ker}(d_n) \cap \mathrm{Ker}(d_k)$, $k \neq n$, *is a free Lie algebra, generated by the standard monomials on $A_{i,n}$ where the letter $A_{k,n}$ has only one entry. In other words the free generators are standard monomials of type (4.7) where each monomial contains only one copy of $A_{k,n}$.*

We recursively define the sets $\mathcal{K}(n)_k$, $1 \leq k \leq n$, in the reverse order as follows:

1) Let $\mathcal{K}(n)_n = \{A_{1,n}, A_{2,n}, \ldots, A_{n-1,n}\}$.
2) Suppose that $\mathcal{K}(n)_{k+1}$ (with $k \leq n-1$) is defined as a subset of Lie monomials on the letters

$$A_{1,n}, A_{2,n}, \ldots, A_{n-1,n}.$$

Let

$$\mathcal{A}_k = \{W \in \mathcal{K}(n)_{k+1} \mid W \text{ does not contain } A_{k,n} \text{ in its entries}\}.$$

3) Define

$$\mathcal{K}(n)_k = \{W' \text{ and } [\cdots [[W', W_1], W_2], \ldots, W_t]\}$$

for $W' \in \mathcal{K}(n)_{k+1} \setminus \mathcal{A}_k$ and $W_1, W_2, \ldots, W_t \in \mathcal{A}_k$ with $t \geq 1$. Note that $\mathcal{K}(n)_k$ is again a subset of Lie monomials on letters $A_{1,n}, A_{2,n}, \ldots, A_{n-1,n}$.

Example 4.8: Let $n = 3$. The set $\mathcal{K}(3)_1$ is constructed by the following steps:

1) $\mathcal{K}(3)_3 = \{A_{1,3}, A_{2,3}\}$.
2) $\mathcal{A}_2 = \{A_{1,3}\}$,
$$\mathcal{K}(3)_2 = \{A_{2,3}, [\cdots[A_{2,3}, A_{1,3}], \ldots, A_{1,3}]\}.$$
3) $\mathcal{A}_1 = \{A_{2,3}\}$,
$\mathcal{K}(3)_1 = \{[\cdots[A_{2,3}, A_{1,3}], \ldots, A_{1,3}], A_{2,3}], \ldots, A_{2,3}]\}$.

Theorem 4.9: *The Lie algebra* $\mathrm{L^P}(\mathrm{Brun_n})$ *is a free Lie algebra generated by* $\mathcal{K}(n)_1$ *as a set of free generators.*

4.5. *The rank of* $L_q^P(\mathrm{Brun}_n)$

Observe that the Lie algebra $L(P)$ is of finite type in the sense that each homogeneous component $L_k(P_n)$ is a free abelian group of finite rank. Thus the subgroup

$$\mathrm{L^P}(\mathrm{Brun_n}) \cap L_k(P_n)$$

is a free abelian group of finite rank. We now give a formula for the rank of $L_q^P(\mathrm{Brun}_n)$.

Let $\mathcal{G} = L_q(\mathbb{P})$. It is a bi-$\Delta$ group. Then $\mathcal{Z}_n(L_q(\mathbb{P})) = L_q^P(\mathrm{Brun}_{n+1})$. $d^i = \mathrm{d}^{i-1} : \mathbb{P}_{n-1} = P_n \to \mathbb{P}_n = P_{n+1}$ is obtained by adding a trivial ith string in front of the other strings ($i = 1, 2, \ldots, n$).

The decomposition formula for bi-Δ groups gives us the following decomposition of Lie algebras.

Proposition 4.10: *There is a decomposition*

$$L_q(P_n) = \bigoplus_{\substack{1 \leq i_1 < \cdots < i_k \leq n \\ 0 \leq k \leq n-1}} d^{i_k} d^{i_{k-1}} \cdots d^{i_1} (L_q^P(\mathrm{Brun}_{n-k}))$$

for each n *and* q. □

Corollary 4.11: *There is a formula*

$$\mathrm{rank}(L_q(P_n)) = \sum_{k=0}^{n-1} \binom{n}{k} \mathrm{rank}(L_q^P(\mathrm{Brun}_{n-k}))$$

for each n and q.

Theorem 4.12:

$$\mathrm{rank}(L_q^P(\mathrm{Brun}_n)) = \sum_{k=0}^{n-1}(-1)^k\binom{n}{k}\mathrm{rank}(L_q(P_{n-k}))$$

for each n and q, where $P_1 = 0$ and, for $m \geq 2$,

$$\mathrm{rank}(L_q(P_m)) = \frac{1}{q}\sum_{k=1}^{m-1}\sum_{d|q}\mu(d)k^{q/d}$$

with μ the Möbius function.

One can see the definition of the Möbius function in the Appendix to Chapter II of the Bourbaki book [14].

5. Virtual braids

5.1. *Virtual knots and braids*

Virtual knots were introduced by L. Kauffman [29] and then studied in particular by M. Goussarov, M. Polyak and O. Viro [23]. One of the motivations lies in the theory of Gauss diagrams and Gauss codes of knots [29, 41]. Namely, for any knot diagram it is possible to construct its Gauss diagram and form its Gauss code. The example of a Gauss diagram of a knot is given on the Figure 5.

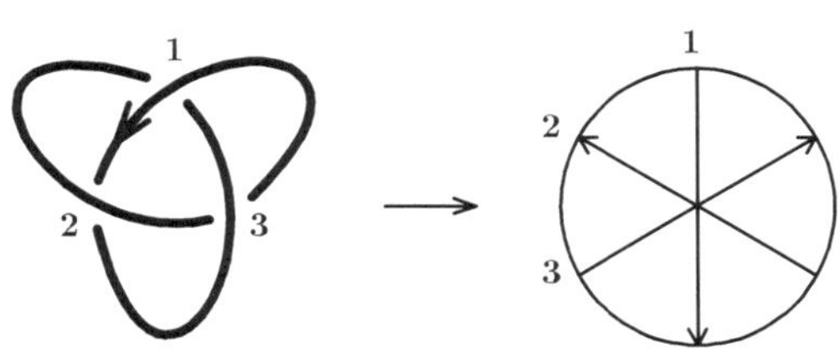

Fig. 5. Gauss diagramm of trefoil

The Gauss code of this Gauss diagram is the following:

O1U2O3U1O2U3,

The problem is that not every Gauss diagram (or Gauss code) corresponds to some knot. Using virtual knots one may escape this difficulty.

Virtual knots are those whose Gauss diagram is not a Gauss diagram of any classical knot.

Many notions from the classical knot theory are generalized to virtual knots, such as fundamental group, rack, quandle, Kauffman and Jones polynomials.

As it follows from works of L. Kauffman [29] and M. Goussarov, M. Polyak and O. Viro [23] two virtually equivalent classical knots are classically equivalent. This means that inclusion of usual knots in the universe of virtual knots does not spoil the classical theory.

Virtual braids are the braid theory analogue of virtual links.

Seichi Kamada [28] proved that any virtual link can be described as the closure of a virtual braid, which is unique up to certain basic moves. This is analogous to the Alexander and Markov theorems for classical braids and links. So, the same way as in classical case virtual braids can be used in the study of virtual links.

5.2. *Virtual braid diagrams*

The simplest way to introduce formally virtual braids is in terms of virtual braid diagrams. We follow here the work of P. Bellingeri, B. A. Cisneros de La Cruz and L. Paris [11].

A *virtual braid diagram* on n strands is an n-tuple $\beta = (b_1, \ldots, b_n)$ of smooth paths in the plane $\mathbb{R}^2$ satisfying the following conditions:

(a) $b_i(0) = (i, 0)$ for all $i \in \{1, \ldots, n\}$.
(b) There exists a permutation $g \in \Sigma_n$ such that $b_i(1) = (g(i), 1)$ for all $i \in \{1, \ldots, n\}$.
(c) $(p_2 \circ b_i)(t) = t$, for all $i \in \{1, \ldots, n\}$ and all $t \in [0, 1]$, where $p_2 : \mathbb{R}^2 \to \mathbb{R}$ denotes the projection on the second coordinate.
(d) The b_i's intersect transversely in a finite number of double points, called the *crossings* of the diagram.

Each crossing is endowed with one of the following attributes: positive, negative or virtual. In the figures they are generally indicated as in Figure 6. Let VBD_n be the set of virtual braid diagrams on n strands, and let $\sim$ be the equivalence relation on VBD_n generated by ambient isotopy and the virtual Reidemeister moves depicted in Figure 7. The concatenation of diagrams induces a group structure on $VBD_n/\sim$. The latter is called *virtual braid group* on n strands, and is denoted by VB_n.

Fig. 6. Crossings in a virtual braid diagram

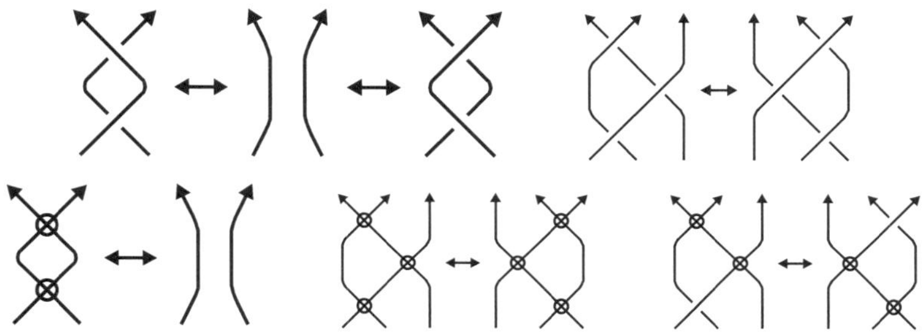

Fig. 7. Virtual Reidemeister moves

5.3. *Virtual braid group*

This group is given by the following set of generators: $\{\sigma_i, \tau_i, \ i = 1, 2, ..., n-1\}$ and relations:

$$\begin{cases} \sigma_i\sigma_j & = \sigma_j\sigma_i, \text{ if } |i-j| > 1, \\ \sigma_i\sigma_{i+1}\sigma_i & = \sigma_{i+1}\sigma_i\sigma_{i+1}. \end{cases}$$

The braid group relations

$$\begin{cases} \tau_i^2 & = 1, \\ \tau_i\tau_j & = \tau_j\tau_i, \quad \text{if} \quad |i-j| > 1, \\ \tau_i\tau_{i+1}\tau_i & = \tau_{i+1}\tau_i\tau_{i+1}. \end{cases}$$

The symmetric group relations

$$\begin{cases} \sigma_i\tau_j & = \tau_j\sigma_i, \text{ if } |i-j| > 1, \\ \tau_i\tau_{i+1}\sigma_i & = \sigma_{i+1}\tau_i\tau_{i+1}. \end{cases}$$

The mixed relations

The generator σ_i corresponds to the canonical generator of the braid group Br_n. The generators τ_i correspond to the virtual crossings.

This presentation was given in [47].

5.4. *Pure virtual braid group*

As for the classical braid groups there exists the canonical epimorphism to the symmetric group $VB_n \to \Sigma_n$ with the kernel called the *pure virtual braid group* PV_n. So we have a short exact sequence

$$1 \to PV_n \to VB_n \to \Sigma_n \to 1.$$

Define the following elements in PV_n

$$\lambda_{i,i+1} = \tau_i\,\sigma_i^{-1}, \quad \lambda_{i+1,i} = \tau_i\,\lambda_{i,i+1}\,\tau_i = \sigma_i^{-1}\,\tau_i, \quad i = 1, 2, \ldots, n-1,$$

$$\lambda_{ij} = \tau_{j-1}\,\tau_{j-2}\ldots\tau_{i+1}\,\lambda_{i,i+1}\,\tau_{i+1}\ldots\tau_{j-2}\,\tau_{j-1},$$

$$\lambda_{ji} = \tau_{j-1}\,\tau_{j-2}\ldots\tau_{i+1}\,\lambda_{i+1,i}\,\tau_{i+1}\ldots\tau_{j-2}\,\tau_{j-1}, \quad 1 \le i < j-1 \le n-1.$$

The group PV_n $(n \ge 2)$ admits a presentation with the generators λ_{ij}, $1 \le i \ne j \le n$, and the following relations:

$$\lambda_{ij}\lambda_{kl} = \lambda_{kl}\lambda_{ij} \tag{5.1}$$

$$\lambda_{ki}\lambda_{kj}\lambda_{ij} = \lambda_{ij}\lambda_{kj}\lambda_{ki}, \tag{5.2}$$

where distinct letters stand for distinct indices [4].

In particular PV_2 is a free group on free generators $\lambda_{1,2}$ and $\lambda_{2,1}$.

Like the usual pure braid groups, groups PV_n admit a semi-direct product decompositions [4]: for $n \ge 2$, the n-th virtual pure braid group can be decomposed as

$$PV_n = V_{n-1}^* \rtimes PV_{n-1}, \tag{5.3}$$

where V_{n-1}^* is a free subgroup of PV_n, PV_1 is supposed to be a trivial group.

These groups appeared in another context under the other name "n-th quasitriangular group $\mathbf{QTr}_n$" in the work of L. Bartholdi, B. Enriquez, P. Etingof and E. Rains [9] as groups associated to the Yang-Baxter equations. By considering the equation (5.2) for $k = 1$, $i = 2$, $j = 3$, we get the quantum Yang-Baxter equation

$$\lambda_{12}\lambda_{13}\lambda_{23} = \lambda_{23}\lambda_{13}\lambda_{12}.$$

In particular L. Bartholdi, B. Enriquez, P. Etingof and E. Rains gave a description of integral homology groups of PV_n: the group $H_r(PV_n, \mathbb{Z})$ is free abelian of rank $\binom{n-1}{r}\frac{n!}{(n-r)!}$ (the number of unordered partitions of the set $[n]$ consisting of n elements in r ordered parts).

5.5. *Group KV_n*

In this and the next section we follow the work [11].

Recall that Σ_n denotes the group of permutations of $\{1, \dots, n\}$. We denote by $\theta : VB_n \to \Sigma_n$ the epimorphism which sends σ_i to 1 and τ_i to $(i, i+1)$ for all $1 \le i \le n-1$, and by KB_n the kernel of θ. Note that θ has a section $\iota : \Sigma_n \to VB_n$ which sends $(i, i+1)$ to τ_i for all $1 \le i \le n-1$, and therefore VB_n is a semi-direct product $VB_n = KB_n \rtimes \Sigma_n$.

For $1 \le i < j \le n$ we set

$$\delta_{i,j} = \tau_i \tau_{i+1} \cdots \tau_{j-2} \sigma_{j-1} \tau_{j-2} \cdots \tau_{i+1} \tau_i \,,$$
$$\delta_{j,i} = \tau_i \tau_{i+1} \cdots \tau_{j-2} \tau_{j-1} \sigma_{j-1} \tau_{j-1} \tau_{j-2} \cdots \tau_{i+1} \tau_i \,.$$

As it is stated in [11] the following proposition is proved in Rabenda's master thesis [42] which is not available anywhere. The proof can be found in [5].

Proposition 5.1: *The group KB_n has a presentation with generating set*

$$\mathcal{S} = \{\delta_{i,j} \mid 1 \le i \ne j \le n\} \,,$$

and relations

$$\delta_{i,j}\delta_{k,\ell} = \delta_{k,\ell}\delta_{i,j} \qquad \text{for } i,j,k,\ell \text{ distinct}$$
$$\delta_{i,j}\delta_{j,k}\delta_{i,j} = \delta_{j,k}\delta_{i,j}\delta_{j,k} \qquad \text{for } i,j,k \text{ distinct}$$

Lemma 5.2: [5] *Let u be a word over $\{\tau_1, \dots, \tau_{n-1}\}$, let $\bar{u}$ be the element of VB_n represented by u, and let $i, j \in \{1, \dots, n\}$, $i \ne j$. Then $\bar{u}\delta_{i,j}\bar{u}^{-1} = \delta_{i',j'}$, where $i' = \theta(\bar{u})(i)$ and $j' = \theta(\bar{u})(j)$.*

Note that $\tau_i^{-1} = \tau_i$, since $\tau_i^2 = 1$, for all $i \in \{1, \dots, n-1\}$. Hence, the letters $\tau_1^{-1}, \dots, \tau_{n-1}^{-1}$ are not needed.

5.5.1. *Group KV_n as an Artin group*

Let S be a finite set. We recall that a *Coxeter matrix* over S is a square matrix $M = (m_{s,t})_{s,t\in S}$, indexed by the elements of S, such that $m_{s,s} = 1$ for all $s \in S$, and $m_{s,t} = m_{t,s} \in \{2, 3, 4, \dots\} \cup \{\infty\}$ for all $s, t \in S$, $s \ne t$. We represent this Coxeter matrix with a labelled graph $\Gamma = \Gamma_M$, called *Coxeter diagram.* The set of vertices of Γ is S. Two vertices $s, t \in S$ are connected by an edge labelled by $m_{s,t}$ if $m_{s,t} \ne \infty$.

If a, b are two letters and m is an integer ≥ 2, we set $\langle a, b\rangle^m = (ab)^{\frac{m}{2}}$ if m is even, and $\langle a, b\rangle^m = (ab)^{\frac{m-1}{2}} a$ if m is odd. In other words, $\langle a, b\rangle^m$

denotes the word $aba\cdots$ of length m. The *Artin group* of Γ is the group $A = A(\Gamma)$ defined by the following presentation.

$$A = \langle S \mid \langle s,t\rangle^{m_{s,t}} = \langle t,s\rangle^{m_{s,t}} \text{ for all } s,t \in S,\ s \neq t \text{ and } m_{s,t} \neq \infty\rangle .$$

The *Coxeter group* of Γ, denoted by $W = W(\Gamma)$, is the quotient of A by the relations $s^2 = 1$, for $s \in S$.

Let $\mathrm{V}\Gamma_n$ be the Coxeter diagram defined as follows. The set of vertices of $\mathrm{V}\Gamma_n$ is $\mathcal{S}$. If $i,j,k,\ell \in \{1,\ldots,n\}$ are distinct, then $\delta_{i,j}$ and $\delta_{k,\ell}$ are connected by an edge labelled by 2. If $i,j,k \in \{1,\ldots,n\}$ are distinct, then $\delta_{i,j}$ and $\delta_{j,k}$ are connected by an edge labelled by 3. There is no other edge in $\mathrm{V}\Gamma_n$. Then, by the presentation given above, KB_n is isomorphic to $A(\mathrm{V}\Gamma_n)$.

Let Γ be a Coxeter diagram. For $X \subset S$, we denote by Γ_X the subdiagram of Γ spanned by X, by A_X the subgroup of $A = A(\Gamma)$ generated by X, and by W_X the subgroup of $W = W(\Gamma)$ generated by X. The group A_X is the Artin group of Γ_X, and W_X is the Coxeter group of Γ_X.

For $\mathcal{X} \subset \mathcal{S}$, we denote by $KB_n(\mathcal{X})$ the subgroup of KB_n generated by $\mathcal{X}$. By the above, $KB_n(\mathcal{X})$ has a presentation with generating set $\mathcal{X}$ and relations

- $st = ts$ if s and t are connected in $\mathrm{V}\Gamma_n$ by an edge labelled by 2,
- $sts = tst$ if s and t are connected in $\mathrm{V}\Gamma_n$ by an edge labelled by 3.

5.5.2. *Full subsets of $\mathcal{S}$*

Definition 5.3: We say that a subset $\mathcal{X}$ of $\mathcal{S}$ is *full* if any two distinct elements s,t of $\mathcal{X}$ are connected by an edge of $\mathrm{V}\Gamma_n$.

We denote by $F_n = F(x_1,\ldots,x_n)$ the free group of rank n freely generated by $x_1,\ldots,x_n$. For $i,j \in \{1,\ldots,n\}$, $i \neq j$, we define $\varphi_{i,j} \in \mathrm{Aut}(F_n)$ by

$$\varphi_{i,j}(x_i) = x_i x_j x_i^{-1}\,,\ \varphi_{i,j}(x_j) = x_i\,,\ \text{and } \varphi_{i,j}(x_k) = x_k \text{ for } k \notin \{i,j\}\,.$$

The map $\mathcal{S} \to \mathrm{Aut}(F_n)$, $\delta_{i,j} \mapsto \varphi_{i,j}$, induces a representation $\varphi : KB_n \to \mathrm{Aut}(F_n)$. For $\mathcal{X} \subset \mathcal{S}$, we denote by $\varphi_{\mathcal{X}} : KB_n(\mathcal{X}) \to \mathrm{Aut}(F_n)$ the restriction of φ to $KB_n(\mathcal{X})$.

Proposition 5.4: [11] *If $\mathcal{X}$ is a full subset of $\mathcal{S}$, then $\varphi_{\mathcal{X}} : KB_n(\mathcal{X}) \to \mathrm{Aut}(F_n)$ is faithful.*

The following fact will be used in Step 4 of the algorithm.

Theorem 5.5: *Let $A_1 *_B A_2$ be an amalgamated product of groups. Let $g_1, \dots, g_\ell$ be a sequence of elements of $A_1 \sqcup A_2$ different from 1 and satisfying the following condition:*

if $g_i \in A_1$ (resp. $g_i \in A_2$), then $g_{i+1} \in A_2 \setminus B$ (resp. $g_{i+1} \in A_1 \setminus B$), for all $i \in \{1, \dots, \ell - 1\}$.

Then $g_1 g_2 \cdots g_\ell$ is different from 1 in $A_1 *_B A_2$.

5.6. *Solution to the word problem for the virtual braid group (after Bellingeri & Cisneros de la Cruz & Paris)*

Solution to the word problem for VB_n is divided into four steps. In Step 1, we give an algorithm (called Algorithm A) which decides whether an element of VB_n belongs to KB_n and, if it does, determines a word over $\mathcal{S}^{\pm 1}$ which represents this element.

For $\mathcal{X} \subset \mathcal{S}$, we denote by $KB_n(\mathcal{X})$ the subgroup of KB_n generated by $\mathcal{X}$. The other three steps provide a solution to the word problem for $KB_n(\mathcal{X})$ which depends recursively on the cardinality of $\mathcal{X}$. Step 2 is the beginning of the induction. More precisely, the algorithm proposed in Step 2 (called Algorithm B) is a solution to the word problem for $KB_n(\mathcal{X})$ when $\mathcal{X}$ is a full subset of $\mathcal{S}$. We notice that singletons are full subsets.

In Step 3 we suppose given a solution to the word problem for $KB_n(\mathcal{X})$, and, for a given subset $\mathcal{Y} \subset \mathcal{X}$, we show an algorithm which solves the membership problem for $KB_n(\mathcal{Y})$ in $KB_n(\mathcal{X})$ (Algorithm C).

In Step 4 we show an algorithm which solves the word problem for $KB_n(\mathcal{X})$ when $\mathcal{X}$ is not a full subset, under the assumption that the group $KB_n(\mathcal{Y})$ has a solvable word problem for any proper subset $\mathcal{Y}$ of $\mathcal{X}$ (Algorithm D).

Step 1

Now, we give an algorithm which, given a word u over $\{\sigma_1^{\pm 1}, \dots, \sigma_{n-1}^{\pm 1}, \tau_1, \dots, \tau_{n-1}\}$, decides whether the element $\bar{u}$ of VB_n represented by u belongs to KB_n. If yes, it also determines a word u' over $\mathcal{S}^{\pm 1} = \{\delta_{i,j}^{\pm} \mid 1 \le i \ne j \le n\}$ which represents $\bar{u}$.

Algorithm A. Let u be a word over $\{\sigma_1^{\pm 1}, \dots, \sigma_{n-1}^{\pm 1}, \tau_1, \dots, \tau_{n-1}\}$. We write u in the form

$$u = v_0 \sigma_{i_1}^{\varepsilon_1} v_1 \cdots v_{\ell-1} \sigma_{i_\ell}^{\varepsilon_\ell} v_\ell \,,$$

where $v_0, v_1, \ldots, v_\ell$ are words over $\{\tau_1, \ldots, \tau_{n-1}\}$, and $\varepsilon_1, \ldots, \varepsilon_\ell \in \{\pm 1\}$. For a word $v = \tau_{j_1} \cdots \tau_{j_k}$ over $\{\tau_1, \ldots, \tau_{n-1}\}$, we set $\theta(v) = (j_1, j_1 + 1) \cdots (j_k, j_k + 1) \in \Sigma_n$. Note that $\theta(\bar{u}) = \theta(v_0)\,\theta(v_1) \cdots \theta(v_\ell)$.

If $\theta(\bar{u}) \neq 1$, then $\bar{u} \notin KB_n$. If $\theta(\bar{u}) = 1$, then $\bar{u} \in KB_n$, and $\bar{u}$ is represented by

$$u' = \delta_{a_1,b_1}^{\varepsilon_1} \delta_{a_2,b_2}^{\varepsilon_2} \cdots \delta_{a_\ell,b_\ell}^{\varepsilon_\ell},$$

where

$$a_k = \theta(v_0 \cdots v_{k-1})(i_k) \text{ and } b_k = \theta(v_0 \cdots v_{k-1})(i_k + 1)$$

for all $k \in \{1, \ldots, \ell\}$.

Step 2

Notation. If u is a word over $\mathcal{S}^{\pm 1}$, then $\bar{u}$ will denote the element of KB_n represented by u.

Algorithm B. Let $\mathcal{X}$ be a full subset of $\mathcal{S}$, and let $u = s_1^{\varepsilon_1} \cdots s_\ell^{\varepsilon_\ell}$ be a word over $\mathcal{X}^{\pm 1}$. We have $\varphi_{\mathcal{X}}(\bar{u}) = \varphi_{\mathcal{X}}(s_1)^{\varepsilon_1} \cdots \varphi_{\mathcal{X}}(s_\ell)^{\varepsilon_\ell}$. If $\varphi(\bar{u}) = \mathrm{Id}$, then $\bar{u} = 1$. Otherwise, $\bar{u} \neq 1$.

Let $\mathcal{Y}$ be a subset of $\mathcal{S}$. For a word $u = s_1^{\varepsilon_1} \cdots s_\ell^{\varepsilon_\ell}$ over $\mathcal{S}^{\pm 1}$, Bellingeri et al. construct a word $\pi_{\mathcal{Y}}(u)$ over $\mathcal{Y}^{\pm 1}$ such that the following statement holds.

Proposition 5.6: *Let $\mathcal{Y}$ be a subset of $\mathcal{S}$. Let u, v be two words over $\mathcal{S}^{\pm 1}$. If $\bar{u} = \bar{v}$, then $\overline{\pi_{\mathcal{Y}}(u)} = \overline{\pi_{\mathcal{Y}}(v)}$. Moreover, we have $\bar{u} \in KB_n(\mathcal{Y})$ if and only if $\bar{u} = \overline{\pi_{\mathcal{Y}}(u)}$.*

Step 3

Algorithm C. Take two subsets $\mathcal{X}$ and $\mathcal{Y}$ of $\mathcal{S}$ such that $\mathcal{Y} \subset \mathcal{X}$, and assume we are given a solution to the word problem for $KB_n(\mathcal{X})$. Let u be a word over $\mathcal{X}^{\pm 1}$. We calculate $v = \pi_{\mathcal{Y}}(u)$. If $\overline{uv^{-1}} \neq 1$, then $\bar{u} \notin KB_n(\mathcal{Y})$. If $\overline{uv^{-1}} = 1$, then $\bar{u} \in KB_n(\mathcal{Y})$ and v is a word over $\mathcal{Y}^{\pm 1}$ which represents the same element as u.

Step 4

We assume that $\mathcal{X}$ is a non-full subset of $\mathcal{S}$, and that we have a solution to the word problem for $KB_n(\mathcal{Y})$ for any proper subset $\mathcal{Y}$ of $\mathcal{X}$ (induction hypothesis). We can and do choose two proper subsets $\mathcal{X}_1, \mathcal{X}_2 \subset \mathcal{X}$ satisfying the following properties.

(a) $\mathcal{X} = \mathcal{X}_1 \cup \mathcal{X}_2$.

(b) Let $\mathcal{X}_0 = \mathcal{X}_1 \cap \mathcal{X}_2$. There is no edge in $\mathrm{V}\Gamma_n$ connecting an element of $\mathcal{X}_1 \setminus \mathcal{X}_0$ to an element of $\mathcal{X}_2 \setminus \mathcal{X}_0$.

It follows from the presentations of the $KB_n(\mathcal{X}_i)$'s that we have the amalgamated product

$$KB_n(\mathcal{X}) = KB_n(\mathcal{X}_1) *_{KB_n(\mathcal{X}_0)} KB_n(\mathcal{X}_2)\,.$$

Algorithm D. Let u be a word over $\mathcal{X}^{\pm 1}$. We write u in the form $u_1 u_2 \cdots u_\ell$, where

- u_i is either a word over $\mathcal{X}_1^{\pm 1}$, or a word over $\mathcal{X}_2^{\pm 1}$,
- if u_i is a word over $\mathcal{X}_1^{\pm 1}$ (resp. over $\mathcal{X}_2^{\pm 1}$), then u_{i+1} is a word over $\mathcal{X}_2^{\pm 1}$ (resp. over $\mathcal{X}_1^{\pm 1}$).

We decide whether $\bar{u}$ is trivial by induction on ℓ. Suppose that $\ell = 1$ and $u = u_1 \in KB_n(\mathcal{X}_j)$ ($j \in \{1, 2\}$). Then we apply the solution to the word problem for $KB_n(\mathcal{X}_j)$ to decide whether $\bar{u}$ is trivial or not. Suppose that $\ell \geq 2$. For all i we set $v_i = \pi_{\mathcal{X}_0}(u_i)$. If $\overline{u_i v_i^{-1}} \neq 1$ for all i, then $\bar{u} \neq 1$. Suppose that there exists $i \in \{1, \dots, \ell\}$ such that $\overline{u_i v_i^{-1}} = 1$. Let $u_i' = v_1 u_2$ if $i = 1$, $u_i' = u_{\ell-1} v_\ell$ if $i = \ell$, and $u_i' = u_{i-1} v_i u_{i+1}$ if $2 \leq i \leq \ell - 1$. Set $v = u_1 \cdots u_{i-2} u_i' u_{i+2} \cdots u_\ell$. Then $\bar{u} = \bar{v}$ and, by induction, we can decide whether v represents 1 or not.

5.7. *Pure virtual braid group PV_3*

Till the end of this section we follow the work [7].

We recall that the group PV_3 admits the following presentation:

$$\begin{aligned}\langle \lambda_{12}, \lambda_{21}, \lambda_{13}, \lambda_{31}, \lambda_{23}, \lambda_{32} \mid\ & \lambda_{12}\lambda_{13}\lambda_{23}\lambda_{12}^{-1}\lambda_{13}^{-1}\lambda_{23}^{-1} = 1\\ & \lambda_{21}\lambda_{23}\lambda_{13}\lambda_{21}^{-1}\lambda_{23}^{-1}\lambda_{13}^{-1} = 1,\ \lambda_{13}\lambda_{12}\lambda_{32}\lambda_{13}^{-1}\lambda_{12}^{-1}\lambda_{32}^{-1} = 1,\\ & \lambda_{31}\lambda_{32}\lambda_{12}\lambda_{31}^{-1}\lambda_{32}^{-1}\lambda_{12}^{-1} = 1,\ \lambda_{23}\lambda_{21}\lambda_{31}\lambda_{23}^{-1}\lambda_{21}^{-1}\lambda_{31}^{-1} = 1,\\ & \lambda_{32}\lambda_{31}\lambda_{21}\lambda_{32}^{-1}\lambda_{31}^{-1}\lambda_{21}^{-1} = 1\rangle. \end{aligned} \tag{5.4}$$

5.7.1. *Residual nilpotence of a group*

We recall that a group G is *nilpotent* if its lower central series (4.1) terminates at the trivial subgroup $\{e\}$ in finitely many steps.

A group is *residually nilpotent* if one of the following equivalent statements holds.

- The intersection of all terms of the lower central series is trivial.
- Every non-identity element avoids some member of the lower central series.
- For every non-identity element there is a normal subgroup of G not containing it with a nilpotent quotient group.
- For every non-identity element g of G, there exists a nilpotent group H and a homomorphism $\phi : G \to H$ such that $\phi(g) \neq e$ in H.

The elements of finite order in a nilpotent group H form a normal subgroup Tors (H), the torsion subgroup of H, so it makes sense to define the subgroups $\overline{\Gamma_i(G)} \leq G$ by the formula

$$\overline{\Gamma_i(G)}/\Gamma_i(G) = \text{Tors}\,(G/\Gamma_i(G)).$$

This defines a series of normal subgroups of G

$$G = \overline{\Gamma_1(G)} \geq \overline{\Gamma_2(G)} \geq \cdots \geq \overline{\Gamma_i(G)} \geq \overline{\Gamma_{i+1}(G)} \geq \cdots. \tag{5.5}$$

The group $\overline{\Gamma_i(G)}/\overline{\Gamma_{i+1}(G)}$ is a torsion-free abelian group.

A group G is *residually torsion-free nilpotent* if one of the following equivalent statements holds.

- The intersection of all terms $\overline{\Gamma_i(G)}$ of the series (5.5) is trivial.
- Every non-identity element avoids some member of the series (5.5).
- For every non-identity element, there is a normal subgroup of G not containing it, with a torsion-free nilpotent quotient group.
- For every non-identity element g of G, there exists a torsion-free nilpotent group H and a homomorphism $\phi : G \to H$, such that $\phi(g) \neq e$ in H.

Free groups give a canonical example of residually torsion-free nilpotent groups.

5.7.2. *Residual nilpotence of PV_3*

Pure braid groups are residually torsion-free nilpotent. So a natural question arises: are the pure virtual braid groups PV_n residually torsion-free nilpotent?

We prove that $PV_3 \cong G_3 * \mathbb{Z}$ for some group G_3, and then that G_3 injects into a residually torsion free nilpotent group Cb_3; this implies that G_3 is itself residually torsion free nilpotent, and by a theorem of Malcev, the same can be said of PV_3.

It follows from (5.3) that

$$PV_3 = V_2^* \rtimes PV_2,$$

where $PV_2 = \langle \lambda_{12}, \lambda_{21} \rangle$ is a free 2-generated group and V_2^* is the normal closure of group $V_2 = \langle \lambda_{13}, \lambda_{23}, \lambda_{31}, \lambda_{32} \rangle$ in PV_3. The group V_2 is a free of rank 4.

Proposition 5.7: *There exists a group G_3, such that the group PV_3 decomposes as the following free product: $PV_3 \cong G_3 * \mathbb{Z}$.*

Proposition 5.8: *There exists a group Q_3, such that the group G_3 decomposes as the following semi-direct product:*

$$G_3 \cong Q_3 \rtimes \mathbb{Z}.$$

Theorem 5.9: *The pure virtual braid group PV_3 is residually torsion free nilpotent.*

5.7.3. *Sketch of the proof of the Theorem*

Consider the subgroup of $\mathrm{Aut}(F_n)$, $n \geq 2$, generated by automorphisms of the form

$$\varepsilon_{ij} : \begin{cases} x_i \longmapsto x_j^{-1} x_i x_j, & \text{if} \quad i \neq j; \\ x_l \longmapsto x_l, & \text{if} \quad l \neq i, \end{cases}$$

and denote this subgroup by Cb_n. This is the group *basis conjugating automorphisms* of a free group. It is also called the *McCool group* [39]. It has topological interpretations. It is the pure group of motions of n unlinked circles in S^3 [22, 26]. On the other hand it is the pure braid-permutation group. It is residually torsion-free nilpotent.

McCool gave the following presentation for it [39]:

$$\langle \varepsilon_{ij},\ 1 \leq i \neq j \leq n \mid \varepsilon_{ij}\varepsilon_{kl} = \varepsilon_{kl}\varepsilon_{ij}, \\ \varepsilon_{ij}\varepsilon_{kj} = \varepsilon_{kj}\varepsilon_{ij}, \varepsilon_{ij}\varepsilon_{kj}\varepsilon_{ik} = \varepsilon_{ik}\varepsilon_{ij}\varepsilon_{kj} \rangle$$

where distinct letters stand for distinct indices.

There exists a homomorphism $f : PV_3 \longrightarrow Cb_3$, defined by

$$f(\lambda_{ij}) = \varepsilon_{ij}, \quad 1 \leq i \neq j \leq 3.$$

We prove that the group G_3 is embedded in Cb_3 by the map f, hence it is residually torsion free nilpotent as a subgroup of the group Cb_3. Recall the following result of Malcev: free product of residually torsion-free nilpotent groups is residually torsion-free nilpotent. Hence, PV_3 is residually torsion free nilpotent.

5.7.4. *Remark on the finite type invariants*

Recall the notion of the finite type invariants for virtual braids. Let J be the two-sided ideal in the integral group ring of virtual braids $\mathbb{Z}[VB_n]$ generated by the elements $\{\sigma_i - \tau_i,\ \sigma_i^{-1} - \tau_i \mid i = 1, \dots, n-1\}$. The filtration of the group ring

$$\mathbb{Z}[VB_n] \supset J \supset J^2 \supset \cdots$$

is called *Goussarov-Polyak-Viro filtration.*

Let A be an abelian group and $n \geq 2$. Consider a map (not necessary a homomorphism)

$$v : VB_n \to A.$$

We say that v in an *A-valued invariant of virtual n-braids of degree d* if its linear extension $\mathbb{Z}[VB_n] \to A$ vanishes on J^{d+1}.

As usual the augmentation ideal for the pure virtual braids is defined by the formula $\Delta(PV_n) = \mathrm{Ker}\{\mathbb{Z}[PV_n] \to \mathbb{Z}\}$. The Goussarov-Polyak-Viro filtration for VB_n corresponds to the filtration by powers of augmentation ideal of the group ring of the pure virtual braids $\mathbb{Z}[PV_n]$:

$$\mathbb{Z}[PV_n] \supset \Delta \supset \Delta^2 \supset \cdots .$$

Residual torsion free nilpotence implies that the intersection of augmentation powers of the group ring $\mathbb{Z}[PV_3]$ is zero. Hence, we have the following:

Proposition 5.10: *The set of invariants of finite degree is complete for virtual pure braids on three strands.*

5.7.5. *Subgroup Q_3*

We establish that the subgroup Q_3 which is essential in the structure of PV_3 is an extension of $\mathbb{Z}^4$ by a free group on countably many generators.

Theorem 5.11: *The kernel of the homomorphism of abelianization* ab *from Q_3 onto $\mathbb{Z}^4$ is free and there is a short exact sequence*

$$1 \longrightarrow F_\infty \longrightarrow Q_3 \xrightarrow{\mathrm{ab}} \mathbb{Z}^2 \times \mathbb{Z}^2 \longrightarrow 1,$$

where F_∞ is a free subgroup of countable rank.

5.7.6. *Asphericity*

Let G be a group and M a $\mathbb{Z}[G]$-module. Recall that M_G is defined in Brown's book [17], p. 34, as a quotient-module of M by the submodule generated by differences $m - gm$, $m \in M$, $g \in G$.

Lemma 5.12: *Let a $\mathbb{Z}[G]$-module M be a submodule of a free $\mathbb{Z}[G]$-module. Suppose that $M \neq 0$ and $M_G = 0$. Then the group G is not residually torsion-free nilpotent.*

Theorem 5.13: *The presentation (5.4) is aspherical.*

Proof: Let K be the standard 2-complex of the presentation (5.4). Consider the following exact sequence [17, Exercise 1, p. 44]:

$$0 \to H_3(PV_3) \to \pi_2(K)_{PV_3} \to H_2(K) \to H_2(PV_3) \to 0. \tag{5.6}$$

As it was calculated in [9] the homology $H_r(PV_n)$ is the free abelian group of rank $\binom{n-1}{r}\frac{n!}{(n-r)!}$ (the number of unordered partitions of the set $[n]$ consisting of n elements in r ordered parts) So, in our case $H_2(PV_3) = \mathbb{Z}^{\oplus 6}$ and $H_3(PV_3)$ is zero. Now observe that $H_2(K)$ is a free abelian group of rank less than or equal to six, since there are only six 2-dimensional cells in the 2-complex K. Hence, the map $H_2(K) \to H_2(PV_3)$ in (5.6) is an isomorphism and therefore, $\pi_2(K)_{PV_3} = 0$. We have the following monomorphism of $\mathbb{Z}[PV_3]$-modules:

$$\pi_2(K) = H_2(\tilde{K}) \hookrightarrow C_2(\tilde{K}) \cong \mathbb{Z}[PV_3]^{\oplus 6},$$

where $C_2(\tilde{K})$ is the second term of the chain complex $C_*(\tilde{K})$ of the universal covering space $\tilde{K}$ of K. Apply now Lemma 5.12 to the case $G = PV_3$, $M = \pi_2(K)$. Hence, the presentation (5.4) is aspherical. □

5.7.7. *Cohomology ring of PV_3*

The homology of PV_n as abelean groups was calculated in [9]. We describe the multiplicative structure (cup-product) for $H^*(PV_3, \mathbb{Z})$. Everywhere the coefficient ring is $\mathbb{Z}$, so, it will be omitted in notations.

Let λ_{ij}^* be the elements of $H^1(PV_3)$ dual to the abelianizations of λ_{ij}.

Theorem 5.14: *The cohomology ring $H^*(PV_3)$ is the quotient algebra of the exterior algebra*

$$E(\lambda_{12}^*, \lambda_{21}^*, \lambda_{13}^*, \lambda_{31}^*, \lambda_{23}^*, \lambda_{32}^*)$$

subject to the following relations:

1) $\lambda^*_{ij}\lambda^*_{ji} \equiv 0$ *for* $1 \le i < j \le 3$;

2) $(\lambda^*_{13} - \lambda^*_{31})\lambda^*_{ij} \equiv (\lambda^*_{12} - \lambda^*_{21})\lambda^*_{ij} + (\lambda^*_{23} - \lambda^*_{32})\lambda^*_{ij}$ *for* $1 \le i \ne j \le 3$;

3) $\lambda^*_{21}\lambda^*_{31} \equiv \lambda^*_{21}\lambda^*_{32} + \lambda^*_{23}\lambda^*_{31}$.

5.7.8. *The associated graded Lie algebra for* PV_3

For the definition of the Lie algebra of a group see Subsection 4.1.

Theorem 5.15: *The Lie algebra* $L(PV_3)$ *is the quotient of the free Lie algebra* $L(A_1, B_1, A_2, B_2, C_1, C_2)$ *by the following relations*

$$[A_1, B_1] = 0$$
$$[A_2, B_2] = 0,$$

$$[C_1, B_1] = [A_2, B_1]$$
$$[C_1, A_1] = [B_2, A_1]$$
$$[C_1, B_2] = [A_1, B_2]$$
$$[C_1, A_2] = [B_1, A_2].$$

6. Braids and Thompson groups

The standard reference about the Thompson group is the survey of J. W. Cannon, W. J. Floyd and W. R. Parry [19]. We use also the thesis of J. M. Belk [10].

6.1. *Geometrical definition of the Thompson group F*

By *dyadic rational numbers* we mean rational numbers of the form $p2^q$, where $p, q \in \mathbb{Z}$.

Let F be the set of piecewise linear homeomorphisms from the closed unit interval $[0, 1]$ to itself that are differentiable except at finitely many dyadic rational numbers, and such that on intervals of differentiability, the derivatives are powers of 2.

Since derivatives are positive where they exist, elements of F preserve orientation.

Let $f \in F$, and let

$$0 = x_0 < x_1 < x_2 < \cdots < x_n = 1$$

be the points at which f is not differentiable. Then since $f(0) = 0$,

$$f(x) = a_1 x, \quad \text{for } x_0 < x < x_1,$$

where a_1 is a power of 2.

Likewise, since $f(x_1)$ is a dyadic rational number,

$$f(x) = a_2 x + b_2 \text{ for } x_1 < x < x_2,$$

where a_2 is a power of 2 and b_2 is a dyadic rational number. It follows inductively that

$$f(x) = a_i x + b_i \quad \text{for} \quad x_i < x < x_{i+1}$$

and $i = 1, ..., n$, where a_i is a power of 2 and b_i is a dyadic rational number.

It follows that $f^{-1} \in F$ and that f maps the set of dyadic rational numbers bijectively to itself. From this it follows that F is closed under composition of functions. Thus F is a subgroup of the group of all homeomorphisms from $[0, 1]$ to $[0, 1]$. This group is Thompson's group F.

As we are dealing with functions words in the group F we are reading from right to left.

Example 6.1: Let A and B be the following functions:

$$A(x) = \begin{cases} \frac{x}{2}, & 0 \leq x \leq \frac{1}{2}, \\ x - \frac{1}{4}, & \frac{1}{2} \leq x \leq \frac{3}{4}, \\ 2x - 1, & \frac{3}{4} \leq x \leq 1, \end{cases}$$

$$B(x) = \begin{cases} x, & 0 \leq x \leq \frac{1}{2}, \\ \frac{x}{2} + \frac{1}{4}, & \frac{1}{2} \leq x \leq \frac{3}{4}, \\ x - \frac{1}{8}, & \frac{3}{4} \leq x \leq \frac{7}{8}, \\ 2x - 1, & \frac{7}{8} \leq x \leq 1. \end{cases}$$

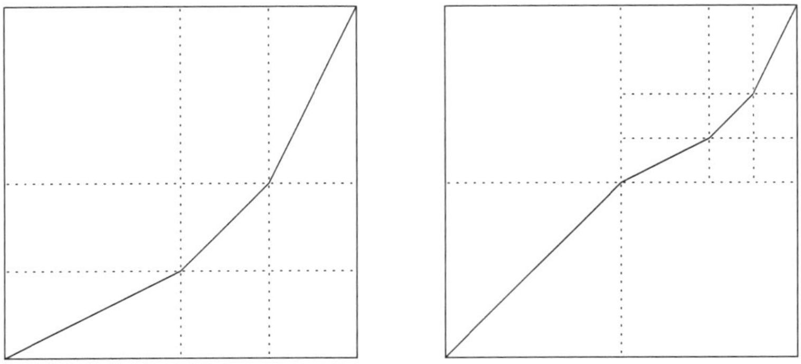

Fig. 8. The generators A and B

Define functions $X_0, X_1, X_2, \ldots$ in F so that

$$X_0 = A$$

and

$$X_n = A^{-(n-1)}\, B\, A^{n-1} \quad \text{for} \quad n \geq 1.$$

6.2. *Tree diagrams*

Define an *ordered rooted binary tree* to be a tree S such that

i) S has a root v_0,

ii) if S consists of more than v_0, then v_0 has valence 2,

iii) if v is a vertex in S with valence greater than 1, then there are exactly two edges $e_{v,L}$, $e_{v,R}$ which contain v and are not contained in the shortest path from v_0 to v.

The edge $e_{v,L}$ is called a *left edge* of S, and $e_{v,R}$ is called a *right edge* of S. Vertices with valence 0 (in case of the trivial tree) or 1 in S will be called leaves of S. There is a canonical left-to-right linear ordering on the leaves of S. The right side of S is the maximal arc of right edges in S which begins at the root of S. The left side of S is defined analogously.

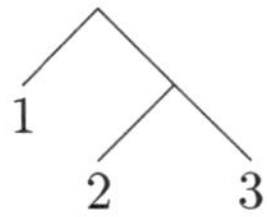

Example of an ordered rooted binary tree

An *ordered rooted binary sub-tree* T of an ordered rooted binary tree S is an ordered rooted binary tree which is a sub-tree of S whose left edges are left edges of S, whose right edges are right edges of S, but whose root need not be the root of S.

Define a *standard dyadic interval* in $[0,1]$ to be an interval of the form $[\frac{a}{2^n}, \frac{a+1}{2^n}]$, where a, n are non-negative integers with $a \leq 2^n - 1$.

There is a tree of standard dyadic intervals, $\mathcal{T}$, which is defined as follows. The vertices of $\mathcal{T}$ are the standard dyadic intervals in $[0,1]$. An edge of $\mathcal{T}$ is a pair (I, J) of standard dyadic intervals I and J such that either I is the left half of J, in which case (I, J) is a left edge, or I is the right half of J, in which case (I, J) is a right edge. The tree $\mathcal{T}$ is an ordered rooted binary tree (see Figure 9).

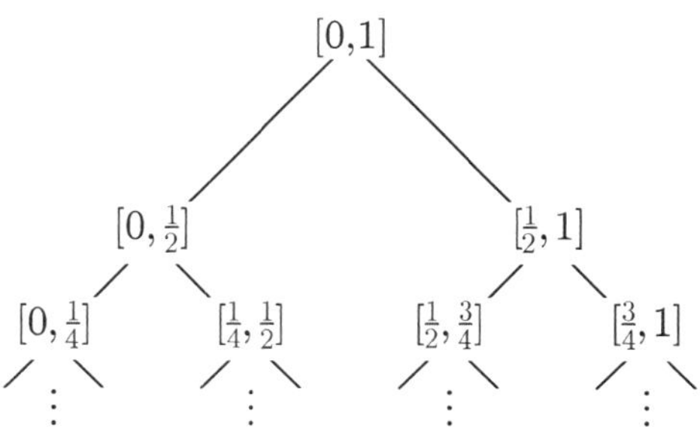

Fig. 9. Tree of standard dyadic intervals $\mathcal{T}$

Define a *$\mathcal{T}$-tree* to be a finite ordered rooted binary subtree of $\mathcal{T}$ with root $[0,1]$. Call the $\mathcal{T}$-tree with just one vertex the *trivial* $\mathcal{T}$-tree. For every non-negative integer n, let $\mathcal{T}_n$ be the $\mathcal{T}$-tree with $n+1$ leaves whose right side has length n.

Define a *caret* to be an ordered rooted binary subtree of $\mathcal{T}$ with exactly two edges:

A partition of $[0,1]$ is called a *standard dyadic partition* if and only if the intervals of the partition are standard dyadic intervals.

There is a canonical bijection between standard dyadic partitions and $\mathcal{T}$-trees.

Lemma 6.2: *Let $f \in F$. Then there exists a standard dyadic partition $0 = x_0 < x_1 < x_2 < \cdots < x_n = 1$ such that f is linear on every interval of the partition and*

$$0 = f(x_0) < f(x_1) < f(x_2) < \cdots < f(x_n) = 1$$

is a standard dyadic partition.

A *tree diagram* is an ordered pair (R, S) of $\mathcal{T}$-trees such that R and S have the same number of leaves:

$$R \to S.$$

The tree R is called the *domain tree* of the diagram, and S is called the *range tree* of the diagram.

Given $f \in F$ there exist standard dyadic partitions P and Q such that f is linear on the intervals of P and maps them to the intervals of Q. To f is associated the tree diagram (R, S), where R is the $\mathcal{T}$-tree corresponding to P and S is the $\mathcal{T}$-tree corresponding to Q.

For $f \in F$ intervals P and Q are not unique. Given one tree diagram (R, S) for f, another can be constructed by adjoining carets to R and S at the leaf with the same number i. Conversely such carets can be deleted.

If there do not exist such carets in R, S, then the tree diagram (R, S) is said to be *reduced*.

There is a canonical bijection between F and the set of reduced tree diagrams.

Recall that the element A sends intervals of the subdivision:

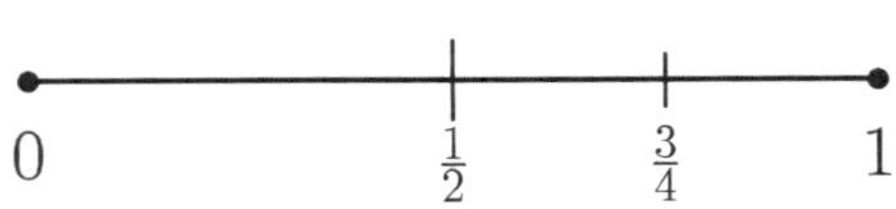

linearly onto intervals of the subdivision:

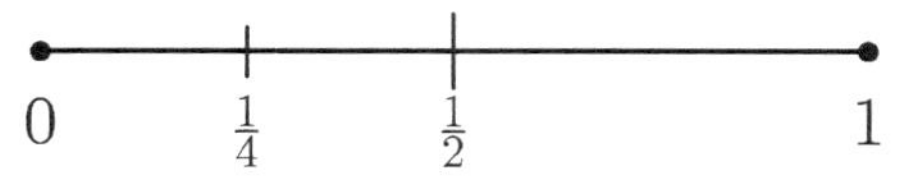

The tree pair diagram for the generator A of F

This tree pair diagram can be interpreted as a law of associativity of a product $*$:

$$a * (b * c) \to (a * b) * c.$$

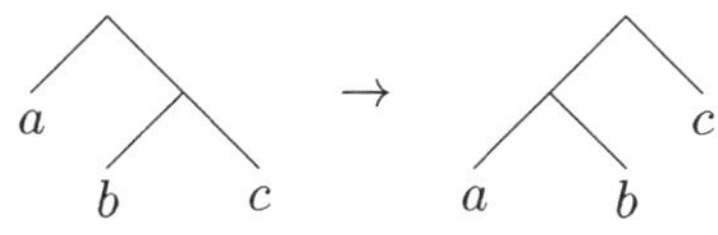

The tree pair diagram for the law of associativity

Tree diagrams can also be depicted in this way:

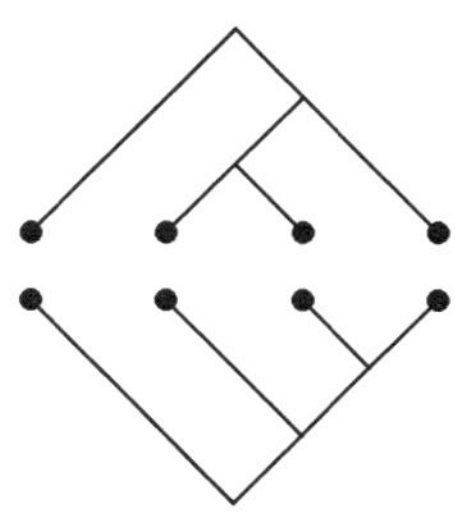

Then reduction of a tree diagram consists of removing an opposing pair of carets, like this:

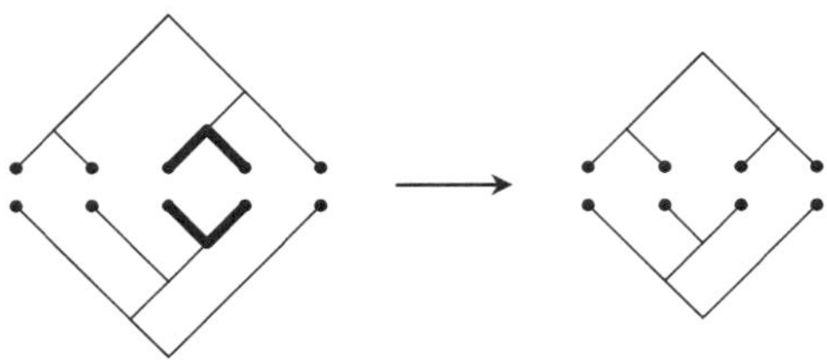

Let $I_0, ..., I_n$ be the leaves of $\mathcal{T}$-tree S in order. For every integer k with $0 \leq k \leq n$, let a_k be the length of the maximal arc of left edges in S which begins at I_k and which does not reach the right side of S. Then a_k is the k-th *expotent* of S.

Theorem 6.3: *Let R, S be $\mathcal{T}$-trees with $n+1$ leaves for some non-negative integer n. Let $a_0, ..., a_n$ be the exponents of R, and let $b_0, \ldots, b_n$ be the exponents of S. Then the function in F with tree diagram (R, S) is*

$$X_0^{b_0} X_1^{b_1} \ldots X_n^{b_n} X_n^{-a_n} \ldots X_1^{-a_1} X_0^{-a_0}.$$

The tree diagram (R,S) is reduced if and only if

i) if the last two leaves of R lie in a caret, then the last two leaves of S do not lie in a caret and

ii) for every integer k with $0 \leq k < n$, if $a_k > 0$ and $b_k > 0$ then either $a_{k+1} > 0$ or $b_{k+1} > 0$.

Corollary 6.4: *Thompson's group F is generated by A and B.*

Corollary 6.5: *Every non-trivial element of F can be expressed in unique form*

$$X_0^{b_0} X_1^{b_1} \ldots X_n^{b_n} X_n^{-a_n} \ldots X_1^{-a_1} X_0^{-a_0}$$

where $n, a_0, \ldots, a_n, b_0, \ldots, b_n$ are non-negative integers such that

i) exactly one of a_n and b_n is non-zero,

ii) if $a_k > 0$ and $b_k > 0$ for some integer k with $0 \leq k < n$, then $a_{k+1} > 0$ or $b_{k+1} > 0$. Furthermore, every such function in F is non-trivial.

The form of the above lemma of an element of F is called the *normal form* of this element.

The functions in F of the form

$$X_0^{b_0} X_1^{b_1} \dots X_n^{b_n}$$

with $b_k \geq 0$ for $k = 0, \dots, n$ will be called *positive*. The positive elements of F are exactly those with tree diagrams having domain tree $\mathcal{T}_n$ for some non-negative integer n. Inverses of positive elements will be called *negative*.

Lemma 6.6: *The set of positive elements of F is closed under multiplication.*

6.3. *Presentations for F*

Let $[y, z] = yzy^{-1}z^{-1}$.

Let F_1 be a group generated by letters α and β, and given by the presentation:

$$\langle \alpha, \beta \,|\, [\alpha\beta^{-1}, \alpha^{-1}\beta\alpha],\ [\alpha\beta^{-1}, \alpha^{-2}\beta\alpha^2] \rangle. \tag{6.1}$$

We rewrite the first relation of F_1 in detail:

$$\alpha\beta^{-1}\alpha^{-1}\beta\alpha\beta\alpha^{-1}\alpha^{-1}\beta^{-1}\alpha = e.$$

This is the same as

$$\beta^{-1}\alpha^{-1}\beta\alpha\beta = \alpha^{-2}\beta\alpha^2. \tag{6.2}$$

We rewrite the second relation of F_1 in detail:

$$\alpha\beta^{-1}\alpha^{-2}\beta\alpha^2\beta\alpha^{-1}\alpha^{-2}\beta^{-1}\alpha^2 = e.$$

This is the same as

$$\beta^{-1}\alpha^{-2}\beta\alpha^2\beta = \alpha^{-3}\beta\alpha^3. \tag{6.3}$$

We define the following notation:

$$x_0 = \alpha,$$

$$x_1 = \beta,$$

$$x_i = x_0^{-i+1} x_1 x_0^{i-1},\ i \geq 2.$$

Using this notation equations (6.2) and (6.3) give us that presentation (6.1) is equivalent to

$$\langle x_0, x_1 \mid x_1^{-1}x_2x_1 = x_3,\ x_1^{-1}x_3x_1 = x_4 \rangle. \tag{6.4}$$

Theorem 6.7: *Group F_1 admits a presentation:*

$$\langle x_0, x_1, ..., x_n, ... \mid x_k^{-1}x_nx_k = x_{n+1} \ \textit{ for } \ k < n \rangle. \tag{6.5}$$

Proof: Let us consider the presentation

$$\langle x_0, x_1, x_2, \ldots, x_n, \ldots \mid x_1^{-1}x_2x_1 = x_3,\ x_1^{-1}x_3x_1 = x_4,\\ x_i = x_0^{-i+1}x_1x_0^{i-1},\ i \geq 2\rangle. \quad (6.6)$$

It is equivalent to presentation (6.4) and so it is equivalent to presentation (6.1). We prove that it is also equivalent to presentation (6.5).

In presentation (6.6) we change the last relation by the following one

$$x_{i+1} = x_0^{-1}x_ix_0,\ i \geq 1.$$

Then by induction we prove relations

$$x_{i+1} = x_{i-1}^{-1}x_ix_{i-1},\ i \geq 3.$$

The case $1 \leq k \leq n-2$ in (6.5) is reduced by induction to the case of relations

$$x_{m+1} = x_1^{-1}x_mx_1,\ m \geq 4, \quad (6.7)$$

where $m = n - k + 1$. We prove (6.7) also by induction starting with $x_1^{-1}x_3x_1 = x_4$:

$$x_{m+1} = x_{m-1}^{-1}x_mx_{m-1} = x_1^{-1}x_{m-2}^{-1}x_1x_1^{-1}x_{m-1}x_1x_1^{-1}x_{m-2}^{-1}x_1\\ = x_1^{-1}x_{m-2}^{-1}x_{m-1}x_{m-2}^{-1}x_1 = x_1^{-1}x_mx_1. \ \square$$

Let us define a map ϕ of the $\{\alpha, \beta\}$ to the Thompson group F by the formulae

$$\phi(\alpha) = A, \quad \phi(\beta) = B.$$

This map defines a homomorphism of the free group generated by α and β to F. We verify directly that the relators of F_1 are sent to e by ϕ, so we have a homomorphism of groups

$$\phi : F_1 \to F.$$

By definition we also have

$$\phi(x_i) = X_i.$$

Theorem 6.8: *The homomorphism*

$$\phi : F_1 \to F$$

is an isomorpism. So formulae (6.1) and (6.5) give two presentations of the Thompson group.

Proof: Surjectivity follows from the fact that F is generated by A and B.

Let us rewrite relations (6.5) in the form

$$x_k^{-1}x_n = x_{n+1}x_k^{-1}, \quad x_n^{-1}x_k = x_k x_{n+1}^{-1},$$

$$x_n x_k = x_k x_{n+1}, \quad x_k^{-1}x_n^{-1} = x_{n+1}^{-1}x_k^{-1}.$$

With the help of these relations we can put every element of the group F_1 in the same normal form as F. This gives injectivity. □

6.4. *Further properties of F*

A discrete group G is *amenable* if there is a left-invariant measure μ on G which is **finitely** additive and has total measure 1.

That is, G is amenable if there is a function

$$\mu : \{\text{subsets of } G\} \to [0, 1]$$

such that

(1) $\mu(gA) = \mu(A)$ for all $g \in G$ and all subsets A of G,
(2) $\mu(G) = 1$,
(3) $\mu(AUB) = \mu(A) + \mu(B)$ if A and B are disjoint subsets of G.

Problem: Is the Richard Thompson group F amenable?

From Mark Sapir's blog (https://marksapir.wordpress.com/): Another false proof of non-amenability of the R. Thompson group F.

It is clear from the definition that a finite group is amenable.

The class of all amenable groups is closed under quotients, subgroups, extensions, and directed unions with respect to inclusion.

Abelian groups are amenable.

We call a group an *elementary amenable* group if it is in the smallest class of groups that contains all Abelian and finite groups and is closed under quotients, subgroups, extensions, and directed unions with respect to inclusion.

Theorem 6.9: *The free group of two variables $\langle a, b\rangle$ is not amenable.*

Proof: Suppose otherwise that μ is a finitely additive, left invariant total measure on $\langle a, b\rangle$. Then $\mu(1) = 0$ as $\langle a, b\rangle$ is infinite. Let

$$g* = \{h \in \langle a, b\rangle : h \text{ has a freely reduced representative beginning with } g\}.$$

Then

$$\langle a,b\rangle = \{1\} \cup a * \cup b * \cup a^{-1} * \cup b^{-1} * .$$

On the other hand we have $a^{-1}a* = \{1\} \cup a * \cup b * \cup\ b^{-1}*$, and so

$$\mu(a*) = \mu(a*) + \mu(b*) + \mu(b^{-1}*),$$

and hence $\mu(b*) = \mu(b^{-1}*) = 0$. The same way $\mu(a*) = \mu(a^{-1}*) = 0$ and $\mu(\langle a,b\rangle) = 0$. □

Theorem 6.10: *The commutator subgroup $[F,F]$ of F consists of all elements in F which are trivial in neighbourhoods of 0 and 1. Furthermore,*

$$H_1(F,\mathbb{Z}) = F/[F,F] \cong \mathbb{Z} \oplus \mathbb{Z}.$$

Proof: Define a group homomorphism

$$\psi : F \to \mathbb{Z} \oplus \mathbb{Z}$$

such that if $f \in F$, then

$$\psi(f) = (a,b),$$

where the right derivative of f at 0 is 2^a and the left derivative of f at 1 is 2^b. Since $\psi(A) = (-1,1)$ and $\psi(B) = (0,1)$, ψ is surjective. The group F is generated by A and B, so ψ is a homomorphism of abelianization and its kernel is $[F,F]$. □

Lemma 6.11: *If $0 = x_0 < x_1 < x_2 < \cdots < x_n = 1$ and $0 = y_0 < y_1 < y_2 < \cdots < y_n = 1$ are partitions of $[0,1]$ consisting of dyadic rational numbers, then there exists $f \in F$ such that $f(x_i) = y_i$ for $i = 0,...,n$. Furthermore, if $x_{i-1} = y_{i-1}$ and $x_i = y_i$ for some i with $0 < i \leq n$, then f can be taken to be trivial on the interval $[x_i, x_{i+1}]$.*

Theorem 6.12: *Every proper quotient group of F is Abelian.*

Lemma 6.13: *Let a,b be dyadic rational numbers with $0 < a < b < 1$ such that $b-a$ is a power of 2. Then the subgroup of F consisting of all functions with support in $[a,b]$ is isomorphic to F by means of the linear conjugation.*

Theorem 6.14: *The commutator subgroup $[F,F]$ of F is a simple group.*

Theorem 6.15: *The sub-monoid of F generated by A, B, B^{-1} is the free product of the sub-monoid generated by A and the subgroup generated by B.*

Corollary 6.16: *Thompson's group F has exponential growth.*

Theorem 6.17: *Every non-Abelian subgroup of F contains a free Abelian subgroup of infinite rank.*

Corollary 6.18: *Thompson's group F does not contain a non-Abelian free group.*

Theorem 6.19: *Thompson's group F is not an elementary amenable group.*

Theorem 6.20: *Thompson's group F is a totally ordered group.*

6.5. *Thompson's group T*

Consider S^1 as the interval $[0,1]$ with the endpoints identified.

Then T is the group of piecewise linear homeomorphisms from S^1 to itself that map images of dyadic rational numbers to images of dyadic rational numbers and that are differentiable except at finitely many images of dyadic rational numbers and, on intervals of differentiability, the derivatives are powers of 2. Just as we proved that F is a group, it is easy to see that T is indeed a group.

The elements A and B of F induce elements of T, which will still be denoted by A and B.

Example 6.21: Let C be the following function:

$$C(x) = \begin{cases} \frac{x}{2} + \frac{3}{4}, & 0 \leq x \leq \frac{1}{2}, \\ 2x - 1, & \frac{1}{2} \leq x \leq \frac{3}{4}, \\ x - \frac{1}{4}, & \frac{3}{4} \leq x \leq 1. \end{cases}$$

We can associate tree diagrams and unique reduced tree diagrams to elements of T almost exactly as we did to elements of F. The only difference is the following. Elements of F map leftmost leaves of domain trees to leftmost leaves of range trees. When an element of T does not do this, we denote the image in its range tree of the leftmost leaf of its domain tree with a small circle.

Lemma 6.22: *The elements $A, B,$ and C generate T and satisfy the* *following relations:*

$$\begin{aligned}
&1)\ [AB^{-1}, A^{-1}BA] = 1,\\
&2)\ [AB^{-1}, A^{-2}BA^2] = 1,\\
&3)\ C = B(A^{-l}CB),\\
&4)\ (A^{-l}CB)(A^{-l}BA) = B(A^{-2}CB^2),\\
&5)\ CA = (A^{-1}CB)^2,\\
&6)\ C^3 = 1.
\end{aligned}$$

Let T_1 be a group generated by A, B and C as formal symbols and relations of the previous lemma.

Lemma 6.23: *The homomorphism*

$$T_1 \to T$$

that maps the formal symbols A, B, and C to the functions A, B, and C in T is a surjection.

Lemma 6.24: *The subgroup of T_1 generated by A and B is isomorphic to F.*

Define the elements C_n , $n \geq 1$, of T_1 by

$$C_n = A^{-(n-1)}CB^{n-1}.$$

For convenience we define $C_0 = 1$.

Theorem 6.25: *The group T_1 is simple.*

Corollary 6.26: *The group T_1 is isomorphic to T.*

6.6. *Thompson's group V*

To define the group V we admit bijections of the circle which can be discontinuous in a finite number of dyadic points. So, V is the group of right-continuous bijections of S^1 that map images of dyadic rational numbers to images of dyadic rational numbers, that are differentiable except at finitely many images of dyadic rational numbers, and such that, on each maximal interval on which the function is differentiable, the function is linear with derivative a power of 2.

We can associate tree diagrams with elements of V as we did for F and T, except that now we need to label the leaves of the domain and range trees to indicate the correspondence between the leaves.

Define

$$\pi_0 : S^1 \to S^1$$

by

$$\pi_0(x) = \begin{cases} \frac{x}{2} + \frac{1}{2}, & 0 \le x \le \frac{1}{2}, \\ 2x - 1, & \frac{1}{2} \le x \le \frac{3}{4}, \\ x, & \frac{3}{4} \le x \le 1. \end{cases}$$

Define $\pi_1 = C_2^{-1} \pi_0 C_2$ and $\pi_n = A^{-n+1} \pi_1 A^{n-1}$ for $n > 1$.

Lemma 6.27: *The elements A, B, C and π_0 generate V and satisfy the following relations:*

1) $[AB^{-1}, A^{-1}BA] = 1$, 2) $[AB^{-1}, A^{-2}BA^2] = 1$,
3) $C = B(A^{-l}CB)$, 4) $(A^{-l}CB)(A^{-l}BA) = B(A^{-2}CB^2)$,
5) $CA = (A^{-1}CB)^2$, 6) $C^3 = 1$,
7) $\pi_1^2 = 1$,
8) $\pi_1 \pi_3 = \pi_3 \pi_1$,
9) $(\pi_2 \pi_1)^3 = 1$,
10) $X_3 \pi_1 = \pi_1 X_3$,
11) $\pi_1 X_2 = B \pi_2 \pi_1$,
12) $\pi_2 B = B \pi_3$,
13) $\pi_1 C_3 = C_3 \pi_2$,
14) $(\pi_1 C_3)^3 = 1$.

Let V_1 be a group generated by A, B, C and π_0 as formal symbols and relations of the previous lemma.

Lemma 6.28: *In V_1 the elements π_i, $i = 0, 1, 2, ..., n, \dots$, satisfy the following relations:*

$$\begin{cases} \pi_i \pi_j & = \pi_j \pi_i, \quad \textit{if } \; |i - j| > 1, \\ \pi_i \pi_{i+1} \pi_i & = \pi_{i+1} \pi_i \pi_{i+1}, \\ \pi_i^2 = 1. \end{cases}$$

For each positive integer n, let $\Pi(n)$ be the subgroup of V_1 generated by π_i, $i = 0, 1, 2, ..., n$, and let $\Pi = \cup_{n \in \mathbb{N}} \Pi(n)$.

Let Σ be the infinite group given by its canonical presentation with generators s_i, $i = 0, 1, 2, ..., n, \ldots$ and relations:

$$\begin{cases} s_i s_j & = s_j\, s_i, \quad \text{if } |i-j| > 1, \\ s_i s_{i+1} s_i & = s_{i+1} s_i s_{i+1}, \\ s_i^2 = 1. \end{cases}$$

Proposition 6.29: *The subgroup* Π *is isomorphic to* Σ.

Theorem 6.30: *The group* V_1 *is simple and isomorphic to* V.

Corollary 6.31: *The group* V *contains all finite groups.*

6.7. *Classifying space for the Thompson group F*

We recall that a space X is a *classifying space* for a group G if $\pi_1(X) \cong G$ and the universal cover $\widetilde{X}$ of X is contractible. This is equivalent to saying X is a $K(G,1)$ space, meaning that $\pi_k(X)$ is trivial when $k \neq 1$ and is G when $k = 1$. This condition that the higher homotopy groups vanish is called being *aspherical.*

We now define the space $\mathcal{C}F$ [43]. Let $\mathcal{C}F_n$ denote the space of all n-tuples of real numbers, $(t_1, \ldots, t_n)$, such that:

(1) the entries are non-decreasing, i.e. $t_1 \leq \cdots \leq t_n$, and
(2) for all $i = 1, \ldots, n-2$, $t_{i+2} - t_i \geq 1$.

This second condition should be thought of as saying that no three distinct entries are too close together. For example, $(1, 1, 2)$ is a point in $\mathcal{C}F_3$, but $(1, 1, 3/2)$ is not.

Let

$$\mathcal{C}F = \left(\coprod_{n=1}^{\infty} \mathcal{C}F_n \right) / \sim$$

denote the disjoint union of the spaces $\mathcal{C}F_n$, subject to the identifications

$$(t_1, \ldots, t_i, \ldots, t_n) \sim (t_1, \ldots, t_i, t_i, \ldots, t_n)$$

whenever $(t_1, \ldots, t_i, t_i, \ldots, t_n) \in \mathcal{C}F_{n+1}$, that is whenever t_i is at least 1 away from its neighbors.

For each n, $\mathcal{C}F_n$ is contractible, for instance to the point $p_n = (1, \ldots, n)$ in $\mathcal{C}F_n$. The contraction is given by the homotopy that at time $0 \leq t \leq 1$ takes the point $\vec{t} = (t_1, \ldots, t_n)$ to $p_n t + (1-t)\vec{t}$.

The fundamental group of $\mathcal{C}F$ (which will be F) will come from the identifications arising from $\sim$. Points in $\mathcal{C}F_n$ are identified with points in $\mathcal{C}F_{n+1}$ in at most n ways, once for each allowed bifurcation of an entry in $\mathcal{C}F_n$, so non-trivial elements of $\pi_1(\mathcal{C}F)$ can arise, for example, by splitting one entry and then merging into a different entry.

Then non-trivial relations in $\pi_1(\mathcal{C}F)$ arise from the fact that splits and merges that are far enough apart may happen in any order.

Theorem 6.32: *The space $\mathcal{C}F$ is a classifying space for F.*

6.8. *Informal introduction to the braided Thompson group BV*

Here we follow the work [15].

As we saw elements of Thompson group V can be coded by pairs of labeled binary trees. For example:

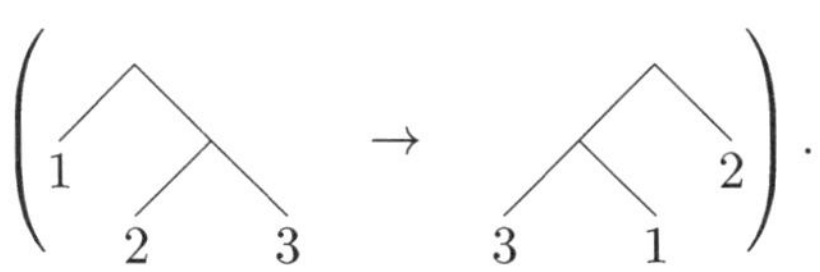

Elements of Thompson group V can also be described with the help of the standard Georg Cantor set K.

Let us recall the standard "deleted middle thirds" description of the Cantor set K. The set K is a limit of a sequence of collections of closed intervals in the unit interval $[0,1]$.

The first few collections are

$$\begin{gathered}
\{[0,1]\} \\
\{[0,\tfrac{1}{3}]\ ,\ [\tfrac{2}{3},1]\} \\
\{[0,\tfrac{1}{9}]\ ,\ [\tfrac{2}{9},\tfrac{1}{3}]\ ,\ [\tfrac{2}{3},\tfrac{7}{9}]\ ,\ [\tfrac{8}{9},1]\} \\
\{[0,\tfrac{1}{27}]\ ,\ [\tfrac{2}{27},\tfrac{1}{9}]\ ,\ [\tfrac{2}{9},\tfrac{7}{27}]\ ,\ [\tfrac{8}{27},\tfrac{1}{3}]\ ,\ [\tfrac{2}{3},\tfrac{19}{27}]\ ,\ [\tfrac{20}{27},\tfrac{7}{9}]\ ,\ [\tfrac{8}{9},\tfrac{25}{27}]\ ,\ [\tfrac{26}{27},1]\} \\
\vdots
\end{gathered} \tag{6.8}$$

Elements of V are defined using covers of K by pairwise disjoint intervals chosen from the collections in (6.8).

Given a pair of such covers with the same number of intervals (necessarily finite) and a one-to-one correspondence between them, an element of

V is created by taking each interval in the first cover affinely in an orientation preserving way onto the corresponding interval in the second cover and restricting this map to K.

The restriction is a homeomorphism of K. The group V is the set of all such maps. Below, we show one the map f where we indicate the bijection by numbering the intervals.

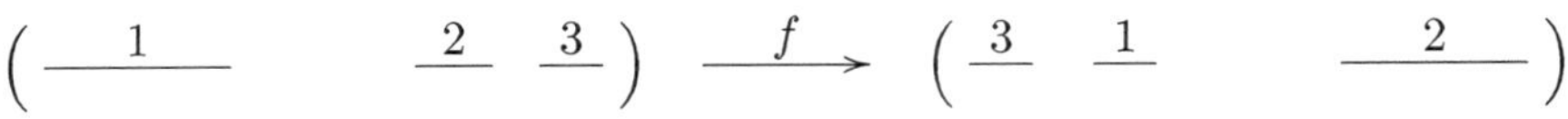

The map f is the restriction of the following affine surjections

$$[0, \tfrac{1}{3}] \to [\tfrac{2}{9}, \tfrac{1}{3}], \qquad [\tfrac{2}{3}, \tfrac{7}{9}] \to [\tfrac{2}{3}, 1], \qquad [\tfrac{8}{9}, 1] \to [0, \tfrac{1}{9}]$$

to the portions of K contained in the given intervals.

The structure of the left tree indicates that the interval $[0, 1]$ (corresponding to the root at the top of the tree) is to be split, and the resulting right interval $[\frac{2}{3}, 1]$ is to be split again. This describes the intervals in the domain of f. The right tree codes the splittings needed to describe the intervals in the range of f.

To obtain an element of BV, we embed K in the plane $\mathbb{R}^2$ as a subset of the x-axis.

Let K be covered by a collection I of pairwise disjoint intervals from (6.8) and also a similar collection J with the same number of intervals.

An element of BV will take intervals in I to intervals in J exactly as described above, but the move will be accomplished by an isotopy of $\mathbb{R}^2$ with compact support.

That is, the move will be accomplished by braiding if we view the isotopy as a level preserving homeomorphism from $\mathbb{R}^2 \times [0, 1]$ to itself and letting the braid strands be the images of the components of $K \times [0, 1]$.

A restriction that must be observed is that during the isotopy, each interval in I must have its image parallel to the x-axis at all times.

Isotopies u and v are equivalent if there is a level preserving isotopy of $\mathbb{R}^2 \times [0, 1]$ from u to v (adhering to the restriction that the images of intervals from I be kept parallel to the x-axis throughout) that are fixed on the Cantor set at the 0 and 1 levels.

Thus BV is seen to be a subgroup of a braid group on a Cantor set of strands.

The surjection from BV to V is obtained by taking each element of BV to the homeomorphism of K obtained at the end of the isotopy.

An element of BV can also be coded by pairs of binary trees, but now the connection from the leaves of the first to the leaves of the second is given by a braid and not a bijection. This is most easily pictured by drawing the second tree upside down below the first and drawing the braid connecting the leaves between the two trees.

As an example, the following is one element of BV that maps to the element f of V in the example above. We draw both the trees and braid encoding of the element as well as a picture of a braiding of intervals.

6.9. *Formal definition to the braided Thompson group BV*

The group BV has the following presentation [16]. The equalities $v_i = v_0^{1-i}v_1v_0^{i-1}$ and $\bar{\pi}_i = v_0^{1-i}\bar{\pi}_1 v_0^{i-1}$ for all $i \geq 2$, and $\pi_i = \bar{\pi}_i v_i \bar{\pi}_{i+1}^{-1}$ for all $i \in \mathbb{N}$ are taken as definitions.

$$
\begin{aligned}
BV = \langle v_0, v_1, \bar{\pi}_0, \bar{\pi}_1 \,|\, v_2v_1 &= v_1v_3, & v_3v_1 &= v_1v_4,\\
\bar{\pi}_2 v_1 &= v_1\bar{\pi}_3, & \bar{\pi}_3 v_1 &= v_1\bar{\pi}_4,\\
\pi_0 v_0 &= v_1\pi_0\pi_1, & \pi_1 v_1 &= v_2\pi_1\pi_2,\\
\pi_0^{-1} v_0 &= v_1\pi_0^{-1}\pi_1^{-1}, & \pi_1^{-1} v_1 &= v_2\pi_1^{-1}\pi_2^{-1},\\
\pi_0 v_2 &= v_2\pi_0, & \pi_0 v_3 &= v_3\pi_0,\\
\pi_1 v_3 &= v_3\pi_1, & \pi_1 v_4 &= v_4\pi_1,\\
\pi_0 &= \bar{\pi}_1^{-1} v_0^{-1}\bar{\pi}_0, & \pi_1 &= \bar{\pi}_2^{-1} v_1^{-1}\bar{\pi}_1,
\end{aligned}
$$

$$\begin{aligned}
&\pi_0\pi_2 = \pi_2\pi_0, \qquad \pi_0\pi_3 = \pi_3\pi_0,\\
&\pi_1\pi_3 = \pi_3\pi_1, \qquad \pi_1\pi_4 = \pi_4\pi_1,\\
&\pi_0\pi_1\pi_0 = \pi_1\pi_0\pi_1, \pi_1\pi_2\pi_1 = \pi_2\pi_1\pi_2,\\
&\bar\pi_2\pi_0 = \pi_0\bar\pi_2, \qquad \bar\pi_3\pi_0 = \pi_0\bar\pi_3,\\
&\bar\pi_3\pi_1 = \pi_1\bar\pi_3, \qquad \bar\pi_4\pi_1 = \pi_1\bar\pi_4,\\
&\pi_0\bar\pi_1\pi_0 = \bar\pi_1\pi_0\bar\pi_1, \pi_1\bar\pi_2\pi_1 = \bar\pi_2\pi_1\bar\pi_2\rangle.
\end{aligned}$$

A presentation for the group V is obtained from that for BV by adding the relations $\bar\pi_0^2 = \bar\pi_1^2 = \pi_0^2 = \pi_1^2 = 1$.

6.10. *The group $\widehat{BV}$. Another model for the braided Thompson group*

Definition 6.33: Let $\widehat{BV}$ [20] be a group given by generators a_i, σ_j; $i, j = 1, 2, \ldots,$ and the following relations: for $i \geq 1$ and $j \geq i+2$,

$$\begin{cases}\sigma_i\sigma_j = \sigma_j\sigma_i, \quad \sigma_i a_j = a_j\sigma_i, \quad a_i a_{j-1} = a_j a_i, \quad a_i\sigma_{j-1} = \sigma_j a_i,\\ \sigma_i\sigma_{i+1}\sigma_i = \sigma_{i+1}\sigma_i\sigma_{i+1}, \ \sigma_{i+1}\sigma_i a_{i+1} = a_i\sigma_i, \ \sigma_i\sigma_{i+1}a_i = a_{i+1}\sigma_i.\end{cases} \tag{6.9}$$

Theorem 6.34: [20] *The subgroup of $\widehat{BV}$ generated by σ_* is (a copy of) the braid group B_∞, and the subgroup generated by a_* is (a copy of) Thompson's group F. These subgroups generate $\widehat{BV}$, and their intersection is $\{1\}$.*

6.11. *Some well-known questions about Thompson groups*

Question. (Matt Brin) Explore the group ring structure of the Thompson groups and their various submonoids

Question. Is F automatic?

Question. Let $\gamma(n)$ be the growth rate of F with respect to the $\{x_0, x_1\}$ generating set, and let:

$$\Gamma(t) = \sum_{n=0}^{\infty} \gamma(n)t^n.$$

Is $\Gamma(t)$ a rational function?

Appendix A. Bi-Δ-structures

We recollect here necessary facts about bi-Δ-structures, which are taken from the work [49].

The sequence of sets $\{X_{n+1}\}_{n\geq 0}$ with faces denoted by $\{d_0, d_1, \ldots\}$ and cofaces denoted by $\{d^0, d^1, \ldots\}$ forms a bi-Δ-set if the following identities hold:

(1) $d_j d_i = d_i d_{j+1}$ for $j \geq i$;
(2) $d^j d^i = d^{i+1} d^j$ for $j \leq i$;
(3) $d_j d^i = \begin{cases} d^{i-1} d_j & \text{if } j < i, \\ \mathrm{id} & \text{if } j = i, \\ d^i d_{j-1} & \text{if } j > i. \end{cases}$

A sequence of groups $\mathcal{G}$ is called a bi-Δ-group if $\mathcal{G}$ is a bi-Δ-set such that all faces and co-faces are group homomorphisms.

Let $\mathbb{P}_n = P_{n+1}$, the pure braid group. The sequence of groups $\mathbb{P} = \{\mathbb{P}_n\}_{n\geq 0}$ with faces relabeled as $\{\mathbb{d}_0, \mathbb{d}_1, \ldots\}$ and co-faces relabeled as $\{\mathbb{d}^0, \mathbb{d}^1, \ldots\}$ forms a bi-Δ-group structure. Where the face operation $\mathbb{d}_i : \mathbb{P}_n \to \mathbb{P}_{n-1} = d_{i+1} : P_{n+1} \to P_n$ is obtained by deleting the $i+1$st string, the co-face operation $\mathbb{d}^i : \mathbb{P}_n \to \mathbb{P}_{n+1}$ is obtained by adding a trivial $i+1$st string in front of the other strings ($i = 0, 1, 2, \ldots, n$).

A.1. *A decomposition formula on bi-Δ-groups*

Let $\mathcal{G} = \{G_n\}_{n\geq 0}$ be a bi-Δ-group. Define

$$\mathcal{Z}_n(\mathcal{G}) = \bigcap_{i=0}^{n} \mathrm{Ker}(d_i \colon G_n \to G_{n-1}).$$

Theorem A.1: (Decomposition Theorem of bi-Δ-groups). *Let $\mathcal{G} = \{G_n\}_{n\geq 0}$ be a bi-Δ-group. Then G_n is the (iterated) semi-direct product of the subgroups*

$$d^{i_k} d^{i_{k-1}} \cdots d^{i_1}(\mathcal{Z}_{n-k}(\mathcal{G})),$$

$0 \leq i_1 < \cdots < i_k \leq n$, $0 \leq k \leq n$, with lexicographic ordering from the right.

Corollary A.2: *Let $\mathcal{G} = \{G_n\}_{n\geq 0}$ be a bi-Δ-group such that each G_n is* *an Abelian group. Then there is direct sum decomposition*

$$G_n = \bigoplus_{\substack{0 \le i_1 < \cdots < i_k \le n \\ 0 \le k \le n}} d^{i_k} d^{i_{k-1}} \cdots d^{i_1}(\mathcal{Z}_{n-k}(\mathcal{G}))$$

for each n.

Acknowledgments

The author is thankful to the IMS, National University of Singapore, for their hospitality and excellent conditions for work, to all local organizers of the Young Topologist Seminar, and the International conference in IMS, Singapore, August 2015, and in particular to Jie Wu. At the last stage of the work on these notes, the author was partially supported by CNRS PICS project of cooperation with Georgia, No. 237647.

The author is thankful to Vlad Sergiescu for the discussions of the subject of the chapter and to Paolo Belligeri who attracted his attention to the work of P. Bellingeri, B. A. Cisneros de La Cruz and L. Paris [11].

References

1. S. Andreadakis, "On the automorphisms of free groups and free nilpotent groups," *Proc. London Math. Soc.*, Ser 3, 1965, **15**, 239–268.
2. E. Artin, "Theorie der Zöpfe," *Abh. Math. Semin. Univ. Hamburg*, 1925, vol. 4, pp. 47–72.
3. E. Artin, "Theory of braids," *Ann. Math.*, 1947, vol. 48, no 1, pp. 101–126.
4. V. G. Bardakov, "The virtual and universal braids," *Fund. Math.* vol. 184 (2004), pp. 1–18.
5. V. G. Bardakov, P. Bellingeri, "Combinatorial properties of virtual braids," *Topology Appl.* **156** (2009), no 6, pp. 1071–1082.
6. V. G. Bardakov, R. Mikhailov, V. V. Vershinin and J. Wu, "Brunnian Braids on Surfaces," *AGT*, 12, 2012, pp. 1607–1648.
7. V. G. Bardakov, R. Mikhailov, V. V. Vershinin and J. Wu, "On the pure virtual braid group PV_3," *Comm. Algebra*, 44 (2016), no. 3, pp. 1350-1378.
8. V. G. Bardakov, V. V. Vershinin, J. Wu, "On Cohen braids. Proc. Steklov Inst. Math." 286 (2014), no. 1, pp. 16-32.
9. L. Bartholdi, B. Enriquez, P. Etingof and E. Rains, "Groups and Lie algebras corresponding to the Yang-Baxter equations," *J. Algebra*, vol. 305 (2006), no. 2, pp. 742–764.
10. J. M. Belk, "Thompson group F," *arXiv:0708.3609v1*.
11. P. Bellingeri, B. A. Cisneros de La Cruz and L. Paris, "A simple solution to the word problem for virtual braid groups," *Pacific J. Math.* vol. 283 (2016), no. 2, pp. 271–287.

12. J. A. Berrick, F. R. Cohen, Y. L. Wong and J. Wu, "Configurations, braids, and homotopy groups", *J. Amer. Math. Soc.* vol. 19 2006, pp. 265–326.
13. J. S. Birman, *Braids, Links, and Mapping Class Groups*, Ann. Math. Stud., no 82, 1974.
14. N. Bourbaki, *Groupes et algèbres de Lie, Chaps. 2–3*, Hermann, Paris, 1972.
15. M. Brin, "The algebra of strand splitting. I. A braided version of Thompson's group V," *J. Group Theory*, vol. 10 (2007), no. 6, pp. 757–788.
16. M. Brin, "The algebra of strand splitting. II. A presentation for the braid group on one strand," *Internat. J. Algebra Comput.* vol. 16 (2006), no. 1, pp. 203–219.
17. K. S. Brown, *Cohomology of groups*, Graduate Texts in Mathematics, 87. Springer-Verlag, New York, 1994. x+306 pp.
18. W. Burau, "Über Zopfinvarianten" (German) *Abh. Math. Semin. Hamb. Univ.*, vol. 9 (1932, pp. 117–124.
19. J. W. Cannon, W. J. Floyd and W. R. Parry, "Introductory notes on Richard Thompson's groups," *Enseign. Math.* (2) vol. 42 (1996), no. 3-4, pp. 215–256.
20. P. Dehornoy, "The group of parenthesized braids," *Adv. Math.*, vol. 205 (2006), no. 2, pp. 354–409.
21. *F. A. Garside*, "The braid group and other groups," *Quart. J. Math. Oxford Ser.*, 1969, vol. 20, pp. 235–254.
22. D. L. Goldsmith, "The theory of motion groups," *Michigan Math. J.* vol. 28 (1981), pp. 3–17.
23. M. Goussarov, M. Polyak and O. Viro, "Finite Type Invariants of Classical and Virtual Knots," *Topology*, vol. 39 (2000), no. 5, pp. 1045–1068.
24. G. G. Gurzo, "The group of smooth braids," *16th All-Union Algebra Conference, Abstract II*, Leningrad, 1981, pp. 39-40.
25. M. Hall "A basis for free Lie rings and higher commutators in free groups," *Proc. Amer. Math. Soc.*, vol. 1 (1950), pp. 575–581.
26. C. Jensen, J. McCammond, and J. Meier, "The integral cohomology of the group of loops," *Geom. Topol.*, vol. 10 (2006), pp. 759–784.
27. D. L. Johnson, "Towards a characterization of smooth braids," *Math. Proc. Cambridge Philos. Soc.* vol. 92 (1982), pp. 425–427.
28. S. Kamada, "Braid presentation of virtual knots and welded knots," *Osaka J. Math.*, vol. 44 (2007), no. 2, pp. 441–458.
29. L. Kauffman, "Virtual Knot theory," *European J. Combin.*, vol. 20 (1999), no. 7, pp. 663–690.
30. T. Kohno, "Série de Poincaré-Koszul associée aux groupes de tresses pure," *Invent. Math.*, vol. 82 (1985), pp. 57–75.
31. Ch. Kassel and V. Turaev, *Braid groups. With the graphical assistance of Olivier Dodane* Graduate Texts in Mathematics, 247. Springer, New York, 2008. xii+340 pp.
32. H. Levinson, "Decomposable braids and linkages", *Trans. Amer. Math. Soc.* vol. 178 (1973), pp. 111–126.
33. H. Levinson, "Decomposable braids as subgroups of braid groups," *Trans. Amer. Math. Soc.* vol. 202, (1975), pp. 51–55.

34. J. Y. Li, V. Vershinin and J. Wu, "Brunnian braids and Lie algebras," *J. Algebra*, vol. 439 (2015), pp. 270–293.
35. J. Y. Li and J. Wu "Artin braid groups and homotopy groups," *Proc. London Math. Soc.*, vol¡; 99 (2009), pp. 521–556.
36. J. Y. Li and J. Wu "On symmetric commutator subgroups, braids, links and homotopy groups," *Trans. Amer. Math. Soc.*, vol. 363 (2011), pp. 3829–3852.
37. G. S. Makanin, Kourovka Notebook (Unsolved Questions in Group Theory) Seventh edition. Novosibirsk. 1980 Question 6.23. p. 78.
38. A. A. Markoff, "Foundations of the Algebraic Theory of Tresses," *Trudy Mat. Inst. Steklova*, no 16, 1945 (Russian, English summary).
39. J. McCool, "On basis-conjugating automorphisms of free groups," *Canadian J. Math.*, vol. 38,**12**(1986), 1525-1529.
40. K. Murasugi, B. Kurpita, *A study of braids.* Mathematics and its Applications, 484. Kluwer Academic Publishers, Dordrecht, 1999. x+272 pp.
41. M. Polyak, O. Viro, "Gauss diagram formulas for Vassiliev invariants," *Internat. Math. Res. Notices*, 1994, no. 11, pp. 445–453.
42. L. Rabenda. Mémoire de DEA (Master thesis), Université de Bourgogne, 2003.
43. L. Sabalka, M. C. B. Zaremsky "On Belk's classifying space for Thompson's group F," *Forum Math.*, vol. 29 (2017), no. 3, pp. 681–691.
44. J.-P. Serre, *Lie algebras and Lie groups*, 1964, Lectures given at Harvard University. Corrected fifth printing of the second (1992) edition. Lecture Notes in Mathematics, 1500. Springer-Verlag, Berlin, 2006. viii+168 pp.
45. V. V. Vershinin, "Homology of braid groups and their generalizations," *Knot theory (Warsaw, 1995), Banach Center Publ.*, 42, Polish Acad. Sci., Warsaw, 1998, pp. 421–446.
46. V. V. Vershinin, "Braid groups and loop spaces," *Russ. Math. Surv.* vol. 54, no.2, pp. 273-350 (1999); translation from *Usp. Mat. Nauk*, vol. 54, no. 2, pp. 3-84 (1999).
47. V. V. Vershinin, "On homology of virtual braids and Burau representation" *Knots in Hellas '98*, vol. 3. *J. Knot Theory Ramifications* 10 (2001), no. 5, 795–812.
48. V. V. Vershinin, "Braid groups, their Properties and Generalizations" *Handbook of Algebra*, vol. **4**, Elsevier, Amsterdam a.o. 2006, pp. 427–465.
49. J. Wu, "On maps from loop suspensions to loop spaces and the shuffle relations on the Cohen groups", *Memoirs Amer. Math. Soc.* Vol. **180** No. **851** (2006).

HIGHER LIMITS, HOMOLOGY THEORIES AND fr-CODES

Sergei O. Ivanov
Chebyshev Laboratory, St. Petersburg State University
14th Line, 29b, Saint Petersburg, 199178 Russia
ivanov.s.o.1986@gmail.com

Roman Mikhailov
Chebyshev Laboratory, St. Petersburg State University
14th Line, 29b, Saint Petersburg, 199178 Russia
and St. Petersburg Department of Steklov Mathematical Institute, Russia
rmikhailov@mail.ru

This text contains an introduction to the theory of limits over the category of presentations, with examples of different well-known functors like homology or derived functors of non-additive functors in the form of derived limits. The theory of so-called **fr**-codes is also developed. This is a method that shows how different functors from the category of groups to the category of abelian groups, such as group homology and tensor products of abelianization, can be coded as sentences in the alphabet with two symbols **f** and **r**.

Contents

1. Introduction

The main idea of the theory presented in this text is the following: given a group (or analogously an associative algebra, Lie algebra over a ring, abelian group etc.) G, we consider the category $\mathsf{Pres}(G)$. The objects of $\mathsf{Pres}(G)$ are surjective homomorphisms $\pi : F \twoheadrightarrow G$, where F is a free group, and the morphisms $f : (\pi_1 : F_1 \twoheadrightarrow G) \to (\pi_2 : F_2 \twoheadrightarrow G)$ are homomorphisms $f : F_1 \to F_2$ such that $\pi_1 = \pi_2 f$. The category $\mathsf{Pres}(G)$ has coproducts and hence is contractible. However, the cohomology group of $\mathsf{Pres}(G)$, with coefficients in functors, define invariants of G. This provides an approach to functors in the category of groups (or associative algebras, Lie algebras, or in general, categories with enough projectives). Here is an illustrative example from [4]: let G be a 2-torsion-free group. For a given $\pi : F \twoheadrightarrow G$, let $R := \ker(\pi)$. Then there is a natural isomorphism

$$H^1(B\mathsf{Pres}(G), \frac{[R,R]}{[[R,R],F]}) = \mathsf{lim}^1 \frac{[R,R]}{[[R,R],F]} = H_3(G;\mathbb{Z}/2),$$

where $B\mathsf{Pres}(G)$ is the classifying space of the category $\mathsf{Pres}(G)$. That is, one can obtain the third homology of G with $\mathbb{Z}/2$-coefficients as the first derived limit of the presentation $\frac{[R,R]}{[[R,R],F]}$ over the category $\mathsf{Pres}(G)$. A lot of different functors like group homology, cyclic and Hochschild homology of algebras, homology of Lie algebras and derived functors in the sense of

Dold-Puppe can be obtained in this way. We refer to [12], [4], [3], [10] for discussion of different aspects of this approach.

One of the subjects of these notes is the theory of so-called **fr**-codes. We consider an alphabet with two symbols $\mathbf{f}, \mathbf{r}$ and form *sentences* with words in this alphabet, for example, (**rfr frr ffff**) or (**rr frf**). Now, looking at the object $\pi : F \twoheadrightarrow G$, we construct an ideal in $\mathbb{Z}[F]$ given by the sum of words in a sentence where we replace **f** by the augmentation ideal of F and **r** by the ideal $(\ker(\pi) - 1)\mathbb{Z}[F]$. The *translation* of a sentence (or **fr**-code) on the language of functors in the category of groups corresponds to that ideal lim^1-functor. For example, the above sentences give natural functors:

$$(\mathbf{rfr\ frr\ ffff}) \rightsquigarrow \mathsf{lim}^1(\mathbf{rfr} + \mathbf{frr} + \mathbf{ffff}) = \mathsf{Tor}(G_{ab} \otimes G_{ab}, G_{ab}),$$

$$(\mathbf{rr\ frf}) \rightsquigarrow \mathsf{lim}^1(\mathbf{rr} + \mathbf{frf}) = H_3(G).$$

Here derived limits are taken over $\mathsf{Pres}(G)$. We give a table of different **fr**-codes at the end of these notes.

2. Background

2.1. *Tarski-Grothendieck set theory*

Through the paper we use the advanced version of axiomatic set theory called Tarski-Grothendieck set theory. It is an extension of the standard Zermelo-Fraenkel set theory with the axiom of choice (ZFC) by the Tarski's axiom of universes (U). Here we recall what it is (for details see [14], [2], [13], [1]).

A **Grothendieck universe** is a set U with the following properties:

(1) if $x \in U$ and $y \in x$, then $y \in U$;
(2) $\emptyset \in U$;
(3) if $x \in U$, then $\mathcal{P}(x) \in U$ (where $\mathcal{P}(x)$ is the power-set);
(4) if $\{x_i\}_{i \in I}$ is a family such that $x_i \in U$ and $I \in U$, then $\bigcup_{i \in I} x_i \in U$.

It is easy to prove that any Grothendieck universe U contains:

- all singletons of each of its elements;
- all couples of its elements;
- all products of all families of elements of U indexed by an element of U;
- all disjoint unions of all families of elements of U indexed by an element of U;
- all intersections of all families of elements of U indexed by an element of U;

- all functions between any two elements of U;
- all subsets of U whose cardinal is an element of U.

The Tarski's axiom (U) is the following.

(U) For each set x there is a Grothendieck universe U such that $x \in U$.

A Grothendieck universe is a set closed under all set-theoretical constructions. It follows that, if $\mathbb{N} \in U$, then $\mathbb{Z}, \mathbb{Q}, \mathbb{R}, \mathbb{C}, \mathbb{R}^n, \mathcal{P}(\mathbb{R}^n), (\mathbb{R}^n)^{\mathbb{R}^m}$ and all their subsets are in U. Roughly speaking, if $\mathbb{N} \in U$, then all classical mathematics is in U. More precisely, if $\mathbb{N} \in U$ then U is a model of ZFC set theory. Moreover, if we fix a couple of universes U, U' such that $U \in U'$, and start to call elements of U by sets and elements of U' by classes we get a model for NBG class theory. So the consideration of universes allows us to avoid considering classes. Moreover, it is more flexible because we can consider a sequence of universes $U_0 \in U_1 \in U_2 \in \dots$ and call elements of U_n classes of the nth level and work with them as with sets because they are sets in the Tarski-Grothendieck set theory.

A set x is called U-small for some Grothedieck universe U if $x \in U$. A topological space (group, ring, module,...) is called U-small if its underlying set is U-small. If U is fixed, we just call them small. A category is called U-small if the set of objects and the set of morphisms are U-small. In order to avoid using classes we will assume that **all categories are small.** It follows that any category is U-small for some Grothendieck universe U. We denote by Sets^U, ($\mathsf{Top}^U, \mathsf{Gr}^U, \mathsf{Rings}^U, \mathsf{Mod}(R)^U$, Cat^U ...) categories of U-small sets (topological spaces, groups, rings, R-modules, categories ...). If $U \in U'$, where U' is another Grothendieck universe, then all these categories are U'-small. Furthermore, we will avoid the superscript U and assume that U is a Grothendieck universe such that $\mathbb{N} \in U$.

2.2. *Simplicial sets*

In this subsection we recall some basic information about simplicial sets. For more information and proofs see [9], [5], [6].

The **simplicial indexing category** is the category Δ whose objects are posets

$$[n] = \{0 < 1 < \cdots < n\}$$

for $n \geq 0$ and morphisms are order-preserving maps. For $n \geq 1$ and $0 \leq i \leq n$ we define ith **coface Map** $d^i : [n-1] \to [n]$ as the unique injective order-preserving map whose image does not contain i. For $n \geq 0$ and $0 \leq i \leq n$

we define ith **codegeneracy Map** $s^i : [n+1] \to [n]$ as the unique sujective order-preserving map such that i has two-element preimage. It is easy to check that they satisfy the following relations: (i) $d^j d^i = d^i d^{j-1}$ if $i < j$; (ii) $s^j s^i = s^{i-1} s^j$ if $i > j$; (iii) $s^j d^i = d^i s^{j-1}$ if $i < j$; (iv) $s^i d^i = \mathsf{id} = s^{i+1} d^i$; (v) $s^j d^i = d^{i-1} s^j$ if $i > j+1$. It follows that any order-preserving map $\alpha : [m] \to [n]$ can be written in one and only one way as

$$\alpha = d^{j_l} \dots d^{j_1} s^{i_k} \dots s^{i_1},$$

where $m > i_1 > \dots > i_k \geq 0$ and $0 \leq j_1 < \dots < j_l \leq n$. Using this one can prove that Δ is the path category of the quiver

$$\cdots \qquad [2] \; \substack{\xleftarrow{d^0} \\ \xleftarrow{d^1} \\ \xleftarrow{d^2} \\ \xrightarrow{s^0} \\ \xrightarrow{s^1}} \; [1] \; \substack{\xleftarrow{d^0} \\ \xleftarrow{d^1} \\ \xrightarrow{s^0}} \; [0]$$

with relations (i)-(v).

Let $\mathcal{C}$ be a category. A **simplicial object** of $\mathcal{C}$ is a functor $X : \Delta^{\mathrm{op}} \to \mathcal{C}$. We set $X_n = X([n])$, $d_i = X(d^i)$ and $s_i = X(s^i)$. A morphism of simplicial objects is a natural transformation. The category of simplicial objects is denoted by $\mathsf{s}\mathcal{C}$. Using that Δ is a path category with relations, we can define a simplicial set as a sequence of objects $X_0, X_1, \dots$ together with **face morphisms** $d_i : X_n \to X_{n-1}$ and **degeneracy morphisms** $s_i : X_n \to X_{n+1}$, for $0 \leq i \leq n$, satisfying identities:

(s1) $d_i d_j = d_{j-1} d_i$ if $i < j$;
(s2) $s_i s_j = s_j s_{i-1}$ if $i > j$;
(s3) $d_i s_j = s_{j-1} d_i$ if $i < j$;
(s4) $d_i s_i = \mathsf{id} = d_i s_{i+1}$;
(s5) $d_i s_j = s_j d_{i-1}$ if $i > j+1$.

The most important case of $\mathcal{C}$ is Sets. Simplicial objects of Sets are called **simplicial sets**. If X is a simplicial set, elements of X_n are called n**-simplexes** of X. An n-simplex x is **degenerate** if $x = s_i(x')$ for some i and $x' \in X_{n-1}$.

The simplicial set $\Delta[m] = \Delta(-, [m])$ is called simplicial m-simplex. It defines a functor

$$\Delta[-] : \Delta \longrightarrow \mathsf{sSets}$$

$$[m] \longmapsto \Delta[m]$$

which is the Yoneda embedding for the category Δ. For simplicial sets X, Y we denote by $X \times Y$ their dimension-wise product (which is the product in the categorical sense) and by $\mathsf{Map}(X,Y)$ the **function complex** (or **internal hom**) which is the simplicial set defined by

$$\mathsf{Map}(X,Y) = \mathsf{sSets}(\Delta[-] \times X, Y).$$

It is easy to check that there is a natural isomorphism $\mathsf{sSets}(X \times Y, Z) \cong \mathsf{sSets}(X, \mathsf{Map}(Y,Z))$.

A (simplicial) **homotopy** between two simplicial maps $f, g : X \to Y$ is an element $H \in \mathsf{Map}(X,Y)_1$ such that $d_1(H) = f$ and $d_0(H) = g$. In other words, it is a simplicial map $H : \Delta[1] \times X \to Y$ such that $H(\Delta[d^1] \times \mathsf{id}) = f$ and $H(\Delta[d^0] \times \mathsf{id}) = g$. Here we identify $\Delta[0] \times X$ with X.

Let $\Delta_{\mathsf{Top}}[n]$ denotes the standard n-simplex in $\mathbb{R}^{n+1}$. Then any order-preserving $\alpha : [m] \to [n]$ induces a linear map $\mathbb{R}^{m+1} \to \mathbb{R}^{n+1}$ that sends e_{i+1} to $e_{\alpha(i)+1}$. This map can be restricted to the map $\Delta_{\mathsf{Top}}[\alpha] : \Delta_{\mathsf{Top}}[m] \to \Delta_{\mathsf{Top}}[n]$. Hence, we have defined a functor to the category of topological spaces

$$\Delta_{\mathsf{Top}}[-] : \Delta \longrightarrow \mathsf{Top}.$$

There is a pair of adjoint functors between categories Top and sSets that make these categories very similar to each other. The first of these functors is the **singular simplicial set** functor

$$\mathsf{Sing} : \mathsf{Top} \longrightarrow \mathsf{sSets}$$

given by

$$\mathsf{Sing}(X) = \mathsf{Top}\left(\Delta_{\mathsf{Top}}[-], X\right).$$

The second is the geometric realisation functor

$$|-| : \mathsf{sSets} \longrightarrow \mathsf{Top}.$$

Its constructive definition is a little bit more complicated but it can be reconstructed from the equality $|\Delta[-]| = \Delta_{\mathsf{Top}}[-]$ and the fact that it commutes with all colimits because any simplicial set is a colimit of simplcial simplexes $\Delta[m]$. Moreover, the geometric realisation functor is left adjoint to Sing, i.e. for a simplicial set X and a topological space T there is a natural isomorphism

$$\mathsf{Top}(|X|, T) \cong \mathsf{sSets}(X, \mathsf{Sing}(T)).$$

An important property of geometric realisations is that they commute with certain products. For example, if X and Y are simplicial sets and one of them is finite (i.e. there are only a finite number of non-degenerate simplexes), then there is a natural isomorphism of topological spaces

$$|X \times Y| \cong |X| \times |Y|. \tag{2.1}$$

In particular, any simplicial homotopy $H : \Delta[1] \times X \to Y$ defines a homotopy $|H| : I \times |X| \to |Y|$, where $I = \Delta_{\mathsf{Top}}[1]$ is the closed interval.

Let X be a simplicial set. Consider the complex $\mathsf{C}_\bullet(X)$ whose nth term $\mathsf{C}_n(X)$ is the free abelian group generated by X_n and $d = \sum_i (-1)^i d_i$. Then the homology and cohomology of X with coefficients in an abelian group A are defined as follows:

$$H_n(X, A) = H_n(\mathsf{C}_\bullet(X) \otimes A), \qquad H^n(X, A) = H^n(\mathsf{Hom}(\mathsf{C}_\bullet(X), A)).$$

It is well known that

$$H_n(X, A) \cong H_n(|X|, A), \qquad H^n(X, A) \cong H^n(|X|, A). \tag{2.2}$$

A simplicial set X is **pointed** if all sets X_n are pointed and the face and degeneracy maps preserve base points. In other words, a pointed simplicial set is a simplicial object in the category of pointed sets. The first item of the following proposition can be found in [1] . The second item can be proved similarly.

Proposition 2.1: ***(Contractible criterion)*** *Let X be a pointed simplicial set.*

(1) If there exist point preserving maps $\tilde{s}_{-1} : X_n \to X_{n+1}$ satisfying the simplicial identities (s2)-(s5), then $|X|$ is contractible.
(2) If there exist point preserving maps $\tilde{s}_{n+1} : X_n \to X_{n+1}$ satisfying the simplicial identities (s2)-(s5), then $|X|$ is contractible.

3. Classifying space of a category

3.1. *Nerve of a category*

We consider posets as categories in the usual way with order-preserving maps as functors. Thus the simplicial indexing category is a full subcategory of the category of small categories $\Delta \subset \mathsf{Cat}$. The **nerve** of a category $\mathcal{C}$ is a simplicial set $\mathsf{N}\mathcal{C} : \Delta^{\mathrm{op}} \to \mathsf{Sets}$ given by

$$\mathsf{N}\mathcal{C} = \mathsf{Cat}(-, \mathcal{C}) \mid_{\Delta^{\mathrm{op}}} .$$

For example $\Delta[n] = \mathsf{N}([n])$. It is easy to see that there is a natural bijection between the set $(\mathsf{N}\mathcal{C})_n = \mathsf{Cat}([n], \mathcal{C})$ and the set of all sequences of morphisms $(\alpha_1, \alpha_2, \dots, \alpha_n)$ such that $\mathsf{codom}(\alpha_{i+1}) = \mathsf{dom}(\alpha_i)$ in the category $\mathcal{C}$:

$$\bullet \xleftarrow{\alpha_1} \bullet \xleftarrow{\alpha_2} \bullet \xleftarrow{\alpha_3} \dots \xleftarrow{\alpha_n} \bullet$$

The face maps act as follows: $d_i(\alpha_n, \dots, \alpha_1) = (\alpha_n, \dots, \alpha_{i+1}\alpha_i, \dots, \alpha_1)$ for $1 \leq i < n$, $d_0(\alpha_n, \dots, \alpha_1) = (\alpha_n, \alpha_{n-1}, \dots, \alpha_2)$ and $d_n(\alpha_n, \alpha_{n-1}, \dots, \alpha_1) = (\alpha_{n-1}, \dots, \alpha_1)$. Degeneracies act as follows $s_i(\alpha_n, \dots, \alpha_1) = (\alpha_n, \dots, \alpha_{i+1}, \mathsf{id}_{c_i}, \alpha_i, \dots, \alpha_1)$, where $c_i = \mathsf{codom}(\alpha_i)$. Roughly speaking, $(\mathsf{N}\mathcal{C})_n$ consists of commutative n-simplexes in $\mathcal{C}$. The nerve $\mathsf{N}\mathcal{C}$ is functorial by $\mathcal{C}$, i.e. it defines a functor $\mathsf{N} : \mathsf{Cat} \to \mathsf{sSets}$.

If $\mathcal{C}, \mathcal{D}$ are categories, we denote by $\mathcal{D}^{\mathcal{C}}$ the category of functors whose objects are functors and morphisms are natural transformations. It is easy to see and well known that there is a natural isomorphism $\mathsf{Cat}(\mathcal{C} \times \mathcal{D}, \mathcal{E}) \cong \mathsf{Cat}(\mathcal{C}, \mathcal{E}^{\mathcal{D}})$.

Proposition 3.1: *Let $\mathcal{C}$ and $\mathcal{D}$ be categories. Then $\mathsf{N} : \mathsf{Cat} \to \mathsf{sSets}$ is a fully faithful functor, i.e. it induces a bijection*

$$\mathsf{Cat}(\mathcal{C}, \mathcal{D}) \cong \mathsf{sSets}(\mathsf{N}\mathcal{C}, \mathsf{N}\mathcal{D}).$$

Moreover, there are natural isomorphisms

$$\mathsf{N}(\mathcal{C} \times \mathcal{D}) \cong \mathsf{N}\mathcal{C} \times \mathsf{N}\mathcal{D}, \qquad \mathsf{N}(\mathcal{D}^{\mathcal{C}}) \cong \mathsf{Map}(\mathsf{N}\mathcal{C}, \mathsf{N}\mathcal{D}).$$

Proof: Let $f : \mathsf{N}\mathcal{C} \to \mathsf{N}\mathcal{D}$ be a simplicial map. We need to prove that there is a unique functor $\mathcal{F} : \mathcal{C} \to \mathcal{D}$ such that $f = \mathsf{N}\mathcal{F}$. If $f = \mathsf{N}\mathcal{F}$, then $\mathcal{F}(c) = f_0(c)$ and $\mathcal{F}(\alpha) = f_1(\alpha)$ for all objects c and morphisms α. Hence, if it exists, it is unique. To prove that it exists we begin by proving, by induction on n, that $f_n(\alpha_n, \dots, \alpha_1) = (f_1(\alpha_n), \dots, f_1(\alpha_1))$ for $n \geq 1$. The basis is obvious. Let $f_n(\alpha_n, \dots, \alpha_1) = (\beta_n, \dots, \beta_1)$. Then $(\beta_n, \dots, \beta_2) = d_0(\beta_n, \dots, \beta_1) = d_0 f_n(\alpha_n, \dots, \alpha_1) = f_{n-1}(\alpha_n, \dots, \alpha_2) = (f_1(\alpha_n), \dots, f_1(\alpha_2))$ and $(\beta_{n-1}, \dots, \beta_1) = d_0(\beta_{n-1}, \dots, \beta_1) = d_0 f_n(\alpha_{n-1}, \dots, \alpha_1) = f_{n-1}(\alpha_{n-1}, \dots, \alpha_1) = (f_1(\alpha_{n-1}), \dots, f_1(\alpha_1))$. Hence $(\beta_n, \dots, \beta_1) = (f_1(\alpha_n), \dots, f_1(\alpha_1))$. It follows that f is defined by f_0 and f_1. By definition we set $\mathcal{F}(c) = f_0(c)$ and $\mathcal{F}(\alpha) = f_1(\alpha)$. Nowt $\mathcal{F}$ is a functor, i.e. it respects identity morphisms and composition, as $\mathcal{F}(\mathsf{id}_c) = f_1(\mathsf{id}_c) = f_1(s_0(c)) = s_0(f_0(c)) = \mathsf{id}_{\mathcal{F}(c)}$ and $\mathcal{F}(\beta\alpha) = f_1(\beta\alpha) = f_1(d_1(\beta, \alpha)) = d_1 f_2(\beta, \alpha) = d_1(\mathcal{F}(\beta), \mathcal{F}(\alpha)) = \mathcal{F}(\beta)\mathcal{F}(\alpha)$. It follows that $\mathcal{F}$ is the functor such that $f = \mathsf{N}\mathcal{F}$.

The isomorphism $\mathsf{N}(\mathcal{C} \times \mathcal{D}) \cong \mathsf{N}\mathcal{C} \times \mathsf{N}\mathcal{D}$ is obvious. The second isomorphism follows from $\mathsf{N}(\mathcal{D}^{\mathcal{C}}) = \mathsf{Cat}(-, \mathcal{D}^{\mathcal{C}}) \cong \mathsf{Cat}(- \times \mathcal{C}, \mathcal{D}) \cong \mathsf{sSets}(\Delta[-] \times \mathsf{N}\mathcal{C}, \mathsf{N}\mathcal{D}) = \mathsf{Map}(\mathsf{N}\mathcal{C}, \mathsf{N}\mathcal{D})$. $\square$

Remark 3.2: This proposition implies that we can consider Cat as a full subcategory in sSets identifying $\mathcal{C}$ with its nerve. Moreover this embedding respects products and inner hom's.

Let $\mathcal{F}, \mathcal{G} : \mathcal{C} \to \mathcal{D}$ be two functors and $\varphi : \mathcal{F} \to \mathcal{G}$ be a natural transformation. Then $\varphi \in \mathsf{N}(\mathsf{Fun}(\mathcal{C}, \mathcal{D}))_1 \cong \mathsf{Map}(\mathsf{N}\mathcal{C}, \mathsf{N}\mathcal{D})_1 = \mathsf{Cat}(\Delta[1] \times \mathsf{N}\mathcal{C}, \mathsf{N}\mathcal{D})$. Hence, the natural transfomation $\varphi : \mathcal{F} \to \mathcal{G}$ defines a simplicial homotopy between $\mathsf{N}\mathcal{F}$ and $\mathsf{N}\mathcal{G}$ that we denote by

$$\mathsf{N}\varphi : \Delta[1] \times \mathsf{N}\mathcal{C} \longrightarrow \mathsf{N}\mathcal{D}, \qquad \mathsf{N}\mathcal{F} \overset{\mathsf{N}\varphi}{\sim} \mathsf{N}\mathcal{G}. \tag{3.1}$$

Here we use the symbol '$\sim$' for simplicial homotopy relation, however, it is not an equivalence relation for arbitrary simplicial sets.

3.2. *Classifying space of a category*

The **classifying space** $\mathsf{B}\mathcal{C}$ of a category $\mathcal{C}$ is the geometric realisation of its nerve

$$\mathsf{B}\mathcal{C} = |\mathsf{N}\mathcal{C}|.$$

A category $\mathcal{C}$ is said to be **contractible** if $\mathsf{B}\mathcal{C}$ is contractible. If we consider a group G as a category with one object, the classifying space $\mathsf{B}G$ is the classifying space of this group. This construction defines a functor

$$\mathsf{B} : \mathsf{Cat} \longrightarrow \mathsf{Top}.$$

Using (2.1) and (3.1) we get that if $\mathcal{F}, \mathcal{G} : \mathcal{C} \to \mathcal{D}$ are functors, then a natural transformation $\varphi : \mathcal{F} \to \mathcal{G}$ defines a homotopy

$$\mathsf{B}\varphi : I \times \mathsf{B}\mathcal{C} \longrightarrow \mathsf{B}\mathcal{D}, \qquad \mathsf{B}\mathcal{F} \overset{\mathsf{B}\varphi}{\sim} \mathsf{B}\mathcal{G}. \tag{3.2}$$

3.3. *Classifying space and adjoint functors*

An **adjunction** between categories $\mathcal{C}$ and $\mathcal{D}$ is a pair of functors $\mathcal{F} : \mathcal{C} \rightleftarrows \mathcal{D} : \mathcal{G}$ with a natural isomorphism

$$\mathcal{D}(\mathcal{F}c, d) \cong \mathcal{C}(c, \mathcal{G}d).$$

In this case the functor $\mathcal{F}$ is called a left adjoint functor, the functor $\mathcal{G}$ is called a right adjoint functor and the adjunction is denoted by $\mathcal{F} \dashv \mathcal{G}$. Then there is an natural isomorphism $\mathcal{D}(\mathcal{F}c, \mathcal{F}c) \cong \mathcal{C}(c, \mathcal{G}\mathcal{F}c)$ and the image of $\mathsf{id}_{\mathcal{F}c}$ under this isomorphism is called the **unit of adjunction** $\eta_c : c \to \mathcal{G}\mathcal{F}c$.

Similarly the isomorphism $\mathcal{C}(\mathcal{G}d, \mathcal{G}d) \cong \mathcal{D}(\mathcal{F}\mathcal{G}d, d)$ defines the **counit of adjuntion** $\varepsilon_d : \mathcal{F}\mathcal{G}d \to d$. These morphisms define natural transformations

$$\eta : \mathsf{Id}_{\mathcal{C}} \longrightarrow \mathcal{G}\mathcal{F}, \qquad \varepsilon : \mathcal{F}\mathcal{G} \longrightarrow \mathsf{Id}_{\mathcal{D}}.$$

Proposition 3.3: *If there is an adjunction between categories $\mathcal{C}$ and $\mathcal{D}$, then their classifying spaces $\mathsf{B}\mathcal{C}$ and $\mathsf{B}\mathcal{D}$ are homotopy equivalent.*

Proof: If $\mathcal{F} : \mathcal{C} \rightleftarrows \mathcal{D} : \mathcal{G}$ is an adjunction, then $\mathsf{B}\eta$ and $\mathsf{B}\varepsilon$, constructed in (3.2), are homotopies $\mathsf{id}_{\mathsf{B}\mathcal{C}} \sim (\mathsf{B}\mathcal{G})(\mathsf{B}\mathcal{F})$ and $(\mathsf{B}\mathcal{F})(\mathsf{B}\mathcal{G}) \sim \mathsf{id}_{\mathsf{B}\mathcal{D}}$. □

Corollary 3.4: *If a category $\mathcal{C}$ has an initial or a terminal object, it is contractible.*

Proof: Let $*$ be a category with one object and one morphism. Then $\mathcal{C}$ has a initial object if and only if there is an adjunction $\mathcal{F} : * \rightleftarrows \mathcal{C} : \mathcal{G}$. Dually, it has a terminal object if and only if there is an adjunction $\mathcal{F} : \mathcal{C} \rightleftarrows * : \mathcal{G}$. □

3.4. *Classifying space of a category with pairwise coproducts*

A **category with pairwise coproducts** is a category $\mathcal{C}$ such that for any objects $c_1, c_2 \in \mathcal{C}$ there exists the coproduct $c_1 \xrightarrow{i_1} c_1 \sqcup c_2 \xleftarrow{i_2} c_2$ in $\mathcal{C}$ ($\mathcal{C}$ does not necessarily have an initial or terminal object). A category with pairwise products is defined dually.

Proposition 3.5: *A category $\mathcal{C}$ with pairwise (co)products is contractible.*

Proof: Assume that $\mathcal{C}$ is a category with pairwise coproducts. Choose an object $c_0 \in \mathcal{C}$ and consider the functor $\mathcal{F} : \mathcal{C} \to \mathcal{C}$ given by the formula $\mathcal{F}(-) = - \sqcup c_0$. By the definition of coproduct $c \to c \sqcup c_0 \leftarrow c_0$, we obtain natural transformations $\mathsf{Id}_{\mathcal{C}} \to \mathcal{F} \leftarrow \mathsf{Const}_{c_0}$. Using (3.2), we get that these natural transformations induce homotopies $\mathsf{id}_{\mathsf{B}\mathcal{C}} \sim \mathsf{B}\mathcal{F} \sim \mathsf{const}_{c_0}$. Hence, $\mathcal{C}$ is contractible. The case of a category with pairwise products is dual. □

4. Representations of categories and higher limits

4.1. *Limits*

Recall [8] that the limit $\lim \mathcal{F}$ of a functor $\mathcal{F} : \mathcal{C} \to \mathcal{D}$ is an object of $\mathcal{D}$ together with a universal collection of morphisms $\{\varphi_c : \lim \mathcal{F} \to \mathcal{F}(c)\}_{c \in \mathcal{C}}$ such that $\mathcal{F}(f)\varphi_c = \varphi_{c'}$ for any morphism $f : c \to c'$. Universality means that for any object $d \in \mathcal{D}$, and any collection of morphisms $\{\psi_c : d \to \mathcal{F}(c)\}_{c \in \mathcal{C}}$, such that $\mathcal{F}(f)\psi_c = \psi_{c'}$ for any morphism $f : c \to c'$, there exists a unique morphism $\alpha : d \to \mathsf{lim}\,\mathcal{F}$ such that $\psi_c = \varphi_c \alpha$. The limit is not always exists but if it exists it is unique up to a unique isomorphism that commutes with morphisms φ_c.

Denote by $\mathsf{diag} : \mathcal{D} \to \mathcal{D}^{\mathcal{C}}$ the diagonal functor given by $\mathsf{diag}(d)(c) = d$ and $\mathsf{diag}(d)(\alpha) = \mathsf{id}_d$. For the sake of simplicity we will denote $\mathsf{diag}(d)$ by the symbol d. If limits of all functors $\mathcal{C} \to \mathcal{D}$ exist, they define the functor

$$\mathsf{lim} : \mathcal{D}^{\mathcal{C}} \longrightarrow \mathcal{D}$$

which is right adjoint to the diagonal functor $\mathsf{diag} \dashv \mathsf{lim}$. In other words there is a natural isomorphism

$$\mathcal{D}^{\mathcal{C}}(d, \mathcal{F}) \cong \mathcal{D}(d, \mathsf{lim}\,\mathcal{F}).$$

4.2. *Limits over strongly connected categories*

Throughout the paper k denotes a commutative ring. Let $\mathcal{F} : \mathcal{C} \to \mathsf{Mod}(k)$ be a functor. Consider the subfunctor $\mathsf{inv}\,\mathcal{F} \subseteq \mathcal{F}$ of 'invariant elements' given by the formula

$$\mathsf{inv}\,\mathcal{F}(c) = \{x \in \mathcal{F}(c) \mid \forall c', \forall f_1, f_2 : c \to c' \quad \mathcal{F}(f_1)(x) = \mathcal{F}(f_2)(x)\}.$$

A category $\mathcal{C}$ is said to be **strongly connected** if for any $c, c' \in \mathcal{C}$ the hom-set $\mathcal{C}(c, c')$ is non-empty. For example, a category with one object is always strongly connected but the category of the ordered set $\{0 < 1\}$ is not strongly connected because $\mathcal{C}(1, 0) = \emptyset$.

Proposition 4.1: *Let $\mathcal{C}$ be a strongly connected category and $\mathcal{F} : \mathcal{C} \to \mathsf{Mod}(k)$ be a functor. Then $\mathsf{lim}\,\mathcal{F}$ exists and for any $c \in \mathcal{C}$ the morphism $\varphi_c : \mathsf{lim}\,\mathcal{F} \to \mathcal{F}(c)$ is a monomorphism and it induces the isomorphism*

$$\mathsf{lim}\,\mathcal{F} \cong \mathsf{inv}\,\mathcal{F}(c).$$

Proof: Fix an object $c_0 \in \mathcal{C}$ and consider a morphism $f : c_0 \to c$. Since $\mathcal{C}$ is strongly connected, there exists a morphism $g : c \to c_0$. Using the definition of $\mathsf{inv}\mathcal{F}$, we get $\mathsf{inv}\mathcal{F}(f) = \mathsf{inv}\mathcal{F}(g)^{-1}$. Hence, $\mathsf{inv}\mathcal{F}(f)$ is an isomorphism for any f. Moreover, for any two morphisms $f_1, f_2 : c_0 \to c$ we have $\mathsf{inv}\,\mathcal{F}(f_1) = \mathsf{inv}\,\mathcal{F}(g)^{-1} = \mathsf{inv}\,\mathcal{F}(f_2)$. It follows that the isomorphism $\mathsf{inv}\,\mathcal{F}(c_0) \cong \mathsf{inv}\,\mathcal{F}(c)$ is independent of the choice of f. We denote by $\varphi_c : \mathsf{inv}\,\mathcal{F}(c_0) \to \mathcal{F}(c)$ the composition of this isomorphism with the embedding $\mathsf{inv}\,\mathcal{F}(c) \subseteq \mathcal{F}(c)$. It is easy to see that $\mathcal{F}(f)\varphi_c = \varphi_{c'}$ for any morphism $f : c \to c'$. To prove the universal property of this collection of morphisms we assume that we have a module M and a collection $\{\psi_c : M \to \mathcal{F}(c)\}$ with the property $\mathcal{F}(f)\psi_c = \psi_{c'}$ for any morphism $f : c \to c'$. Thus $\mathcal{F}(f_1)\psi_{c_0} = \psi_c = \mathcal{F}(f_2)\psi_{c_0}$ for any two morphisms $f_1, f_2 : c_0 \to c$, and hence, $\mathrm{Im}\, \psi_{c_0} \subseteq \mathsf{inv}\, \mathcal{F}(c_0)$. Denote by $\alpha : M \to \mathsf{inv}\, \mathcal{F}(c_0)$ the homomorphism induced by ψ_{c_0}. It is easy to check that α is the unique homomorphism satisfying the property $\psi_c = \varphi_c\alpha$. Hence $\lim \mathcal{F} = \mathsf{inv}\, \mathcal{F}(c_0)$. □

Example 4.2: Let G be a group considered as a category with one object. Then a functor $M : G \to \mathsf{Mod}(k)$ is a kG-module. Denote by M^G submodule of invariants of M. Then Proposition 4.1 implies

$$\lim M = M^G.$$

4.3. *Higher limits of representations of categories*

A *representation* of a category $\mathcal{C}$ is a functor $\mathcal{F} : \mathcal{C} \to \mathsf{Mod}(k)$. Here we have to consider the category of 'big enough' modules. More precisely, if $\mathcal{C}$ is U-small for a Grothendieck universe U, then by $\mathsf{Mod}(k)$ we denote the category of U-small modules. It is well-known that the category of representations $\mathsf{Mod}(k)^{\mathcal{C}}$ is an abelian category with enough projective and injective objects, and a sequence in this category is exact if and only if it is exact objectwise.

For any $c \in \mathcal{C}$ we denote by P_c the representation given by $P_c = k[\mathcal{C}(c, -)]$, where $k[-]$ denotes the functor of free k-module. The following isomorphism is a k-linear analogue of the Yoneda lemma:

$$\mathsf{Hom}_{\mathsf{Mod}(k)^{\mathcal{C}}}(P_c, \mathcal{F}) \cong \mathcal{F}(c). \tag{4.1}$$

It follows that the functor $\mathsf{Hom}_{\mathsf{Mod}(k)^{\mathcal{C}}}(P_c, -)$ is exact, and hence, P_c is a projective representation.

All limits of all representations $\mathcal{C} \to \mathsf{Mod}(k)$ exist. Therefore, we have a functor $\lim : \mathsf{Mod}(k)^{\mathcal{C}} \to \mathsf{Mod}(k)$ with the natural isomorphism

$$\mathsf{Hom}_{\mathsf{Mod}(k)^{\mathcal{C}}}(M, \mathcal{F}) \cong \mathsf{Hom}_k(M, \lim \mathcal{F}) \tag{4.2}$$

for any representation $\mathcal{F}$ and any k-module M. In particular, we have a natural isomorphism

$$\lim \mathcal{F} = \mathsf{Hom}_{\mathsf{Mod}(k)^{\mathcal{C}}}(k, \mathcal{F}). \tag{4.3}$$

Since $\lim$ is a right adjoint functor, it is a left exact additive functor. Hence it has right derived functors.

$$\lim{}^n = \mathbf{R}^n \lim. \tag{4.4}$$

Using (4.3) we get

$$\mathsf{lim}^n \mathcal{F} = \mathsf{Ext}^n_{\mathsf{Mod}(k)^{\mathcal{C}}}(k, \mathcal{F}). \tag{4.5}$$

It is well-known that if G is a group and M is a k-module, then $\mathsf{lim}^n M = \mathsf{Ext}^n_{kG}(k, M) = H^n(G, M) = H^n(\mathsf{B}G, M)$. This can be generalised as follows:

Proposition 4.3: *Let $\mathcal{C}$ be a category and M be a k-module. Then*

$$\mathsf{lim}^n M \cong H^n(\mathsf{B}\mathcal{C}, M).$$

Proof: Denote by $E : \Delta^{\mathrm{op}} \to \mathsf{Sets}^{\mathcal{C}}$ the simplicial functor whose $E_n(c)$ consists of sequences $(\beta, \alpha_n, \dots, \alpha_1)$ with $\mathsf{dom}(\alpha_{i+1}) = \mathsf{codom}(\alpha_i)$, $\mathsf{dom}(\beta) = \mathsf{codom}(\alpha_n)$ and $\mathsf{codom}(\beta) = c$. Face maps act as follows: $d_i(\beta, \alpha_n, \dots, \alpha_1) = (\beta, \alpha_n, \dots, \alpha_{i+1}\alpha_i, \dots, \alpha_1)$, for $1 \leq i < n$, $d_0(\beta, \alpha_n, \dots, \alpha_1) = (\alpha_n, \alpha_{n-1}, \dots, \alpha_2)$ and $d_n(\beta, \alpha_n, \alpha_{n-1}, \dots, \alpha_1) = (\beta\alpha_n, \alpha_{n-1}, \dots, \alpha_1)$. Degeneracies act as follows: $s_i(\beta, \alpha_n, \dots, \alpha_1) = (\beta, \alpha_n, \dots, \alpha_{i+1}, \mathsf{id}_{c_i}, \alpha_i, \dots, \alpha_1))$, where $c_i = \mathsf{codom}(\alpha_i)$. For $f : c \to c'$ the map $E_n(f) : E_n(c) \to E_n(c')$ is given by $E_n(f)(\beta, \alpha_n, \dots, \alpha_1) = (f\beta, \alpha_n, \dots, \alpha_1)$. For any n consider the map $\tilde{s}_{n+1} : E_n(c) \to E_{n+1}(c)$ given by $\tilde{s}_{n+1}(\beta, \alpha_n, \dots, \alpha_1) = (\mathsf{id}_c, \beta, \alpha_n, \dots, \alpha_1)$. These maps satisfy the following identities $d_i\tilde{s}_{n+1} = \tilde{s}_n d_i$ and $d_{n+1}\tilde{s}_{n+1} = \mathsf{id}$, $\tilde{s}_{n+1}s_i = s_i\tilde{s}_n$. By Proposition 2.1 we obtain that the simplicial set $|E(c)|$ is contractible for any c. It follows that the homology of the complex $\mathsf{C}_\bullet(E(c)) \otimes k$ is concentrated in degree zero and $H_0(\mathsf{C}_\bullet(E(c)) \otimes k) = k$. Note that there is an isomoprhism of representations

$$\mathsf{C}_n(E) \otimes k = \bigoplus_{(\alpha_n, \dots, \alpha_1) \in (\mathsf{N}\mathcal{C})_n} P_{\mathsf{codom}(\alpha_n)}.$$

Therefore, the complex presentations $\mathsf{C}_\bullet(E(-))$ are a projective resolution of the trivial presentation k. Moreover, using the formula (4.1) we get

$$\mathsf{Hom}_{\mathsf{Mod}(k)^{\mathcal{C}}}(\mathsf{C}_n(E) \otimes k, M) \cong \bigoplus_{(\alpha_n, \dots, \alpha_1) \in (\mathsf{N}\mathcal{C})_n} M \cong \mathsf{Hom}(\mathsf{C}_n(\mathsf{N}\mathcal{C}), M).$$

One can check that this isomorphism gives an isomorphism of complexes

$$\mathsf{Hom}_{\mathsf{Mod}(k)^{\mathcal{C}}}(\mathsf{C}_\bullet(E) \otimes k, M) \cong \mathsf{Hom}(\mathsf{C}_\bullet(\mathsf{N}\mathcal{C}), M).$$

Then applying homology H^n to both sides we get $\mathsf{lim}^n M \cong H^n(\mathsf{N}\mathcal{C}, M)$. Using (2.2), we obtain the required statement. □

Corollary 4.4: *If a category $\mathcal{C}$ is contractible and M is a k-module then*

$$\mathsf{lim}^i M = 0,$$

for $i > 0$.

4.4. *Preserving of higher limits*

A category $\mathcal{C}$ is said to be *k-acyclic* if the reduced homology groups $\tilde{H}_*(\mathsf{B}\mathcal{C}, k)$ vanish. Of course, a contractible category is k-acyclic.

Let $\Phi : \mathcal{C} \to \mathcal{D}$ be a functor. For an object $d \in D$ we denote by $\Phi{\downarrow}d$ the comma category (see [8]) whose objects are pairs (c, α), where $c \in \mathcal{C}$ and $\alpha \in \mathcal{D}(\Phi(c), d)$. A morphism in the comma category from (c, α) to (c', α') is given by $f : c \to c'$, which satisfies $\alpha'\Phi(f) = \alpha$. If $\mathcal{C} = \mathcal{D}$ and $\Phi = \mathsf{Id}_{\mathcal{D}}$, the corresponding comma category is denoted by $\mathcal{D}{\downarrow}d$.

Proposition 4.5: *(see [7]). Let $\Phi : \mathcal{C} \to \mathcal{D}$ be a functor such that $\Phi{\downarrow}d$ is k-acyclic for any $d \in D$. Then for any representation $\mathcal{F}$ of $\mathcal{D}$ there is an isomorphism*

$$\theta_\Phi : \mathsf{lim}^n \mathcal{F} \xrightarrow{\cong} \mathsf{lim}^n(\mathcal{F}\Phi), \tag{4.6}$$

which is natural in $\mathcal{F}$ and Φ.

Remark 4.6: In the previous proposition by naturality in Φ we mean that for a natural transformation $\alpha : \Phi_1 \to \Phi_2$ of functors satisfying the condition of the proposition we have $\alpha_*\theta_{\Phi_1} = \theta_{\Phi_2}$, where $\alpha_* : \mathsf{lim}^n(\mathcal{F}\Phi_1) \to \mathsf{lim}^n(\mathcal{F}\Phi_2)$ is the map induced by α. In [7] the fact that the isomorphism is natural is not stated explicitly, but it follows easily from the proof and the fact that the obvious map $\theta^0_\Phi : \mathsf{lim}\, \mathcal{F} \to \mathsf{lim}(\mathcal{F}\Phi)$ is natural in $\mathcal{F}$ and Φ.

Corollary 4.7: *Let $\Phi_1, \Phi_2 : \mathcal{C} \to \mathcal{D}$ be functors satisfying the condition of Proposition 4.5 and $\alpha : \Phi_1 \to \Phi_2$ be a natural transformation. The induced homomorphism $\alpha_* : \mathsf{lim}^n(\mathcal{F}\Phi_1) \to \mathsf{lim}^n(\mathcal{F}\Phi_2)$ is an isomorphism.*

Proof: Since the isomorphism in Proposition 4.5 is natural in Φ, there is a commutative diagram

$$\begin{array}{ccc} & \mathsf{lim}^n\mathcal{F} & \\ {\scriptstyle \theta_{\Phi_1}\,\cong}\swarrow & & \searrow{\scriptstyle \theta_{\Phi_2}\,\cong} \\ \mathsf{lim}^n(\mathcal{F}\Phi_1) & \xrightarrow{\alpha_*} & \mathsf{lim}^n(\mathcal{F}\Phi_2). \end{array}$$

Hence, α_* is an isomorphism. □

Lemma 4.8: *Let $\mathcal{F}$ be a representation of a category $\mathcal{C}$ with coproducts. Then the morphisms $i_1, i_2 : c \to c \sqcup c$ induce isomorphisms*

$$(i_k)_* : \mathsf{lim}^n\, \mathcal{F}(c) \xrightarrow{\cong} \mathsf{lim}^n\, \mathcal{F}(c \sqcup c). \tag{4.7}$$

Proof: We denote by sq the functor $\mathcal{C} \to \mathcal{C}$ given by $c \mapsto c \sqcup c$. An object of the comma category $\mathsf{sq} \downarrow c_0$ is a pair $(c, \alpha : c \sqcup c \to c_0)$. An arrow $\alpha : c \sqcup c \to c_0$ is defined by the pair of arrows $\alpha_1 = \alpha \circ i_1 : c \to c_0$ and $\alpha_2 = \alpha \circ i_2 : c \to c_0$. Thus the category $\mathsf{sq}\downarrow c_0$ is isomorphic to the category, whose objects are triples $(c, \alpha_1 : c \to c_0, \alpha_2 : c \to c_0)$ and morphisms $f : (c, \alpha_1, \alpha_2) \to (c', \alpha_1', \alpha_2')$ are morphisms $f : c \to c'$ such that $\alpha_1' f = \alpha_1$ and $\alpha_2' f = \alpha_2$. It is easy to see that the object $(c \sqcup c', \alpha_1 + \alpha_1', \alpha_2 + \alpha_2')$ is the coproduct of objects (c, α_1, α_2) and $(c', \alpha_1', \alpha_2')$ in this category. Hence the category $\mathsf{sq} \downarrow c_0$ has coproducts and by Lemma 3.5 it is contractible. Therefore, the functor sq satisfies the condition of Proposition 4.5. Finally, applying Corollary 4.7 to the natural transformation $i_k : \mathsf{Id}_\mathcal{C} \to \mathsf{sq}$, we obtain the claimed isomorphism. □

4.5. *Monoadditive representations*

Let $\mathcal{C}, \mathcal{D}$ be categories with pairwise coproducts. For $c_1, c_2 \in \mathcal{C}$ we denote by $i_k = i_k(c_1, c_2) : c_k \to c_1 \sqcup c_2$ the universal morphisms $c_1 \xrightarrow{i_1} c_1 \sqcup c_2 \xleftarrow{i_2} c_2$. For any functor $\mathcal{F} : \mathcal{C} \to \mathcal{D}$ there is a morphism

$$(\mathcal{F}(i_1), \mathcal{F}(i_2)) : \mathcal{F}(c_1) \sqcup \mathcal{F}(c_2) \longrightarrow \mathcal{F}(c_1 \sqcup c_2). \tag{4.8}$$

The functor $\mathcal{F}$ is said to be **additive** (resp. **monoadditive, split monoadditive**) if (4.8) is an isomorphism (resp. monomorphism, split monomorphism in the category of bifunctors). Consider the functors $\mathsf{sq} : \mathcal{C} \to \mathcal{C}$ and $\mathsf{sq} : \mathcal{D} \to \mathcal{D}$ given by $\mathsf{sq}(x) = x \sqcup x$. Then (4.8) gives a natural transformation

$$\mathsf{T}_\mathcal{F} : \mathsf{sq} \circ \mathcal{F} \longrightarrow \mathcal{F} \circ \mathsf{sq}.$$

Let $\mathcal{D} = \mathsf{Mod}(k)$. Starting from a monoadditive representation $\mathcal{F} : \mathcal{C} \to \mathsf{Mod}(k)$ we construct the representation $\Sigma\mathcal{F} := \operatorname{coker}(\mathsf{T}_\mathcal{F})$ and get the short exact sequence

$$0 \longrightarrow \mathcal{F} \oplus \mathcal{F} \xrightarrow{\mathsf{T}_\mathcal{F}} \mathcal{F} \circ \mathsf{sq} \longrightarrow \Sigma\mathcal{F} \longrightarrow 0. \tag{4.9}$$

A monoadditive representation $\mathcal{F}$ is said to be ***n*-monoadditive**, if $\Sigma\mathcal{F}$ is $(n-1)$-monoadditive, where a 1-monoadditive representation is just a monoadditive representation. For an n-monoadditive representation $\mathcal{F}$

we can define $\Sigma^m\mathcal{F}$ for $m \le n$ by induction. If a representation $\mathcal{F}$ is n-monoadditive, for any $n \ge 1$, $\mathcal{F}$ is called **∞-monoadditive**. For example, additive representations are ∞-monoadditive.

Proposition 4.9: *Let $\mathcal{C}$ be a category with pairwise coproducts and $\mathcal{F}$ be a monoadditive representation of $\mathcal{C}$. Then there is an isomorphism*

$$\mathsf{lim}^n\, \mathcal{F} \cong \mathsf{lim}^{n-1}\, \Sigma\mathcal{F}, \tag{4.10}$$

for any $n \ge 0$.

Proof: Consider the long exact sequence of higher limits associated with the short exact sequence $0 \longrightarrow \mathcal{F} \oplus \mathcal{F} \longrightarrow \mathcal{F} \circ \mathsf{sq} \longrightarrow \Sigma\mathcal{F} \longrightarrow 0$. Using Lemma 4.8, we obtain that the homomorphism $((i_1)_*, (i_2)_*) : \mathsf{lim}^n\, \mathcal{F} \oplus \mathsf{lim}^n\, \mathcal{F} \longrightarrow \mathsf{lim}^n \mathcal{F} \circ \mathsf{sq}$ is an epimorphism, and hence the map $\mathsf{lim}^n \mathcal{F} \circ \mathsf{sq} \longrightarrow \mathsf{lim}^n\, \Sigma\mathcal{F}$ vanishes. Therefore we get the short exact sequences

$$0 \longrightarrow \mathsf{lim}^n\, \Sigma\mathcal{F} \longrightarrow \mathsf{lim}^{n+1}\, \mathcal{F} \oplus \mathsf{lim}^{n+1}\, \mathcal{F} \longrightarrow \mathsf{lim}^{n+1}\, \mathcal{F} \circ \mathsf{sq} \longrightarrow 0, \tag{4.11}$$

which are the totalisations of the bicartesian squares

$$\begin{array}{ccc} \mathsf{lim}^n\, \Sigma\mathcal{F} & \xrightarrow{\quad s_1 \quad} & \mathsf{lim}^{n+1}\, \mathcal{F} \\ \Big\downarrow{\scriptstyle s_2} & & \cong\Big\downarrow{\scriptstyle (i_1)_*} \\ \mathsf{lim}^{n+1}\, \mathcal{F} & \xrightarrow[\cong]{\quad (i_2)_* \quad} & \mathsf{lim}^{n+1} \mathcal{F} \circ \mathsf{sq}. \end{array}$$

By Lemma 4.8 $(i_1)_*$ and $(i_2)_*$ are isomorphisms, and hence s_1 and s_2 are isomorphisms. □

Corollary 4.10: *If $\mathcal{F}$ is an n-monoadditive representation of a category with pairwise coproducts, then $\mathsf{lim}^i\, \mathcal{F} = 0$, for $0 \le i < n$, and $\mathsf{lim}^i\, \mathcal{F} = \mathsf{lim}^{i-n}\Sigma^n\mathcal{F}$.*

Corollary 4.11: *If $\mathcal{F}$ is an ∞-monoadditive representation of a category with pairwise coproducts, then $\mathsf{lim}^i\mathcal{F} = 0$ for any $i \ge 0$.*

Lemma 4.12: *Let $\mathcal{C}, \mathcal{D}$ be categories with pairwise coproducts and $\mathcal{F} : \mathcal{C} \to \mathcal{D}$ be a split monoadditive functor. Then:*

(1) if $\Phi : \mathcal{D} \to \mathcal{E}$ is an additive functor, then $\Phi \circ \mathcal{F}$ is a split monoadditive functor;
(2) if $\mathcal{F}' : \mathcal{C} \to \mathcal{D}$ is a retract of $\mathcal{F}$, then $\mathcal{F}'$ is a split monoadditive functor.

Proof: The proof is obvious. □

Proposition 4.13: *If $\mathcal{F}$ is a split monoadditive representation, then $\Sigma\mathcal{F}$ is a split monoadditive representation.*

Proof: Let $r_{c_1,c_2} : \mathcal{F}(c_1 \sqcup c_2) \to \mathcal{F}(c_1) \sqcup \mathcal{F}(c_2)$ be a retraction of the natural transformation $(\mathcal{F}(i_1), \mathcal{F}(i_2))$. Then $r_{c_1 \sqcup c_1, c_2 \sqcup c_2}$ is a retraction for the natural transformation $((\mathcal{F} \circ \mathsf{sq})(i_1), (\mathcal{F} \circ \mathsf{sq})(i_2))$. It follows that $\mathcal{F} \circ \mathsf{sq}$ is a split monoadditive functor. The short exact sequence $\mathsf{sq} \circ \mathcal{F} \rightarrowtail \mathcal{F} \circ \mathsf{sq} \twoheadrightarrow \Sigma\mathcal{F}$ splits in the category of representations. It follows that $\Sigma\mathcal{F}$ is a retract of $\mathcal{F} \circ \mathsf{sq}$. Then Lemma 4.12 implies that $\Sigma\mathcal{F}$ is a split monoadditive representation. □

Corollary 4.14: *A split monoadditive representation is ∞-monoadditive.*

Hence, we have the following gradation:

split monoadditive $\Rightarrow$ ∞-monoadditive $\Rightarrow$ monoadditive.

4.6. *Spectral sequence of higher limits*

Proposition 4.15: *Let $\Phi : \mathcal{A} \to \mathcal{B}$ be a left exact functor between abelian categories with enough injectives and $a^\bullet \in \mathsf{Com}^+(\mathcal{A})$ be a complex bounded below with Φ-acyclic cohomologies. Then there exists a cohomological spectral sequence E of objects of $\mathcal{B}$ so that*

$$E \Rightarrow \Phi(H^n(a^\bullet)) \quad \textit{and} \quad E_1^{pq} = \mathbf{R}^q\Phi(a^p). \tag{4.12}$$

Proof: Consider an injective Cartan-Eilenberg resolution $I^{\bullet\bullet}$ of $a^\bullet$ with an injection $a^\bullet \rightarrowtail I^{\bullet,0}$ and differentials $d_{\mathrm{I}}^{pq} : I^{pq} \to I^{p+1,q}$ and $d_{\mathrm{II}}^{pq} : I^{pq} \to I^{p,q+1}$. Then there are two spectral sequences ${}_{\mathrm{I}}E$ and ${}_{\mathrm{II}}E$ associated with the bicomplex $B^{\bullet\bullet} = \Phi(I^{\bullet\bullet})$ which converge to the cohomology of the totalisation $H^n(\mathsf{Tot}(B^{\bullet\bullet}))$ so that ${}_{\mathrm{I}}E_1^{pq} = H_{\mathrm{II}}^q(B^{p\bullet})$, ${}_{\mathrm{I}}E_2^{pq} = H_{\mathrm{I}}^p(H_{\mathrm{II}}^q(B^{\bullet\bullet}))$ and ${}_{\mathrm{II}}E_1^{pq} = H_{\mathrm{I}}^p(B^{\bullet,q})$, ${}_{\mathrm{II}}E_2^{pq} = H_{\mathrm{II}}^q(H_{\mathrm{I}}^p(B^{\bullet\bullet}))$. Since $I^{\bullet\bullet}$ is a Cartan-Eilenberg resolution, the complex $I^{\bullet,q}$ is homotopy equivalent to the complex of its cohomologies $H_{\mathrm{I}}^*(I^{\bullet,q})$ with zero differentials, and hence $H_{\mathrm{I}}^p(B^{\bullet,q}) = \Phi(H_{\mathrm{I}}^p(I^{\bullet,q}))$. The complex $H_{\mathrm{I}}^p(I^{\bullet\bullet})$ is an injective resolution of $H^p(a^\bullet)$. It follows that $H_{\mathrm{II}}^q(H_{\mathrm{I}}^p(B^{\bullet\bullet})) = \mathbf{R}^q\Phi(H^p(a^\bullet))$. Using the fact that the objects $H^p(a^\bullet)$ are Φ-acyclic, we obtain

$${}_{\mathrm{II}}E_2^{pq} = \begin{cases} 0, & q \neq 0 \\ \Phi(H^p(a^\bullet)), & q = 0. \end{cases} \tag{4.13}$$

Hence we get $H^n(\mathsf{Tot}(B^{\bullet\bullet})) = \Phi(H^n(a^\bullet))$. Therefore the spectral sequence $E := {}_{\mathrm{I}}E$ converges to $\Phi(H^n(a^\bullet))$, and since $I^{p,\bullet}$ is an injective resolution of a^p, we have $E_1^{pq} = H^q_{\mathrm{II}}(\Phi(I^{p,\bullet})) = \mathbf{R}^q\Phi(a^\bullet)$. □

Corollary 4.16: *Let $\mathcal{F}^\bullet$ be a complex of representations of a category $\mathcal{C}$ bounded below with* lim*-acyclic cohomologies. Then there exists a cohomological spectral sequence E such that*

$$E \Rightarrow \mathsf{lim}\, H^n(\mathcal{F}^\bullet) \quad \textit{and} \quad E_1^{pq} = \mathsf{lim}^q \mathcal{F}^p. \tag{4.14}$$

5. Limits over $\mathsf{Pres}(G)$

5.1. *The category of presentations of a group*

In this section, we assume that $k = \mathbb{Z}$ and we consider the category $\mathsf{Pres}(G)$ of free presentations of a group G. The objects of $\mathsf{Pres}(G)$ are surjective homomorphisms $\pi : F \twoheadrightarrow G$ where F is a free group and morphisms $f : (\pi_1 : F_1 \twoheadrightarrow G) \to (\pi_2 : F_2 \twoheadrightarrow G)$ are homomorphisms $f : F_1 \to F_2$ such that $\pi_1 = \pi_2 f$. The category $\mathsf{Pres}(G)$ has coproducts given by

$$(\pi_1 : F_1 \twoheadrightarrow G) \sqcup (\pi_2 : F_2 \twoheadrightarrow G) = ((\pi_1, \pi_2) : F_1 * F_2 \twoheadrightarrow G),$$

and hence it is contractible. In particular, for any abelian group A higher limits $\mathsf{lim}^i A$ vanish for $i > 0$. It is easy to check that the category $\mathsf{Pres}(G)$ is strongly connected, and hence, the limit of a functor $\mathcal{F} : \mathsf{Pres}(G) \to \mathsf{Ab}$ coincides with the set of invariant elements of $\mathcal{F}(\pi)$ for any presentation π. All the limits considered in this section are taken over the category $\mathsf{Pres}(G)$.

We will always denote by R the kernel of an epimorphism $\pi : F \twoheadrightarrow G$. Therefore an object of $\mathsf{Pres}(G)$ defines a short exact sequence of groups

$$1 \longrightarrow R \longrightarrow F \xrightarrow{\pi} G \longrightarrow 1,$$

and R can be considered as a functor $R : \mathsf{Pres}(G) \to \mathsf{Gr}$.

5.2. *Representations $\mathcal{M} \otimes_{\mathbb{Z}[F]} \mathbf{f}$*

Consider the functor $\mathbb{Z}[F] : \mathsf{Pres}(G) \to \mathsf{Rings}$ that sends $(F \twoheadrightarrow G)$ to $\mathbb{Z}[F]$. Then the representation

$$\mathbf{f} = I(F) : \mathsf{Pres}(G) \longrightarrow \mathsf{Ab}$$

has the structure of a $\mathbb{Z}[F]$-bimodule. So if $\mathcal{M} : \mathsf{Pres}(G) \to \mathsf{Ab}$ is a right $\mathbb{Z}[F]$-module and $\mathcal{N} : \mathsf{Pres}(G) \to \mathsf{Ab}$ is a left $\mathbb{Z}[F]$-module, we can define a tensor product

$$\mathcal{M} \otimes_{\mathbb{Z}[F]} \mathbf{f} \otimes_{\mathbb{Z}[F]} \mathcal{N} : \mathsf{Pres}(G) \longrightarrow \mathsf{Ab}.$$

Proposition 5.1: *Let $\mathcal{M}$ be a right $\mathbb{Z}[F]$-module and $\mathcal{N}$ be a left $\mathbb{Z}[F]$-module. Then the representations $\mathcal{M}\otimes_{\mathbb{Z}[F]}\mathbf{f}$ and $\mathbf{f}\otimes_{\mathbb{Z}[F]}\mathcal{N}$ are split monoadditive. In particular*

$$\mathsf{lim}^i\, \mathcal{M} \otimes_{\mathbb{Z}[F]} \mathbf{f} = 0.$$

for any $i \geq 0$.

Proof: First we note that since the category $\mathsf{Pres}(G)$ is strongly connected, that for any two presentations $\pi_1 : F_1 \twoheadrightarrow G$ and $\pi_2 : F_2 \twoheadrightarrow G$, the embeddings $\pi_k \to \pi_1 \sqcup \pi_2$ split. Indeed, for any morphism $f : \pi_2 \to \pi_1$, the morphism $(\mathsf{id}_{\pi_1}, f) : \pi_1 \sqcup \pi_2 \to \pi_1$ is a retraction of $i_1 : \pi_1 \to \pi_1 \sqcup \pi_2$. It follows that for any representation $\mathcal{F}$ the monomorphism $\mathcal{F}(\pi_k) \to \mathcal{F}(\pi_1 \sqcup \pi_2)$ splits.

It is well known [15] and easy to prove that for any groups G and H there is an isomorphism of right $\mathbb{Z}[G * H]$-modules

$$I(G * H) \cong (I(G) \otimes_{\mathbb{Z}[G]} \mathbb{Z}[G * H]) \oplus (I(H) \otimes_{\mathbb{Z}[G]} \mathbb{Z}[G * H])$$

and an isomorphism of left $\mathbb{Z}[G * H]$-modules

$$I(G * H) \cong (\mathbb{Z}[G * H] \otimes_{\mathbb{Z}[G]} I(G)) \oplus (\mathbb{Z}[G * H] \otimes_{\mathbb{Z}[H]} I(H)).$$

It follows that

$$I(F_1 * F_1) \otimes_{\mathbb{Z}[F_1 * F_2]} \mathcal{N}(\pi_1 \sqcup \pi_2) = (I(F_1) \otimes_{\mathbb{Z}[F_1]} \mathcal{N}(\pi_1 \sqcup \pi_2)) \oplus (I(F_2) \otimes_{\mathbb{Z}[F_2]} \mathcal{N}(\pi_1 \sqcup \pi_2)).$$

Since for any $k \in \{1, 2\}$ the monomorphism $\mathcal{N}(\pi_k) \to \mathcal{N}(\pi_1 \sqcup \pi_2)$ splits, the monomirphism

$$I(F_k) \otimes_{\mathbb{Z}[F_k]} \mathcal{N}(\pi_k) \longrightarrow I(F_k) \otimes_{\mathbb{Z}[F_k]} \mathcal{N}(\pi_1 \sqcup \pi_2)$$

splits too. It follows that $\mathbf{f}\otimes_{\mathbb{Z}[F]}\mathcal{N}$ is split monoadditive. Similarly $\mathcal{M}\otimes_{\mathbb{Z}[F]}\mathbf{f}$ is split monoadditive. □

Corollary 5.2: *For any functor $\mathcal{F} : \mathsf{Pres} \to \mathsf{Ab}$ and any $i \geq 0$, we have $\mathsf{lim}^i\, \mathbf{f} \otimes \mathcal{F} = 0$ and $\mathsf{lim}^i\, F_{ab} \otimes \mathcal{F} = 0$.*

Proof: It follows from the equalities $\mathbf{f} \otimes \mathcal{F} = \mathbf{f} \otimes_{\mathbb{Z}[F]} (\mathbb{Z}[F] \otimes \mathcal{F})$ and $F_{ab} \otimes \mathcal{F} = \mathbf{f} \otimes_{\mathbb{Z}[F]} \mathcal{F}$. □

5.3. *Technical lemmas*

Lemma 5.3: *If* $\mathbf{a}, \mathbf{b}$ *are two right ideals of a ring* Λ *which are free as right* Λ*-modules with bases* $\{x_i\}$ *and* $\{y_j\}$*, then* $\mathbf{ab}$ *is a free* Λ*-module with basis* $\{x_i y_j\}$.

Proof: Obvious. □

Corollary 5.4: *Let* $R_1, \dots, R_n$ *be normal subgroups of* F*,* $\{y_{i,j}\}$ *be a free basis of the group* R_i *and* $\mathbf{r}_i = (R_i - 1)\mathbb{Z}[F]$*. Then* $\mathbf{r}_1\mathbf{r}_2 \dots \mathbf{r}_n$ *is a free right (or left)* F*-module and products* $(y_{1,j_1} - 1) \dots (y_{n,j_n} - 1)$ *form its basis.*

Lemma 5.5: *If* $R \rightarrowtail F \twoheadrightarrow G$ *is a presentation of* G *then*

$$\dots \to \frac{\mathbf{fr}^2}{\mathbf{fr}^3} \to \frac{\mathbf{r}^2}{\mathbf{r}^3} \to \frac{\mathbf{fr}}{\mathbf{fr}^2} \to \frac{\mathbf{r}}{\mathbf{r}^2} \to \frac{\mathbf{f}}{\mathbf{fr}} \to \frac{\mathbb{Z}[F]}{\mathbf{r}} \to \mathbb{Z} \to 0$$

is a free resolution of the right trivial G*-module* $\mathbb{Z}$*, whose* $2n$*th term is* $\mathbf{r}^n/\mathbf{r}^{n+1}$*,* $(2n+1)$*th term is* $\mathbf{fr}^n/\mathbf{fr}^{n+1}$ *and morphisms are induced by embeddings.*

Proof: Using Corollary 5.4 we get that $\mathbf{fr}^n$ and $\mathbf{r}^n$ are free F-modules. Hence $\mathbf{fr}^n/\mathbf{fr}^{n+1} = \mathbf{fr}^n \otimes_{\mathbb{Z}[F]} \mathbb{Z}G$ and $\mathbf{r}^n/\mathbf{r}^{n+1} = \mathbf{r}^n \otimes_{\mathbb{Z}[F]} \mathbb{Z}G$ are free G-modules. Exactness is obvious. □

Lemma 5.6: *The map* $r_1 \otimes \dots \otimes r_n \mapsto (r_1 - 1) \dots (r_n - 1)$ *induces an isomorphism*

$$R_{ab}^{\otimes n} \cong \mathbf{r}^n/\mathbf{r}^n\mathbf{f}.$$

Proof: Let $\{y_i\}$ is a free basis of R. Corollary 5.4 implies that $\mathbf{r}^n$ is a free F-module and products $(y_{i_1} - 1) \dots (y_{i_n} - 1)$ form its basis. It follows that $\mathbf{r}^n/\mathbf{r}^n\mathbf{f} = \mathbf{r}^n \otimes_{\mathbb{Z}[F]} \mathbb{Z}$ is a free abelian group and products $(y_{i_1} - 1) \dots (y_{i_n} - 1)$ form its basis. $R_{ab}^{\otimes n}$ is a free abelian group with the basis given by tensors $y_{i_1} \otimes \dots \otimes y_{i_n}$. Hence the map $r_1 \otimes \dots \otimes r_n \mapsto (r_1 - 1) \dots (r_n - 1)$ induces a bijection on the bases of free abelian groups, and hence, it is an isomorphism. □

Lemma 5.7: *If* Λ *is a ring,* $\mathbf{a}$ *is its right ideal and* $\mathbf{b}$ *is its left ideal, then there is a short exact sequence*

$$0 \longrightarrow \mathsf{Tor}_2^{\Lambda}(\Lambda/\mathbf{a}, \Lambda/\mathbf{b}) \longrightarrow \mathbf{a} \otimes_{\Lambda} \mathbf{b} \longrightarrow \mathbf{ab} \longrightarrow 0,$$

where the map $\mathbf{a} \otimes_{\Lambda} \mathbf{b} \to \mathbf{ab}$ *is given by multiplication.*

Proof: Tensoring by $\mathbf{b}$ the short exact sequence $\mathbf{a} \rightarrowtail \Lambda \twoheadrightarrow \Lambda/\mathbf{a}$ we get the exact sequence

$$0 \longrightarrow \mathsf{Tor}_1^\Lambda(\Lambda/\mathbf{a}, \mathbf{b}) \longrightarrow \mathbf{a} \otimes_\Lambda \mathbf{b} \longrightarrow \mathbf{b} \longrightarrow \mathbf{b}/\mathbf{ab} \longrightarrow 0.$$

Since the kernel of $\mathbf{b} \to \mathbf{b}/\mathbf{ab}$ is $\mathbf{ab}$ we get the short exact sequence

$$0 \longrightarrow \mathsf{Tor}_1^\Lambda(\Lambda/\mathbf{a}, \mathbf{b}) \longrightarrow \mathbf{a} \otimes_\Lambda \mathbf{b} \longrightarrow \mathbf{ab} \longrightarrow 0.$$

Using the long exact sequence of $\mathsf{Tor}_*^\Lambda(\Lambda/\mathbf{a}, -)$ with respect to the short exact sequence $\mathbf{b} \rightarrowtail \Lambda \twoheadrightarrow \Lambda/\mathbf{b}$ we obtain $\mathsf{Tor}_1^\Lambda(\Lambda/\mathbf{a}, \mathbf{b}) \cong \mathsf{Tor}_2^\Lambda(\Lambda/\mathbf{a}, \Lambda/\mathbf{b})$. □

Lemma 5.8: *Let* $\mathbf{a}$ *and* $\mathbf{b}$ *be ideals of* $\mathbb{Z}[F]$ *such that* $\mathsf{Tor}(\mathbb{Z}[F]/\mathbf{a}, \mathbb{Z}[F]/\mathbf{b}) = 0$. *Then the map* $x \otimes y \mapsto xy$ *induces*

$$\mathbf{a} \otimes_{\mathbb{Z}[F]} \mathbf{b} \cong \mathbf{ab}.$$

Proof: For any $\mathbb{Z}[F]$-modules M, N there is an isomorphism $M \otimes_{\mathbb{Z}[F]} N = (M \otimes N)_F$. Then there is a spectral sequence of composition $H_*(F, \mathsf{Tor}_*(M, N)) \Rightarrow \mathsf{Tor}_*^{\mathbb{Z}[F]}(M, N)$. Using this spectral sequence we obtain that $\mathsf{Tor}(\mathbb{Z}[F]/\mathbf{a}, \mathbb{Z}[F]/\mathbf{b}) = 0$ and $H_2(F, -) = 0$ implying $\mathsf{Tor}_2^{\mathbb{Z}[F]}(\mathbb{Z}[F]/\mathbf{a}, \mathbb{Z}[F]/\mathbf{b}) = 0$. Hence Lemma 5.7 implies the claimed isomorphism. □

Lemma 5.9: *Let* $\mathbf{a}' \subset \mathbf{a}$ *and* $\mathbf{b}' \subset \mathbf{b}$ *be ideals of* $\mathbb{Z}[F]$ *such that* $\mathsf{Tor}(\mathbb{Z}[F]/\mathbf{a}, \mathbb{Z}[F]/\mathbf{b}) = 0$. *Then the map* $x \otimes y \mapsto xy$ *induces*

$$(\mathbf{a}/\mathbf{a}') \otimes_{\mathbb{Z}[F]} (\mathbf{b}/\mathbf{b}') \cong \mathbf{ab}/(\mathbf{a}'\mathbf{b} + \mathbf{ab}').$$

Proof: Tensoring on $\mathbf{b}$ the short exact sequence $\mathbf{a}' \rightarrowtail \mathbf{a} \twoheadrightarrow \mathbf{a}/\mathbf{a}'$ we get the exact sequence

$$\mathbf{a}' \otimes_{\mathbb{Z}[F]} \mathbf{b} \longrightarrow \mathbf{ab} \longrightarrow (\mathbf{a}/\mathbf{a}') \otimes_{\mathbb{Z}[F]} \mathbf{b} \longrightarrow 0.$$

The image of the map $\mathbf{a}' \otimes_{\mathbb{Z}[F]} \mathbf{b} \to \mathbf{ab}$ is $\mathbf{a}'\mathbf{b}$. Hence $(\mathbf{a}/\mathbf{a}') \otimes_{\mathbb{Z}[F]} \mathbf{b} \cong \mathbf{ab}/\mathbf{a}'\mathbf{b}$. Tensoring the short exact sequence $\mathbf{b}' \rightarrowtail \mathbf{b} \twoheadrightarrow \mathbf{b}/\mathbf{b}'$ on $\mathbf{a}/\mathbf{a}'$ we get the exact sequence

$$(\mathbf{a}/\mathbf{a}') \otimes_{\mathbb{Z}[F]} \mathbf{b}' \longrightarrow \mathbf{ab}/\mathbf{a}'\mathbf{b} \longrightarrow (\mathbf{a}/\mathbf{a}') \otimes_{\mathbb{Z}[F]} (\mathbf{b}/\mathbf{b}') \longrightarrow 0.$$

The image of the map $(\mathbf{a}/\mathbf{a}') \otimes_{\mathbb{Z}[F]} \mathbf{b}' \to \mathbf{ab}/\mathbf{a}'\mathbf{b}'$ is $(\mathbf{a}'\mathbf{b} + \mathbf{ab}')/\mathbf{a}'\mathbf{b}$ which implies the claimed isomorphism. □

A **functorial ideal** of $\mathbb{Z}[F]$ is a functor $\mathbf{x} : \mathsf{Pres}(G) \to \mathsf{Ab}$ such that $\mathbf{x}$ is an ideal of $\mathbb{Z}[F]$ for any $\pi \in \mathsf{Pres}(G)$. For example, $\mathbf{x} = \mathbf{r}^n + \mathbf{f}^m$.

Lemma 5.10: *If* $\mathbf{x}$ *is a functorial ideal of* $\mathbb{Z}[F]$ *such that* $\mathbb{Z}[F]/\mathbf{x}$ *is torsion free, then* $\mathbf{x}$ *is a flat* $\mathbb{Z}[F]$*-module.*

Proof: Using the spectral sequence $H_*(F, \mathsf{Tor}_*(M,N)) \Rightarrow \mathsf{Tor}^{\mathbb{Z}[F]}(M,N)$, we get

$$\mathsf{Tor}_1^{\mathbb{Z}[F]}(\mathbf{x}, -) = \mathsf{Tor}_2^{\mathbb{Z}[F]}(\mathbb{Z}[F]/\mathbf{x}, -) = H_1(F, \mathsf{Tor}(\mathbb{Z}[F]/\mathbf{x}, -)).$$

So if $\mathbb{Z}[F]/\mathbf{x}$ is torsion free, then $\mathbf{x}$ is flat over $\mathbb{Z}[F]$. □

Lemma 5.11: *If* $\mathbf{x}$ *is a functorial ideal of* $\mathbb{Z}[F]$*, then for* $i \geq 0$

$$\mathsf{lim}^i\, \mathbf{xf} = 0$$

and

$$\mathsf{lim}^i\, \mathbf{fx} = 0.$$

Proof: It follows from Lemmas 5.8, 5.10 and Proposition 5.1. □

Lemma 5.12: *The abelian group* $\mathbb{Z}[F]/\mathbf{r}^n$ *is a free abelian group for any* $n \geq 1$.

Proof: It is enough to prove that $\mathbf{r}^k/\mathbf{r}^{k+1}$ is free abelian. Since $\mathbf{r}^k$ is a free F-module, $\mathbf{r}^k/\mathbf{r}^{k+1} = \mathbf{r}^k \otimes_{\mathbb{Z}[F]} \mathbb{Z}G$ is a free G-module, and hence, it is a free abelian group. □

Lemma 5.13: *For any abelian group* A

$$\mathsf{lim}^i\, \mathbf{r}^n \otimes A = \mathsf{lim}^i\, R_{ab}^{\otimes n} \otimes A = \begin{cases} I^{\otimes n} \otimes A, & i = n, \\ 0, & i \neq n. \end{cases}$$

Proof: Lemma 5.6 gives the short exact sequence:

$$0 \longrightarrow R_{ab} \longrightarrow \mathbf{f}/\mathbf{rf} \longrightarrow I \longrightarrow 0.$$

Since the short exact sequence consists of free abelian groups, tensoring it on $I^{\otimes k} \otimes R_{ab}^{\otimes n-k-1} \otimes A$ we get the short exact sequence

$$0 \longrightarrow I^{\otimes k} \otimes R_{ab}^{\otimes n-k} \otimes A \longrightarrow P \longrightarrow I^{\otimes k+1} \otimes R_{ab}^{\otimes n-k-1} \otimes A \longrightarrow 0,$$

where

$$P = I^{\otimes k} \otimes R_{ab}^{\otimes n-k-1} \otimes A \otimes \mathbf{f}/\mathbf{rf} = (I^{\otimes k} \otimes R_{ab}^{\otimes n-k-1} \otimes A \otimes \mathbb{Z}G) \otimes_{\mathbb{Z}[F]} \mathbf{f}.$$

Using Proposition 5.1 we obtain $\mathsf{lim}^i\, P = 0$ for all i. Hence

$$\mathsf{lim}^i I^{\otimes k} \otimes R_{ab}^{\otimes n-k} \otimes A = \mathsf{lim}^{i-1} I^{\otimes k+1} \otimes R_{ab}^{\otimes n-k-1} \otimes A.$$

By induction we get $\mathsf{lim}^i\, R_{ab}^{\otimes n} \otimes A = \mathsf{lim}^{i-n}\ I(G)^{\otimes n} \otimes A$. It follows that

$$\mathsf{lim}^i\, R_{ab}^{\otimes n} \otimes A = \begin{cases} I^{\otimes n} \otimes A, & i = n, \\ 0, & i \neq n. \end{cases}$$

Now we prove that $\mathsf{lim}^i\, \mathbf{r}^n \otimes A = \mathsf{lim}^i\, R_{ab}^{\otimes n} \otimes A$. Lemma 5.6 gives a short exact sequence $\mathbf{r}^n\mathbf{f} \rightarrowtail \mathbf{r}^n \twoheadrightarrow R_{ab}^{\otimes n}$. Since it consists of free abelian groups, we get the following short exact sequence:

$$0 \longrightarrow \mathbf{r}^n\mathbf{f} \otimes A \longrightarrow \mathbf{r}^n \otimes A \longrightarrow R_{ab}^{\otimes n} \otimes A \longrightarrow 0.$$

Using Lemma 5.8 and Lemma 5.12 we get $\mathbf{r}^n\mathbf{f} \otimes A = (A \otimes \mathbf{r}^n) \otimes_{\mathbb{Z}[F]} \mathbf{f}$. Proposition 5.1 implies that $\mathsf{lim}^*\mathbf{r}^n\mathbf{f} \otimes A = 0$. □

5.4. *Group homology as higher limits*

Theorem 5.14: *For any group G and abelian group A*

$$H_{2n-i}(G, A) = \mathsf{lim}^i\, \mathbf{r}^n/(\mathbf{fr}^n + \mathbf{r}^n\mathbf{f}) \otimes A, \qquad 0 \leq i \leq n-1,$$

$$H_{2n+1-i}(G, A) = \mathsf{lim}^i\, \mathbf{fr}^n/(\mathbf{r}^{n+1} + \mathbf{fr}^n\mathbf{f}) \otimes A, \qquad 0 \leq i \leq n-1.$$

Proof: We prove the first formula, the second formula can be proved similarly. Consider the '$2n$-truncated Gruenberg resolution':

$$\mathsf{t}_{2n}P: \quad (0 \to \mathbf{r}^n/\mathbf{fr}^n \rightarrowtail \mathbf{fr}^{n-1}/\mathbf{fr}^n \to \ldots \to \mathbf{r}/\mathbf{r}^2 \to \mathbf{f}/\mathbf{fr} \to \mathbb{Z}[F]/\mathbf{r} \to 0).$$

Then $H_i(\mathsf{t}_{2n}P \otimes_{\mathbb{Z}G} A) = H_i(G, A)$ for $0 \leq i \leq 2n$. Since A has trivial action of G, we get $M \otimes_{\mathbb{Z}G} A = M/M\mathbf{f} \otimes A$. Hence, $\mathsf{t}_{2n}P \otimes_{\mathbb{Z}G} A$ is equal to

$$\mathbf{r}^n/(\mathbf{fr}^n + \mathbf{r}^n\mathbf{f}) \otimes A \to (\mathbf{fr}^{n-1}/\mathbf{fr}^{n-1}\mathbf{f}) \otimes A \to \ldots \to \mathbf{r}/\mathbf{rf} \otimes A \to \mathbf{f}/\mathbf{f}^2 \otimes A \\ \to \mathbb{Z}[F]/\mathbf{f} \otimes A.$$

Apply the spectral sequence of higher limits to this complex:

$$\mathsf{lim}^i\, \mathsf{t}_{2n}P_j \otimes_{\mathbb{Z}G} A \Rightarrow H_{j-i}(G, A).$$

The short exact sequence

$$\mathbf{fr}^k\mathbf{f} \otimes A \rightarrowtail \mathbf{fr}^k \otimes A \twoheadrightarrow (\mathbf{fr}^k/\mathbf{fr}^k\mathbf{f}) \otimes A$$

implies $\mathsf{lim}^i\, (\mathbf{fr}^k/\mathbf{fr}^k\mathbf{f}) \otimes A = 0$. Recall $(\mathbf{r}^k/\mathbf{r}^k\mathbf{f}) \otimes A = R_{ab}^{\otimes k} \otimes A$. Set $\mathcal{F} = (\mathbf{r}^n/(\mathbf{fr}^n + \mathbf{r}^n\mathbf{f})) \otimes A$. Then the first page of the spectral sequence looks as follows:

$\vdots$	$\vdots$	$\vdots$	$\vdots$	$\vdots$	$\vdots$	$\vdots$	$\vdots$
n	$\mathsf{lim}^n\,\mathcal{F}$	0	0	$\dots$	0	0	0
$n-1$	$\mathsf{lim}^{n-1}\,\mathcal{F}$	0	$I^{\otimes n-1}\otimes A$	$\dots$	0	0	0
$\vdots$	$\vdots$	$\vdots$	$\vdots$	$\dots$	$\vdots$	$\vdots$	$\vdots$
2	$\mathsf{lim}^2\,\mathcal{F}$	0	0	$\dots$	0	0	0
1	$\mathsf{lim}^1\,\mathcal{F}$	0	0	$\dots$	$I\otimes A$	0	0
0	$\mathsf{lim}^0\,\mathcal{F}$	0	0	$\dots$	0	0	A
	$2n$	$2n-1$	$2n-2$	$\dots$	2	1	0

By the lacunary reason we get the isomorphisms. □

Corollary 5.15: *For any group G, abelian group A and $n \geq 1$*

$$H_{2n}(G, A) = \mathsf{lim}^0\, \mathbf{r}^n/(\mathbf{fr}^n + \mathbf{r}^n\mathbf{f}) \otimes A,$$

$$H_{2n+1}(G, A) = \mathsf{lim}^0\, \mathbf{fr}^n/(\mathbf{r}^{n+1} + \mathbf{fr}^n\mathbf{f}) \otimes A.$$

Theorem 5.16: [4] *For any group G and a G-module M there are isomorphisms*

$$H_{2n-i}(G, M) \cong \mathsf{lim}^i\, R_{ab}^{\otimes n} \otimes_{\mathbb{Z}G} M, \qquad 0 \leq i \leq n-1.$$

Moreover,

$$\mathsf{lim}^i\, R_{ab}^{\otimes n} \otimes_{\mathbb{Z}G} M = 0, \qquad n+1 \leq i,$$

and there is a short exact sequence:

$$0 \longrightarrow H_n(G, M) \longrightarrow \mathsf{lim}^n\, R_{ab}^{\otimes n} \otimes_{\mathbb{Z}G} M \longrightarrow (I^{\otimes n-1} \otimes M) \cdot I \longrightarrow 0.$$

6. Hochschild and cyclic homology as higher limits

In this section we state results that are proved in [4] and [12]. Let k be a field and A be an algebra. Consider the category $\mathsf{Pres}(A)$ of free presentations of an algebra A. The objects of $\mathsf{Pres}(A)$ are surjective homomorphisms $\pi : F \twoheadrightarrow A$ where F is a free algebra and morphisms $f : (\pi_1 : F_1 \twoheadrightarrow A) \to (\pi_2 : F_2 \twoheadrightarrow A)$ are homomorphisms $f : F_1 \to F_2$ such that $\pi_1 = \pi_2 f$. The limits considered in this section are taken over this category.

Theorem 6.1: [4] *For an algebra A and an A-bimodule M, for $n \geq 1$, there are natural isomorphisms*

$$H_{2n-i}(A, M) \simeq \mathsf{lim}^i\, (I^n/I^{n+1}) \otimes_{A^e} M \ \ \textit{for}\ \ 0 \leq i < n.$$

For an F-bimodule M we set

$$M_\natural = \frac{M}{[M,F]} = H_0(A,M),$$

where $[M,F]$ is the vector space generated by elements $mf - fm$.

Corollary 6.2: *For an algebra A and $n \geq 1$, there are natural isomorphisms*

$$HH_{2n-i}(A) \simeq \mathsf{lim}^i\ (I^n/I^{n+1})_\natural \quad \textit{for} \ \ 0 \leq i < n.$$

Theorem 6.3: *(Quillen [12]) Let A be an algebra over a field k of characteristic* 0. *Then there are isomorphisms*

$$HC_{2n}(A) \cong \mathsf{lim}\ (F/I^{n+1})_\natural. \tag{6.1}$$

Theorem 6.4: [4] *Let A be an augmented algebra over a field k of characteristic* 0. *Then there are isomorphisms*

$$HC_{2n-1}(A) \cong \mathsf{lim}^1\ (F/I^{n+1})_\natural. \tag{6.2}$$

7. fr-codes of functors Gr → Ab

Denote by Pres the category whose objects are presentations of a group $F \overset{\pi}{\twoheadrightarrow} G$, and whose morphisms are commutative squares

$$\begin{array}{ccc} F_1 & \longrightarrow & F_2 \\ \downarrow{\scriptstyle \pi_1} & & \downarrow{\scriptstyle \pi_2} \\ G_1 & \longrightarrow & G_2. \end{array}$$

Consider the forgetful functor

$$\mathsf{Pres} \longrightarrow \mathsf{Gr},$$

that sends the presentation $F \twoheadrightarrow G$ to G. The fibre of this functor over a group G is the category $\mathsf{Pres}(G)$. The following ideals of $\mathbb{Z}[F]$

$$\mathbf{f} = I(F) = (F-1)\mathbb{Z}[F], \qquad \mathbf{r} = (R-1)\mathbb{Z}[F]$$

depend on F and can be considered as functors $\mathbf{f}, \mathbf{r} : \mathsf{Pres} \to \mathsf{Ab}$. Using the operations of sum, product and intersection of ideals we can get a lot of different functors. For example

$$\mathbf{fr}^n + \mathbf{r}^n\mathbf{f}, \mathbf{r}^{n+1} + \mathbf{fr}^n\mathbf{f} : \mathsf{Pres} \longrightarrow \mathsf{Ab}.$$

Denote by $\mathsf{ML}(\mathbf{f}, \mathbf{r})$ the set of all functorial ideals of $\mathbb{Z}[F]$ that can be received from $\mathbf{r}, \mathbf{f}$ using the operations of sum, product and intersection, including $\mathbb{Z}[F]$ itself. Then for any $\mathbf{x} \in \mathsf{ML}(\mathbf{f}, \mathbf{r})$ and $i \geq 0$ we get the functor

$$^{i}[\mathbf{x}] : \mathsf{Gr} \longrightarrow \mathsf{Ab}$$

given by

$$^{i}[\mathbf{x}](G) = \mathsf{lim}^{i}_{\mathsf{Pres}(G)}\ \mathbf{x}.$$

Moreover, if $\mathbf{x} \subseteq \mathbf{y}$ then we get a morphism $^{i}[\mathbf{x}] \longrightarrow {}^{i}[\mathbf{y}]$. We will prove that $^{0}[\mathbf{x}] = 0$ for any $\mathbf{x}$, and the most important case is $i = 1$. So we set $^{1}[\mathbf{x}] = [\mathbf{x}]$.

Definition 7.1: An **fr-code** of a functor $\mathcal{F} : \mathsf{Gr} \to \mathsf{Ab}$ is an isomorphism

$$\mathcal{F} \cong {}^{i}[\mathbf{x}].$$

A **polynomial fr-code** is an **fr**-code $\mathcal{F} \cong {}^{i}[\mathbf{x}]$, such that $\mathbf{x} = \sum_j \mathbf{a}_{j,1}\mathbf{a}_{j,2}\dots\mathbf{a}_{j,m_j}$, where $\mathbf{a}_{j,k} \in \{\mathbf{f}, \mathbf{r}\}$.

Example 7.2: Consider the functor $G \mapsto I(G)$. There is a short exact sequence $\mathbf{r} \rightarrowtail \mathbf{f} \twoheadrightarrow I(G)$. Using the long exact sequence of higher limits and the fact that $\mathsf{lim}^{i}\mathbf{f} = 0$, we get the equality $\mathsf{lim}^{i}\ \mathbf{r} = \mathsf{lim}^{i-1}\ I(G)$. Thus we obtain a polynomial **fr**-code for $I(G)$:

$$I(G) \cong [\mathbf{r}].$$

The main point of this section is the fact that a lot of useful functors $\mathsf{Gr} \to \mathsf{Ab}$ and transformations between them have polynomial **fr**-codes. In particular, we will prove the following polynomial **fr**-codes:

$$I(G)/I(G)^n = [\mathbf{r} + \mathbf{f}^n],$$
$$G_{ab} = [\mathbf{r} + \mathbf{f}^2],$$
$$I(G)^{\otimes_{\mathbb{Z}[F]} n} = [\sum_{i=1}^{n} \mathbf{f}^{i-1}\mathbf{r}\mathbf{f}^{n-i}],$$
$$G_{ab}^{\otimes n} = [\sum_{i=1}^{n} \mathbf{f}^{i-1}\mathbf{r}\mathbf{f}^{n-i} + \mathbf{f}^{n+1}],$$
$$H_{2n}(G) = [\mathbf{f}\mathbf{r}^n + \mathbf{r}^n\mathbf{f}],$$
$$H_{2n-1}(G) = [\mathbf{r}^n + \mathbf{f}\mathbf{r}^{n-1}\mathbf{f}],$$
$$L_{n-i}\bigotimes^{n} G_{ab} = {}^{i}[\mathbf{r}^n + \mathbf{f}^{n+1}],$$
$$I(G)^{\otimes n} = {}^{n}[\mathbf{r}^n].$$

There are also non-polynomial codes for some functors:

$$I(G)^n = [\mathbf{r} \cap \mathbf{f}^n].$$

7.1. fr-*codes for homology of groups*

Proposition 7.3: *There are polynomial* **fr**-*codes for functors of homology* $H_* : \mathsf{Gr} \to \mathsf{Ab}$ *except* H_2:

$$H_{2n+2}(G) \cong [\mathbf{fr}^{n+1} + \mathbf{r}^{n+1}\mathbf{f}], \qquad H_{2n-1}(G) \cong [\mathbf{r}^n + \mathbf{fr}^{n-1}\mathbf{f}]$$

for $n \geq 1$.

Proof: Theorem 5.14 implies $H_{2n+2}(G) \cong \mathsf{lim}^0\, \mathbf{r}^{n+1}/(\mathbf{fr}^{n+1} + \mathbf{r}^{n+1}\mathbf{f})$ and $H_{2n+1}(G) = \mathsf{lim}^0\, \mathbf{fr}^n/(\mathbf{r}^{n+1} + \mathbf{fr}^n\mathbf{f})$. Applying the long exact sequence of limits and using $\mathsf{lim}^{0,1}\, \mathbf{r}^{n+1} = 0$ and $\mathsf{lim}^*\, \mathbf{fr}^n = 0$ we get the claimed isomorphisms. □

7.2. fr-*codes for* $L_i \bigotimes^n G_{ab}$

Recall that if $\mathcal{F} : \mathsf{Ab} \to \mathsf{Ab}$ and A is abelian simplicial group, then the Dold-Puppe derived functors $L_i\, \mathcal{F} : \mathsf{Ab} \to \mathsf{Ab}$ can be defined using a free abelian simplicial resolution $X_\bullet$ of A as follows

$$L_i\mathcal{F}(A) = \pi_i(\mathcal{F}(X_\bullet)).$$

Here we consider the derived functors of the functor of tensor power $\bigotimes^n$: $\mathsf{Ab} \to \mathsf{Ab}$ that sends A to $A^{\otimes n}$.

Let $A_1, \ldots, A_n$ be abelian groups. Consider a free resolution $P_i = (0 \to P_{i,1} \to P_{i,0} \to 0)$ of A_i. Then $\mathsf{Tor}_i(A_1, \ldots, A_n)$ is defined as follows

$$\mathsf{Tor}_i(A_1, \ldots, A_n) = H_i(P_1 \otimes \cdots \otimes P_n).$$

The Eilenberg-Zilber Lemma implies that

$$L_i \bigotimes^n A = \mathsf{Tor}_i(A, \ldots, A).$$

Proposition 7.4: *The functors* $L_i \bigotimes^{\otimes n} G_{ab}$ *have the following* **fr**-*codes*

$$L_{n-i} \bigotimes^n G_{ab} \cong {}^i[\mathbf{r}^n + \mathbf{f}^{n+1}].$$

Proof: Consider the resolution $\bar{R} \to F_{ab}$ of G_{ab}, where $\bar{R} = R[F,F]/[F,F]$. Then $H_i((\bar{R} \to F_{ab})^{\otimes n}) = L_i \bigotimes^n G_{ab}$. Consider the acyclic complex $(F_{ab} \to F_{ab})^{\otimes n}$ and cokernel of the embedding

$$(\bar{R} \to F_{ab})^{\otimes n} \rightarrowtail (F_{ab} \to F_{ab})^{\otimes n} \twoheadrightarrow C_\bullet.$$

Then $H_i(C_\bullet) = L_{i-1} \bigotimes^n G_{ab}$. Consider the spectral sequence of higher limits (Corollary 4.16) $\mathsf{lim}^i C_k \Rightarrow L_{k-i-1} \bigotimes^n G_{ab}$. If $0 \leq k < n$, then C_k is a direct sum of quotients of $F_{ab}^{\otimes n}$ by an abelian group of the form $A \otimes F_{ab} \otimes B$. Using Lemma 5.2 and long exact sequences of higher limits, we get $\mathsf{lim}^i C_k = 0$ for all i and $k \neq n$. It follows that $\mathsf{lim}^i C_n = L_{n-i-1} \bigotimes^n G_{ab}$. Note that $F_{ab} = \mathbf{f}/\mathbf{f}^2$, $\bar{R} = (\mathbf{r} + \mathbf{f}^2)/\mathbf{f}^2$, and hence,

$$F_{ab}^{\otimes n} \cong \mathbf{f}^n/\mathbf{f}^{n+1}, \ \bar{R}^{\otimes n} \cong (\mathbf{r}^n + \mathbf{f}^{n+1})/\mathbf{f}^{n+1}.$$

Then $C_n = F_{ab}^{\otimes n}/\bar{R}^{\otimes n} = \mathbf{f}^n/(\mathbf{r}^n + \mathbf{f}^{n+1})$. Since ${}^{i}[\mathbf{f}^n] = 0$, we obtain $\mathsf{lim}^i C_n = {}^{i+1}[\mathbf{r}^n + \mathbf{f}^{n+1}]$. □

7.3. *Other computations*

Proposition 7.5: *For any $n \geq 1$ there is a polynomial* **fr***-code for $I^{\otimes_{\mathbb{Z}G} n}$:*

$$I^{\otimes_{\mathbb{Z}G} n} \cong \left[\sum\nolimits_{i=1}^{n} \mathbf{f}^{i-1}\mathbf{r}\mathbf{f}^{n-i}\right].$$

Moreover, ${}^{i}[\sum \mathbf{f}^{i-1}\mathbf{r}\mathbf{f}^{n-i}] = 0$ for $i \neq 1$.

Proof: Since $I \cong \mathbf{f}/\mathbf{r}$, we get $I^{\otimes_{\mathbb{Z}[F]} n} \cong \mathbf{f}^n/(\sum \mathbf{f}^{i-1}\mathbf{r}\mathbf{f}^{n-i})$. Then using the short exact sequence $\sum \mathbf{f}^{i-1}\mathbf{r}\mathbf{f}^{n-i} \rightarrowtail \mathbf{f}^n \twoheadrightarrow I^{\otimes n}$ we obtain the required isomorphism. □

Proposition 7.6: *There is a polynomial* **fr***-code for the functor I/I^n:*

$$I/I^n \cong [\mathbf{r} + \mathbf{f}^n].$$

Moreover, ${}^{i}[\mathbf{r} + \mathbf{f}^n] = 0$ for $i \neq 1$, and the projection $I \twoheadrightarrow I/I^n$ is induced by the embedding $\mathbf{r} \subset \mathbf{r} + \mathbf{f}^n$.

Proof: This follows from the long exact sequence of higher limits applied to rows of the diagram

$$\begin{array}{ccccccccc} 0 & \longrightarrow & \mathbf{r} & \longrightarrow & \mathbf{f} & \longrightarrow & I(G) & \longrightarrow & 0 \\ & & \downarrow & & \| & & \downarrow & & \\ 0 & \longrightarrow & \mathbf{r} + \mathbf{f}^n & \longrightarrow & \mathbf{f} & \longrightarrow & I(G)/I(G)^n & \longrightarrow & 0. \end{array}$$

□

Proposition 7.7: *There is a polynomial* **fr***-code for the functor $G_{ab}^{\otimes n}$:*

$$G_{ab}^{\otimes n} = [\sum\nolimits_{i=1}^{n} \mathbf{f}^{i-1}\mathbf{r}\mathbf{f}^{n-i} + \mathbf{f}^{n+1}].$$

Moreover, ${}^{i}[\sum_{i=1}^{n} \mathbf{f}^{i-1}\mathbf{r}\mathbf{f}^{n-i} + \mathbf{f}^{n+1}] = 0$ for $i \neq 1$, and the map $I(G)^{\otimes n} \to G_{ab}^{\otimes n}$ is induced by the embedding $\sum_{i=1}^{n} \mathbf{f}^{i-1}\mathbf{r}\mathbf{f}^{n-i} \subset \sum_{i=1}^{n} \mathbf{f}^{i-1}\mathbf{r}\mathbf{f}^{n-i} + \mathbf{f}^{n+1}$.

Proof: Since $G_{ab} = \mathbf{f}/(\mathbf{r}+\mathbf{f}^2)$, we get $G_{ab}^{\otimes n} = \mathbf{f}^n/(\sum_{i=1}^n \mathbf{f}^{i-1}\mathbf{r}\mathbf{f}^{n-i} + \mathbf{f}^{n+1})$. Then the statement follows from the diagram

$$\begin{array}{ccccccccc} 0 & \longrightarrow & \sum_{i=1}^n \mathbf{f}^{i-1}\mathbf{r}\mathbf{f}^{n-i} & \longrightarrow & \mathbf{f}^n & \longrightarrow & I(G)^{\otimes_{\mathbb{Z}[F]}n} & \longrightarrow & 0 \\ & & \downarrow & & \| & & \downarrow & & \\ 0 & \longrightarrow & \sum_{i=1}^n \mathbf{f}^{i-1}\mathbf{r}\mathbf{f}^{n-i} + \mathbf{f}^{n+1} & \longrightarrow & \mathbf{f}^n & \longrightarrow & G_{ab}^{\otimes n} & \longrightarrow & 0. \end{array}$$ □

Now let's consider more non-standard example, namely that mentioned in the introduction. There is a **fr**-code

$$\mathsf{Tor}(G_{ab} \otimes G_{ab}, G_{ab}) = [\mathbf{rfr} + \mathbf{frr} + \mathbf{ffff}]. \tag{7.1}$$

To see (7.1), observe first that these is a natural isomorphism

$$G_{ab}^{\otimes 2} = \frac{\mathbf{f}^2}{\mathbf{rf} + \mathbf{fr} + \mathbf{f}^3}.$$

Hence, there is a short exact sequence with free abelian two left terms:

$$0 \to \frac{\mathbf{rf} + \mathbf{fr} + \mathbf{f}^3}{\mathbf{f}^3} \to \frac{\mathbf{f}^2}{\mathbf{f}^3} \to G_{ab}^{\otimes 2} \to 0.$$

Tensoring with $G_{ab} = \frac{\mathbf{f}}{\mathbf{r}+\mathbf{f}^2}$, we get the exact sequence

$$0 \to \mathsf{Tor}(G_{ab}^{\otimes 2}, G_{ab}) \to \frac{\mathbf{rf} + \mathbf{fr} + \mathbf{f}^3}{\mathbf{f}^3} \otimes \frac{\mathbf{f}}{\mathbf{r} + \mathbf{f}^2} \to \frac{\mathbf{f}^2}{\mathbf{f}^3} \otimes G_{ab} \to G_{ab}^{\otimes 3} \to 0$$

By Lemma 5.9,

$$\frac{\mathbf{rf} + \mathbf{fr} + \mathbf{f}^3}{\mathbf{f}^3} \otimes \frac{\mathbf{f}}{\mathbf{r} + \mathbf{f}^2} = \frac{\mathbf{rff} + \mathbf{frf} + \mathbf{f}^4}{\mathbf{rfr} + \mathbf{frr} + \mathbf{f}^4}$$

Now observe that $\mathsf{lim}^i \frac{\mathbf{f}^2}{\mathbf{f}^3} \otimes G_{ab} = \mathsf{lim}^i F_{ab}^{\otimes 2} \otimes G_{ab} = 0,\ i \geq 0$. Hence

$$\mathsf{Tor}(G_{ab}^{\otimes 2}, G_{ab}) = \mathsf{lim} \frac{\mathbf{rff} + \mathbf{frf} + \mathbf{f}^4}{\mathbf{rfr} + \mathbf{frr} + \mathbf{f}^4}.$$

Now consider the exact sequence

$$0 \to \frac{\mathbf{rff} + \mathbf{frf} + \mathbf{f}^4}{\mathbf{rfr} + \mathbf{frr} + \mathbf{f}^4} \to \frac{\mathbf{f}^3}{\mathbf{rfr} + \mathbf{frr} + \mathbf{f}^4} \to \frac{\mathbf{f}^3}{\mathbf{rff} + \mathbf{frf} + \mathbf{f}^4} \to 0.$$

By Lemma 5.9,

$$\frac{\mathbf{f}^3}{\mathbf{rff} + \mathbf{frf} + \mathbf{f}^4} = \frac{\mathbf{f}^2}{\mathbf{rf} + \mathbf{fr} + \mathbf{f}^3} \otimes \frac{\mathbf{f}}{\mathbf{f}^2} = G_{ab}^{\otimes 2} \otimes F_{ab},$$

hence

$$\mathsf{lim}^i \frac{\mathbf{f}^3}{\mathbf{rff}+\mathbf{frf}+\mathbf{f}^4} = 0,\ i \geq 0.$$

Therefore,

$$\mathsf{Tor}(G_{ab}^{\otimes 2}, G_{ab}) = \mathsf{lim} \frac{\mathbf{rff}+\mathbf{frf}+\mathbf{f}^4}{\mathbf{rfr}+\mathbf{frr}+\mathbf{f}^4} = \mathsf{lim}^1(\mathbf{rff}+\mathbf{frf}+\mathbf{f}^4)$$

and the required isomorphism (7.1) is proved.

Some other examples of description of limits can be found in [4] and [11]. Here we give some examples (see [4], [11])[a]:

$$\begin{aligned}
\mathsf{lim} \frac{[F,F]}{[R,R][F,F,F]} &= L_1 S^2(G_{ab}),\\
\mathsf{lim} \frac{[F,F,F]}{[R,R,R][F,F,F,F]} &= L_2 \mathfrak{L}_s^3(G_{ab}),\\
\mathsf{lim} \frac{[F,F,F]}{[R,R,F][F,F,F,F]} &= L_1 S^3(G_{ab}),\\
\mathsf{lim} \frac{[F,F,F]}{[F,F,R][F,F,F,F]} &= 0,\\
\mathsf{lim} \frac{[R,R]}{[R,R,F]} &= H_2(G, S^2(I(G))),
\end{aligned}$$

where $S^2, S^3, \mathfrak{L}_s^3$ are functors of symmetric square, cube and super-Lie-cube respectively. If G is 2-torsion free then, as it is shown in [4],

$$\begin{aligned}
\mathsf{lim} \frac{[R,R]}{[R,R,F]} &= H_4(G;\mathbb{Z}/2),\\
\mathsf{lim}^1 \frac{[R,R]}{[R,R,F]} &= H_3(G;\mathbb{Z}/2).
\end{aligned}$$

8. Intersection games

Let $\mathbf{a}_1, \dots, \mathbf{a}_n$, $n \geq 2$ be ideals in $\mathbb{Z}[F]$ which can be presented as monomials in $\mathbf{f}, \mathbf{r}$. There is a natural problem: define a minimal set $\mathbf{b}_1, \dots, \mathbf{b}_m$, $m \geq 1$, of monomials in $\mathbf{f}, \mathbf{r}$ such that:

1) All of them lie in the intersection $\mathbf{a}_1 \cap \dots \cap \mathbf{a}_n$;

[a] We assume that the commutator brackets are left-normalized. For example, $[R,R,F] := [[R,R],F]$.

2) Any monomial $\mathbf{b}$ in $\mathbf{f}, \mathbf{r}$ which lies in $\mathbf{a}_1 \cap \cdots \cap \mathbf{a}_n$ also lies in some $\mathbf{b}_i$ as an ideal.

After constructing the set of monomials $\mathbf{b}_1, \ldots, \mathbf{b}_m$, if $m \geq 2$, one can repeat this process. In this way we obtain what we will call *generations* of collections of ideals.

Here is our main example. Take two ideals $\mathbf{r}, \mathbf{f}^2$. One can easily check that any monomial written in terms of $\mathbf{f}$ and $\mathbf{r}$ which lies in $\mathbf{r} \cap \mathbf{f}^2$, lies either in $\mathbf{rf}$ or in $\mathbf{fr}$ (or in both). Now we take monomials from the intersection $\mathbf{rf} \cap \mathbf{fr}$ and see that the minimal set of monomials which covers all of them is $\mathbf{r}^2, \mathbf{frf}$. Repeating this process, we obtain the following *generations*:

$$(\mathbf{r}, \mathbf{f}^2) \to (\mathbf{rf}, \mathbf{fr}) \to (\mathbf{r}^2, \mathbf{frf}) \to (\mathbf{r}^2\mathbf{f}, \mathbf{fr}^2) \to (\mathbf{r}^3, \mathbf{fr}^2\mathbf{f}) \to \ldots \tag{8.1}$$

In this example, we obtain the following isomorphism which follows from Gruenberg resolution:

$$\frac{\text{intersection of ideals in } i\text{th generation}}{\text{sum of ideals in } (i+1)\text{st generation}} = H_{i+1}(G).$$

We collect some examples of origins and generators in the following table:

origin	r, ff	rr, fff	r, fff
2 generation	rf, fr	rrf, rfr, frr	rff, frf, ffr
3 generation	rr, frf	rrr, frfrf	rrr, frrf, rfrf, frfr, ffrff
4 generation	rrf, frr	frrrf, frfrr, rrfrf	rrrfr, rfrrr, frrrf, rrfrff, rfrfrf, frfrfr, ffrfrr
5 generation	rrr, frrf	rrrrr, rrfrrf, rrfrrf, frrrrf, frrfrr	rrrrrr, rrrfrfr, rrfrfrr, rfrrrfr, rfrfrrr, frrrrrf, rfrfrfrf, frfrfrfr, ffrfrrfrff

In the above example (8.1), we see that

$$\begin{aligned} &\mathsf{lim}^{j+1} \text{ (sum of ideals in the } (i+1)\text{th generation)} \\ &\quad = \mathsf{lim}^{j} \text{ (sum of ideals in the } i\text{th generation)} \end{aligned} \tag{8.2}$$

for $j \geq 1$. The reason for that connection is simple. All generations consist of two ideals, say $(\mathbf{a}_i, \mathbf{b}_i)$, $i = 1, 2, \ldots$ and there are natural short exact sequences:

$$0 \to \frac{\mathbf{a}_i \cap \mathbf{b}_i}{\mathbf{a}_{i+1} + \mathbf{b}_{i+1}} \to \frac{\mathbf{f}}{\mathbf{a}_{i+1} + \mathbf{b}_{i+1}} \to \frac{\mathbf{f}}{\mathbf{a}_i} \oplus \frac{\mathbf{f}}{\mathbf{b}_i} \to \frac{\mathbf{f}}{\mathbf{a}_i + \mathbf{b}_i} \to 0$$

Now (8.2) follows from the following properties:

$$\mathsf{lim}^j \frac{\mathbf{a}_i \cap \mathbf{b}_i}{\mathbf{a}_{i+1} + \mathbf{b}_{i+1}} = 0, \ j > 0, \ \mathsf{lim}^j \left(\frac{\mathbf{f}}{\mathbf{a}_i} \oplus \frac{\mathbf{f}}{\mathbf{b}_i} \right) = 0, \ j \geq 0.$$

It would be interesting to explore general connections between derived limits and the above intersection game.

9. Table

In the next table we summarize the computations of derived limits for certain **fr**-sentences. In this table we will denote the augmentation ideal of G by $\mathfrak{g}$.

fr-code	$\lim^1$	$\lim^2$	$\lim^3$	$\lim^4$
f	0	0	0	0
r	$\mathfrak{g}$	0	0	0
rr	0	$\mathfrak{g} \otimes \mathfrak{g}$	0	0
rrr	0	0	$\mathfrak{g} \otimes \mathfrak{g} \otimes \mathfrak{g}$	0
rrrr	0	0	0	$\mathfrak{g} \otimes \mathfrak{g} \otimes \mathfrak{g} \otimes \mathfrak{g}$
fr+rf	$\mathfrak{g} \otimes_{\mathbb{Z}[G]} \mathfrak{g}$	0	0	0
ffr+frf+rff	$\mathfrak{g} \otimes_{\mathbb{Z}[G]} \mathfrak{g} \otimes_{\mathbb{Z}[G]} \mathfrak{g}$	0	0	0
r+ff	G_{ab}	0	0	0
r+fff	$\mathfrak{g}/\mathfrak{g}^3$	0	0	0
rf+ffr	$\mathfrak{g}^2 \otimes_{\mathbb{Z}[G]} \mathfrak{g}$	0	0	0
rf+fffr	$\mathfrak{g}^3 \otimes_{\mathbb{Z}[G]} \mathfrak{g}$	0	0	0
rfr+frr+ffff	$\mathsf{Tor}(G_{ab} \otimes G_{ab}, G_{ab})$	0	0	0
fr+rf+fff	$G_{ab} \otimes G_{ab}$	0	0	0
rff+frf+rff+ffff	$G_{ab} \otimes G_{ab} \otimes G_{ab}$	0	0	0
rr+fff	$\mathsf{Tor}(G_{ab}, G_{ab})$	$G_{ab} \otimes G_{ab}$	0	0
rrr+ffff	$L_2 \otimes^3 (G_{ab})$	$L_1 \otimes^3 (G_{ab})$	$G_{ab} \otimes G_{ab} \otimes G_{ab}$	0
rrrr+fffff	$L_3 \otimes^4 (G_{ab})$	$L_2 \otimes^4 (G_{ab})$	$L_1 \otimes^4 (G_{ab})$	$G_{ab}^{\otimes 4}$
rr+frf	$H_3(G)$	$\mathfrak{g} \otimes_{\mathbb{Z}[G]} \mathfrak{g}$	0	0
rrf+frr	$H_4(G)$	$\mathfrak{g} \otimes_{\mathbb{Z}[G]} \mathfrak{g} \otimes_{\mathbb{Z}[G]} \mathfrak{g}$	0	0
rrr+frrf	$H_5(G)$	$H_4(G)$	$\mathfrak{g} \otimes_{\mathbb{Z}[G]} \mathfrak{g} \otimes_{\mathbb{Z}[G]} \mathfrak{g}$	0
rf+ffr+ffff	$\mathfrak{g}^2/\mathfrak{g}^3 \otimes G_{ab}$	0	0	0
rfff+rfr+rrf	0	$\mathfrak{g} \otimes G_{ab} \otimes G_{ab}$	0	0
rrfff+rrfr+rrrf	0	0	$\mathfrak{g} \otimes \mathfrak{g} \otimes G_{ab} \otimes G_{ab}$	0

Acknowledgements

This text is based on lectures given by the authors in the summer of 2015. The authors thank F. Pavutnitsky for discussions related to the subject of these notes. The research is supported by Saint-Petersburg State University research grant N 6.37.208.2016 and by JSC "Gazprom Neft".

References

1. A. Jon Berrick, F. R. Cohen, E. Hanbury, Y.-L. Wong, J. Wu, *Braids: Introductory lectures on braids, configurations and their applications*, volume 19

of Lecture Notes Series. Institute for Mathematical Sciences. National University of Singapore. World Scientific Publishing Co. Pte. Ltd., Hackensack, NJ, 2010.
2. N. Bourbaki, "Univers", Appendix to *Expose I* in SGA 4.
3. I. Emmanouil and R. Mikhailov, "A limit approach to group homology", *J. Algebra* **319** (2008), 1450–1461.
4. S. O. Ivanov and R. Mikhailov, "A higher limit approach to homology theories", *J. Pure Appl. Algebra*, **219** (2015), 1915–1939.
5. P. Gabriel and M. Zisman, *Calculus of fractions and homotopy theory*, Springer Verlag, New York, 1967.
6. P. Goerss, J. F. Jardine, *Simplicial homotopy theory*, Progress in Mathematics, Vol. 174, Birkhauser, Basel, 1999.
7. S. Jackowski, J. Słomińska, "G-functors, G-posets and homotopy decompositions of G-spaces", *Fund. Math.* **169** (2001), 249–287.
8. S. Mac Lane, *Categories for the working mathematician*, (Graduate Texts in Mathematics 5), Springer Verlag, 2nd edition (1998).
9. J. P. May, *Simplicial Objects in Algebraic Topology*, Van Nostrand, Princeton, 1968.
10. R. Mikhailov and I. B. S. Passi, "Limits over categories of extensions", *Ind. J. Pure Appl. Math.* **41** (2010), 113–131, arXiv: 0904.0634.
11. R. Mikhailov and I. B. S. Passi, "Generalized dimension subgroups and derived functors", to appear *J. Pure Appl. Alg.*, arxiv:1506.04400.
12. D. Quillen, "Cyclic cohomology and algebra extensions", *K-Theory* **3**, (1989), 205–246.
13. M. Shulman, "Set theory for category theory", arxiv:0810.1279.
14. A. Tarski "On well-ordered subsets of any set", *Fundamenta Mathematicae*, 32:176-183, (1939).
15. C. A. Weibel, *An introduction to homological algebra*, Cambridge Stud. Adv. Math. **38**, Cambridge University Press, Cambridge, 1994.

CONFIGURATION SPACES AND ROBOT MOTION PLANNING ALGORITHMS

Michael Farber

School of Mathematical Sciences
Queen Mary, University of London
Mile End Road, London E1 4NS, UK
m.farber@qmul.ac.uk

These notes survey topological problems relevant to the motion planning problem of robotics and include some new results and constructions. First we analyse the notion of topological complexity of configuration spaces which is responsible for discontinuities in algorithms for robot navigation. Then we present explicit motion planning algorithms for coordinated collision free control of many particles moving in Euclidean spaces or on graphs. These algorithms are optimal in the sense that they have minimal number of regions of continuity. Moreover, we describe in full detail the topology of configuration spaces of two particles on a tree and use it to construct some top-dimensional cohomology classes in configuration spaces of n particles on a tree.

Contents

1. Introduction

These notes start with a survey of the topological approach to the motion planning problem complementing [13] and chapter 4 of [16]. In §2 – §7 we present a general description of the method and some basic results.

In §8 and §9 we analyse in full detail motion planning algorithms for collision free motion of many particles moving in the Euclidean spaces $\mathbb{R}^d$. Problems of this kind appear in many areas of engineering when multiple objects have to be moved in a coordinated way from one state to another avoiding collisions. The motion planning algorithms presented here are *optimal* in the sense that they have minimal topological complexity (equal to $2n - 1$ or $2n - 2$ depending on the parity of the dimension d, where n is the number of moving objects). The motion planning algorithms suggested in [13] had topological complexity quadratic in n. A recent paper [25] proposed a motion planning algorithm for n particles moving on the plane $\mathbb{R}^2$ having complexity $2n - 1$. The algorithms presented here are inspired by the construction of [25].

In §10 we analyse the topology of configuration spaces of graphs and present (following [10]) a motion planning algorithm for collision free control of n particles on a tree. In §11 we describe explicitly the configuration space $F(\Gamma, 2)$ of two particles on a tree proving the main Theorem 11.1 in full detail; this theorem was stated in [10] without proof.

Theorem 10.3 claims that the topological complexity of collision free motion of many particles on a tree is independent of the number of moving particles and depends only on the number of essential vertices of Γ. This fact contrasts the corresponding result for Euclidean spaces where the complexity is linear in n. The proof of Theorem 10.3 is completed in §13 after an analysis of the top-dimensional cohomology of configuration spaces of trees which is carried out in §12. Theorem 10.3 was stated in [10] without proof. Recently S. Scheirer [27] published a detailed proof of a similar result under some additional assumptions.

In §14 we give some further comments and most recent literature references.

2. Motion planning algorithms and topological complexity of configuration spaces

2.1. *Motion planning algorithms*

Any mechanical system S possesses a variety of states $C(S)$, called *the configuration space.* As an example, one may imagine a robot whose state consists of its location in the 3-space as well as the mutual positions of all its body parts such as elbows, knees, fingers etc.

We want to programme our system S so that it is capable of moving *autonomously* from any initial state $A \in C(S)$ to any final state $B \in C(S)$. Such programme is *a motion planning algorithm.* Once a motion planning algorithm has been specified, we may simply *order* our system to move to a new state B, and the motion planning algorithm will prescribe *how* the system will implement the motion departing from the current state A.

A state of the system is typically described by a collection of numerical parameters which can be interpreted as coordinates of a point in $\mathbb{R}^N$. The variety of all states of the system is then represented by a subset $C(S) \subset \mathbb{R}^N$; we see that the configuration space of the system comes naturally with a topology. The topology of the configuration space $C(S)$ is important since motions of the system are represented by *continuous* paths in $C(S)$.

We refer to [24] and [29] for additional information about motion planning in robotics.

2.2. *The concept of* $\mathsf{TC}(X)$

We shall study a topological invariant $\mathsf{TC}(X)$ of a topological space X, originally introduced in [8], see also [9] and [13]. It is a numerical homotopy invariant inspired by the robot motion problem, similar in spirit to the classical Lusternik–Schnirelmann category $\mathsf{cat}(X)$. Intuitively, $\mathsf{TC}(X)$ is a measure of the navigational complexity of X viewed as the configuration space of a system. $\mathsf{TC}(X)$, as well as $\mathsf{cat}(X)$, are special cases of a more general notion of the genus of a fibration introduced by A. Schwarz [28].

Next we give the formal definitions. Let X denote a topological space thought of as the configuration space of a mechanical system. The states of the system are represented by the points of X, and continuous motions of the system are represented by continuous paths $\gamma : [0,1] \to X$. Here the point $A = \gamma(0)$ represents the initial state and $\gamma(1) = B$ represents the final state of the system. The space X is path connected if and only if the system can be brought to an arbitrary state from any given state by a continuous motion.

Denote by $PX = X^I$ the space of all continuous paths $\gamma : I = [0,1] \to X$. The space PX is supplied with the compact-open topology, see [30], which is characterised by the property that a map $Z \to X^I$ is continuous if and only if the associated map $Z \times I \to X$ is continuous. Let

$$\pi : PX \to X \times X \tag{2.1}$$

be the map which assigns to a path γ the pair $(\gamma(0), \gamma(1)) \in X \times X$ of the initial-final configurations. It is easy to see that π is a fibration in the sense of Hurewicz, see [30], Chapter 2, §8, Corollary 3.

A motion planning algorithm is a section of the fibration π. In other words a motion planning algorithm is a map (not necessarily continuous)

$$s : X \times X \to PX$$

satisfying $\pi \circ s = 1_{X \times X}$.

A motion planning algorithm $s : X \times X \to PX$ is *continuous* if the suggested route $s(A, B)$ of going from A to B depends continuously on the states A and B. A continuous motion planning algorithm in X exists if and only if the space X is contractible, see [16], Lemma 4.2. Thus, for a system with a non-contractible configuration space any motion planning algorithm must be discontinuous.

Definition 2.1: Given a path-connected topological space X, we define the topological complexity of X as the minimal number $\mathsf{TC}(X) = k$ such that the Cartesian product $X \times X$ may be covered by k open subsets $X \times X = U_1 \cup U_2 \cup \cdots \cup U_k$, such that for any $i = 1, 2, ..., k$, there exists a continuous section $s_i : U_i \to PX$, $\pi \circ s_i = \mathsf{id}$ over U_i. If no such k exists we will set $\mathsf{TC}(X) = \infty$.

Example 2.2: Suppose we are to construct a motion planning algorithm on the circle $X = S^1$. Given two points $A, B \in S^1$, which are not antipodal, i.e. $B \neq -A$, we may move from A to B along the shortest geodesic curve $s_1(A, B)$ which is unique and depends continuously on A and B. This defines a continuous section $s_1 : U_1 \to (S^1)^I$, where $U_1 \subset S^1 \times S^1$ denotes the set $U_1 = \{(A, B) \in S^1 \times S^1; A \neq -B\}$.

However, if the points A and B are antipodal then there are two distinct shortest geodesic curves from A to B so that the section s_1 does not extend to a continuous section over the whole product $S^1 \times S^1$.

Denote $U_2 = \{(A, B) \in S^1 \times S^1; A \neq B\} \subset S^1 \times S^1$. We may define a continuous section $s_2 : U_2 \to (S^1)^I$ by setting $s_2(A, B)$ to be the path moving from A to B in the clockwise direction along the circle with constant velocity. Again, we observe that the section s_2 cannot be extended to a continuous section on the whole space $S^1 \times S^1$.

The open sets U_1, U_2 cover $S^1 \times S^1$ and therefore $\mathsf{TC}(S^1) \leq 2$ according to Definition 2.1. On the other hand, since the circle S^1 is not contractible we know that $\mathsf{TC}(S^1) > 1$. Therefore $\mathsf{TC}(S^1) = 2$.

2.3. *Homotopy invariance*

Next we show that the topological complexity $\mathsf{TC}(X)$ depends only on the homotopy type of X. We start with the following auxiliary statement.

Theorem 2.3: *Let X and Y be topological spaces. Suppose that X dominates Y, i.e. there exist continuous maps $f : X \to Y$ and $g : Y \to X$ such that $f \circ g \sim \mathsf{id}_Y$. Then $\mathsf{TC}(Y) \leq \mathsf{TC}(X)$.*

Proof: Assume that $U \subset X \times X$ is an open subset such that there exists a continuous section $s : U \to PX$ of (2.1) over U. Define $V = (g \times g)^{-1}(U) \subset Y \times Y$. We may construct a continuous section $\sigma : V \to PY$ over V as follows. Fix a homotopy $h_t : Y \to Y$ with $h_0 = \mathsf{id}_Y$ and $h_1 = f \circ g$; here $t \in [0, 1]$. For $(A, B) \in V$ and $t \in [0, 1]$ set

$$\sigma(A,B)(\tau) = \begin{cases} h_{3\tau}(A), & \text{for} \quad 0 \leq \tau \leq 1/3, \\ \\ f(s(gA, gB)(3\tau - 1)), & \text{for} \quad 1/3 \leq \tau \leq 2/3, \\ \\ h_{3(1-\tau)}(B), & \text{for} \quad 2/3 \leq \tau \leq 1. \end{cases}$$

Thus we obtain that for $k = \mathsf{TC}(X)$, any open cover $U_1 \cup \cdots \cup U_k = X \times X$ with a continuous section $s_i : U_i \to PX$ over each U_i defines an open cover

$V_1 \cup \cdots \cup V_k$ of $Y \times Y$, where each $V_i = (g \times g)^{-1}(U_i)$ admits a continuous section $\sigma_i : V_i \to PY$. This proves that $\mathsf{TC}(Y) \leq \mathsf{TC}(X)$. □

Corollary 2.4: *If X and Y are homotopy equivalent then* $\mathsf{TC}(X) = \mathsf{TC}(Y)$.

3. Upper and lower bounds for $\mathsf{TC}(X)$

3.1. *The upper bound*

We start with a dimensional upper bound.

Theorem 3.1: *For any path-connected paracompact locally contractible topological space X one has*

$$\mathsf{TC}(X) \leq \dim(X \times X) + 1. \tag{3.1}$$

Here $\dim(X \times X)$ denotes the covering dimension of $X \times X$.

Recall that $\dim(Y) \leq n$ if any open cover of Y has a locally finite open refinement such that no point of Y belongs to more than $n+1$ open sets of the refinement. If Y is a polyhedron, then $\dim(Y)$ coincides with the maximum of the dimensions of the simplices of Y.

A topological space Y is called *locally contractible* if any point of Y has an open neighbourhood $U \subset Y$ such that the inclusion $U \to Y$ is null-homotopic.

Proof: Suppose $\dim(X \times X) = n$. Let $\mathcal{U} = \{U_i\}_{i \in I}$ be an open cover of $X \times X$ such that each open set $U_i \subset X \times X$ admits a continuous section $s_i : U_i \to PX$, where $i \in I$. Such a cover exists since X is locally contractible. Let $\mathcal{V} = \{V_j\}_{j \in J}$ be a refinement of $\mathcal{U}$ having multiplicity $\leq n+1$, i.e. for every $(x,y) \in X \times X$ there exist at most $n+1$ values of $j \in J$ such that $(x,y) \in V_j$. Construct a partition of unity $\{h_j\}_{j \in J}$ subordinate to $\mathcal{V}$, i.e. each $h_j : X \times X \to [0,1]$ is continuous, $\mathrm{supp}(h_j) \subset V_j$ and $\sum_{j \in J} h_j = 1_{X \times X}$. Given a subset $S \subset J$ define

$$W(S) \subset X \times X$$

as the set of all pairs (x,y) such that $h_j(x,y) > h_k(x,y)$ for all $j,k \in J$ satisfying $j \in S$ and $k \notin S$. Each set $W(S)$ is open and admits a continuous section $W(S) \to PX$. Besides, $W(S) = \emptyset$ for $|S| > n+1$ and the family $\{W(S); |S| \leq n+1\}$ is an open cover of $X \times X$. If $S, S' \subset J$ are two subsets such that none of them contains the other, i.e. there is $j \in S$, $j \notin S'$

and there exists $k \in S'$, $k \notin S$, then the intersection $W(S) \cap W(S') = \emptyset$. Therefore the union

$$W_k = \bigcup_{|S|=k} W(S),$$

is open and admits a continuous section $W_k \to PX$, where $k = 1, 2, \ldots, n+1$. We obtain an open cover $\{W_1, \ldots, W_{n+1}\}$ of $X \times X$ with the desired properties implying that $\mathsf{TC}(X) \leq n+1$. □

3.2. *The lower bound*

Next we give a lower bound for $\mathsf{TC}(X)$ which depends on the structure of the cohomology algebra of X.

Let $\mathbf{k}$ be a field. The singular cohomology $H^*(X;\mathbf{k})$ is a graded $\mathbf{k}$-algebra with the multiplication

$$\cup : H^*(X;\mathbf{k}) \otimes H^*(X;\mathbf{k}) \to H^*(X;\mathbf{k}) \tag{3.2}$$

given by the cup-product, see [23]. For two cohomology classes $u \in H^i(X;\mathbf{k})$ and $v \in H^j(X;\mathbf{k})$ we shall denote their cup-product by

$$u \cup v = uv \in H^{i+j}(X;\mathbf{k}).$$

The tensor product $H^*(X;\mathbf{k}) \otimes H^*(X;\mathbf{k})$ is also a graded $\mathbf{k}$-algebra with the multiplication

$$(u_1 \otimes v_1) \cdot (u_2 \otimes v_2) = (-1)^{|v_1| \cdot |u_2|}\, u_1 u_2 \otimes v_1 v_2. \tag{3.3}$$

Here $|v_1|$ and $|u_2|$ denote the degrees of cohomology classes v_1 and u_2 correspondingly. The cup-product (3.2) is an algebra homomorphism.

Definition 3.2: The kernel of homomorphism (3.2) is called *the ideal of the zero-divisors* of $H^*(X;\mathbf{k})$. The *zero-divisors-cup-length* of $H^*(X;\mathbf{k})$ is the length of the longest nontrivial product under the multiplication (3.3) in the ideal of the zero-divisors of $H^*(X;\mathbf{k})$.

Theorem 3.3: *The topological complexity of motion planning* $\mathsf{TC}(X)$ *is greater than the zero-divisors-cup-length of* $H^*(X;\mathbf{k})$.

Proof: Let $\Delta_X \subset X \times X$ denote the diagonal. First we observe that the kernel of the induced homomorphism $\pi^* : H^j(X \times X;\mathbf{k}) \to H^j(PX;\mathbf{k})$ coincides with the set of cohomology classes $u \in H^j(X \times X;\mathbf{k})$ such that

$$u|_{\Delta_X} = 0 \in H^j(X;\mathbf{k}).$$

Let $\alpha : X \to PX$ be the map which associates to any point $x \in X$ the constant path $[0,1] \to X$ at this point. Note that α is a homotopy equivalence and the composition $\pi \circ \alpha : X \to X \times X$ is the inclusion onto the diagonal Δ_X and thus our statement follows.

Next we note that the composition

$$H^*(X;\mathbf{k}) \otimes H^*(X;\mathbf{k}) \simeq H^*(X \times X;\mathbf{k}) \xrightarrow{\pi^*} H^*(PX;\mathbf{k}) \underset{\simeq}{\xrightarrow{\alpha^*}} H^*(X;\mathbf{k})$$

coincides with the cup-product homomorphism (3.2) where the homomorphism on the left is the Künneth isomorphism.

Combining these two remarks we obtain that a cohomology class

$$u = \sum_r a_r \times b_r \in H^*(X \times X;\mathbf{k})$$

satisfies $\pi^* u = 0$ if and only if the tensor

$$\sum_r a_r \otimes b_r \in H^*(X;\mathbf{k}) \otimes H^*(X;\mathbf{k})$$

is a zero-divisor.

Suppose that $u_1, \ldots, u_s \in H^*(X \times X;\mathbf{k})$ are cohomology classes satisfying $\pi^*(u_j) = 0$, for $j = 1, \ldots, s$, and such that their cup-product

$$0 \neq u_1 \cup \cdots \cup u_s \in H^*(X \times X;\mathbf{k})$$

is nonzero. We claim that the topological complexity $\mathsf{TC}(X)$ must satisfy $\mathsf{TC}(X) \geq s+1$. Indeed, suppose that $\mathsf{TC}(X) \leq s$, i.e. one may find an open cover $U_1, \ldots, U_s$ of $X \times X$ with each open set U_i admitting a continuous section $\sigma_i : U_i \to PX$. We have

$$u_i|_{U_i} = \sigma_i^* \circ \pi^*(u_i) = 0$$

and from the long exact cohomology sequence of the pair $(X \times X, U_i)$ one obtains that there exists a relative cohomology class $\tilde{u}_i \in H^*(X \times X, U_i;\mathbf{k})$ such that

$$u_i = \tilde{u}_i|_{\Delta_X}, \quad i = 1, \ldots, s. \tag{3.4}$$

Thus we see that the nontrivial product $u_1 u_2 \ldots u_s$ equals $(\tilde{u}_1 \tilde{u}_2 \ldots \tilde{u}_s)|_{\Delta_X}$; however the product $\tilde{u}_1 \tilde{u}_2 \ldots \tilde{u}_s$ lies in the trivial group $H^*(X \times X, \cup_i U_i;\mathbf{k}) = H^*(X \times X, X \times X;\mathbf{k}) = 0$ contradicting our assumption that $u_1 u_2 \ldots u_s \neq 0$. □

Example 3.4: Let $X = S^n$. Let $u \in H^n(S^n;\mathbf{k})$ be the fundamental class, and let $1 \in H^0(S^n;\mathbf{k})$ be the unit. Then the class $a = 1 \otimes u - u \otimes 1 \in$

$H^*(S^n;\mathbf{k}) \otimes H^*(S^n;\mathbf{k})$ is a zero-divisor, since applying the homomorphism (3.2) to it we obtain $1 \cdot u - u \cdot 1 = 0$. Another zero-divisor is $b = u \otimes u$, since $u^2 = 0$. Computing $a^2 = a \cdot a$ by means of rule (3.3) we find

$$a^2 = ((-1)^{n-1} - 1) \cdot u \otimes u.$$

Hence $a^2 = -2b$ for n even and $a^2 = 0$ for n odd; the product ab vanishes for any n. We conclude that *the zero-divisors-cup-length of $H^*(S^n;\mathbb{Q})$ is greater than or equal to 1 for n odd and is greater than or equal to 2 for n even.*

Applying Theorem 3.3 we find that $\mathsf{TC}(S^n) > 1$ for n odd and $\mathsf{TC}(S^n) > 2$ for n even. This means that any motion planner on the sphere S^n must have at least two open sets U_i; moreover, if n is even, any motion planner on the sphere S^n must have at least three open sets U_i. It is easy to construct a motion planner of S^n having two open sets for n odd and three open sets for n even; compare Example 6.3. This gives

$$\mathsf{TC}(S^n) = \begin{cases} 2, & \text{for } n \text{ odd}, \\ 3, & \text{for } n \text{ even}. \end{cases} \tag{3.5}$$

Example 3.5: Let X be a connected finite graph with $b_1(X) > 1$. Then there exist two linearly independent cohomology classes $u_1, u_2 \in H^1(X;\mathbb{Q})$. So, for $i = 1, 2$, the tensors $1 \otimes u_i - u_i \otimes 1$ are zero-divisors and their product equals $u_2 \otimes u_1 - u_1 \otimes u_2 \neq 0$. Hence by Theorem 3.3 we have $\mathsf{TC}(X) \geq 3$. On the other hand, applying Theorem 3.1 we obtain $\mathsf{TC}(X) \leq 3$. Therefore, $\mathsf{TC}(X) = 3$.

4. Simultaneous control of several objects

Suppose that we have a system which is a union of two independent systems S_1 and S_2 such that S_1 and S_2 can move independently without interaction. For example one may imagine the situation that an operator has to control two robots confined to two different rooms in the house simultaneously. If X_i denotes the configuration space of the system S_i, where $i = 1, 2$, then the configuration space of our entire system is the Cartesian product $X_1 \times X_2$, the variety of all pairs of states (x_1, x_2) where $x_1 \in X_1$ and $x_2 \in X_2$.

Note that in the case of two robots operating in the same room we would have to exclude from the product $X_1 \times X_2$ the set of all pairs of configurations (x_1, x_2) where the robots collide; thus, in this case the actual configuration space will be a suitable subspace of the product $X_1 \times X_2$.

4.1. *The product inequality*

Theorem 4.1: *For path–connected metric spaces X and Y one has*

$$\mathsf{TC}(X \times Y) \leq \mathsf{TC}(X) + \mathsf{TC}(Y) - 1. \tag{4.1}$$

Proof: Suppose $\mathsf{TC}(X) = n$, $\mathsf{TC}(Y) = m$. Let $U_1, \ldots, U_n$ be an open cover of $X \times X$ with a continuous section $s_i : U_i \to PX$ for $i = 1, \ldots, n$. Let $f_i : X \times X \to \mathbb{R}$, where $i = 1, \ldots, n$, be a partition of unity subordinate to the cover $\{U_i\}$. Similarly, let $V_1, \ldots, V_m$ be an open cover of $Y \times Y$ with a continuous section $\sigma_j : V_j \to PY$ for $j = 1, \ldots, m$, and let $g_j : Y \times Y \to \mathbb{R}$, where $j = 1, \ldots, m$, be a partition of unity subordinate to the cover $\{V_j\}$.

For any pair of nonempty subsets $S \subset \{1, \ldots, n\}$ and $T \subset \{1, \ldots, m\}$, let

$$W(S,T) \subset (X \times Y) \times (X \times Y)$$

denote the set of all 4-tuples $(A, B, C, D) \in (X \times Y) \times (X \times Y)$, such that for any $(i, j) \in S \times T$

$$f_i(A, C) \cdot g_j(B, D) > 0$$

is satisfied, and for any $(i', j') \notin S \times T$,

$$f_i(A, C) \cdot g_j(B, D) > f_{i'}(A, C) \cdot g_{j'}(B, D).$$

One easily checks that:

(a) each set $W(S,T) \subset (X \times Y) \times (X \times Y)$ is open;

(b) $W(S,T)$ and $W(S',T')$ are disjoint if neither $S \times T \subset S' \times T'$ nor $S' \times T' \subset S \times T$;

(c) if $(i, j) \in S \times T$, then $W(S,T)$ is contained in $U_i \times V_j$; therefore there exists a continuous motion planning algorithm over each $W(S,T)$ (it can be described explicitly in terms of s_i and σ_j);

(d) the sets $W(S,T)$ (with all possible nonempty S and T) form a cover of $(X \times Y) \times (X \times Y)$.

Let us prove (d). Suppose that $(A, B, C, D) \in (X \times Y) \times (X \times Y)$. Let S be the set of all indices $i \in \{1, \ldots, n\}$, such that $f_i(A, C)$ equals the maximum of $f_k(A, C)$, where $k = 1, 2, \ldots, n$. Similarly, let T be the set of all $j \in \{1, \ldots, m\}$, such that $g_j(B, D)$ equals the maximum of $g_\ell(B, C)$, where $\ell = 1, \ldots, m$. Then clearly (A, B, C, D) belongs to $W(S,T)$.

Let $W_k \subset (X \times Y) \times (X \times Y)$ denote the union of all sets $W(S,T)$, where $|S| + |T| = k$. Here $k = 2, 3, \ldots, n + m$. The sets $W_2, \ldots, W_{n+m}$ form an open cover of $(X \times Y) \times (X \times Y)$. If $|S| + |T| = |S'| + |T| =$

k, then the corresponding sets $W(S,T)$ and $W(S',T')$ either coincide (if $S = S'$ and $T = T'$) or are disjoint. Hence we see (using (c)) that there exists a continuous motion planning algorithm over each open set W_k. This completes the proof. □

4.2. *The reduced topological complexity*

Theorem 4.1 suggests the notation

$$\widetilde{\mathsf{TC}}(X) = \mathsf{TC}(X) - 1,$$

which is called *the reduced topological complexity.* Then we have:

Corollary 4.2: *For path-connected metric spaces $X_1, \ldots, X_k$ one has*

$$\widetilde{\mathsf{TC}}(X_1 \times X_2 \times \cdots \times X_k) \leq \sum_{i=1}^{k} \widetilde{\mathsf{TC}}(X_i). \tag{4.2}$$

We shall use the following notation. For a topological space X we shall denote by $\mathsf{zcl}(X)$ the largest integer k such that that there exist k zero-divisors $u_1, u_2, \ldots, u_k \in H^*(X;\mathbb{Q}) \otimes H^*(X;\mathbb{Q})$ having a nontrivial product

$$u_1 u_2 \ldots u_k \neq 0 \in H^*(X;\mathbb{Q}) \otimes H^*(X;\mathbb{Q}).$$

Theorem 3.3 can be restated as the inequality

$$\widetilde{\mathsf{TC}}(X) \geq \mathsf{zcl}(X).$$

By Example 3.4 we have

$$\mathsf{zcl}(S^n) \geq \begin{cases} 2, \text{ if } n \text{ is even,} \\ \\ 1, \text{ if } n \text{ is odd.} \end{cases}$$

Lemma 4.3: *One has* $\mathsf{zcl}(X \times Y) \geq \mathsf{zcl}(X) + \mathsf{zcl}(Y)$.

Proof: See [16], Lemma 4.52. □

Example 4.4: Suppose that each space X_i is the n-dimensional sphere S^n. Then using (3.5), Corollary 4.2 and Lemma 4.3 one has

$$\mathsf{TC}(\prod_{i=1}^{k} S^n) \leq \begin{cases} 2k+1, \text{ for } n \text{ even,} \\ \\ k+1, \quad \text{for } n \text{ odd.} \end{cases}$$

On the other hand we have

$$\begin{aligned}\widetilde{\mathsf{TC}}(\prod_{i=1}^k S^n) &\geq \mathsf{zcl}(\prod_{i=1}^k S^n)\\ &\geq \sum_{i=1}^k \mathsf{zcl}(S^n)\\ &= \begin{cases} 2k, & \text{if } n \text{ is even,}\\ k, & \text{if } n \text{ is odd.}\end{cases}\end{aligned}$$

Thus:

$$\mathsf{TC}(\prod_{i=1}^k S^n) = \begin{cases} 2k+1, & \text{for } n \text{ even,}\\ k+1, & \text{for } n \text{ odd.}\end{cases} \tag{4.3}$$

5. Centralised and distributed controls for large systems

Consider a large system S consisting of many independently moving parts $S_1, \dots, S_k$. As we discussed earlier, the configuration space of this system is the Cartesian product $X_1 \times X_2 \times \cdots \times X_k$ of the configuration spaces X_i of individual parts S_i. One may compare the distributed and centralised motion planning algorithms for S.

In the case of *distributed motion planning algorithms*, one controls each system S_i independently of the other systems S_j. The motion planning algorithm for S_i will have at least $\mathsf{TC}(X_i)$ domains of continuity, and therefore a distributed motion planning algorithm for S will have at least

$$\prod_{i=1}^k \mathsf{TC}(X_i) \tag{5.1}$$

domains of continuity.

However, in the case of *centralised control*, when the system S is viewed as a single system, there exists a motion planning algorithm with

$$\mathsf{TC}(\prod_{i+1}^k X_i)$$

domains of continuity. Taking into account inequality (4.2) we obtain that one may find a centralised motion planning algorithm for S having at most

$$1 - k + \sum_{i=1}^{k} \mathsf{TC}(X_i) = 1 + \sum_{i=1}^{k} \widetilde{\mathsf{TC}}(X_i) \tag{5.2}$$

domains of continuity.

In the special case when $\mathsf{TC}(X_i) = a \geq 2$ is independent of i, we obtain that any distributed motion planning algorithm has at least a^k domains of continuity and one can find a centralised motion planning algorithm with at most $k(a-1)+1$ domains of continuity.

In conclusion, *the centralised control has potentially significantly more stability compared to the distributed control.*

6. Tame motion planning algorithms

The definition of $\mathsf{TC}(X)$ (see Definition 2.1) deals with open subsets of $X \times X$ admitting continuous sections of the path fibration (2.1). To construct a motion planning algorithm in practice one partitions the whole space $X \times X$ into pieces and defines a continuous (often smooth or analytic) section over each of the obtained sets. Any such partition necessarily contains sets which are not open and hence we need to be able to operate with subsets of $X \times X$ of a more general nature.

Definition 6.1: A topological space X is a Euclidean Neighbourhood Retract (ENR) if it can be embedded into a Euclidean space $\mathbb{R}^k$ such that for some open neighbourhood $X \subset U \subset \mathbb{R}^k$ there is a retraction $r : U \to X$, $r|_X = 1_X$.

It is known that a subset $X \subset \mathbb{R}^k$ is an ENR if and only if it is locally compact and locally contractible, see [6], Chapter 4, §8. This implies that all finite-dimensional polyhedra, manifolds and semi-algebraic sets are ENRs.

Definition 6.2: Let X be an ENR. A motion planning algorithm $s : X \times X \to PX$ is said to be tame if $X \times X$ can be split into finitely many sets

$$X \times X = F_1 \cup F_2 \cup \cdots \cup F_k$$

such that

(1) Each restriction $s|F_i : F_i \to PX$ is continuous, where $i = 1, \ldots, k$;
(2) $F_i \cap F_j = \emptyset$, for $i \neq j$;
(3) Each F_i is an ENR.

It is known that for an ENR X, the minimal number of domains of continuity $F_1, \dots, F_k$ in tame motion planning algorithms $s : X \times X \to PX$ equals $\mathsf{TC}(X)$, see [13], Theorem 13.1.

Example 6.3: Here we construct a tame motion planning algorithm on the sphere S^n. Let $F_1 \subset S^n \times S^n$ be the set of all pairs (A, B) such that $A \neq -B$. We may construct a continuous section $s_1 : F_1 \to PS^n$ by moving A towards B along the shortest geodesic arc. Consider now the set $F_2 \subset S^n \times S^n$ of pairs of antipodal points $(A, -A)$. If n is odd we may construct a continuous section $s_2 : F_2 \to PS^n$ as follows. Fix a non-vanishing tangent vector field v on S^n; such a v exists for n odd. Move A towards the antipodal point $-A$ along the semi-circle tangent to the vector $v(A)$.

In the case when n is even the above procedure has to be modified since for n even any vector field v tangent to S^n has at least one zero. We may find a tangent vector field v having a single zero $A_0 \in S^n$. Denote $F_2 = \{(A, -A); A \neq A_0\}$ and define $s_2 : F_2 \to PS^n$ as in the previous paragraph. The set $F_3 = \{(A_0, -A_0)\}$ consists of a single pair; define $s_3 : F_3 \to PS^n$ by choosing an arbitrary path from A_0 to $-A_0$.

Example 6.4: Constructing sections via deformations. Let $A \subset X \times X$ be a subset. A continuous section

$$s_A : A \to X^I$$

of the fibration (2.1) can be viewed as a continuous function of three variables $s_A(a_1, a_2, t) \in X$, where $a_1, a_2 \in X$ are such that $(a_1, a_2) \in A$ and $t \in I = [0, 1]$; this function must satisfy $s_A(a_1, a_2, 0) = a_1$ and $s_A(a_1, a_2, 1) = a_2$.

Suppose that a subset $B \subset X \times X$ can be continuously deformed inside $X \times X$ into the subset A. In other words, assume that there exists a continuous map

$$h : B \times I \to X \times X \tag{6.1}$$

such that $h(b, 0) = b$ and $h(b, 1) \in A$ for any $b \in B$. We may write

$$h(b, t) = (h_1(b, t), h_2(b, t))$$

where h_1 and h_2 are the compositions of h with the projections. The path $s_A(h(b, 1), \tau)$, where $\tau \in [0, 1]$, connects the points $h_1(b, 1)$ and $h_2(b, 1)$, i.e.

$s_A(h(b,1),0) = h_1(b,1)$ and $s_A(h(b,1),1) = h_2(b,1)$. Thus, the formula

$$s_B(b,\tau) = \begin{cases} h_1(b,3\tau), & \text{for } \tau \in [0,1/3], \\ s_A(h(b,1),3\tau-1), & \text{for } \tau \in [0,1/3], \\ h_2(b,3-3\tau), & \text{for } \tau \in [2/3,1], \end{cases} \tag{6.2}$$

defines a path from b_1 to b_2 which depends continuously on $(b,t) = (b_1,b_2,t)$. Hence we obtain a continuous section

$$s_B : B \to X^I$$

of the path fibration (2.1) over B. Thus a deformation of B into A and a continuous section s_A over A define a continuous section s_B over B.

We shall often use the above remark in the case when A is the diagonal $\Delta_X \subset X \times X$, i.e. $\Delta_X = \{(x,y) \in X \times X; x = y\}$. There exists an obvious section $s : \Delta_X \to X^I$ over the diagonal and hence any deformation (6.1) of a subset $B \subset X \times X$ into the diagonal Δ_X will automatically give a motion planning section over B, using (6.2).

7. Topological complexity of collision free motion planning in $\mathbb{R}^d$

Consider a system consisting of n small objects moving in the Euclidean space $\mathbb{R}^d$ without collisions. Mathematically we may assume that each of the objects is a point and hence our configuration space is

$$F(\mathbb{R}^d,n) = \{(z_1,\dots,z_n) \in \mathbb{R}^d \times \cdots \times \mathbb{R}^d; z_i \neq z_j \quad \text{for} \quad i \neq j\}.$$

Here the vectors $z_i \in \mathbb{R}^d$ represent the locations of the moving objects and the condition $z_i \neq z_j$ reflects the requirement that the objects must move without collisions.

A motion planning algorithm in $F(\mathbb{R}^d,n)$ assigns to any pair of configurations

$$(z_1,\dots,z_n),\ (z'_1,\dots,z'_n) \in F(\mathbb{R}^d,n)$$

a continuous curve of configurations

$$(z_1(t),z_2(t),\dots,z_n(t)) \in F(\mathbb{R}^d,n), \quad t \in [0,1],$$

such that $(z_1(0),\dots,z_n(0)) = (z_1,\dots,z_n)$ and $(z_1(1),\dots,z_n(1)) = (z'_1,\dots,z'_n)$.

The following theorem gives the topological complexity of this motion planning problem:

Theorem 7.1: *See* [12], [15]. *For* $n \geq 2$, *one has*

$$\mathsf{TC}(F(\mathbb{R}^d, n)) = \begin{cases} 2n-1, \text{ for } d \geq 3 \text{ odd}, \\ \\ 2n-2, \text{ for } d \geq 2 \text{ even}. \end{cases} \tag{7.1}$$

We see that the topological complexity of collision free motion planning in the Euclidean space is roughly $\sim 2n$ where n is the number of controlled objects. One naturally wants to know explicit motion planning algorithms for $F(\mathbb{R}^d, n)$ with optimal topological complexity as given by Theorem 7.1. Such algorithms will be given in the following two sections.

The paper [13] suggested a motion planning algorithm in $F(\mathbb{R}^d, n)$ having topological complexity quadratic in n. In [25], Hugo Mas-Ku and Enrique Torres-Giese suggested a motion planning algorithm for $F(\mathbb{R}^2, n)$ with complexity $2n-1$. They also briefly indicated how their algorithm may work for $F(\mathbb{R}^d, n)$ with any $d \geq 2$; such an algorithm would be optimal for $d \geq 3$ odd.

The paper [25] also contains the useful observation that the lower bound of Theorem 7.1, for the case $d \geq 3$, odd can be deduced from the fact that the configuration space $F(\mathbb{R}^d, n)$ contains the product $\prod_{i=1}^{n-1} S^{d-1}$ of $n-1$ copies of the sphere S^{d-1} as a retract. We repeat this argument below.

Let us describe the maps

$$\prod_{i=1}^{n-1} S^{d-1} \xrightarrow{\alpha} F(\mathbb{R}^d, n) \xrightarrow{\beta} \prod_{i=1}^{n-1} S^{d-1}; \tag{7.2}$$

we want β to be a retraction onto the image of α. We think of $S^{d-1} \subset \mathbb{R}^d$ as being the unit sphere with its centre at the origin. For

$$(u_1, \ldots, u_{n-1}) \in \prod_{i=1}^{n-1} S^{d-1}$$

we set

$$\alpha(u_1, u_2, \ldots, u_{n-1}) = (z_1, z_2, \ldots, z_n) \in F(\mathbb{R}^d, n),$$

where $z_1 = 0$ and for $i = 1, \ldots, n-1$ one has

$$z_{i+1} = z_i + 3^{i-1} u_i.$$

Clearly, α is injective. We have for $k \geq 1$,

$$z_{i+k} - z_i = 3^{i+k-2} u_{i+k-1} + 3^{i+k-3} u_{i+k-2} + \cdots + 3^{i-1} u_i$$

and

$$|z_{i+k} - z_i| \geq 3^{i+k-2} - 3^{i+k-3} - 3^{i+k-4} - \cdots - 3^{i-1} = \frac{1}{2}3^{i-1}(3^{k-1} + 1).$$

This shows that $z_i \neq z_j$ for $i \neq j$, i.e. the map α indeed takes its values in the configuration space $F(\mathbb{R}^d, n)$.

Next we define the second map $\beta : F(\mathbb{R}^d, n) \to \prod_{i=1}^{n-1} S^{d-1}$,

$$\beta(z_1, z_2, \ldots, z_n) = (u_1, u_2, \ldots, u_{n-1})$$

where

$$u_i = \frac{z_{i+1} - z_i}{|z_{i+1} - z_i|} \in S^{d-1}, \qquad i = 1, 2, \ldots, n-1.$$

It is obvious that $\beta \circ \alpha$ is the identity map, i.e. β is a retraction of the image of α.

Assuming that $d \geq 3$ is odd and $n \geq 2$, one applies Theorem 2.3 and formula (4.3) to obtain

$$\mathsf{TC}(F(\mathbb{R}^d, n)) \geq \mathsf{TC}(\prod_{i=1}^{n-1} S^{d-1}) = 2n - 1.$$

8. A motion planning algorithm in $F(\mathbb{R}^d, n)$

In this section we present a tame motion planning algorithm in $F(\mathbb{R}^d, n)$ with $2n - 1$ regions of continuity. This algorithm works for any $d \geq 2$ and $n \geq 2$; it is optimal in the case when $d \geq 3$ is odd. In the following section we suggest a modification of this algorithm which works in the case of $d \geq 2$ even and has $2n - 2$ regions of continuity; this algorithm is optimal for $d \geq 2$ even in the sense that it has the smallest possible number of regions of continuity.

The algorithm we describe in this section can be used in designing practical systems controlling motion of many objects moving in space without collisions.

8.1. *The space $F(L, n)$*

Fix a line $L \subset \mathbb{R}^d$ and denote by $p : \mathbb{R}^d \to L$ the orthogonal projection. Let $e \in \mathbb{R}^d$ denote a unit vector in the direction of L. The vector e determines an order on L: we say that for $a, b \in L$ one has $a \leq b$ if the scalar product $\langle b - a, e \rangle$ is non-negative.

Note that $F(L, n)$ is naturally a subset of $F(\mathbb{R}^d, n)$ and while the configuration space $F(\mathbb{R}^d, n)$ is connected, the configuration space $F(L, n)$ is disconnected. More precisely, the space $F(L,n)$ contains $n!$ connected components and each of the components of $F(L,n)$ is contractible. Indeed, a configuration

$$C = (z_1, \ldots, z_n) \in F(L,n)$$

determines a permutation

$$\tau = (i_1, \ldots, i_n) \in \Sigma_n$$

of the set $\{1, 2, \ldots, n\}$ where

$$z_{i_1} < z_{i_2} < \cdots < z_{i_n}.$$

This permutation describes the order in which the points z_i appear on the line L. Clearly, two configurations of $F(L,n)$ lie in the same connected component of $F(L,n)$ if and only if they have the same ordering, i.e. they determine the same permutation.

For a permutation $\tau \in \Sigma_n$ we denote by $F(L,n,\tau)$ the set of all configurations $C = (z_1, \ldots, z_n) \in F(L,n)$ such that the order of points z_i on the line L is described by the permutation τ. We have

$$F(L,n) = \bigsqcup_{\tau \in \Sigma_n} F(L,n,\tau). \tag{8.1}$$

To show that each space $F(L,n,\tau)$ is contractible we note that for two configurations $C, C' \in F(L,n,\tau)$ where $C = (z_1, \ldots, z_n)$ and $C' = (z'_1, \ldots, z'_n)$ we may define the linear deformation

$$z_i(t) = (1-t)z_i + tz'_i, \quad t \in [0,1], \quad i = 1, \ldots, n, \tag{8.2}$$

which represents a continuous path in $F(L,n,\tau)$. Clearly, if $z_i < z_j$ and $z'_i < z'_j$, then $z_i(t) < z_j(t)$ for any $t \in [0,1]$. In other words, we have a continuous section

$$\sigma_\tau : F(L,n,\tau) \times F(L,n,\tau) \to F(L,n,\tau)^I \tag{8.3}$$

of the path fibration

$$F(L,n,\tau)^I \to F(L,n,\tau) \times F(L,n,\tau).$$

Fix a specific configuration $C_\tau \in F(L,n,\tau)$ for each permutation $\tau \in \Sigma_n$. Since $d > 1$, each pair of configurations C_τ and $C_{\tau'}$ can be connected by a continuous path

$$\gamma_{\tau,\tau'} : [0,1] \to F(\mathbb{R}^d, n), \quad \gamma_{\tau,\tau'}(0) = C_\tau, \quad \gamma_{\tau,\tau'}(1) = C_{\tau'}.$$

The family of paths $\{\gamma_{\tau,\tau'}\}$ gives a continuous section of the path fibration over the (discrete) subset

$$\Sigma_n \times \Sigma_n \subset F(L,n) \times F(L,n).$$

Since (8.2) gives a deformation of the set $F(L,n,\tau)\times F(L,n,\tau')$ to the single point $\{(C_\tau, C_{\tau'})\} \subset \Sigma_n \times \Sigma_n$, we obtain via concatenation (as explained in Example 6.4) a continuous section

$$\sigma : F(L,n) \times F(L,n) \to F(\mathbb{R}^d, n)^I \tag{8.4}$$

of the path fibration, i.e. such that the composition

$$F(L,n) \times F(L,n) \xrightarrow{\sigma} F(\mathbb{R}^d, n)^I \xrightarrow{\pi} F(\mathbb{R}^d, n) \times F(\mathbb{R}^d, n) \tag{8.5}$$

coincides with the inclusion $F(L,n)\times F(L,n) \subset F(\mathbb{R}^d,n)\times F(\mathbb{R}^d,n)$. Recall that π denotes the path fibration (2.1).

8.2. *Sets A_i*

For a configuration $C \in F(\mathbb{R}^d, n)$, where $C = (z_1, \dots, z_n)$ with $z_i \in \mathbb{R}^d$, $z_i \neq z_j$ for $i \neq j$, consider the set of projection points

$$p(C) = \{p(z_1), \dots, p(z_n)\}, \qquad p(z_i) \in L, \quad i = 1, \dots, n.$$

The cardinality of this set will be denoted $\mathrm{cp}(C)$. Here the symbol "cp"stands for *"cardinality of projection"*. Note that $\mathrm{cp}(C)$ can be any number $1, 2, \dots, n$. Let A_i denote the set of all configurations $C \in F(\mathbb{R}^d, n)$ with $\mathrm{cp}(C) = i$. Clearly, A_i is an ENR.

The set A_n is open and dense in $F(\mathbb{R}^d, n)$. If $C = (z_1, \dots, z_n) \in A_n$, then $p(C) \in F(L,n)$ and the formula

$$z_i(t) = z_i + t(p(z_i) - z_i), \quad i = 1, \dots, n \tag{8.6}$$

defines a continuous deformation of A_n onto $F(L,n)$.

In general, the closure of each set A_i is contained in the union of the sets A_j with $j \leq i$, i.e.

$$\overline{A}_i \subset \bigcup_{j \leq i} A_j.$$

For a configuration $C \in A_i$, where $i \geq 2$, $C = (z_1, \dots, z_n)$ denote

$$\epsilon(C) = \frac{1}{n} \min\{|p(z_r) - p(z_s)|; p(z_r) \neq p(z_s)\}.$$

For $i = 1$ the formula above makes no sense and we set $\epsilon(C) = 1$ for any $C \in A_1$.

For $C \in A_i$ and $t \in [0,1]$, where $C = (z_1, \dots, z_n)$, define

$$F_i(t)(C) = (z_1(t), \dots, z_n(t)), \text{ where } z_j(t) = z_j + t(j-1)\epsilon(C)e, \ j = 1, \dots, n.$$

This defines a continuous deformation of A_i into A_n inside $F(\mathbb{R}^d, n)$; we shall call the deformation $F_i : A_i \times I \to F(\mathbb{R}^d, n)$ *"desingularization"*.

8.3. *Sections* σ_{ij}

We have constructed several deformations and a section over $F(L, n)$; applying iteratively the construction of Example 6.4 we obtain a continuous section

$$\sigma_{ij} : A_i \times A_j \to F(\mathbb{R}^d, n)^I, \quad i, j = 1, 2, \dots, n, \tag{8.7}$$

of the path fibration, i.e. such that the composition

$$A_i \times A_j \stackrel{\sigma_{ij}}{\to} F(\mathbb{R}^d, n)^I \stackrel{\pi}{\to} F(\mathbb{R}^d, n) \times F(\mathbb{R}^d, n) \tag{8.8}$$

coincides with the inclusion $A_i \times A_j \to F(\mathbb{R}^d, n) \times F(\mathbb{R}^d, n)$. Indeed, the desingularization deformation $F_i \times F_j$ takes $A_i \times A_j$ into $A_n \times A_n$; then we apply the deformation (8.6) which takes $A_n \times A_n$ into $F(L, n) \times F(L, n)$; and finally we apply section (8.4). Let us emphasise that the above description of σ_{ij} is totally algorithmic and practically implementable.

8.4. *Combining the regions of continuity*

The sets $A_i \times A_j$ where $i, j = 1, \dots, n$, are mutually disjoint and cover the whole product $F(\mathbb{R}^d, n) \times F(\mathbb{R}^d, n)$. Over each of these sets we have a continuous section σ_{ij}; in total we have n^2 of these sets. In this subsection we observe that one may combine these sets into $2n - 1$ sets W_k, where $k = 2, \dots, 2n$, such that the sections σ_{ij} determine a continuous section over each W_k.

Define

$$W_k = \bigcup_{i+j=k} A_i \times A_j, \quad \text{where} \quad k = 2, 3, \dots, 2n. \tag{8.9}$$

We know that the closure of each set A_i is contained in the union of the sets A_r with $r \leq i$. This implies that for any two distinct pairs (i, j) and (i', j'), with $i + j = k = i' + j'$, one has

$$\overline{A_i \times A_j} \cap (A_{i'} \times A_{j'}) = \emptyset.$$

Therefore no limit point of $A_i \times A_j$ lies in $A_{i'} \times A_{j'}$ for $i + j = i' + j'$. Hence the sections σ_{ij}, see (8.7), jointly define a continuous section of the path fibration $\pi : F(\mathbb{R}^d, n)^I \to F(\mathbb{R}^d, n) \times F(\mathbb{R}^d, n)$ over each set W_k. Thus, we have constructed a tame motion planning algorithm in $F(\mathbb{R}^d, n)$ having $2n-1$ domains of continuity $W_2, W_3, \ldots, W_{2n}$.

9. A motion planning algorithm in $F(\mathbb{R}^d, n)$ with $d \geq 2$ even

In this section we improve the motion planning algorithm in $F(\mathbb{R}^d, n)$ of the previous section under the assumption that $d \geq 2$ is even. This motion planning algorithm will have $2n-2$ domains of continuity.

For a configuration $C = (z_1, \ldots, z_n) \in F(\mathbb{R}^d, n)$ consider the line $L' = L'_C$ through the origin which is parallel to the affine line $L = L_C$ connecting the points z_1 and z_2. The line L_C has a natural orientation from z_1 to z_2 and we denote by $e = e_C \in L'_C$ the unit vector

$$e_C = \frac{z_2 - z_1}{|z_2 - z_1|}.$$

Let $p_C : \mathbb{R}^d \to L_C$ denote the orthogonal projection. For a configuration $C = (z_1, \ldots, z_n) \in F(\mathbb{R}^d, n)$ we denote by $\operatorname{cp}(C)$ the cardinality of the set $\{p_C(z_1), \ldots, p_C(z_n)\}$ of the projection points; note that $\operatorname{cp}(C) \in \{2, \ldots, n\}$.

9.1. *Desingularization*

For a configuration $C \in F(\mathbb{R}^d, n)$, $C = (z_1, \ldots, z_n)$ with $\operatorname{cp}(C) = i$, where $i \geq 2$, denote

$$\epsilon(C) = \frac{1}{n} \min\{|p_C(z_r) - p_C(z_s)|; p_C(z_r) \neq p_C(z_s)\}.$$

For $t \in [0, 1]$ and C as above define $F_i(t)(C) = (z_1(t), \ldots, z_n(t))$, where

$$z_j(t) = z_j + t(j-1)\epsilon(C)e_C$$

for $j = 1, \ldots, n$. This gives a "desingularization" deformation $F_i(t)(C)$ with $F_i(0)(C) = C$ and

$$\operatorname{cp}(F_i(t)(C)) = n \quad \text{for} \quad t \in (0, 1].$$

Note that the lines L_C and L'_C do not change under the desingularization, i.e. $L_{F_i(t)(C)} = L_C$ and $L'_{F_i(t)(C)} = L'_C$. Besides, the desingularization $F_i(t)(C)$ is continuous as a function of (t, C) if we restrict it to the set of configurations C with $\operatorname{cp}(C) = i$, where i is fixed.

9.2. *Colinear configurations*

For $i, j = 2, \ldots, n$ we denote by A_{ij} the set of all pairs of configurations (C, C'), where $C, C' \in F(\mathbb{R}^d, n)$, such that $e_C \neq -e_{C'}$, $\mathrm{cp}(C) = i$ and $\mathrm{cp}(C') = j$. Similarly, for $i, j = 2, \ldots, n$ we denote by B_{ij} the set of all pairs of configurations (C, C'), where $C, C' \in F(\mathbb{R}^d, n)$, such that $e_C = -e_{C'}$, $\mathrm{cp}(C) = i$ and $\mathrm{cp}(C') = j$.

Clearly,

$$\overline{B_{ij}} \subset \bigcup_{\substack{r \leq i \\ s \leq j}} B_{rs}, \qquad \overline{A_{ij}} \subset \bigcup_{\substack{r \leq i \\ s \leq j}} A_{rs} \cup \bigcup_{\substack{r \leq i \\ s \leq j}} B_{rs}. \tag{9.1}$$

Denote by $X \subset F(\mathbb{R}^d, n) \times F(\mathbb{R}^d, n)$ the set of all pairs (C, C') of configurations such that (a) the vectors e_C and $e_{C'}$ are not opposite to each other, i.e. $e_C \neq -e_{C'}$, and (b) the configurations C and C' are *colinear*, i.e. $C \in F(L_C, n)$ and $C' \in F(L_{C'}, n)$.

Consider also the subset $X' \subset X$ consisting of pairs of colinear configurations (C, C') with $e_C = e_{C'}$ and $L_C = L_{C'}$.

Besides, we shall denote by $Y \subset F(\mathbb{R}^d, n) \times F(\mathbb{R}^d, n)$ the set of all pairs of colinear configurations (C, C') such that the vectors e_C and $e_{C'}$ are opposite to each other, i.e. $e_C = -e_{C'}$. Note that in this case $L_C = L_{C'}$.

The union $X \cup Y$ is the set of all pairs of colinear configurations.

9.3. *Deformations* σ_{ij}

Next we define the deformations

$$\begin{aligned} \sigma_{ij} &: A_{ij} \to (F(\mathbb{R}^d, n) \times F(\mathbb{R}^d, n))^I, \\ \sigma'_{ij} &: B_{ij} \to (F(\mathbb{R}^d, n) \times F(\mathbb{R}^d, n))^I, \end{aligned} \tag{9.2}$$

deforming A_{ij} into X and B_{ij} into Y correspondingly, i.e. such that

(1) $\sigma_{ij}(C, C')(0) = (C, C')$ and $\sigma_{ij}(C, C')(1) \in X$,
(2) $\sigma'_{ij}(C, C')(0) = (C, C')$ and $\sigma'_{ij}(C, C')(1) \in Y$.

Given a pair $(C, C') \in A_{ij}$, we apply first the desingularization deformations $F_i(t)(C)$ and $F_j(t)(C')$ taking the pair (C, C') to a pair of configurations (C_1, C'_1), with $\mathrm{cp}(C_1) = n$, $L_{C_1} = L_C$ and $\mathrm{cp}(C'_1) = n$, $L_{C'_1} = L_{C'}$. Next we apply the linear deformation (8.6) taking the pair (C_1, C'_1) to a pair of colinear configurations (C_2, C'_2), where $C_2 \in F(L_C, n)$ and $C'_2 \in F(L_{C'}, n)$. The deformation σ_{ij} is the concatenation of the two deformations described above; the deformation σ'_{ij} is defined similarly.

Next we deform X into X' by a deformation $X \times I \to F(\mathbb{R}^d, n) \times F(\mathbb{R}^d, n)$ as follows. Suppose we are given two colinear configurations $C = (z_1, \dots, z_n)$ and $C' = (z'_1, \dots, z'_n)$ with vectors e_C and $e_{C'}$ satisfying $e_C \neq -e_{C'}$. Making parallel translation, we may assume that both lines L_C and $L_{C'}$ pass through the origin $0 \in \mathbb{R}^d$. We may now view e_C and $e_{C'}$ as points of the unit sphere $S^{d-1} \subset \mathbb{R}^d$ and, since they are not antipodal, there exists a unique geodesic path $e(t) \in S^{d-1}$ of minimal length connecting them. We obtain a continuous path V_t of orthogonal transformations of $V_t : \mathbb{R}^d \to \mathbb{R}^d$, which is identical on the orthogonal complement to the subspace spanned by the vectors e_C and $e_{C'}$, and such that $V_t(e_C) = e(t)$. Applying V_t to the configuration $C = (z_1, \dots, z_n)$ we get a path $(V_t(z_1), \dots, V_t(z_n))$ in $F(\mathbb{R}^d, n)$ taking C to a colinear configuration C" such that $e_{C"} = e_{C'}$.

Finally we observe that there exist continuous sections

$$\sigma_{X'} : X' \to F(\mathbb{R}^d, n)^I \quad \text{and} \quad \sigma_Y : Y \to F(\mathbb{R}^d, n)^I \tag{9.3}$$

of the path space fibration

$$\pi : F(\mathbb{R}^d, n)^I \to F(\mathbb{R}^d, n) \times F(\mathbb{R}^d, n) \tag{9.4}$$

over the sets X' and Y correspondingly. Here we will use our assumption that $d \geq 2$ is even. Let us start with $\sigma_{X'}$. Suppose we are given two colinear configurations $C = (z_1, \dots, z_n)$ and $C' = (z'_1, \dots, z'_n)$ with $L = L_C = L_{C'}$ and $e_C = e_{C'}$. The points $z_1, \dots, z_n, z'_1, \dots, z'_n$ lie on the oriented line L and their "ordering" determines two permutations $(i_1, i_2, \dots, i_n)$ and $(j_1, j_2, \dots, j_n)$ such that $z_{i_1} < z_{i_2} < \dots < z_{i_n}$ and $z'_{j_1} < z'_{j_2} < \dots < z'_{j_n}$. Since $d \geq 2$ is even, the unit sphere S^{d-1} admits a continuous and nowhere zero tangent vector field. This means that we may continuously choose a unit vector $e'_C \in S^{d-1}$ perpendicular to e_C for any colinear configuration C. Now we define the following path in $F(\mathbb{R}^d, n)$ which takes C onto C' and is continuous as a function of (C, t); we set $C^t = (z^t_1, z^t_2, \dots, z^t_n)$ where

$$z^t_{i_k} = \begin{cases} z_{i_k} + 3tke'_C, & \text{for } t \in [0, 1/3], \\ z_{i_k} + ke'_C + (3t-1)(z_{j_k} - z_{i_k}), & \text{for } t \in [1/3, 2/3], \\ z_{j_k} + k(3-3t)e'_C & \text{for, } t \in [2/3, 1]. \end{cases}$$

This formula defines a continuous section of (9.4) over X' which we shall denote by $\sigma_{X'}$. The section σ_Y, see (9.3), is defined by the similar formulae.

Now we may concatenate (as explained in Example 6.4) the above deformations and the section $\sigma_{X'}$ to obtain a continuous section

$$s_{ij} : A_{ij} \longrightarrow F(\mathbb{R}^d, n)^I$$

of the path fibration over each A_{ij}, where $i, j = 2, \ldots, n$. Similarly, concatenating the deformation σ'_{ij} and the section σ_Y we obtain a continuous section

$$s'_{ij} : B_{ij} \longrightarrow F(\mathbb{R}^d, n)^I, \quad i, j = 2, \ldots, n.$$

9.4. *Repackaging the regions of continuity*

The sets A_{ij}, B_{rs} are pairwise disjoint and their union is $F(\mathbb{R}^d, n) \times F(\mathbb{R}^d, n)$; each of these sets is an ENR and on each of these sets we have constructed a continuous section s_{ij} or s'_{ij}.

We can define the sets W_k (repackaging) as follows

$$W_k = \bigcup_{i+j=k} A_{ij} \cup \bigcup_{r+s=k+1} B_{rs},$$

where $k = 3, \ldots, 2n$. From (9.1) we see that for $i+j = k$ and $r+s = k+1$ no limit point of A_{ij} may be contained in B_{rs}. The sections s_{ij} and s'_{rs} define a continuous section of the path fibration over each W_k, where $k = 3, \ldots, 2n$. As the result we obtain $2n-2$ regions of continuity $W_3, W_4, \ldots, W_{2n}$; note that $W_3 = B_{22}$.

10. Configuration spaces of graphs

Let Γ be a connected finite graph. The symbol $F(\Gamma, n)$ denotes the configuration space of n distinct particles on Γ. In other words, $F(\Gamma, n)$ is the subset of the Cartesian product

$$\underbrace{\Gamma \times \Gamma \times \cdots \times \Gamma}_{n \text{ times}} = \Gamma^n$$

consisting of configurations $C = (z_1, z_2, \ldots, z_n)$ where $z_i \in \Gamma$ and $z_i \neq z_j$ for $i \neq j$. The topology of $F(\Gamma, n)$ is induced from its embedding into Γ^n.

Configuration spaces of graphs were studied by R. Ghrist, D. Koditschek and A. Abrams, see [1], [2], [18], [19]. To illustrate the importance of these configuration spaces for robotics one may mention the control problems where a number of automated guided vehicles (AGV) have to move along a network of floor wires [19]. The motion of the vehicles must be safe: it should be organized so that collisions do not occur. If n is the number of AGV then the natural configuration space of this problem is the space $F(\Gamma, n)$ where Γ is a graph describing the network of floor wires. Here we idealise reality by assuming that the vehicles have size 0 (i.e. they are points).

The first question to ask is whether the configuration space $F(\Gamma, n)$ is connected. Clearly $F(\Gamma, n)$ is disconnected if $\Gamma = [0,1]$ is a closed interval (and $n \geq 2$) or if $\Gamma = S^1$ is the circle and $n \geq 3$. These are the only examples of this kind as the following simple lemma claims:

Lemma 10.1: *Let Γ be a connected finite graph having at least one essential vertex. Then the configuration space $F(\Gamma, n)$ is connected.*

An *essential vertex* is a vertex of the graph which is incident to at least 3 edges. We denote the number of essential vertexes of Γ by $m(\Gamma)$.

10.1. *Motion planning algorithm in $F(\Gamma, n)$*

The algorithm presented here was first described in [10]. We assume below that Γ is a tree having an essential vertex. Fix a univalent vertex $u_0 \in \Gamma$ which will be called *the root.* Any point in Γ can be connected by a simple path to the root u_0 and this connecting path is unique up to homotopy. The choice of the root determines a partial order on Γ: we say that $x \succeq y$, where $x, y \in \Gamma$ if any path from x to the root u_0 passes through y. Of course, $\succeq$ is only a partial order, i.e. there may exist pairs $x, y \in \Gamma$ such that neither $x \succeq y$, nor $y \succeq x$. On the following picture we see $u \succeq v$ and $w \succeq v$ however u and w are not comparable.

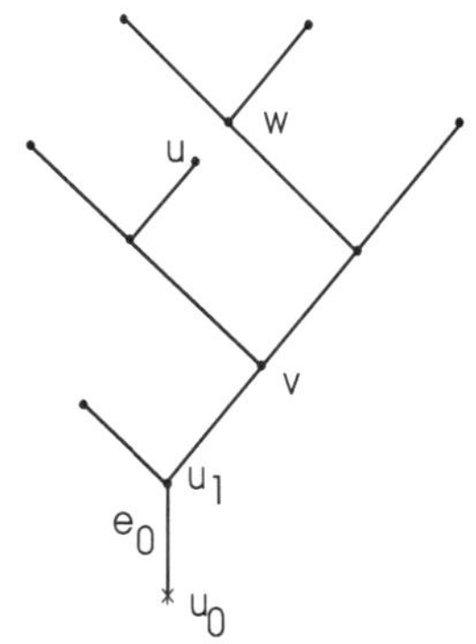

Fig. 1. A partial order between the points of the tree.

Let $e_0 \subset \Gamma$ denote the root edge of Γ. For a configuration $C = (z_1, \ldots, z_n) \in F(e_0, n) \subset F(\Gamma, n)$, such that $z_i \in e_0$ for all $i = 1, \ldots, n$, one has

$$z_{i_1} \preceq z_{i_2} \preceq \cdots \preceq z_{i_n} \tag{10.1}$$

for some permutation $\tau = (i_1, i_2, \ldots, i_n) \in \Sigma_n$. The space $F(e_0, n)$ consists of $n!$ connected components labeled by permutations $\tau \in \Sigma_n$,

$$F(e_0, n) = \bigsqcup_{\tau \in \Sigma_n} F(e_0, n, \tau),$$

where $F(e_0, n, \tau)$ is the set of all configurations $C = (z_1, \ldots, z_n) \in F(e_0, n)$ with the order of the points z_j described by the permutation τ as in (10.1). The fact that each space $F(e_0, n, \tau)$ is contractible follows similarly to the arguments of Section 8.1. Using the connectivity of $F(\Gamma, n)$ and the contractibility of $F(e_0, n, \tau)$ we may construct a continuous section

$$\sigma_0 : F(e_0, n) \times F(e_0, n) \to F(\Gamma, n)^I \tag{10.2}$$

of the path fibration

$$\pi : F(\Gamma, n)^I \to F(\Gamma, n) \times F(\Gamma, n)$$

which is similar to (8.4). The section σ_0 is a continuous motion planning algorithm moving any configuration of n points lying on the root edge e_0 to any other such configuration avoiding collisions. Note that under this motion some points will have to leave the root edge before returning to it.

Our algorithm works as follows. Let

$$A = (A_1, \ldots, A_n) \in F(\Gamma, n) \quad \text{and} \quad B = (B_1, \ldots, B_n) \in F(\Gamma, n)$$

be two given configurations of n distinct points on Γ. Let $A_{i_1}, \ldots, A_{i_r}$ be all the minimal elements (with respect to the order $\succeq$) of the set of points of A. Here we assume that the indices satisfy $i_1 < i_2 < \cdots < i_r$. First we move the point A_{i_1} down to an interior point of the root edge e_0. Next we move A_{i_2} to the root edge e_0 and we continue moving similarly the remaining points $A_{i_3}, \ldots, A_{i_r}$ in order of their indices. As the result, after this first stage of the algorithm, all the minimal points of A are transferred into the root edge e_0. On the second stage we find the minimal set among the remaining points of A and move them down, one after another, to the edge e_0. Iterating this procedure we find a continuous collision free motion of all the points of A moving them onto the interior of the root edge e_0. We obtain a configuration of points $A' = (A'_1, \ldots, A'_n) \in F(e_0, n)$ which all lie in the interior of the root edge e_0, in a certain order.

Applying a similar procedure to the configuration B we obtain a configuration $B' = (B'_1, \ldots, B'_n) \in F(e_0, n) \subset F(\Gamma, n)$ connected with B by a continuous collision free motion.

Next we apply the section σ_0 giving a continuous collision free motion from A' to B'.

Finally, the output of the algorithm is the concatenation of

(1) the motion from A to A';
(2) the motion from A' to B' fiven by σ_0;
(3) the reverse motion from B to B'.

The above algorithm has discontinuities: if one of the points A_j is a vertex $v \in \Gamma$ then a small perturbation of A_j inside Γ may lead to a different set of minimal points (see Figure 2) and hence to a completely different ultimate motion. Note that the vertices of Γ which have valence one or two do not cause discontinuity, i.e. we only need to worry about the essential vertexes of Γ.

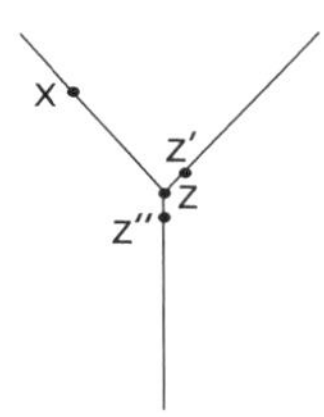

Fig. 2. Perturbations z' and z'' of the vertex point z lead to different sets of minimal points of the set $\{x, z\}$

Let $S_i \subset F(\Gamma, n)$ denote the set of all configurations $A = (A_1, \ldots, A_n)$ such that precisely i points among the points A_j are essential vertices of Γ. If we restrict the above algorithm to the set of pairs $(A, B) \in S_i \times S_j$ with fixed i, j, then the result of the algorithm is a continuous function of the input; in other words we have a continuous section

$$\sigma_{ij} : S_i \times S_j \to F(\Gamma, n)^I, \quad \text{where} \quad i, j = 0, 1, \ldots, m(\Gamma). \tag{10.3}$$

Recall that $m(\Gamma)$ denotes the number of essential vertices of Γ.

We observe that the closure of S_i satisfies

$$\overline{S}_i \subset \bigcup_{j \geq i} S_j. \tag{10.4}$$

It follows that for two distinct pairs (i,j) and (i',j') with $i+j=i'+j'$ one has

$$\overline{S_i \times S_j} \cap (S_{i'} \times S_{j'}) = \emptyset.$$

Hence we obtain that the continuous sections σ_{ij} constructed above define a continuous section of the path fibration over each set

$$W_k = \bigcup_{i+j=k} S_i \times S_j, \quad k = 0, 1, \dots, 2m(\Gamma).$$

The sets $W_0, W_1, \dots, W_{2m(\Gamma)}$ form a partition of $F(\Gamma,n) \times F(\Gamma,n)$ and each of these sets is an ENR. Hence we have described a tame motion planning algorithm on $F(\Gamma,n)$ with $2m(\Gamma)+1$ regions of continuity.

Corollary 10.2: *Let Γ be a tree having an essential vertex. Then the topological complexity of the configuration space $F(\Gamma,n)$ satisfies*

$$\mathsf{TC}(F(\Gamma,n)) \leq 2m(\Gamma) + 1. \tag{10.5}$$

Our goal in the following sections will be to prove the following result:

Theorem 10.3: *Let Γ be a tree not homeomorphic to the interval $[0,1]$ and let n be an integer satisfying $n \geq 2m(\Gamma)$; in the case when $n=2$ we shall additionally assume that Γ is not homeomorphic to the letter Y. Then*

$$\mathsf{TC}(F(\Gamma,n)) = 2m(\Gamma) + 1. \tag{10.6}$$

In other words the upper bound of Corollary 10.2 is exact assuming that $n \geq 2m(\Gamma)$ and hence the motion planning algorithm described above in this section is optimal. There is however one exception: if Γ is homeomorphic to the letter Y then $F(\Gamma,2)$ is homotopy equivalent to the circle S^1 as follows from Theorem 11.1 below. Hence in this case $\mathsf{TC}(F(\Gamma,2)) = 2$, see Example 2.2; the inequality (10.5) is strict in this case.

Theorem 10.3 was stated in [10] without proof. A similar (but slightly different) theorem appears also in a recent preprint [27].

11. The space $F(\Gamma, 2)$ for a tree Γ

In this section (which can be read independently of the rest of the notes) we describe the $\mathbb{Z}_2$-equivariant homotopy type of the configuration space $F(\Gamma,2)$ of two distinct particles of a tree Γ. The involution

$$\tau : F(\Gamma,2) \to F(\Gamma,2)$$

acts by permutting the particles, i.e. $\tau(x,y) = (y,x)$, where $(x,y) \in F(\Gamma,2)$.

Recall that *the degree* of a vertex v (denoted by $\eta(v)$) is the number of edges of Γ incident to v. A vertex v is *essential* if $\eta(v) \geq 3$. Fix a univalent *root vertex* $u_0 \in \Gamma$, $\eta(u_0) = 1$. Then any vertex $v \neq u_0$ has a unique *descending* edge e incident to it; the minimal path connecting v to the root vertex passes through e. The other $\eta(v) - 1$ edges incident to v will be called *ascending*. We associate with a tree Γ a 1-dimensional cell

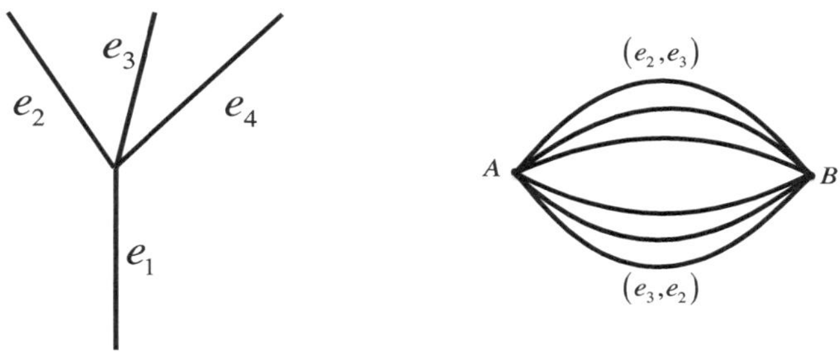

Fig. 3. An essential vertex v, the descending edge e_1 and the ascending edges e_2, e_3, e_4 (left). The graph Q_Γ (right).

complex Q_Γ which is constructed as follows. The complex Q_Γ has two 0-cells (vertices) A and B and an even number

$$\sum_v (\eta(v) - 1)(\eta(v) - 2)$$

of 1-dimensional cells connecting A to B, each labelled by a triple (v, e, e') where v is an essential vertex of Γ and e, e' is an ordered pair of distinct ascending edges of Γ incident to v. The complex Q_Γ has a free involution $T : Q_\Gamma \to Q_\Gamma$ which maps A to B and homeomorphically maps each edge with the label (v, e, e') onto the edge with the label (v, e', e). See Figure 3.

Theorem 11.1: *For a tree Γ having an essential vertex the configuration space $F(\Gamma, 2)$ is $\mathbf{Z}_2$-equivariantly homotopy equivalent to the complex Q_Γ.*

This theorem was stated in [10] without proof. The configuration spaces $F(\Gamma, 2)$ for various classes of graphs Γ complementing the class of trees were explicitly described in [3, 17].

Proof: [Proof of Theorem 11.1] We repeat the standard arguments (compare Proposition 4G.2 from [23]) emphasising the equivariant features we are dealing with.

First we describe an open cover $F(\Gamma, 2) = U \cup V$. Since Γ is a tree, for any two points $x, x' \in \Gamma$ there exists a unique simple path in Γ connecting x to x'. Fix an interior point u_0' of the edge incident to the root vertex u_0. Denote by U the set of all configurations $(x, y) \in F(\Gamma, 2)$ such that the simple path connecting x to u_0 does not pass through y. Similarly, we denote by $V \subset F(\Gamma, 2)$ the set of all configurations (x, y) such that the simple path connecting y to the root u_0 does not pass through x. It is obvious that U and V are open and cover $F(\Gamma, 2)$.

The set U is contractible. Indeed, if $(x, y) \in U$ then we may move the configuration (x, y) continuously to the configuration (u_0, u_0') by first moving x along the minimal path to u_0 and then moving y along the minimal path to u_0'. We obtain a path $(x(t), y(t)) \in F(\Gamma, 2)$ (where $t \in [0, 1]$) with $(x(0), y(0)) = (x, y)$ and $(x(1), y(1)) = (u_0, u_0')$, which is not only continuous as function of t, but it also depends continuously on the initial pair (x, y). Therefore we obtain a continuous deformation retraction of the set U to the point (u_0, u_0').

Similarly, the set V is contractible.

A configuration $(x, y) \in F(\Gamma, 2)$ lies in the intersection $U \cap V$ if the minimal path connecting x to u_0 does not pass through y and the minimal path connecting y to u_0 does not pass through x. Initially, these two minimal paths have distinct routes before they meet at an essential vertex v and then they coincide and follow the minimal path connecting v to u_0 (see Figure 4). We see that the intersection $U \cap V$ has many connected components which can be labelled by triples (v, e, e') where v is an essential vertex of Γ and e, e' is an ordered pair of ascending edges incident to v. We denote by $W_{v,e,e'}$ the set of configurations $(x, y) \in F(\Gamma, 2)$ such that the minimal path connecting x to the root u_0 contains an internal point of e and the minimal path connecting y to the root u_0 contains an internal point of e'. The sets $W_{v,e,e'}$ corresponding to different triples v, e, e' are disjoint and

$$U \cap V = \bigsqcup_{v,e,e'} W_{v,e,e'}.$$

Each set $W_{v,e,e'}$ is contractible since one may continuously move any configuration $(x, y) \in W_{v,e,e'}$ into a fixed configuration (x_0, y_0), where $x_0 \in e$ and $y_0 \in e'$, by moving x and y along the minimal paths connecting these points to the root u_0; this motion is continuous both as a function of time t and as a function of the initial conditions (x, y).

The involution $\tau : F(\Gamma, 2) \to F(\Gamma, 2)$, where $\tau(x, y) = (y, x)$, maps U

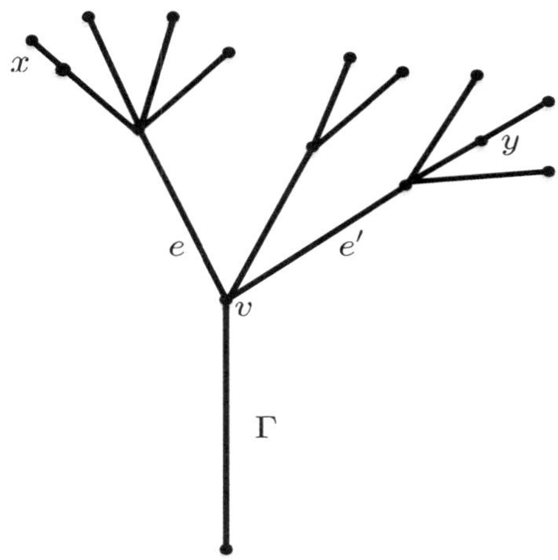

Fig. 4. A configuration (x, y) lying in the set $W_{v,e,e'}$.

onto V and vice versa. Besides, τ maps each set $W_{v,e,e'}$ homeomorphically onto the set $W_{v,e',e}$.

Consider the quotient Q of the disjoint union

$$U \sqcup V \sqcup ((U \cap V) \times [0,1]),$$

where for each configuration $(x, y) \in U \cap V$ we identify the point $((x, y), 0)$ with $(x, y) \in U$ and the point $((x, y), 1)$ with $(x, y) \in V$. The quotient Q carries a natural involution

$$\tau[(x, y), t] = [(y, x), 1 - t],$$

where the square brackets denote the equivalence class in Q. The projection map $P : Q \to F(\Gamma, 2)$ is $\mathbb{Z}_2$-equivariant. Here for $(x, y) \in U \cap V$ one has $P[(x, y), t] = (x, y)$; similarly, for $(x, y) \in U$ or $(x, y) \in V$ one has $P(x, y) = (x, y)$.

Next we show that there exists a continuous $\mathbb{Z}_2$-equivariant section $S : F(\Gamma, 2) \to Q$. Indeed, fix a partition of unity subordinate to the cover U, V; it is a pair of continuous functions $\phi_U, \phi_V : F(\Gamma, 2) \to [0, 1]$ such that $\operatorname{supp}(\phi_U) \subset U$, $\operatorname{supp}(\phi_V) \subset V$ and for each configuration $(x, y) \in F(\Gamma, 2)$ one has

$$\phi_U(x, y) + \phi_V(x, y) = 1.$$

Define the functions $\psi_U, \psi_V : F(\Gamma, 2) \to [0,1]$ by

$$\psi_U(x,y) = \frac{1}{2} \cdot (\phi_U(x,y) + \phi_V(y,x)),$$

$$\psi_V(x,y) = \frac{1}{2} \cdot (\phi_V(x,y) + \phi_U(y,x)).$$

This is a partition of unity satisfying the additional property $\psi_U(x,y) = \psi_V(y,x)$. We may define the section $S : F(\Gamma, 2) \to Q$ by setting

$$S(x,y) \;=\; [(x,y), \psi_U(x,y)].$$

We have $S(y,x) = [(y,x), 1 - \psi_U(x,y)]$, i.e. S is $\mathbb{Z}_2$-equivariant. Clearly, $P \circ S$ is the identity map. Besides, the homotopy $\Phi_\tau : Q \to Q$ given by

$$\Phi_\tau\left[(x,y), t\right] = \left[(x,y), (1-\tau)\psi_U(x,y) + \tau t\right]$$

connects the identity map and the composition $S \circ P$. This shows that P is an equivariant deformation retraction.

Let Q' denote the quotient of Q where the set U is collapsed to a single point (denoted A) and the set V is collapsed to another single point denoted B. The space Q' is the unreduced suspension

$$\Sigma(U \cap V) \;=\; \Sigma(\bigsqcup_{(v,e,e')} W_{v,e,e'}).$$

Since U and V are contractible, we see that the quotient map $Q \to Q'$ is an equivariant homotopy equivalence.

Next we use the fact that each set $W_{v,e,e'}$ is contractible, see above. Hence we obtain that Q' equivariantly deformation retracts onto the suspension $\Sigma(\{(v,e,e')\})$, where $\{(v,e,e')\}$ is viewed as a discrete set of labels. Clearly, the suspension $\Sigma(\{(v,e,e')\})$ coincides with what we denoted by Q_Γ. Thus we have constructed a chain of equivariant homotopy equivalences $F(\Gamma, 2) \simeq Q \simeq Q' \simeq Q_\Gamma$. This completes the proof. □

Example 11.2: Consider the graph Γ of the letter Y which has a single essential vertex of degree 3. Applying Theorem 11.1 we obtain that the configuration space $F(\Gamma, 2)$ is equivariantly homotopy equivalent to the circle S^1 with the standard antipodal involution.

The following result is a straightforward corollary of Theorem 11.1.

Corollary 11.3: *Let Γ, Γ' be trees such that $m(\Gamma') > 0$ and let $\alpha : \Gamma' \to \Gamma$ be a topological embedding. Then the natural inclusion $\alpha' : F(\Gamma, 2) \to F(\Gamma, 2)$ induces a monomorphism*

$$\alpha'_* : H_1(F(\Gamma', 2)) \to H_1(F(\Gamma, 2)).$$

Proof: Let v_0 be a root vertex of Γ', and let u_0 be a root vertex of Γ such that the path connecting u_0 to $\alpha(v_0)$ is disjoint from $\alpha(\Gamma' - \{v_0\})$. We see that the complex $Q_{\Gamma'}$ is naturally a subcomplex of Q_Γ which implies our statement due to Theorem 11.1. □

Corollary 11.4: *If Γ' is the graph homeomorphic to the letter Y then $H_1(F(\Gamma', 2)) = \mathbb{Z}$ and each topological embedding $\alpha : \Gamma' \to \Gamma$ determines a generator of the group $\mathbb{Z} \simeq \alpha_*(H_1(F(\Gamma', 2)) \subset H_1(F(\Gamma, 2))$, unique up to a sign. The homology classes corresponding to all such embeddings α generate the group $H_1(F(\Gamma, 2))$ although not freely.*

12. Top-dimensional cohomology of $F(\Gamma, n)$

In this section we utilise the results of §11 to construct useful cohomology classes of $F(\Gamma, n)$ of the top dimension. The results of this section will be used in the proof of Theorem 10.3.

Let Γ be a tree with $m = m(\Gamma)$ essential vertices. It is known for any n that the configuration space $F(\Gamma, n)$ has the homotopy type of a cell complex of dimension $\leq m$; in particular $H^i(F(\Gamma, n)) = 0$ for $i > m$, see [18]. In this section we shall consider the m-dimensional cohomology classes of $F(\Gamma, n)$ assuming that $n \geq 2m$.

We start from the following general remark which will be useful in the sequel.

Lemma 12.1: *Let Γ be a connected graph having a univalent vertex. Then for any $n' > n$ the natural projection*

$$p : F(\Gamma, n') \to F(\Gamma, n),$$

where $(x_1, \ldots, x_{n'}) \mapsto (x_1, \ldots, x_n)$, is a domination, i.e. there exists a continuous map $q : F(\Gamma, n) \to F(\Gamma, n')$ such that $p \circ q \sim \mathsf{id}$. In particular p induces a monomorphism $p^* : H^*(F(\Gamma, n)) \to H^*(F(\Gamma, n'))$.

Proof: Let u_0 be a univalent vertex of Γ. Let $U \subset \Gamma$ be a small open neighbourhood of u_0 in Γ. The graph $\Gamma' = \Gamma - U$ is homeomorphic to Γ. Define the map $s : F(\Gamma', n) \to F(\Gamma, n')$ as follows: fix a set of $n' - n$ pairwise distinct points $a_{n+1}, a_{n+2}, \ldots, a_{n'} \in U$, and for any configuration $C = (z_1, \ldots, z_n) \in F(\Gamma', n)$ define $s(C) = (z'_1, \ldots, z'_{n'}) \in F(\Gamma, n')$ by

$$z'_i = \begin{cases} z_i, & \text{for } i \leq n; \\ a_i, & \text{for } i > n. \end{cases}$$

The composition

$$F(\Gamma', n) \xrightarrow{s} F(\Gamma, n') \xrightarrow{p} F(\Gamma, n)$$

is a homeomorphism induced by the inclusion $\Gamma' \to \Gamma$. We observe that there is a homotopy of injective maps $h_t : \Gamma \to \Gamma$, with $t \in [0,1]$, such that $h_0 = 1_\Gamma$ and $h_1(\Gamma) = \Gamma'$. Then the composition

$$F(\Gamma, n) \xrightarrow{h_1} F(\Gamma', n) \xrightarrow{s} F(\Gamma, n') \xrightarrow{p} F(\Gamma, n)$$

is homotopic to the identity (through the homotopy $h_t : F(\Gamma, n) \to F(\Gamma, n)$). Thus $p \circ q \sim \mathsf{id}$ where $q = s \circ h_1$. □

From here on, let the symbol Γ denote a tree. For $n \geq 2m$, let

$$\Phi_i : F(\Gamma, n) \longrightarrow F(\Gamma, 2), \quad \text{where} \quad i = 1, \ldots, m,$$

denote the projection

$$\Phi_i(x_1, \ldots, x_n) = (x_{2i-1}, x_{2i}).$$

Denote by $v_1, \ldots, v_m$ the essential vertices of Γ. For each $j = 1, \ldots, m$, fix a topological embedding $\Gamma_j \subset \Gamma$ of a letter Y graph into Γ around the essential vertex v_j. Also, let $\Gamma_0 \subset \Gamma$ be a small interval containing the root vertex. We assume that the subtrees $\Gamma_0, \ldots, \Gamma_m$ are sufficiently small so that $\Gamma_i \cap \Gamma_j = \emptyset$, for $i \neq j$. We shall consider the space $F(\Gamma_i, 2)$ as being a subspace of $F(\Gamma, 2)$ for each i.

Define the embedding

$$\Psi : \prod_{i=1}^{m} F(\Gamma_i, 2) \longrightarrow F(\Gamma, 2m) \tag{12.1}$$

by

$$((x_1, x_2), (x_3, x_4), \ldots, (x_{2m-1}, x_{2m})) \mapsto (x_1, x_2, x_3, \ldots, x_{2m}).$$

We shall denote by $T^m \subset F(\Gamma, 2m)$ the image of Ψ. It is a subset homotopy equivalent to the m-dimensional torus, see Example 11.2.

We have the commutative diagram

$$\begin{array}{ccc} \prod_{j=1}^m F(\Gamma_j, 2) & \xrightarrow{\Psi} & F(\Gamma, 2m) \\ \Big\downarrow{\scriptstyle \Pi_i} & & \Big\downarrow{\scriptstyle \Phi_i} \\ F(\Gamma_i, 2) & \longrightarrow & F(\Gamma, 2) \end{array} \tag{12.2}$$

Here Π_i is the projection on the i-th factor and the lower horizontal map is the inclusion.

For any $j = 1, \ldots, m$ choose a cohomology class

$$\alpha_j \in H^1(F(\Gamma, 2)),$$

which is associated with the vertex v_j via Theorem 11.1; more specifically, we require that

$$\begin{aligned} &\alpha_i | F(\Gamma_i, 2) \neq 0 \in H^1(F(\Gamma_i, 2)) = \mathbb{Z}, \\ &\alpha_j | F(\Gamma_i, 2) = 0, \quad \text{if} \quad i \neq j. \end{aligned} \tag{12.3}$$

Such classes exist due to Theorem 11.1.

We obtain m^2 cohomology classes

$$u_{ij} \in H^1(F(\Gamma, 2m)),$$

defined by

$$u_{ij} = \Phi_i^*(\alpha_j), \quad i, j = 1, \ldots, m.$$

Using the commutative diagram (12.2) we obtain

$$\begin{aligned} &u_{ii}|_{T^m} \neq 0 \in H^1(T^m), \\ &u_{ij}|_{T^m} = 0, \quad \text{if} \quad i \neq j. \end{aligned} \tag{12.4}$$

Moreover, we see that the cup-product

$$u_{11}u_{22} \ldots u_{mm} = \prod_{i=1}^{m} u_{ii} \in H^m(F(\Gamma, 2m)) \tag{12.5}$$

is nonzero since

$$(\prod_{i=1}^{m} u_{ii})\,|_{T^m} \neq 0.$$

This follows from our remark above that the class u_{ii} is induced from a nonzero class $\alpha_i | F(\Gamma_i, 2)$ under the projection $\Pi_i : T^m \to F(\Gamma_i, 2)$.

If $z \in H_m(F(\Gamma, 2m))$ denotes the homology class realised by T^m then

$$\langle \prod_{i=1}^{m} u_{ii}, \ z \rangle \neq 0. \tag{12.6}$$

Next we consider different m-fold products of the classes u_{ij}. First we observe that $u_{ij}u_{ik} = 0$ for any i, j, k. Indeed, $\alpha_j \alpha_k = 0 \in H^2(F(\Gamma, 2))$ since $F(\Gamma, 2)$ is homotopy equivalent to a graph; hence $u_{ij}u_{ik} = \Phi_i^*(\alpha_j \alpha_k) = 0$.

Let $\sigma = (i_1, i_2, \dots, i_m)$ be a sequence with $i_k \in \{1, 2, \dots, m\}$; we do not require it to be a permutation, i.e. repetitions of the indices are allowed. We associate with σ the top-dimensional cohomology class

$$u_\sigma = u_{1i_1} u_{2i_2} \dots u_{mi_m} \in H^m(F(\Gamma, 2m)).$$

It follows from (12.4) that

$$\langle u_\sigma, z \rangle = 0, \tag{12.7}$$

assuming that σ is distinct from the sequence $(1, 2, \dots, m)$.

For a permutation $\tau = (j_1, j_2, \dots, j_m)$ of the indices $1, 2, \dots, m$ define the homeomorphism

$$L^\tau : F(\Gamma, 2m) \to F(\Gamma, 2m)$$

by $L^\tau(x_1, x_2, \dots, x_{2m}) = (x_{2j_1-1}, x_{2j_1}, \dots, x_{2j_m-1}, x_{2j_m})$. Define also the homology class

$$z^\tau = L^\tau_*(z) \in H_m(F(\Gamma, 2m)).$$

For a sequence $\sigma = (i_1, i_2, \dots, i_m)$ and for a permutation $\tau = (j_1, j_2, \dots, j_m)$, we claim that the evaluation

$$\langle u_\sigma, z^\tau \rangle \neq 0 \tag{12.8}$$

is nonzero if and only if σ and τ coincide. Indeed, we compute

$$\begin{aligned}
\langle u_\sigma, z^\tau \rangle &= \langle u_\sigma, L^\tau_*(z) \rangle \\
&= \langle (L^\tau)^*(u_\sigma), z \rangle = \langle (L^\tau)^*(\prod_{k=1}^m u_{ki_k}), z \rangle \\
&= \langle \prod_{k=1}^m (L^\tau)^*(u_{ki_k}), z \rangle = \langle \prod_{k=1}^m (\Phi_k \circ L^\tau)^*(\alpha_{i_k}), z \rangle \\
&= \langle \prod_{k=1}^m \Phi^*_{j_k}(\alpha_{i_k}), z \rangle = \langle \prod_{k=1}^m u_{j_k i_k}, z \rangle.
\end{aligned}$$

Here we used that $\Phi_k \circ L^\tau = \Phi_{j_k}$. Using (12.4) and (12.5) we obtain that the number $\langle u_\sigma, z^\tau \rangle$ is nonzero iff $j_k = i_k$ for any k, i.e. iff σ and τ are equal.

Corollary 12.2: *The cohomology classes $u_\sigma \in H^m(F(\Gamma, 2m))$ corresponding to various permutations σ are linearly independent. In particular, for $n \geq 2m$ the rank of the group $H^m(F(\Gamma, n))$ is at least $m!$.*

13. Proof of Theorem 10.3

Below, we assume that Γ is a tree and $n \geq 2m$. Let us first assume that $m \geq 2$.

Any degree one cohomology class $u \in H^1(F(\Gamma, n))$ determines a zero-divisor

$$\bar{u} = u \otimes 1 - 1 \otimes u \in H^*(F(\Gamma, n)) \otimes H^*(F(\Gamma, n)).$$

Our goal is to find $2m$ cohomology classes of degree one such that the product of the corresponding zero-divisors is nonzero.

We shall use the notations introduced in the previous section. Consider the classes $u_{11}, u_{22}, \dots, u_{mm}$ and $u_{12}, u_{23}, \dots, u_{(m-1)m}, u_{m1}$ and the corresponding zero-divisors $\bar{u}_{ii}, \bar{u}_{i(i+1)}$. We want to show that the product

$$\prod_{i=1}^{m} \bar{u}_{ii} \times \prod_{i+1}^{m} \bar{u}_{i(i+1)} \tag{13.1}$$

is nonzero. We know that the cohomology of $F(\Gamma, n)$ vanishes in degrees $> m$. Therefore we obtain

$$\prod_{i=1}^{m} \bar{u}_{ii} \times \prod_{i+1}^{m} \bar{u}_{i(i+1)} = \sum_{S} \pm \left(\prod_{i \in S} u_{ii} \times \prod_{i \notin S} u_{i(i+1)} \right) \otimes \left(\prod_{i \notin S} u_{ii} \times \prod_{i \in S} u_{i(i+1)} \right).$$

Here S runs over all subsets $S \subset \{1, 2, \dots, m\}$; the sign $\times$ denotes the cup-product. We shall evaluate the product (13.1) on the tensor product of two homology classes $z \otimes z^{\tau}$ where τ is the permutation $(2, 3, \dots, m, 1)$ and $z \in H_m(F(\Gamma, n))$ is the homology class defined above. Using statements (12.6), (12.7), (12.8) we find that all the terms in the sum

$$\begin{aligned} &\langle \prod_{i} \bar{u}_{ii} \otimes \prod_{i} \bar{u}_{i(i+1)}, z \otimes z^{\tau} \rangle \\ &= \sum_{S} \pm \langle \prod_{i \in S} u_{ii} \times \prod_{i \notin S} u_{i(i+1)}, z \rangle \cdot \langle \prod_{i \notin S} u_{ii} \times \prod_{i \in S} u_{i(i+1)}, z^{\tau} \rangle \end{aligned}$$

vanish except for the term with $S = \{1, 2, \dots, m\}$ which is nonzero. This shows that the product (13.1) is nonzero. Therefore,

$$\mathsf{zcl}(F(\Gamma, n) \geq 2m, \quad \text{for} \quad n \geq 2m.$$

By Theorem 3.3 we have $\mathsf{TC}(F(\Gamma, n)) \geq 2m + 1$ and the inverse inequality is given by Corollary 10.2.

The above arguments fail in the case $m = 1$, i.e. when Γ is a tree with a single essential vertex v, $\eta(v) \geq 3$. In the case $\eta(v) = 3$ the tree

Γ is homeomorphic to the letter Y, hence we shall assume that $\eta(v) \geq 4$. Consider the graph Q_Γ given by Theorem 11.1. By Theorem 11.1 the space $F(\Gamma, 2)$ is homotopy equivalent to a wedge of circles where the number of circles equals

$$b_1(F(\Gamma, 2)) = \sum_{v \in V(\Gamma)} (\eta(v) - 1)(\eta(v) - 2) - 1 \geq 5.$$

Using Example 3.5 we have $\mathsf{TC}(F(\Gamma, 2)) = 3$. This proves our statement for $n = 2$. If $n > 2$ we apply Lemma 12.1 and Theorem 2.3 to conclude $\mathsf{TC}(F(\Gamma, n)) \geq \mathsf{TC}(F(\Gamma, 2)) = 3$. This completes the proof.

14. Further comments

It is interesting to compare Theorems 7.1 and 10.3. The topological complexity $\mathsf{TC}(F(\mathbb{R}^d, n))$ is linear in n but, in contrast, $\mathsf{TC}(F(\Gamma, n))$ equals $2m(\Gamma) + 1$, i.e. it is independent of n. This result may have some practical implications: *to simplify the task of controlling a large number of objects moving in space without collisions one may restrict their motion to a graph.*

In [14] the authors analysed the topological complexity of collision free motion planning of multiple objects in $\mathbb{R}^d$ in the presence of moving obstacles.

The notion of *higher topological complexity* $\mathsf{TC}_s(X)$, where $s = 2, 3, \ldots$ was introduced by Rudyak [26]. The number $\mathsf{TC}_s(X)$ can be defined as the Schwarz genus of the fibration

$$p_s : X^I \to X^s$$

where

$$p_s(\gamma) = \left(\gamma(0), \gamma\left(\frac{1}{s-1}\right), \ldots, \gamma\left(\frac{k}{s-1}\right), \ldots, \gamma(1)\right),$$

compare with (2.1). The invariant $\mathsf{TC}(X)$ which we studied in these notes coincides with $\mathsf{TC}_2(X)$.

The invariant $\mathsf{TC}_s(X)$ is also related to robotics: while in the case of $\mathsf{TC}(X)$ we are dealing with algorithms for a robot to move from an initial state to a final state, in the case of $\mathsf{TC}_s(X)$ with $s > 2$ we require that while moving from the initial state to the final state the robot visits $s - 2$ additional intermediate states. This explains why $\mathsf{TC}_s(X)$ is also called *"the sequential topological complexity"*.

Note that our notation $\mathsf{TC}_s(X)$ stands for what is called *"the unreduced"* topological complexity; *"the reduced"* version is smaller by one.

The sequential topological complexities of configuration spaces $F(\mathbb{R}^d, n)$ were computed in [20]:

$$\mathsf{TC}_s(F(\mathbb{R}^d, n)) = \begin{cases} sn - s + 1, & \text{for } d \text{ odd}, \\ \\ sn - s, & \text{for } d \text{ even}. \end{cases}$$

The topological complexity of a closed orientable surface Σ_g of genus g was computed in the initial paper [8]:

$$\mathsf{TC}(\Sigma_g) = \begin{cases} 3, \text{ for } g = 0 \text{ and } g = 1, \\ \\ 5, \text{ for } g \geq 2. \end{cases}$$

The task of finding $\mathsf{TC}(N_g)$ turned out to be much more difficult; here N_g stands for the closed non-orientable surface of genus g. The case N_1 (the real projective plane) was settled in [11]:

$$\mathsf{TC}(N_1) = 4.$$

A. Dranishnikov [7] proved that $\mathsf{TC}(N_g) = 5$ for any $g \geq 5$; he also mentioned that his method can be pushed to prove that $\mathsf{TC}(N_4) = 5$ as well. While preparing these notes for publication (December 2016) I received information that two independent groups of researchers obtained the full solution to the problem:

$$\mathsf{TC}(N_g) = 5, \quad \text{for any } g \geq 2,$$

see [4] and [5].

Proposition 5.1 from [22] claims that[a]

$$\mathsf{TC}_s(\Sigma) = 2s + 1, \quad s \geq 3,$$

for any closed surface Σ not homeomorphic to the sphere or the torus. See also Proposition 3.2 from [21].

References

1. A. Abrams, *Configuration spaces of colored graphs*, Geom. Dedicata **92** (2002), 185–194.
2. A. Abrams, R. Ghrist, *Finding topology in a factory: configuration spaces.* Amer. Math. Monthly **109** (2002), no. 2, 140–150.
3. K. Barnett, M. Farber, *Topology of configuration space of two particles on a graph.* I. Algebr. Geom. Topol. 9 (2009), no. 1, 593–624.

[a]Note that we are using in these notes the non-reduced topological complexity.

4. D.C. Cohen and L. Vandembroucq, *Topological complexity of the Klein bottle*, preprint, December 2016.
5. D. Davis, *The topological complexity of the Klein bottle equals 5*, preprint arXiv:1612.02747, 2016.
6. A. Dold, *Lectures on Algebraic Topology*, Springer-Verlag, 1972.
7. A. Dranishnikov, *The topological complexity and the homotopy cofiber of the diagonal map for non-orientable surfaces.* Proc. Amer. Math. Soc. 144 (2016), no. 11, 4999–5014.
8. M. Farber, *Topological Complexity of Motion Planning.* Discrete and Computational Geometry **29** (2003), 211–221.
9. M. Farber, *Instabilities of Robot Motion.* Topology and its Applications **140** (2004), 245–266.
10. M. Farber, *Collision free motion planning on graphs,* in: "Algorithmic Foundations of Robotics IV", M. Erdmann, D. Hsu, M. Overmars, A. Frank van der Stappen editors, Springer, 2005, 123–138.
11. M. Farber, S. Tabachnikov, S. Yuzvinsky, *Topological Robotics: Motion Planning in Projective Spaces.* "International Mathematical Research Notices", **34** (2003), 1853–1870.
12. M. Farber, S. Yuzvinsky, *Topological Robotics: Subspace Arrangements and Collision Free Motion Planning.* Transl. of AMS, **212** (2004), 145–156.
13. M. Farber, *Topology of robot motion planning,* In "Morse Theoretic Methods in Nonlinear Analysis and in Symplectic Topology", Paul Biran, Octav Cornea, Francois Lalonde editors, pp. 185–230, 2006.
14. M. Farber, M. Grant, S. Yuzvinsky, *Topological complexity of collision free motion planning algorithms in the presence of multiple moving obstacles,* In "Topology and Robotics" (M. Farber, R. Ghrist *et al.* editors), Contemporary Mathematics AMS, vol. 438, 2007, 75–83.
15. M. Farber and M. Grant, *Topological complexity of configuration spaces*, Proc. of the AMS, **137** (2009), 1841–1847.
16. M. Farber, *Invitation to topological robotics*, Zurich Lectures in Advanced Mathematics, EMS, 2008.
17. M. Farber, E. Hanbury, *Topology of configuration space of two particles on a graph, II.* Algebr. Geom. Topol. 10 (2010), no. 4, 2203–2227.
18. R. Ghrist, *Configuration spaces and braid groups on graphs in robotics*, In "Knots, Braids, and Mapping Class Groups", dedicated to Joan Birman, pages 29–40, AMS/IP Stud. Adv. Math., 24, AMS, Providence, 2001.
19. R.W. Ghrist, D.E. Koditschek, *Safe cooperative robot dynamics on graphs*, SIAM J. Control Optim. **40** (2002), no. 5, 1556–1575.
20. J. González, M. Grant, *Sequential motion planning of non-colliding particles in Euclidean spaces.* Proc. Amer. Math. Soc. 143 (2015), no. 10, 4503–4512.
21. J. González, B. Gutiérrez, A. Guzman, C. Hidber, M. Mendoza, C. Roque, *Motion planning in tori revisited*, Morfismos, **19** (2015), 7–18.
22. J. Gonzĺez, B. Gutiérrez, D. Gutiérrez, A. Lara, *Motion planning in real flag manifolds*, Homology, Homotopy and Applications, **18** (2016), 359–375.
23. A. Hatcher, *Algebraic Topology*, Cambridge, 2002.
24. J.-C. Latombe, *Robot Motion Planning.* Kluwer Academic Publishers, 1991.

25. H. Mas-Ku, E. Torres-Giese, *Motion planning algorithms for configuration spaces*, Bol. Soc. Mex. (2015), 265–274.
26. Y. Rudyak, *On higher analogs of topological complexity*, Topology and its Applications, 157 (2010), 916–920.
27. S. Scheirer, *Topological complexity of n points on a tree*, arXiv:1607.08185, 2016.
28. A.S. Schwarz, *The genus of a fiber space.* Amer. Math. Sci. Transl. **55** (1966), 49–140.
29. M. Sharir, *Algorithmic motion planning.* Handbook of Discrete and Computational Geometry. J.E. Goodman, J. O'Rourke eds. CRC Press, Boca Raton, FL, 1997, 733–754.
30. E. Spanier, *Algebraic Topology,* McGraw-Hill Company, 1966.

CELLULAR STRATIFIED SPACES

Dai Tamaki
Department of Mathematics
Faculty of Science
Shinshu University
3-1-1, Asahi, Matsumoto, 390-8621, Japan
rivulus@shinshu-u.ac.jp

Dedicated to Professor Fred Cohen on the occasion of his 70th birthday.

The notion of cellular stratified spaces was introduced in a joint work of the author with Basabe, González, and Rudyak with the aim of constructing a cellular model of the configuration space of a sphere. Although the original aim was not achieved in the project, the notion of cellular stratified spaces turns out to be useful, at least, in the study of configuration spaces of graphs. In particular, the notion of totally normal cellular stratified spaces was used successfully in a joint work with the former students of the author [1] to study the homotopy type of configuration spaces of graphs with a small number of vertices.

Roughly speaking, totally normal cellular stratified spaces correspond to acyclic categories in the same way regular cell complexes correspond to posets.

In these notes, we extend this correspondence by replacing cells by stellar cells and acyclic categories by topological acyclic categories.

Contents

1. Introduction

The author has been interested in configuration spaces since he was a graduate student at the University of Rochester under the guidance of Professor Fred Cohen. The interest arose when the author tried to understand the global structure of homotopy groups of spheres in his Ph.D. thesis [2, 3, 4]. One of the key aspects are the underlying combinatorial structures in the configuration spaces of Euclidean spaces.

Recently the author renewed his interest in configuration spaces during a joint project with Basabe, González, and Rudyak [5] on higher symmetric topological complexities. The notion of cellular stratified spaces was discovered during discussions with them.

It turns out that cellular stratified spaces have already appeared in many areas in topology. For example, the stratification on the complement of a complexified hyperplane arrangement used in the construction of the Salvetti complex [6, 7, 8] is one of the motivating examples.

1.1. *Cellular stratified spaces everywhere*

Before we state the main results, let us take a look at examples of cellular stratified spaces. We begin with configuration spaces. The configuration space of n distinct points in a topological space X is defined by

$$\begin{aligned}\mathrm{Conf}_n(X) &= \{(x_1,\ldots,x_n)\in X^n \mid x_i\neq x_j \text{ for } i\neq j\}\\ &= X^n\setminus\Delta_n(X),\end{aligned}$$

where

$$\Delta_n(X)=\bigcup_{1\le i<j\le n}\{(x_1,\ldots,x_n)\in X^n\mid x_i=x_j\}.$$

The starting point of this work is the following problem:

Problem 1.1: *Given a space X, construct a combinatorial model for the homotopy type of the configuration space* $\mathrm{Conf}_k(X)$ *of k distinct points in X. In other words, find a regular cell complex or a simplicial complex $C_k(X)$ embedded in* $\mathrm{Conf}_k(X)$ *as a Σ_k-equivariant deformation retract.*

Several solutions are known in special cases.

Example 1.2: For a finite CW-complex X of dimension 1, namely a graph, Abrams constructed a subspace $C_k^{\mathrm{Abrams}}(X)$ contained in $\mathrm{Conf}_k(X)$ in his thesis [9] and proved that there is a homotopy equivalence

$$C_k^{\mathrm{Abrams}}(X)\simeq\mathrm{Conf}_k(X)$$

as long as the following two conditions are satisfied:

(1) each path connecting vertices in X of valency more than 2 has length at least $k+1$, and
(2) each homotopically essential path connecting a vertex to itself has length at least $k+1$.

Here a path means a 1-dimensional subcomplex homeomorphic to a closed interval.

Example 1.3: Consider the case $X = \mathbb{R}^n$. For $1 \le i < j \le k$, the hyperplane

$$H_{i,j} = \left\{(x_1, \dots, x_k) \in \mathbb{R}^k \,\middle|\, x_i = x_j\right\}$$

in $\mathbb{R}^k$ defines a linear subspace $H_{i,j} \otimes \mathbb{R}^n$ in $\mathbb{R}^k \otimes \mathbb{R}^n = \underbrace{\mathbb{R}^n \times \cdots \times \mathbb{R}^n}_{k} = X^k$ and we have

$$\mathrm{Conf}_k(\mathbb{R}^n) = \mathbb{R}^k \otimes \mathbb{R}^n \setminus \bigcup_{1 \le i < j \le k} H_{i,j} \otimes \mathbb{R}^n.$$

The collection $\{H_{i,j} \mid 1 \le i < j \le k\}$ is called the *braid arrangement* of rank $k-1$ and is denoted by $\mathcal{A}_{k-1}$.

When $n = 2$, the construction due to Salvetti [6] gives us a regular cell complex $\mathrm{Sal}(\mathcal{A}_{k-1})$ embedded in $\mathrm{Conf}_k(\mathbb{R}^2)$ as a Σ_k-equivariant deformation retract.

More generally, the construction sketched at the end of [7] by Björner and Ziegler and elaborated in [8] by De Concini and Salvetti gives us a regular cell complex $\mathrm{Sal}^{(n)}(\mathcal{A}_{k-1})$ embedded in $\mathrm{Conf}_k(\mathbb{R}^n)$ as a Σ_k-equivariant deformation retract.

This construction is a special case of the construction of a regular cell complex whose homotopy type represents the complement of the subspace arrangement associated with a real hyperplane arrangement.

There are pros and cons in these two constructions. The conditions in Abrams' theorem require us to subdivide a given 1-dimensional CW-complex finely. For example, his construction fails to give the right homotopy type of the configuration space $\mathrm{Conf}_2(S^1)$ of two points in S^1 when it is applied to the minimal cell decomposition: $S^1 = e^0 \cup e^1$. The minimal regular cell decomposition $S^1 = e^0_- \cup e^0_+ \cup e^1_- \cup e^1_+$ is not fine enough either. We need to subdivide S^1 into three 1-cells to use Abrams' model.

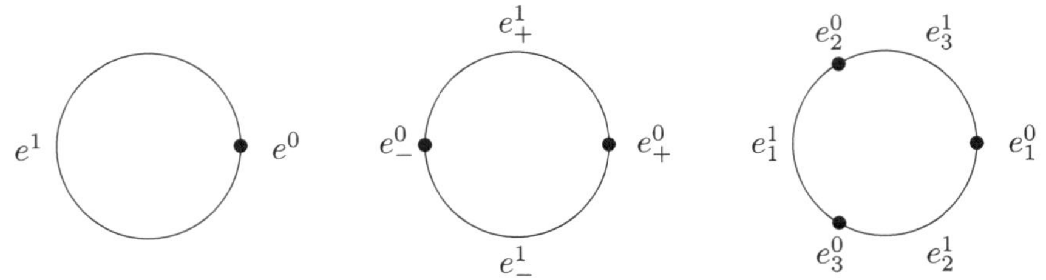

Another problem is that his theorem is restricted to 1-dimensional CW-complexes, although the construction of the model itself works for any cell complex and higher dimensional cases appeared recently in [10].

The second construction suggests that we should consider more general stratifications than cell decompositions. The complex $\mathrm{Sal}^{(n)}(\mathcal{A}_{k-1})$ is constructed from the combinatorial structure of the "cell decomposition" of $\mathbb{R}^k \otimes \mathbb{R}^n$ defined by the hyperplanes in the arrangement $\mathcal{A}_{k-1}$ together with the standard framing in $\mathbb{R}^n$. "Cells" in this decomposition are unbounded regions in $\mathbb{R}^k \otimes \mathbb{R}^n$. Although such a decomposition is not regarded as a cell decomposition in the usual sense, we may extend the definition of face posets to such generalized cell decompositions. And the complex $\mathrm{Sal}^{(n)}(\mathcal{A}_{k-1})$ is constructed in terms of the combinatorial structure of the face poset of the "cell decomposition" of $\mathbb{R}^k \otimes \mathbb{R}^n$. The crucial deficiency of the second construction is, however, that it works only for Euclidean spaces.

One of the motivations of these notes is to find a common framework for working with configuration spaces and complements of arrangements. Although there are many interesting "parallel theories" between configuration spaces and arrangements, e.g. the Fulton-MacPherson-Axelrod-Singer compactification [11, 12] and the De Concini-Procesi wonderful model [13], there is no "Salvetti complex" for configuration spaces in general. A more concrete motivation is, therefore, to solve Problem 1.1 in such a way that it generalizes the Salvetti complex for the braid arrangement.

By analyzing the techniques of combinatorial algebraic topology used in the proof of Salvetti's theorem, the notion of cellular stratified spaces was introduced in [14]. It turns out that, in the case of configuration spaces of spheres, which was the main target of study in the project, it was not easy to use cellular stratified spaces to construct a combinatorial model. The section for cellular stratified spaces was removed from the published version [5].

However, the theory of cellular stratified spaces can be used to study configuration spaces of graphs, as is done in [1], in which the notion of totally normal cellular stratified spaces (Definition 3.6) played a central role.

Definition 1.4: Let X be a normal cellular stratified space. X is called *totally normal* if, for each n-cell e_λ,

(1) there exists a structure of a regular cell complex on S^{n-1} containing ∂D_λ as a stratified subspace, and
(2) for any cell e in ∂D_λ, there exists a cell e_μ in ∂e_λ such that e_μ and e share the same domain and the characteristic map of e_μ factors through

D_λ via the characteristic map of e:

$$\begin{array}{ccccccc} \overline{e} & \hookrightarrow & \partial D_\lambda & \hookrightarrow & D_\lambda & \xrightarrow{\varphi_\lambda} & X \\ \uparrow & & & & & \nearrow_{\varphi_\mu} & \\ D & = & D_\mu . & & & & \end{array}$$

The category $C(X)$ consisting of cells in X as objects and the above maps $D_\mu \to \overline{e} \to D_\lambda$ as morphisms is called the *face category* of X.

The following result (Theorem 2.50 in [1]) says that we can always recover the homotopy type of X from its face poset, if X is totally normal.

Theorem 1.5: *For a totally normal cellular stratified space X, the classifying space $BC(X)$ of the face category $C(X)$ can be embedded in X as a strong deformation retract.*

When $X = \mathbb{R}^k \otimes \mathbb{R}^n$ and the stratification is given by a real hyperplane arrangement $\mathcal{A}$ in $\mathbb{R}^k$ by the method of Björner-Ziegler and De Concini-Salvetti [7, 8], *then $BC(X)$ is homeomorphic to the higher order Salvetti complex* $\mathrm{Sal}^{(n)}(\mathcal{A})$.

Note that, if X is a regular cell complex, it is a fundamental fact in combinatorial algebraic topology that the classifying space $BF(X)$ of the face poset $F(X)$ is the barycentric subdivision of X and is homeomorphic to X. The above result is a generalization of this well-known fact.

Examples of totally normal cellular stratified spaces include

- regular cell complexes,
- the Björner-Ziegler stratification [7] of Euclidean spaces defined by subspace arrangements,
- graphs regarded as 1-dimensional cell complexes,
- the minimal cell decomposition of $\mathbb{R}\mathrm{P}^n$,
- Kirillov's PLCW-complexes [15] satisfying a certain regularity condition, and
- the geometric realization of Δ-sets.

Another source of inspiration for these notes is a preprint [16] of R. Cohen, J.D.S. Jones, and G.B. Segal. Given a Morse-Smale function $f : M \to \mathbb{R}$ on a smooth closed manifold M, they constructed a topological acyclic category $C(f)$ and proved that the classifying space $BC(f)$ is homeomorphic to M. Under a weaker assumption, they also proved that

$BC(f)$ is homotopy equivalent to X. Their results strongly suggest that a "topological acyclic category version" of Theorem 1.5 should exist.

1.2. *Statements of results*

The aim of these notes is to extend these results and develop the theory of cellular stratified spaces. For this purpose, we first introduce the notion of cylindrical structures on cellular stratified spaces.

Definition 1.6: A *cylindrical structure* on a normal cellular stratified space X consists of:

- a normal stratification on S^{n-1} containing ∂D_λ as a stratified subspace for each n-cell $\varphi_\lambda : D_\lambda \to \overline{e_\lambda}$ in X,
- a stratified space $P_{\mu,\lambda}$ and a morphism of stratified spaces

$$b_{\mu,\lambda} : P_{\mu,\lambda} \times D_\mu \longrightarrow \partial D_\lambda$$

  for each pair of cells $e_\mu \subset \partial e_\lambda$,
- a morphism of stratified spaces

$$c_{\lambda_0,\lambda_1,\lambda_2} : P_{\lambda_1,\lambda_2} \times P_{\lambda_0,\lambda_1} \longrightarrow P_{\lambda_0,\lambda_2}$$

  for each sequence $\overline{e_{\lambda_0}} \subset \overline{e_{\lambda_1}} \subset \overline{e_{\lambda_2}}$,

satisfying certain compatibility and associativity conditions. A cellular stratified space equipped with a cylindrical structure is called a *cylindrically normal* cellular stratified space.

Examples of cylindrically normal cellular stratified spaces include:

- totally normal cellular stratified spaces,
- PLCW complexes,
- the minimal cell decomposition of $\mathbb{C}\mathrm{P}^n$,
- the geometric realization of simplicial sets.

The cylindrical structure on $\mathbb{C}\mathrm{P}^n$ can be defined by using the "moduli spaces of flows" of a Morse-Smale function, as is done in the preprint [16] by R. Cohen, J.D.S Jones, and G.B. Segal mentioned above. Alternatively we can construct the same cylindrical structure by identifying $\mathbb{C}\mathrm{P}^n$ with the Davis-Januszkiewicz construction M_{λ_n}. See Example 4.26 for more details. This observation suggests a large class of quasitoric and torus manifolds have cylindrically normal cell decompositions.

Given a cylindrically normal cellular stratified space X, we define a topological acyclic category $C(X)$, called the cylindrical face category of

X. Objects are cells in X and the space of morphisms from e_μ to e_λ is defined to be $P_{\mu,\lambda}$.

Our first result says that the classifying space $BC(X)$ of $C(X)$ can always be embedded in X.

Theorem 1.7: *For any cylindrically normal cellular stratified space X, there exists an embedding*

$$i_X : BC(X) \hookrightarrow X$$

which is natural with respect to morphisms of cellular stratified spaces. Furthermore, when all cells in X are closed, i_X is a homeomorphism.

When X contains non-closed cells, i_X is not a homeomorphism. Our second result says that, under a reasonable condition, those non-closed cells can be collapsed into $BC(X)$.

Theorem 1.8: *For a polyhedral cellular stratified space X (Definition 4.7), the image of the embedding $i_X : BC(X) \hookrightarrow X$ is a strong deformation retract of X. The deformation retraction can be taken to be natural with respect to morphisms of polyhedral cellular stratified spaces.*

It turns out that the theorems still hold when we replace cells by "star-shaped" cells. We introduce the notion of stellar stratified spaces and show that the functor $BC(-)$ transforms cellular stratified spaces to stellar stratified spaces, giving us a dualizing operation $D(-)$. With this structure, Theorem 1.7 (Theorem 5.16) can be rephrased as follows:

Theorem 1.9: *For a totally normal stellar stratified space X, we have an embedding of stellar stratified spaces*

$$D(D(X)) \hookrightarrow X,$$

which is an isomorphism when all cells in X are closed.

It should be worthwhile noting that, when X is the complement of the complexification of a real hyperplane arrangement $\mathcal{A}$, $DD(X)$ coincides with the Salvetti complex of $\mathcal{A}$. This fact and examples of cylindrical structures suggest that we may apply these theorems to the following problems:

(1) Construct combinatorial models for configuration spaces and apply them to study the homotopy types of configuration spaces.
(2) Develop a refinement and an extension of the Cohen-Jones-Segal Morse theory.

(3) Reformulate Forman's discrete Morse theory [17, 18] in terms of topological acyclic categories.

We should also be able to apply the results in these notes to other types of configurations and arrangements. We might be able to apply the results to toric topology.

1.3. *Organization*

The notes are organized as follows:

- §2 is preliminary. We fix notations and terminologies for stratified spaces in §2.1. The definition of cell structures is given in §2.2. Cellular stratified spaces are introduced in §2.3. We introduce stellar stratified spaces in §2.4.
- We extend the regularity and normality for cell complexes in §3. The regularity and normality are extended to cellular stratified spaces in §3.1. In §3.2, the definition and basic properties of totally normal cellular stratified spaces are recalled from [1]. §3.3 is devoted to examples of totally normal cellular stratifies spaces.
- A new structure, called cylindrically normal cellular and stellar stratifications, is introduced and studied in §4. After defininng cylindrically normal cellular stratified spaces in §4.1, we impose a piecewise-linear structure on each cell and introduce polyhedral structures in §4.2. We review examples of cylindrically normal cellular stratified spaces in §4.3.
- Theorems 1.7 and 1.8 are proved in §5.
- For applications to configuration spaces, we need to understand basic operations on cellular stratified spaces, which is the subject of §6. We study three kinds of operations; stratified subspaces in §6.1, products in §6.2, and subdivisions in §6.3.
- As an extension of the barycentric sudivision of regular cell complexes, our framework is suitable for discussing duality, which is the subject of §7.

The notes contain three appendices for the convenience of the reader.

- Understanding the behavior of quotient maps is important in these notes, since cell structure maps are required to be quotient maps. We summerize important properties of quotient maps in Appendix A.
- In Appendix B, we recall definitions and properties of simplicial complexes, simplicial sets, and related structures.

- Appendix C is a summary on the basics of topological categories, including their classifying spaces.

2. Stratifications and cells

This section is preliminary. We introduce the notions of stratified spaces, cell structures, and cellular stratified spaces. Although the term "stratified space" has been used in singularity theory since its beginning and is a well-established concept, we need our own definition of stratified spaces.

2.1. *Stratified spaces*

Before we introduce *cellular* stratified spaces, let us first recall the notion of stratified spaces in general, whose theory has been developed in singularity theory. Unfortunately, however, there seems to be no standard definition of stratified spaces. There are many non-equivalent definitions in the literature. For this reason, we decided to examine several books and extract properties for our needs. As our prototypes, we use definitions in books by Kirwan [19], Bridson and Haefliger [20], Pflaum [21], and Schürmann [22].

Here is our reformulation.

Definition 2.1: Let Λ be a poset. Define $D \subset \Lambda$ to be closed if and only if, for $\lambda \in D$ and $\mu \in \Lambda$, $\mu \leq \lambda$ implies $\mu \in D$. Or $U \subset \Lambda$ is open if and only if, for $\lambda \in U$ and $\mu \in \Lambda$, $\lambda \leq \mu$ implies $\mu \in U$. This topology is called the *Alexandroff topology* on Λ.

Definition 2.2: Let X be a topological space and Λ be a poset. A *stratification* of X indexed by Λ is an open continuous map

$$\pi : X \longrightarrow \Lambda$$

satisfying the condition that, for each $\lambda \in \operatorname{Im}\pi$, $\pi^{-1}(\lambda)$ is connected and locally closed, where Λ is topologized by the Alexandroff topology.

For simplicity, we put $e_\lambda = \pi^{-1}(\lambda)$ and call it a *stratum* with index λ.

Recall that a subset A of a topological space X is said to be *locally closed* if every point $x \in A$ has a neighborhood U in X with $A \cap U$ closed in U. This condition is known to be equivalent to saying that A is an intersection of an open and a closed subset of X, or A is open in $\overline{A}$.

Remark 2.3: We may safely assume that π is surjective. When we define morphisms of stratified spaces and stratified subspaces, however, it is more convenient not to assume the surjectivity.

Given a map $\pi : X \to \Lambda$, we have a decomposition of X, i.e.

(1) $X = \bigcup_{\lambda \in \operatorname{Im}\pi} e_\lambda$.
(2) For $\lambda, \lambda' \in \operatorname{Im}\pi$, $e_\lambda \cap e_{\lambda'} = \emptyset$ if $\lambda \neq \lambda'$.

Thus the image of π in the indexing poset Λ can be identified with the set of strata. The definition of stratification can be rephrased in terms of closures of strata.

Lemma 2.4: *Let $\pi : X \to \Lambda$ be a continuous map from a topological space X to a poset Λ. Then it is open if and only if the condition $e_\lambda \subset \overline{e_{\lambda'}}$ is equivalent to $\lambda \leq \lambda'$ for $\lambda, \lambda' \in \operatorname{Im}\pi$.*

Proof: It is well known that a map π is an open continuous map if and only if $\overline{\pi^{-1}(B)} = \pi^{-1}(\overline{B})$ for any subset $B \subset \Lambda$. Thus, when π is open continuous, $\pi^{-1}(\lambda) \subset \overline{\pi^{-1}(\lambda')}$ if and only if $\lambda \in \overline{\{\lambda'\}}$, which is equivalent to saying $\lambda \leq \lambda'$.

Conversely suppose that $\pi^{-1}(\lambda) \subset \overline{\pi^{-1}(\lambda')}$ is equivalent to $\lambda \leq \lambda'$. For a subset $B \subset \Lambda$, we have $\overline{\pi^{-1}(B)} \subset \pi^{-1}(\overline{B})$ by the continuity of π. For $x \in \pi^{-1}(\overline{B})$, $\pi(x) \in \overline{B}$. By the definition of the Alexandroff topology, there exists $\lambda \in B$ such that $\pi(x) \leq \lambda$. By assumption, this is equivalent to $\pi^{-1}(\pi(x)) \subset \overline{\pi^{-1}(\lambda)}$. Thus $x \in \overline{\pi^{-1}(B)}$ and we have shown that $\overline{\pi^{-1}(B)} \supset \pi^{-1}(\overline{B})$. □

Definition 2.5: For a stratification $\pi : X \to \Lambda$, the image $\operatorname{Im}\pi$ is called the *face poset* and is denoted by $P(X, \pi)$ or simply by $P(X)$.

When $P(X)$ is finite or countable, (X, π) is said to be *finite* or *countable.*

Remark 2.6: The above structure (without the connectivity of $\pi^{-1}(\lambda)$) is called a decomposition in Pflaum's book [21]. Pflaum used the notion of set germ to define a stratification from local decompositions. Furthermore, Pflaum imposed three further conditions:

- If $e_\mu \cap \overline{e_\lambda} \neq \emptyset$, then $e_\mu \subset \overline{e_\lambda}$.
- Each stratum is a smooth manifold.
- The collection $\{e_\lambda\}_{\lambda \in \Lambda}$ is locally finite in the sense that, for any $x \in X$, there exists a neighborhood U of x such that $U \cap e_\lambda \neq \emptyset$ only for a finite number of strata e_λ.

The first condition corresponds to the normality of cell complexes. We would like to separate the third condition as one of the conditions for CW stratifications. For the second condition, as is remarked in his book, we may replace smooth manifolds by any collection of geometric objects such as complex manifolds, real analytic sets, polytopes, and so on. In §2.3, we choose the class of spaces equipped with "cell structures" and define the notion of cellular stratified spaces. As an example of another choice, we use "star-shaped cells" and define the notion of stellar stratified spaces in §2.4. We also impose more structures on cells and introduce notions of cubical and polyhedral structures in Definition 6.20 and Definition 4.7 respectively.

Bridson and Haefliger [20] define stratifications by using closed strata. Their strata correspond to closures of strata in our definition. Furthermore, they also required their stratifications to be normal in the following sense:

Definition 2.7: We say a stratum e_λ in a stratified space (X, π) is *normal* if $e_\mu \subset \overline{e_\lambda}$ whenever $e_\mu \cap \overline{e_\lambda} \neq \emptyset$. When all strata are normal, the stratification π is said to be *normal*.

It is immediate to verify the following:

Lemma 2.8: *A stratum e_λ is normal if and only if $\partial e_\lambda = \overline{e_\lambda} \setminus e_\lambda$ is a union of strata.*

Another difference between our definition and the one by Bridson and Haefliger is that they considered intersections of closed strata.

Lemma 2.9: *Let (X, π) be a normal stratified space. Then for any pair of strata e_μ, e_λ, the intersection $\overline{e_\mu} \cap \overline{e_\lambda}$ is a union of strata.*

Proof: This is obvious since closures of different strata can intersect only on the boundaries, which are unions of strata by the definition of normality. □

The following is a typical example of stratifications we are interested in:

Example 2.10: Let $S_1 = \{-1, 0, 1\}$ with poset structure $0 < \pm 1$. The sign function

$$\mathrm{sign} : \mathbb{R} \longrightarrow S_1$$

given by

$$\mathrm{sign}(x) = \begin{cases} +1, & \text{if } x > 0; \\ 0, & \text{if } x = 0; \\ -1, & \text{if } x < 0, \end{cases}$$

defines a stratification on $\mathbb{R}$:

$$\mathbb{R} = (-\infty, 0) \cup \{0\} \cup (0, \infty).$$

This innocent-looking stratification turns out to be one of the most important ingredients in the theory of real hyperplane arrangements. Let $\mathcal{A} = \{H_1, \dots, H_k\}$ be a real affine hyperplane arrangement in $\mathbb{R}^n$ defined by affine 1-forms $L = \{\ell_1, \dots, \ell_k\}$. Hyperplanes cut $\mathbb{R}^n$ into convex regions that are homeomorphic to the interior of the n-disk. Each hyperplane H_i is cut into convex regions of dimension $n-1$ by other hyperplanes, and so on. These cuttings can be described as a stratification defined by the sign

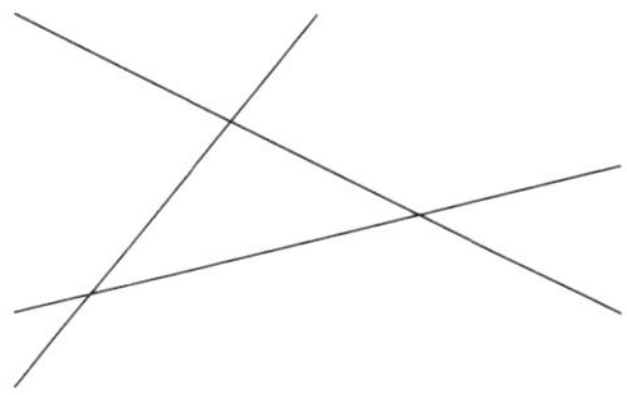

Fig. 1. Hyperplanes cut the ambient Euclidean space.

function as follows. Define a map

$$\pi_{\mathcal{A}} : \mathbb{R}^n \longrightarrow \mathrm{Map}(L, S_1)$$

by

$$\pi_{\mathcal{A}}(a)(\ell_i) = \mathrm{sign}(\ell_i(a)).$$

The partial order in S_1 induces a partial order on $\mathrm{Map}(L, S_1)$ by $\varphi \leq \psi$ if and only if $\varphi(\ell_i) \leq \psi(\ell_i)$ for all i. Then $\pi_{\mathcal{A}}$ is the indexing map for the stratification of $\mathbb{R}^n$ induced by $\mathcal{A}$.

There is a standard way to extend the above construction to a stratification on $\mathbb{C}^n$ by the complexification $\mathcal{A} \otimes \mathbb{C} = \{H_1 \otimes \mathbb{C}, \dots, H_k \otimes \mathbb{C}\}$ as is studied intensively in the theory of hyperplane arrangements. A good reference is the paper [7] by Björner and Ziegler. As is sketched at the end of the above paper, the stratification of $\mathbb{R}^n$ defined above can be extended to a stratification on $\mathbb{R}^n \otimes \mathbb{R}^\ell$ as follows: let $\mathcal{A} = \{H_1, \dots, H_k\}$ and $L = \{\ell_1, \dots, \ell_k\}$ be as above. Then the maps

$$\ell_1 \otimes \mathbb{R}^\ell, \dots, \ell_k \otimes \mathbb{R}^\ell : \mathbb{R}^n \otimes \mathbb{R}^\ell \longrightarrow \mathbb{R}^\ell$$

define a subspace arrangement $\mathcal{A} \otimes \mathbb{R}^n = \{H_1 \otimes \mathbb{R}^\ell, \dots, H_k \otimes \mathbb{R}^\ell\}$ in $\mathbb{R}^n \otimes \mathbb{R}^\ell$.

Let $S_\ell = \{0, \pm e_1, \dots, \pm e_\ell\}$ be the poset with partial ordering $0 < \pm e_1 < \cdots < \pm e_\ell$. Define the ℓ-th order sign function

$$\mathrm{sign}_\ell : \mathbb{R}^\ell \longrightarrow S_\ell$$

by

$$\mathrm{sign}_\ell(\boldsymbol{x}) = \begin{cases} \mathrm{sign}(x_\ell)e_\ell, & x_\ell \neq 0; \\ \mathrm{sign}(x_{\ell-1})e_{\ell-1}, & x_\ell = 0, x_{\ell-1} \neq 0; \\ \vdots & \\ \mathrm{sign}(x_1)e_1, & x_n = \cdots = x_2 = 0, x_1 \neq 0; \\ 0 & \boldsymbol{x} = 0. \end{cases}$$

Define a stratification on $\mathbb{R}^n \otimes \mathbb{R}^\ell$

$$\pi_{\mathcal{A}\otimes\mathbb{R}^\ell} : \mathbb{R}^n \otimes \mathbb{R}^\ell \longrightarrow \mathrm{Map}(L, S_\ell)$$

by

$$\pi_{\mathcal{A}\otimes\mathbb{R}^\ell}(a \otimes x)(\ell_i) = \mathrm{sign}_\ell(\ell_i(a)x).$$

This is a normal stratification on $\mathbb{R}^n \otimes \mathbb{R}^\ell$.

Example 2.11: Consider the standard n-simplex

$$\Delta^n = \left\{(x_0, \ldots, x_n) \in \mathbb{R}^n \mid x_0 + \cdots + x_n = 1, x_i \geq 0\right\}.$$

Define

$$\pi_n : \Delta^n \longrightarrow 2^{[n]}$$

by

$$\pi_n(x_0, \ldots, x_n) = \left\{i \in [n] \mid x_i \neq 0\right\},$$

where $[n] = \{0, \ldots, n\}$ and $2^{[n]}$ is the power set of $[n]$. Under the standard poset structure on $2^{[n]}$, π is a stratification with strata simplices in Δ^n. This is a normal stratification.

Define another stratification $\pi_n^{\max} : \Delta^n \to [n]$ by

$$\pi_n^{\max}(x_0, \ldots, x_n) = \max\left\{i \mid x_i \neq 0\right\},$$

where $[n]$ is equipped with the partial order $0 < 1 < \cdots < n$. The resulting decomposition is

$$\Delta^n = (\Delta^n \setminus \Delta^{n-1}) \cup (\Delta^{n-1} \setminus \Delta^{n-2}) \cup \cdots \cup (\Delta^1 \setminus \Delta^0) \cup \Delta^0.$$

This is also a normal stratification.

Example 2.12: Let G be a compact Lie group acting smoothly on a smooth manifold M. M. Davis [23] defined a stratification on M and on the quotient space M/G as follows: the indexing set $I(G)$ is called the set of normal orbit types and is defined by

$$I(G) = \left\{ (\varphi : H \to \mathrm{GL}(V)) \;\middle|\; \begin{array}{l} H < G \text{ a closed subgroup,} \\ \varphi \text{ a representation,} \\ V^H = \{0\} \end{array} \right\} /\sim$$

where $(\varphi : H \to \mathrm{GL}(V)) \sim (\varphi' : H' \to \mathrm{GL}(V'))$ if and only if there exists an element $g \in G$ and a linear isomorphism $f : V \to V'$ such that $H' = gHg^{-1}$ and the diagram

$$\begin{array}{ccc} H & \xrightarrow{\varphi} & \mathrm{GL}(V) \\ {\scriptstyle g(-)g^{-1}} \downarrow & & \downarrow {\scriptstyle f_*} \\ H' & \xrightarrow[\varphi']{} & \mathrm{GL}(V') \end{array}$$

is commutative.

He defined a map

$$\pi_M : M \longrightarrow I(G)$$

by

$$\pi_M(x) = \left(G_x, S_x/(S_x)^{G_x}\right),$$

where G_x is the isotropy subgroup at x and $S_x = T_xM/T_x(Gx)$. There is a canonical partial order on $I(G)$ and it is proved that π_M is a normal stratification (Theorem 1.6 in [23]). It is also proved that the stratification descends to M/G.

In particular, for a torus manifold in the sense of Hattori-Masuda [24] or a small cover in the sense of Davis-Januszkiewicz [25] M, of dimension $2n$ or n respectively, the quotient M/T^n or $M/\mathbb{Z}_2^n$ has a canonical normal stratification.

When $M = \mathbb{C}\mathrm{P}^n$ or $\mathbb{R}\mathrm{P}^n$, the quotient is Δ^n and the stratification by normal orbit types corresponds to the stratification π_n in Example 2.11. We will see the other stratification $\pi_n^{\max}$ in Example 2.11 corresponds to the minimal cell decompositions of $\mathbb{R}\mathrm{P}^n$ and $\mathbb{C}\mathrm{P}^n$ in Example 4.26.

Example 2.13: The stratification on the quotient M/G has been generalized to the notion of manifolds with faces or more generally manifolds with corners. See §6 of Davis' paper [26], for example. These are also important examples of normal stratifications.

It is easy to verify that the product of two stratifications is again a stratification.

Lemma 2.14: *Let (X, π_X) and (Y, π_Y) be stratified spaces. The map*

$$\pi_X \times \pi_Y : X \times Y \longrightarrow P(X) \times P(Y)$$

defines a stratification on $X \times Y$.

Proof: The product of open maps is again open. □

The following requirements for morphisms of stratified spaces should be reasonable.

Definition 2.15: Let (X, π_X) and (Y, π_Y) be stratified spaces.

- A *morphism of stratified spaces* is a pair $\boldsymbol{f} = (f, \underline{f})$ of a continuous map $f : X \to Y$ and a map of posets $\underline{f} : P(X) \to P(Y)$ making the following diagram commutative:

$$\begin{array}{ccc} X & \xrightarrow{f} & Y \\ {\scriptstyle \pi_X}\downarrow & & \downarrow{\scriptstyle \pi_Y} \\ P(X) & \xrightarrow[\underline{f}]{} & P(Y). \end{array}$$

- When $X = Y$ and f is the identity, $\boldsymbol{f} = (f, \underline{f})$ is called a *subdivision.* We also say that (X, π_X) is a subdivision of (Y, π_Y) or (Y, π_Y) is a *coarsening* of (X, π_X).
- When $f(e_\lambda) = e_{\underline{f}(\lambda)}$ for each λ, it is called a *strict morphism.*
- When $\boldsymbol{f} = (f, \underline{f})$ is a strict morphism of stratified spaces and f is an embedding of topological spaces, $\boldsymbol{f}$ is said to be an *embedding* of X into Y.

For a stratified space $\pi : Y \to \Lambda$ and a continuous map $f : X \to Y$, the composition $f^*(\pi) = \pi \circ f : X \to \Lambda$ may or may not be a stratification.

Definition 2.16: Let $f : (X, \pi_X) \to (Y, \pi_Y)$ be a morphism of stratified spaces. When $\underline{f} : P(X) \to \pi_Y(f(X))$ is an isomorphism of posets

$$\begin{array}{ccccc} X & \xrightarrow{f} & f(X) & \hookrightarrow & Y \\ {\scriptstyle \pi_X}\downarrow & & \downarrow{\scriptstyle \pi_Y} & & \downarrow{\scriptstyle \pi_Y} \\ P(X) & \xrightarrow[\cong]{} & \pi_Y(f(X)) & \hookrightarrow & P(Y), \end{array}$$

we say that π_X is *induced* from π_Y via f. We sometimes denote it by $f^*(\pi_Y)$.

Example 2.17: Consider the double covering map

$$2 : S^1 \longrightarrow S^1.$$

The minimal cell decomposition $\pi_{\min} : S^1 = e^0 \cup e^1$ on S^1 in the range does not induce a stratification on S^1 in the domain, since the inverse images of strata are not connected. But we have a strict morphism of stratified spaces

$$\begin{array}{ccc} S^1 & \xrightarrow{2} & S^1 \\ \downarrow & & \downarrow \\ \{0_+, 0_-, 1_+, 1_-\} & \longrightarrow & \{0, 1\} \end{array}$$

if S^1 in the domain is equipped with the minimal Σ_2-equivariant cell decomposition $S^1 = e^0_- \cup e^0_+ \cup e^1_- \cup e^1_+$.

Consider the complex analogue of the double covering on S^1, i.e. the Hopf bundle

$$\eta : S^3 \longrightarrow S^2.$$

This is a principal fiber bundle with fiber S^1. The minimal cell decomposition on S^2, $S^2 = e^0 \cup e^2$, is a stratification. This stratification induces a stratification on S^3

$$S^3 = \eta^{-1}(e^0) \amalg \eta^{-1}(e^2).$$

Note that we have

$$\eta^{-1}(e^0) \cong e^0 \times S^1;$$
$$\eta^{-1}(e^2) \cong e^2 \times S^1.$$

The face posets of these stratifications are isomorphic to the poset $\{0 < 2\}$ and we have a commutative diagram

$$\begin{array}{ccc} S^3 & \xrightarrow{\eta} & S^2 \\ \downarrow & & \downarrow \\ \{0,2\} & = & \{0,2\}. \end{array}$$

Note that these two examples can be described as

$$\begin{aligned} S^1 &= e^0 \times S^0 \amalg e^1 \times S^0 \\ S^3 &= e^0 \times S^1 \amalg e^2 \times S^1, \end{aligned}$$

suggesting the existence of a common framework for handling them simultaneously. We propose the notion of cylindrical structures as such in §4.1.

As a special case of embeddings of stratified spaces, we have the notion of stratified subspaces.

Definition 2.18: Let (X, π) be a stratified space and A be a subspace of X. If the restriction $\pi|_A$ is a stratification, $(A, \pi|_A)$ is called a *stratified subspace* of (X, π).

When the inclusion $i : A \hookrightarrow X$ is a strict morphism, A is called a *strict stratified subspace* of X.

Remark 2.19: We study stratified subspaces in detail in §6.1.

As is the case of cell complexes, the CW condition is useful.

Definition 2.20: A stratification π on X is said to be *CW* if it satisfies the following two conditions:

(1) (Closure Finite) For each stratum e_λ, ∂e_λ is covered by a finite number of strata.
(2) (Weak Topology) X has the weak topology determined by the covering $\{\overline{e_\lambda} \mid \lambda \in P(X)\}$.

For example, weak topologies are useful when we glue quotient maps.

Lemma 2.21: *Let $\pi_X : X \to \Lambda$ and $\pi_Y : Y \to \Lambda$ be normal CW stratified spaces with $P(X) = P(Y) = \Lambda$ and*

$$\boldsymbol{f} : X \longrightarrow Y$$

be a strict morphism of stratified spaces with $\underline{f} = 1_\Lambda$. Let us denote the strata in X and Y indexed by λ by e^X_λ and e^Y_λ, respectively. Suppose

$$f_\lambda = f|_{\overline{e^Y_\lambda}} : \overline{e^Y_\lambda} \to \overline{e^X_\lambda}$$

is a quotient map for all $\lambda \in \Lambda$. Then $f : X \to Y$ is a quotient map.

Proof: For a subset $U \subset Y$, suppose $f^{-1}(U)$ is open in X. By the weak topology condition, $f^{-1}(U) \cap \overline{e^X_\lambda}$ is open in $\overline{e^X_\lambda}$ for each $\lambda \in \Lambda$. We have

$$f\left(f^{-1}(U) \cap \overline{e^X_\lambda}\right) = U \cap f\left(\overline{e^X_\lambda}\right) = U \cap \overline{e^Y_\lambda},$$

since f is a strict morphism and both X and Y are normal. Thus

$$f_\lambda^{-1}\left(U \cap \overline{e^Y_\lambda}\right) = f^{-1}\left(f\left(f^{-1}(U) \cap \overline{e^X_\lambda}\right)\right) = f^{-1}(U) \cap \overline{e^X_\lambda}$$

is open in $\overline{e^X_\lambda}$ for each λ. By assumption, f_λ is a quotient map and Y is CW. And U is open in Y. □

It is straightforward to verify the following by using the fact that any topological space has the weak topology with respect to a locally finite closed covering.

Proposition 2.22: *Any locally finite stratified space is CW.*

Corollary 2.23: *Let (X, π) be a CW stratified space and (X, π') be a subdivision. If each stratum e_λ in (X, π) is subdivided into a finite number of strata in (X, π'), then (X, π') is CW.*

Proof: For each cell e_λ in (X, π), $\overline{e_\lambda}$ has the weak topology with respect to the covering

$$\overline{e_\lambda} = \bigcup_{\lambda' \in P(X,\pi'), e_{\lambda'} \subset e_\lambda} \overline{e_{\lambda'}}$$

because of the finiteness assumption. Thus X has the weak topology with respect to the covering

$$X = \bigcup_{\lambda' \in P(X,\pi')} \overline{e_{\lambda'}} = \bigcup_{\lambda \in P(X,\pi)} \left(\bigcup_{\lambda' \in P(X,\pi'), e_{\lambda'} \subset e_\lambda} \overline{e_{\lambda'}} \right).$$

The closure finiteness condition also follows from the finiteness of the subdivision of each stratum. □

As is the case of CW complexes, metrizability implies local finiteness.

Lemma 2.24: *Any metrizable CW stratified space is locally finite.*

Proof: This fact is well known for CW complexes. The same argument can be used to prove this generalized statement. We give a proof for the convenience of the reader.

If X is not locally finite, there exists a point $x \in X$ such that, for any open neighborhood U of x, U intersects with infinitely many strata. For each n, let U_n be the $\frac{1}{n}$-neighborhood of x and choose a stratum e_n with $U_n \cap e_n \neq \emptyset$ and $x \notin e_n$. Choose $x_n \in U_n \cap e_n$. Then the set $A = \{x_n\}_{n=1,2,\dots}$ is closed by the CW conditions. This contradicts to the fact that $x \notin A$ and $x \in \overline{A}$. Thus X is locally finite. □

2.2. *Cells*

We would like to define a cellular stratification on a topological space X as a stratification on X whose strata are "cells". As we have seen in Example 2.10, we would like to regard chambers and faces of a real hyperplane arrangement as "cells", suggesting the need of non-closed cells.

Definition 2.25: A *globular n-cell* is a subset D of D^n containing $\mathrm{Int}(D^n)$. We call $D \cap \partial D^n$ the *boundary* of D and denote it by ∂D. The number n is called the *globular dimension* of D.

Remark 2.26: We introduce another dimension, called the stellar dimension, for a more general class of subsets of D^n in §2.4.

We use the following definition of cell structures:

Definition 2.27: Let X be a topological space. For a non-negative integer n, an *n-cell structure* on a subspace $e \subset X$ is a pair (D, φ) of a globular n-cell D and a continuous map

$$\varphi : D \longrightarrow X$$

satisfying the following conditions:

(1) $\varphi(D) = \overline{e}$ and $\varphi : D \to \overline{e}$ is a quotient map.
(2) The restriction $\varphi|_{\mathrm{Int}(D^n)} : \mathrm{Int}(D^n) \to e$ is a homeomorphism.

For simplicity, we denote an n-cell structure (D, φ) on e by e when there is no risk of confusion. The map φ is called the *cell structure map* or the *characteristic map* of e and D is called the *domain* of e. The number n is called the (globular) *dimension* of e.

Example 2.28: The open n-disk $\mathrm{Int}(D^n)$ is a globular n-cell. The standard homeomorphism

$$\mathrm{Int}(D^n) \xrightarrow{\cong} \mathbb{R}^n$$

defines an n-cell structure on $\mathbb{R}^n$. The domain is $\mathrm{Int}(D^n)$.

Example 2.29: Let $X = \mathrm{Int}(D^2) \cup \{(1,0)\}$. The identity map defines a 2-cell structure on X.

There is another choice. Let $D = \mathrm{Int}(D^2) \cup S^1_+$. The deformation retraction of S^1_+ onto $(1,0)$ can be extended to a continuous map

$$\varphi : D \longrightarrow X$$

whose restriction to $\mathrm{Int}(D^n)$ is a homeomorphism. For example, φ is given in polar coordinates by

$$\varphi(re^{i\theta}) = \begin{cases} re^{i(1-r)\theta}, & |\theta| \leq \frac{\pi}{2}; \\ re^{i\{\theta-(\pi-\theta)r\}}, & \frac{\pi}{2} \leq \theta \leq \pi; \\ re^{i\{\theta+(\pi+\theta)r\}}, & -\pi \leq \theta \leq -\frac{\pi}{2}. \end{cases}$$

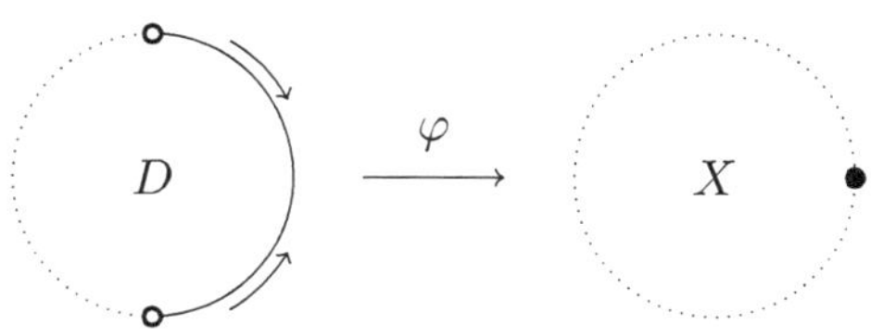

Fig. 2. An exotic cell structure on $\mathrm{Int}(D^2) \cup \{(1,0)\}$

Note that φ is not a quotient map. For example, the image of $\{(x,y) \in D \mid x > 0\}$ under φ is open under the quotient topology, but it is not open under the relative topology on X. Thus this is not a 2-cell structure on X.

In other words, we need to put the quotient topology on X in order for the map φ in the above example to be a cell structure. Fortunately, the quotient topology is a popular choice in many practical examples.

Example 2.30: For any simplicial set X, the geometric realization $|X|$ is known to be a CW complex, whose cells are in one-to-one correspondence with nondegenerate simplices in X.

Consider the simplicial set $X = s(\Delta^2)/s(\Delta^1)$, where Δ^1 and Δ^2 are regarded as ordered simplicial complexes and $s(-)$ is the functor in Example

B.12 which transforms ordered simplicial complexes to simplicial sets. The geometric realization $|X|$ is a cell complex consisting of two 0-cells $[0] = [1]$, $[2]$, two 1-cells $[0, 2]$, $[1, 2]$, and a 2-cell $[0, 1, 2]$. The characteristic map for the 2-cell is given by the composition

$$\psi : D^2 \cong \Delta^2 \times \{[0, 1, 2]\} \hookrightarrow \coprod_{i=0}^{\infty} \Delta^n \times X_n \longrightarrow |X|.$$

Thus it is given by collapsing the right-most arc in the left circle in Figure 3 to $[0] = [1]$. By definition, $|X|$ is equipped with the quotient topology and

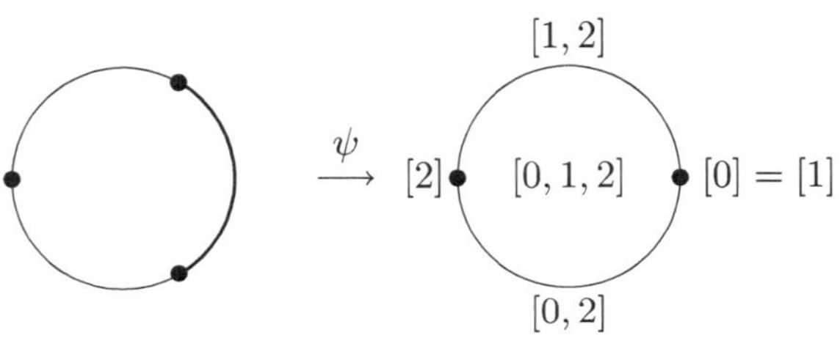

Fig. 3. A 2-cell structure on $s(\Delta^2)/s(\Delta^1)$

ψ defines a 2-cell structure.

Cells satisfying one (or both) of the following conditions appear frequently.

Definition 2.31: Let X be a topological space and $e \subset X$ a subspace. An n-cell structure (D, φ) on e, or simply an n-cell e, is said to be

- *closed* if $D = D^n$,
- *regular* if $\varphi : D \to \overline{e}$ is a homeomorphism.

Example 2.32: Given a cell complex X and its n-cell e, the characteristic map

$$\varphi : D^n \longrightarrow X$$

defines a closed n-cell structure on e. When X is a regular cell complex, it is a regular n-cell structure on e.

Example 2.33: Let $\mathcal{A}$ be an arrangement of a finite number of hyperplanes in $\mathbb{R}^n$. Bounded strata in the stratification $\pi_{\mathcal{A}}$ are convex polytopes and they are closed and regular cells.

We may also define non-closed but regular cell structures on unbounded strata as follows. Suppose $\mathcal{A}$ is essential, namely normal vectors to the hyperplanes span $\mathbb{R}^n$. Then we may choose a closed ball B with center at the origin in $\mathbb{R}^n$ which contains all bounded strata. Hyperplanes also cut

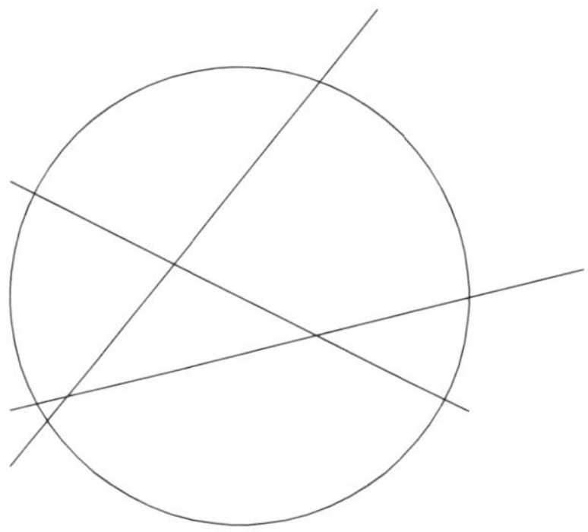

Fig. 4. A bounding sphere for bounded strata.

the boundary sphere and define a stratification $\pi_{\mathcal{A},B}$ on B whose strata are all closed cells. The inclusion $\mathrm{Int}(B) \hookrightarrow \mathbb{R}^n$ is a morphism of stratified spaces under which face posets can be identified

$$\begin{array}{ccc} \mathrm{Int}(B) & \hookrightarrow & \mathbb{R}^n \\ {\scriptstyle \pi_{\mathcal{A},B}} \downarrow & & \downarrow {\scriptstyle \pi_{\mathcal{A}}} \\ P(\pi_{\mathcal{A},B}|_{\mathrm{Int}(B)}) & = & P(\pi_{\mathcal{A}}). \end{array}$$

For each unbounded stratum e in $\pi_{\mathcal{A}}$, the intersection $e \cap \mathrm{Int}(B)$ is a cell in B. Let $\varphi : D^k \to \overline{e \cap B}$ be a characteristic map for $e \cap \mathrm{Int}(B)$ and define $D = \varphi^{-1}(\overline{e} \cap \mathrm{Int}(B))$. We can compress the outside of B into $e \cap \mathrm{Int}(B)$ via a homeomorphism $\psi : \overline{e} \to \overline{e} \cap \mathrm{Int}(B)$. The composition

$$D \xrightarrow{\varphi|_D} \overline{e} \cap \mathrm{Int}(B) \xrightarrow{\psi^{-1}} \overline{e}$$

defines a regular cell structure on e.

Non-closed cells might have a bad topology.

Example 2.34: Let

$$p : D^2 \setminus \{(0,1)\} \longrightarrow \{(x,y) \in \mathbb{R}^2 \mid y \geq 0\}$$

be the homeomorphism given by extending the stereographic projection $S^1 \setminus \{(0,1)\} \to \mathbb{R}$. Let

$$X = \{(x,y) \in \mathbb{R}^2 \mid y > 0\} \cup \{(x,0) \mid x \in \mathbb{Q}\}$$

and define $D = p^{-1}(X)$. Then the restriction

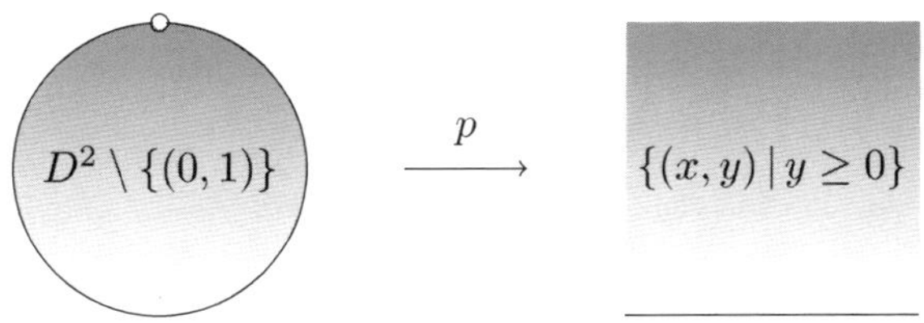

Fig. 5. A 2-cell structure for the closed upper half-plane.

$$p|_D : D \longrightarrow X$$

defines a 2-cell structure on X. X and D are not locally compact. This

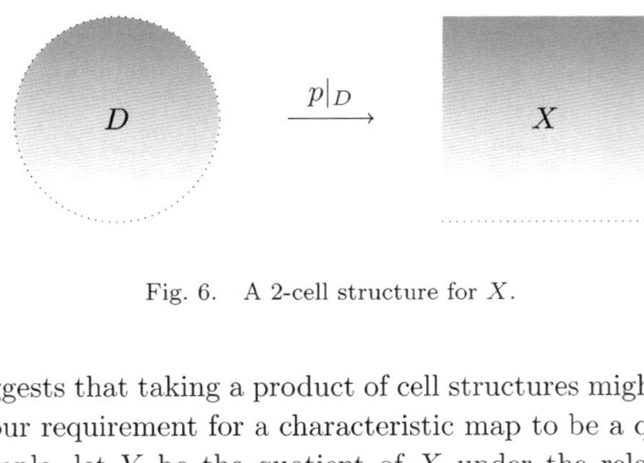

Fig. 6. A 2-cell structure for X.

example suggests that taking a product of cell structures might not be easy because of our requirement for a characteristic map to be a quotient map.

For example, let Y be the quotient of X under the relation $(x,0) \sim (x',0)$ if and only if $x - x' \in \mathbb{Z}$. The composition of $p|_D$ with the canonical projection

$$\varphi_Y : D \xrightarrow{p|_D} X \longrightarrow Y$$

defines a 2-cell structure on Y. But the product with $p|_D$

$$p|_D \times \varphi_Y : D \times D \longrightarrow X \times Y$$

is not a quotient map, since the product $\mathbb{Q}\times\mathbb{Q}\to\mathbb{Q}\times(\mathbb{Q}/\mathbb{Z})$ of the identity map and the quotient map is not a quotient map.

We need to impose certain conditions to take products of cell structures freely. Our solution is to require cellular structures on the boundaries of the domains of cells. See §4.1 for more details.

2.3. *Cellular stratifications*

So far we have defined the notions of stratified spaces and cell structures. Now we are ready to define cellular stratified spaces by combining these two structures.

Definition 2.35: Let X be a Hausdorff space. A *cellular stratification* on X is a pair (π,Φ) of a stratification $\pi: X\to P(X)$ on X and a collection of cell structures $\Phi=\{\varphi_\lambda : D_\lambda\to\overline{e_\lambda}\}_{\lambda\in P(X)}$ satisfying the condition that, for each n-cell e_λ, ∂e_λ is covered by cells of dimension less than or equal to $n-1$.

A *cellular stratified space* is a triple (X,π,Φ) where (π,Φ) is a cellular stratification on X. As usual, we abbreviate it by (X,π) or X, if there is no danger of confusion.

Remark 2.36: The term "cellular stratified space" has been already used in the study of singularities. See, for example, Schürmann's book [22]. We found, however, that his definition is too restrictive for our purposes.

Example 2.37: Consider the "topologist's sine curve"

$$S=\left\{(x,\sin\tfrac{1}{x})\;\middle|\;0<x\le 1\right\}.$$

Its closure in $\mathbb{R}^2$ is given by

$$\overline{S}=S\cup\{(0,t)\mid -1\le t\le 1\}.$$

Since S is homeomorphic to the half interval $(0,1]$ via the function $\sin\frac{1}{x}$, the decomposition

$$\overline{S}=\{(1,\sin 1)\}\cup\{(0,1)\}\cup\{(0,-1)\}\\ \cup\left\{(x,\sin\tfrac{1}{x})\,\middle|\,0<x<1\right\}\cup\{(0,t)\mid -1<t<1\}$$

is a stratification of $\overline{S}$ consisting of five strata. Although the stratum $\left\{(x,\sin\frac{1}{x})\,\middle|\,0<x<1\right\}$ is homeomorphic to $\mathrm{Int}(D^1)$, there is no 1-cell structure on this stratum, since there is no way to extend a homeomorphism $(0,1]\cong S$ to a continuous map $[0,1]\to\overline{S}$.

Furthermore this stratification does not satisfy the dimension condition in Definition 2.35, since $\partial S = \{(0,t) \mid -1 \leq t \leq 1\}$.

Remark 2.38: The above example is borrowed from Pflaum's book [21]. He describes an even more pathological example. See 1.1.12 on page 18 of his book.

Here is another non-example.

Example 2.39: Consider the 2-cell structure on $X = \{(x,y) \in \mathbb{R}^2 \mid y > 0\} \cup \{(x,0) \mid x \in \mathbb{Q}\}$ defined in Example 2.34. Let $e^2 = \{(x,y) \in \mathbb{R}^2 \mid y > 0\}$ and $e^0_x = \{(x,0)\}$ for $x \in \mathbb{Q}$. We have a decomposition

$$X = \bigcup_{x \in \mathbb{Q}} e^0_x \cup e^2$$

and, for each $x \in \mathbb{Q}$, the identification

$$\psi_x : D^0 \cong \{(x,0)\} \hookrightarrow X$$

defines a 0-cell structure on $\{(x,0)\}$. But this is not a cellular stratified space, since e^0_x is not locally closed.

On the other hand, let $Y = \{(x,y) \in \mathbb{R}^2 \mid y > 0\} \cup \mathbb{Z} \times \{0\}$ and consider the decomposition

$$Y = \bigcup_{n \in \mathbb{Z}} e^0_n \cup e^2.$$

Each e^0_n is locally closed in Y and this is a cellular stratification on Y. Note, however, this is not a CW stratification, since we need infinitely many 0-cells to cover the boundary of the 2-cell ∂e^2.

Definition 2.40: We say a cellular stratification is *CW* if its underlying stratification is CW (Definition 2.20).

Lemma 2.41: *A CW cellular stratification (π, Φ) on a space X defines and is defined by the following structure:*

- *a filtration $X_0 \subset X_1 \subset \cdots \subset X_{n-1} \subset X_n \subset \cdots$ on X,*
- *an n-cell structure on each connected component of $X_n \setminus X_{n-1}$*

satisfying the following conditions:

(1) $X = \bigcup_{n=0}^{\infty} X_n$.

(2) For each stratum e in X_n, ∂e is covered by a finite number of strata in X_{n-1}.

(3) X has the weak topology determined by the covering $\{\overline{e} \mid e \in P(X)\}$.

Proof: Suppose (π, Φ) is a CW cellular stratification. Define

$$X_n = \bigcup_{e \in P(X), \dim e \leq n} e.$$

Then we obtain a filtration on X with $X = \bigcup_{n=0}^{\infty} X_n$. By definition, the boundary ∂e of each n-cell e is covered by a finite number of cells of dimension less than or equal to $n-1$. Also by definition, X has the weak topology by the covering $\{\overline{e} \,|\, e \in P(X)\}$.

It remains to show that the n-cells are connected components of the difference $X_n \setminus X_{n-1}$, i.e. each n-cell e is open and closed in $X_n \setminus X_{n-1}$. By the local closedness, e is open in $\overline{e}$. Since $\partial e \subset X_{n-1}$, e is open in $X_n \setminus X_{n-1}$. For any n-cell e', the intersection $e \cap \overline{e'}$ in $X_n \setminus X_{n-1}$ is $e \cap e'$ and is $\emptyset$ or $e = e'$. And thus it is closed in $\overline{e'}$.

It is left to the reader to check the converse. □

Example 2.42: The stratification $\pi_{\mathcal{A}}$ of $\mathbb{R}^n$ defined by a hyperplane arrangement $\mathcal{A}$ (Example 2.10) is a CW cellular stratification.

The Björner-Ziegler stratification $\pi_{\mathcal{A} \otimes \mathbb{R}^\ell}$ on $\mathbb{R}^n \otimes \mathbb{R}^\ell$ is also a CW cellular stratification.

In Definition 2.15, we defined morphisms of stratified spaces as "stratification-preserving maps". Note that, in our definition of cellular stratified spaces, we include characteristic maps as defining data. When we consider maps between cellular stratified spaces, we require them to be compatible with characteristic maps.

Definition 2.43: Let (X, π_X, Φ_X) and (Y, π_Y, Φ_Y) be cellular stratified spaces. A *morphism of cellular stratified spaces* from (X, π_X, Φ_X) to (Y, π_Y, Φ_Y) consists of:

- a morphism $\boldsymbol{f} : (X, \pi_X) \to (Y, \pi_Y)$ of stratified spaces,
- a family of maps

$$f_\lambda : D_\lambda \longrightarrow D_{\underline{f}(\lambda)}$$

indexed by cells $\varphi_\lambda : D_\lambda \to \overline{e_\lambda}$ in X making the diagrams

$$\begin{array}{ccc} X & \xrightarrow{f} & Y \\ \uparrow{\scriptstyle \varphi_\lambda} & & \uparrow{\scriptstyle \psi_{\underline{f}(\lambda)}} \\ D_\lambda & \xrightarrow[f_\lambda]{} & D_{\underline{f}(\lambda)} \end{array}$$

commutative, where $\psi_{\underline{f}(\lambda)} : D_{\underline{f}(\lambda)} \to \overline{e_{\underline{f}(\lambda)}}$ is the characteristic map for $e_{\underline{f}(\lambda)}$.

The category of cellular stratified spaces is denoted by **CSSpaces**.

Remark 2.44: When $\boldsymbol{f} : (X, \pi_X, \Phi_X) \to (Y, \pi_Y, \Phi_Y)$ is a morphism of cellular stratified spaces, the compatibility of f_λ with characteristic maps implies $f_\lambda(\mathrm{Int}(D_\lambda)) \subset \mathrm{Int}(D_{\underline{f}(\lambda)})$.

Remark 2.45: In algebraic topology, the requirement for maps between cell complexes is much weaker. A map $f : X \to Y$ between cell complexes is said to be *cellular*, if $f(X_n) \subset Y_n$ for each n. The author thinks, however, this terminology is misleading.

Definition 2.46: A morphism $(\boldsymbol{f}, \{f_\lambda\}) : (X, \pi_X, \Phi_X) \to (Y, \pi_Y, \Phi_Y)$ of cellular stratified spaces is said to be *strict* if $\boldsymbol{f} : (X, \pi_X) \to (Y, \pi_Y)$ is a strict morphism of stratified spaces and $f_\lambda(0) = 0$ for each $\lambda \in P(X, \pi_X)$.

Once we have morphisms between cellular stratified spaces, we have a notion of equivariant stratification.

Definition 2.47: Let G be a group. A cellular stratified space X equipped with a monoid morphism $G \to$ **CSSpaces**(X, X) or a functor $G \to$ **CSSpaces** is called a *G-cellular stratified space.*

In order to study configuration spaces, products and subdivisions are important. However, the product of two cell structures may not be a cell structure as we have seen in Example 2.34. Subdivisions of cell structures are not easy to handle either. We discuss products and subdivisions of cellular stratified spaces in §6.2 and §6.3, respectively.

2.4. *Stellar stratified spaces*

In Definition 2.27, we required the domain D of an n-cell to contain $\mathrm{Int}(D^n)$. As we will see in the proof of Theorem 5.16, this condition is requiring too much. Furthermore, the definition of the globular dimension of a cell is not appropriate when the cell is not closed. In this section, we introduce stellar cells and study stratified spaces whose strata are stellar cells. Stellar structures also play an essential role in our description of the classifying space of the face category of cellular stratified spaces.

Let us first define "star-shaped cells".

Definition 2.48: A subset S of D^N is said to be an *aster* if $\{0\} * \{x\} \subset S$ for any $x \in S$, where $*$ is the join operation defined by connecting points by line segments. The subset $S \cap \partial D^N$ is called the *boundary* of S and is denoted by ∂S. The complement $S \setminus \partial S$ of the boundary is called the *interior* of S and is denoted by $\mathrm{Int}(S)$.

We say S is *thin* if $S = \{0\} * \partial S$.

We require the existence of a cellular stratification on the boundary in order to define the dimension.

Definition 2.49: A *stellar cell* is an aster S in D^N for some N such that there exists a cellular stratification on ∂D^N containing ∂S as a cellular stratified subspace.

When the (globular) dimension of ∂S is $n-1$, we define the *stellar dimension* of S to be n and call S a *stellar n-cell*.

An n-cell in the sense of §2.2 is stellar if its boundary has a structure of cellular stratified space. However, the dimension as a stellar cell might be smaller than n.

Example 2.50: Consider the globular n-cell $\mathrm{Int}(D^n)$ in Example 2.28. It is a stellar cell with empty boundary. Thus its stellar dimension is 0.

By adding three points to the boundary, for example, we obtain a globular n-cell D whose stellar dimension is 1.

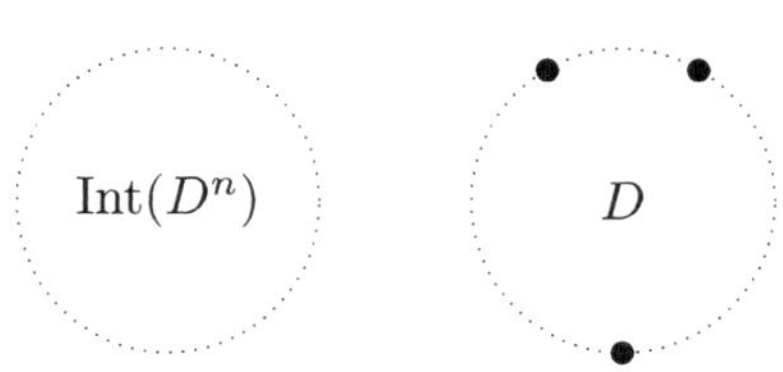

Fig. 7. An open disk and an open disk with three points added.

These two stellar cells are not thin. The first example contains $\{0\}$ as a thin stellar cell. The second example contains a graph of the shape of Y as a thin stellar cell. See Figure 8.

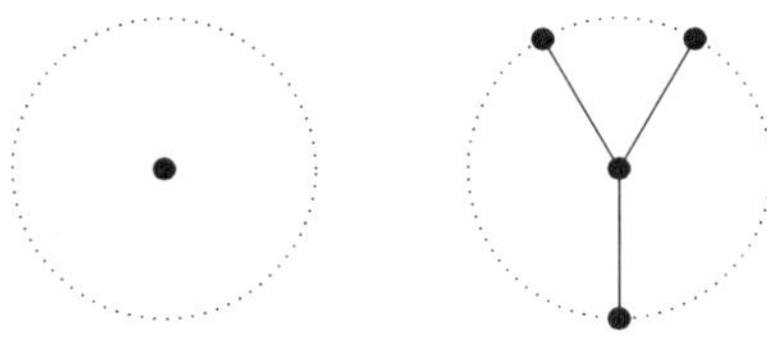

Fig. 8. Thin stellar cells.

Definition 2.51: An *n-stellar structure* on a subset e of a topological space X is a pair (S, φ) of a stellar n-cell S and a quotient map

$$\varphi : S \longrightarrow X$$

satisfying the following conditions:

(1) $\varphi(S) = \overline{e}$.
(2) The restriction of φ to $\mathrm{Int}(S)$ is a homeomorphism onto e.

By replacing cell structures by stellar structures in the definition of cellular stratifications, we obtain the notion of stellar stratifications.

Definition 2.52: A *stellar stratification* on a topological space X consists of a stratification (X, π), and a stellar structure on each $e_\lambda = \pi^{-1}(\lambda)$ for $\lambda \in P(X)$ satisfying the condition that for each stellar n-cell e_λ, ∂e_λ is covered by stellar cells of stellar dimension less than or equal to $n-1$.

A space equipped with a stellar stratification is called a *stellar stratified space*.

The following is a typical example.

Example 2.53: Consider a finite graph X regarded as a 1-dimensional cell complex. Suppose X is regular as a cell complex. Then we may define the barycentric subdivision $\mathrm{Sd}(X)$ of X. This cell complex can be expressed as a union of open stars of vertices of X and the barycenters of 1-cells in X. And we have a structure of stellar stratified space on $\mathrm{Sd}(X)$.

For example, in Figure 9, $\mathrm{Sd}(X)$ consists of three 1-dimensional stellar cells, in which the middle one is not a globular cell.

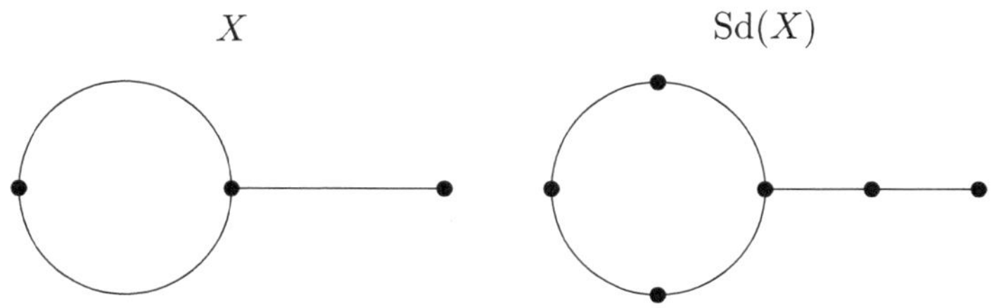

Fig. 9. A stellar stratification on Sd(X).

Remark 2.54: We will extend the definition of the barycentric subdivision to cellular and stellar stratified spaces in §5.2 and investigate more precise relations between stellar stratifications and their barycentric subdivisions.

The definition of morphisms of stellar stratified spaces should be obvious.

Definition 2.55: The category of stellar stratified spaces is denoted by **SSSpaces**.

If a stellar stratified space X is CW, we may describe X as a quotient space of the collection of domains of cell structures.

Lemma 2.56: *For a CW stellar stratified space X with stellar structure $\{\varphi_\lambda : D_\lambda \to \overline{e_\lambda}\}_{\lambda\in\Lambda}$, define*

$$D(X) = \coprod_{\lambda\in\Lambda} D_\lambda$$

and let $\Phi : D(X) \to X$ be the map defined by the stellar structure maps. Then Φ is a quotient map. More explicitly, we have a homeomorphism

$$X \cong D(X)/_{\sim_\Phi},$$

where the relation $\sim_\Phi$ is defined by $x \sim_\Phi y$ if and only if $\varphi_\mu(x) = \varphi_\lambda(y)$ for $x \in D_\mu$ and $y \in D_\lambda$.

This fact is proved for CW cellular stratified spaces as Lemma 2.28 in [1]. The proof can be applied to stellar stratified spaces without a change and is omitted.

3. Totally normal cellular stratified spaces

In order to study their homotopy types, we would like to impose appropriate "niceness conditions" on cellular and stellar stratified spaces.

3.1. *Regularity and normality*

Regularity and normality are frequently used conditions on CW complexes. We have already defined normality for stratified spaces (Definition 2.7) and regularity for cells (Definition 2.31).

Definition 3.1: Let X be a cellular or stellar stratified space. We say X is *normal*, if it is normal as a stratified space. We say X is *regular* if all cells in X are regular.

Example 3.2: The cellular stratification $\pi_{\mathcal{A}\otimes\mathbb{R}^\ell}$ on $\mathbb{R}^n \otimes \mathbb{R}^\ell$ in Example 2.10 defined by a real arrangement $\mathcal{A}$ in $\mathbb{R}^n$ is regular and normal.

Example 3.3: Consider the following cellular stratified space obtained by

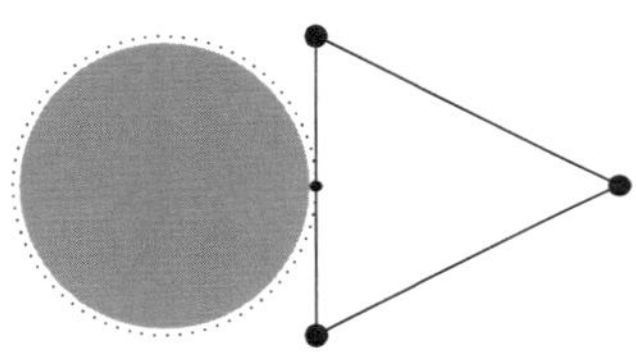

Fig. 10. A non-normal regular cellular stratified space.

gluing $\mathrm{Int}(D^2)$ to the boundary of a 2-simplex at the middle point of an edge. The domain of the characteristic map of the 2-cell is $\mathrm{Int}(D^2) \cup \{(1,0)\}$, whose boundary is mapped into the 1-skeleton. Thus this is a regular cellular stratified space. However, this is not normal.

If we regard the globular 2-cell $\mathrm{Int}(D^2) \cup \{(1,0)\}$ as a stellar cell, its stellar dimension is 1 and this stellar structure does not satisfy the dimensional requirement. Thus this is not a stellar stratified space.

In the case of CW complexes, regularity always implies normality. See Theorem 2.1 in Chapter III of the book [27] by Lundell and Weingram. The above example suggests that the failure of this fact for cellular stratified spaces is partly due to the "wrong definition" of dimensions of globular cells. The right notion of dimension is the stellar one.

Even for stellar stratified spaces, however, regularity does not necessarily imply normality.

Example 3.4: By adding an arc to the previous example, we obtain a stratified space as follows: the globular 2-cell $\mathrm{Int}(D^2) \cup \{(1,0)\} \cup$

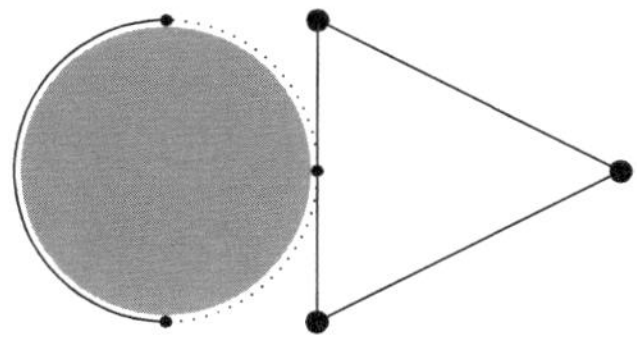

Fig. 11. A non-normal regular stellar stratified space.

$\{(x, y) \in S^1 \mid x \leq 0\}$ is a stellar 2-cell. And this is a regular stellar stratified space. But this is not normal.

Remark 3.5: It seems that, if X is a stellar stratified space in which the boundary of each stellar n-cell is "pure of dimension $n - 1$" and if X is regular, then X is normal.

3.2. *Total normality*

In the case of cellular or stellar stratified spaces, relations among cells are not as rigid as those in cell complexes.

Definition 3.6: Let X be a normal cellular stratified space. Then X is called *totally normal* if, for each n-cell e_λ,

(1) there exists a structure of regular cell complex on S^{n-1} containing ∂D_λ as a strict stratified subspace of S^{n-1};
(2) for any cell e in ∂D_λ, there exists a cell e_μ in ∂e_λ such that e_μ and e share the same domain and the characteristic map φ_μ of e_μ factors through D_λ via the characteristic map of e:

$$\begin{array}{ccccccccc} \overline{e} & \hookrightarrow & \partial D_\lambda & \hookrightarrow & D_\lambda & \xrightarrow{\varphi_\lambda} & \overline{e_\lambda} & \hookrightarrow & X \\ \uparrow & & & & & & & \nearrow & \\ D & = & D_\mu & \xrightarrow[\varphi_\mu]{} & \overline{e_\mu}. & & & & \end{array}$$

(with $\overline{e_\mu} \hookrightarrow X$ the diagonal inclusion.)

The total normality for stellar stratified spaces is defined analogously but we need to use cellular subdivisions.

Definition 3.7: Let X be a normal stellar stratified space. We say X is *totally normal* if, for each stellar n-cell e_λ with domain $D_\lambda \subset D^N$,

(1) there exists a structure of a regular cell complex on S^{N-1} and a structure of a stellar stratified space on ∂D_λ such that each stellar cell in ∂D_λ is a strict stratified subspace of S^{N-1};
(2) for any stellar cell e in ∂D_λ, there exists a stellar cell e_μ in ∂e_λ such that e_μ and e share the same domain and the characteristic map of e_μ factors through D_λ via the characteristic map of e.

Lemma 3.8: *Any totally normal cellular stratified space has a structure of a totally normal stellar stratified space.*

The characteristic maps of totally normal stellar stratified spaces preserve cells.

Lemma 3.9: *For a cell e_λ in a totally normal stellar stratified space X, let $\varphi_\lambda : D_\lambda \to \overline{e_\lambda} \subset X$ be the characteristic map. Then there exists a structure of a stellar stratified space on D_λ under which φ_λ is a strict morphism of stellar stratified spaces.*

Proof: Let e_λ be a cell in a totally normal stellar stratified space X and $\varphi_\lambda : D_\lambda \to \overline{e_\lambda}$ the characteristic map. Let

$$\partial D_\lambda = \bigcup_\nu e'_\nu$$

be the stellar stratification in the definition of total normality. We have

$$\partial e_\lambda = \varphi_\lambda(\partial D_\lambda) = \bigcup_\nu \varphi_\lambda(e'_\nu).$$

By the definition of total normality, for each ν, there exists a cell $e_{\mu'}$ in ∂e_λ whose characteristic map makes the following diagram commutative

$$\begin{array}{ccccc} \overline{e'_\nu} & \hookrightarrow & \partial D_\lambda & \xrightarrow{\varphi_\lambda} & \partial e_\lambda \\ \uparrow{\scriptstyle \psi_\nu} & & & & \uparrow \\ D_\nu & = & D_{\mu'} & \xrightarrow[\varphi_{\mu'}]{} & \overline{e_{\mu'}}, \end{array}$$

where ψ_ν is the characteristic map for e'_ν. This implies that each $\varphi_\lambda(e'_\nu)$ is a cell in X and thus φ_λ is a strict morphism of stellar stratified spaces. □

Corollary 3.10: *Let (π, Φ) be a stellar stratification on X satisfying the first condition of total normality.*

Then it is totally normal if and only if the following condition is satisfied: for each pair $e_\mu < e_\lambda$, let $F(X)(e_\mu, e_\lambda)$ be the set of all maps

$$b : D_\mu \longrightarrow D_\lambda$$

making the diagram

$$\begin{array}{ccc} D_\lambda & \xrightarrow{\varphi_\lambda} & \overline{e_\lambda} \\ {\scriptstyle b}\uparrow & & \uparrow \\ D_\mu & \xrightarrow[\varphi_\mu]{} & \overline{e_\mu} \end{array}$$

commutative, where φ_λ and φ_μ are the characteristic maps of e_λ and e_μ, respectively. Then

$$\partial D_\lambda = \bigcup_{e_\mu < e_\lambda} \bigcup_{b \in F(X)(e_\mu, e_\lambda)} b(\mathrm{Int}(D_\mu)). \tag{3.1}$$

Proof: Suppose (π, Φ) is totally normal. For a pair of cells $e_\mu < e_\lambda$, by Lemma 3.9, there exists a stellar cell e in D_λ such that $\varphi_\lambda(e) = e_\mu$. By the assumption of total normality, the characteristic map $\psi : D \to \overline{e}$ of e makes the following diagram commutative

$$\begin{array}{ccccc} \overline{e} & \hookrightarrow & \partial D_\lambda & \xrightarrow{\varphi_\lambda} & \partial e_\lambda \\ {\scriptstyle \psi}\uparrow & & & & \uparrow \\ D & = & D_\mu & \xrightarrow[\varphi_\mu]{} & \overline{e_\mu} \end{array}$$

The collection of all such characteristic maps ψ is $F(X)(e_\mu, e_\lambda)$ and we have

$$\partial D_\lambda = \bigcup_{e_\mu < e_\lambda} \bigcup_{b \in F(X)(e_\mu, e_\lambda)} b(\mathrm{Int}(D_\mu)).$$

Conversely, the assumption (3.1) implies that, for any stellar cell e in ∂D_λ, there is a corresponding cell e_μ in ∂e_λ whose characteristic map makes the required diagram commutative. □

The lifts $b : D_\mu \to D_\lambda$ of characteristic maps appearing in the above corollary play an essential role when we analyze totally normal stellar stratified spaces. It is easy to see that each b is an embedding.

Lemma 3.11: *Let X be a totally normal stellar stratified space. Then each $b \in F(X)(e_\mu, e_\lambda)$ is an embedding of stellar stratified spaces for each pair $e_\mu < e_\lambda$.*

Proof: This follows from the assumption that the cellular stratification on D_λ is regular. $\square$

3.3. *Examples of totally normal cellular stratified spaces*

Let us take a look at some examples. The first example is borrowed from Kirillov's paper [15].

Example 3.12: Consider $D = \mathrm{Int}(D^2) \cup S^1_+ = e^1 \cup e^2$. Subdivide e^1 into five arcs as is indicated by the left figure of Figure 12. Define X by folding the middle three arcs in e^1 according to the directions indicated by the arrows.

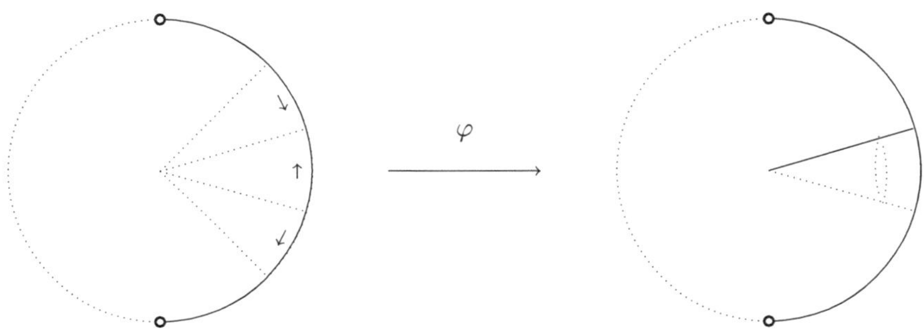

Fig. 12. An example from Kirillov's paper.

Note that $\varphi(e^1)$ is homeomorphic to $\mathrm{Int}(D^1)$. Let

$$\psi : \mathrm{Int}(D^1) \longrightarrow e^1 \subset X$$

be a homeomorphism. Identifications only occur on e^1 and the quotient map

$$\varphi : D \longrightarrow X$$

is a homeomorphism onto its image when restricted to $\mathrm{Int}(D^2)$ and thus defines a characteristic map for the 2-cell.

These maps ψ and φ define a cellular stratification on X. However, there is no way to obtain a map $h : \mathrm{Int}(D^1) \to D$ making the diagram commutative

$$\begin{array}{ccc} D & \xrightarrow{\varphi} & X \\ {\scriptstyle h}\uparrow & \nearrow{\scriptstyle \psi} & \\ \mathrm{Int}(D^1) & & \end{array}$$

and thus this cellular stratification is not totally normal. Note, however, that we obtain a totally normal cellular stratification by an appropriate subdivision of e^1 and $\varphi(e^1)$.

Example 3.13: Consider the minimal cell decomposition

$$S^1 = e^0 \cup e^1 = \{(1,0)\} \cup \left(S^1 \setminus \{(1,0)\}\right).$$

The characteristic map for the 1-cell

$$\varphi_1 : D^1 = [-1,1] \longrightarrow S^1$$

is given by $\varphi(t) = (\cos(2\pi t), \sin(2\pi t))$. There are two lifts b_{-1} and b_1 of the characteristic map φ_0 for the 0-cell.

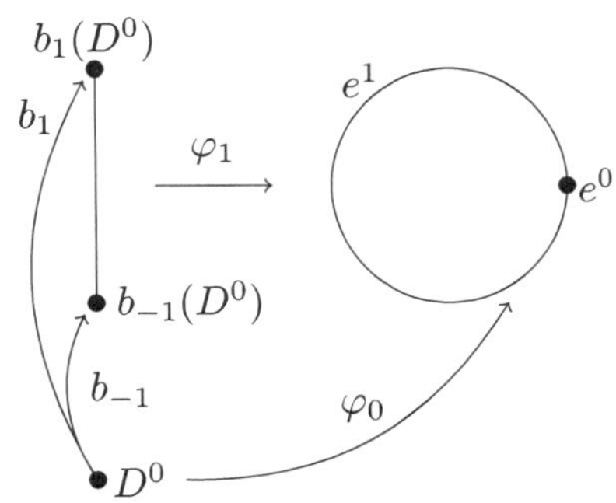

Fig. 13. The minimal cell decomposition of S^1 is totally normal.

Since $\partial D^1 = \{-1,1\} = b_{-1}(D^0) \cup b_1(D^0)$, this is totally normal.

Example 3.14: More generally, any 1-dimensional CW-complex is totally normal. In fact, 0-cells are always regular and there are only two types of 1-cells, i.e. regular cells and cells whose characteristic maps are given by collapsing the boundary of D^1 to a point. By the above example, all 1-cells are totally normal. This fact allows us to apply results of these notes to configuration spaces of graphs [1].

Example 3.15: Consider the punctured torus

$$X = S^1 \times S^1 \setminus e^0 \times e^0 = e^1 \times e^0 \cup e^0 \times e^1 \cup e^1 \times e^1$$

with the stratification induced from the product stratification of the minimal cell decomposition of S^1.

A characteristic map for the 2-cell can be obtained by removing four corners from the domain of the product

$$\varphi_1 \times \varphi_1 : D^1 \times D^1 \setminus \{(1,1),(1,-1),(-1,1),(-1,-1)\} \longrightarrow X$$

of φ_1 in Example 3.13. Let us denote the characteristic maps for $e^1 \times e^0$ and $e^0 \times e^1$ by $\varphi_{1,0}$ and $\varphi_{0,1}$, respectively. The characteristic map of each

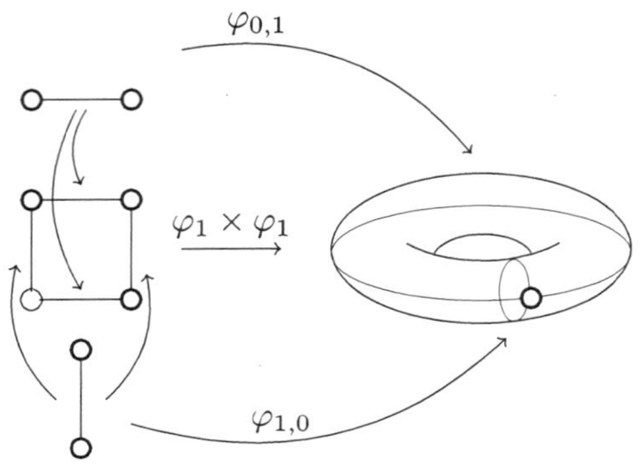

Fig. 14. A totally normal punctured torus.

1-cell has two lifts and the images cover the boundary of the domain of the 2-cell. Thus this is totally normal.

Example 3.16: Let X be a Δ-set. (See Definition B.14.) Define

$$\pi_X : \|X\| \longrightarrow \coprod_{n=0}^{\infty} X_n$$

by $\pi_X(x) = \sigma$ if x is represented by $(\boldsymbol{t}, \sigma) \in \Delta^n \times X_n$ with $t_i \neq 0$ for all $i \in \{0, \ldots, n\}$. Note that $P(\|X\|) = \coprod_{n=0}^{\infty} X_n$ can be made into a poset by defining $\tau \leq \sigma$ if and only if there exists a morphism $u : [m] \to [n]$ in Δ_{inj} with $X(u)(\sigma) = \tau$.

Then the map π_X is a cellular stratification and we have

$$\|X\| = \bigcup_{n=0}^{\infty} \bigcup_{\sigma \in X_n} \operatorname{Int}(\Delta^n) \times \{\sigma\}.$$

Let us denote the n-cell $\operatorname{Int}(\Delta^n) \times \{\sigma\}$ corresponding to $\sigma \in X_n$ by e_σ. The characteristic map φ_σ for e_σ is defined by the composition

$$\varphi_\sigma : D^n \cong \Delta^n \times \{\sigma\} \hookrightarrow \coprod_{n=0}^{\infty} \Delta^n \times X_n \longrightarrow \|X\|.$$

Suppose $\tau < \sigma$ for $\tau \in X_m$ and $\sigma \in X_n$. There exists a morphism $u : [m] \to [n]$ with $X(u)(\sigma) = \tau$. With this morphism, we have the following commutative diagram

$$\begin{array}{ccc} \Delta^n \times \{\sigma\} & \xrightarrow{\varphi_\sigma} & \overline{e_\sigma} \\ {\scriptstyle u_*^{\tau,\sigma}} \uparrow & & \uparrow \\ \Delta^m \times \{\tau\} & \xrightarrow[\varphi_\tau]{} & \overline{e_\tau}, \end{array}$$

where $u_*^{\tau,\sigma} : \Delta^m \times \{\tau\} \to \Delta^n \times \{\sigma\}$ is a copy of the affine map $u_* : \Delta^m \to \Delta^n$ induced by the inclusion of vertices $u : [m] \to [n]$. Hence this cellular stratification on $\|X\|$ is totally normal.

In particular, for any acyclic category C (with discrete topology), the classifying space BC is a totally normal CW complex by Lemma C.16.

Example 3.17: A. Kirillov, Jr. introduced a structure called a PLCW-complex in [15]. A PLCW-complex is defined by attaching PL-disks. The attaching map of an n-cell is required to be a strict morphism of cellular stratified spaces under a suitable PLCW decomposition of the boundary sphere.

Besides the PL requirement in Kirillov's definition, the only difference between totally normal CW complexes and PLCW complexes is that, in totally normal CW complexes, the boundary sphere of each characteristic map is required to have a regular cell decomposition. For example, the cell decomposition of D^3 in Figure 15 is a PLCW complex but is not totally normal.

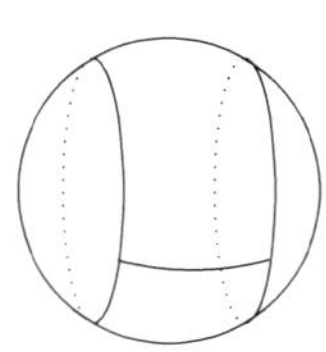

Fig. 15. A PLCW complex which is not totally normal.

Example 3.18: Let $X = \mathbb{R} \times \mathbb{R}_{\geq 0}$ with 0-cells $e_n^0 = \{(n, 0)\}$ for $n \in \mathbb{Z}$, 1-cells $e_n^1 = (n, n+1) \times \{0\}$ for $n \in \mathbb{Z}$, and a 2-cell $e^2 = \mathbb{R} \times \mathbb{R}_{>0}$. The characteristic map φ of the 2-cell is given by extending the stereographic projection $S^1 \backslash \{(0,1)\} \to \mathbb{R}$. The domain is $D = D^2 \backslash \{(0,1)\}$. This is regular

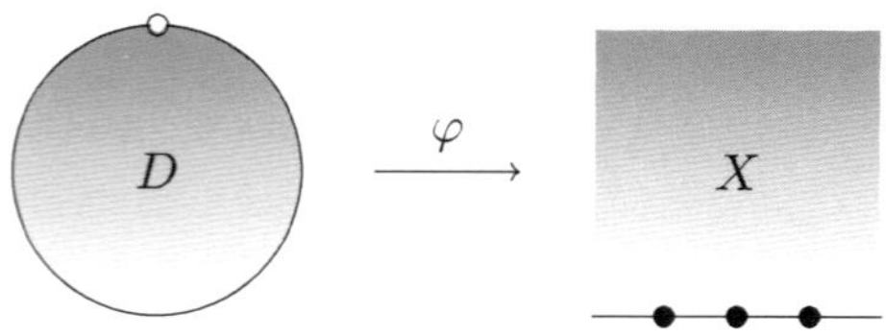

Fig. 16. A cellular stratification on the closed upper half-plane with 0-cells $\mathbb{Z}$.

and normal. It even satisfies the second condition in Definition 3.6. But it is not totally normal, since the corresponding stratification on $\partial D^2 \setminus \{(0,1)\}$ cannot be a strict stratified subspace of a regular cell decomposition of ∂D^2.

Note that this example does not satisfy the closure finiteness condition. Hence it is not CW.

Example 3.19: Consider the minimal cell decomposition

$$S^2 = e^0 \cup e^2.$$

We may choose a regular cell decomposition of ∂D^2 and make it a stellar stratification. For example, we may use the minimal regular cell decomposition $S^1 = e^0_+ \cup e^0_- \cup e^1_+ \cup e^1_-$.

In order to make S^2 totally normal, however, this stratification is not fine enough. There are infinitely many lifts of the characteristic map of the 0-cell parametrized by points in S^1. See Figure 17. Although ∂D^2 is covered

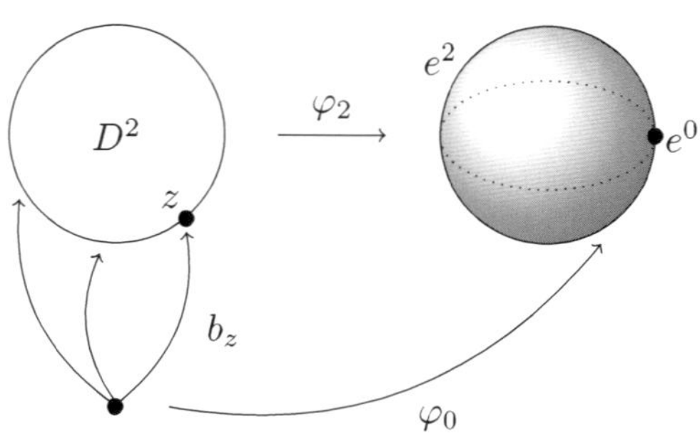

Fig. 17. A minimal cell decomposition of S^2.

by the images of b_z's

$$\partial D^2 = \bigcup_{z \in S^1} b_z(D^0),$$

this is not a cell decomposition of ∂D^2. Hence this is not totally normal.

4. Cylindrically normal cellular stratified spaces

4.1. *Cylindrical structures*

In Example 3.19, lifts of φ_0 are parametrized by points in $S^1 = \partial D^2$. This example suggests that we need to topologize the set of all lifts. Inspired by the work of Cohen, Jones and Segal on Morse theory [16] and this example, we introduce the following definition:

Definition 4.1: A *cylindrical structure* on a normal cellular stratified space (X, π) consists of

- a normal stratification on ∂D^n containing D_λ as a strict stratified subspace for each n-cell $\varphi_\lambda : D_\lambda \to \overline{e_\lambda}$,
- a stratified space $P_{\mu,\lambda}$ and a strict morphism of stratified spaces

$$b_{\mu,\lambda} : P_{\mu,\lambda} \times D_\mu \longrightarrow \partial D_\lambda$$

  for each pair of cells $e_\mu \subset \partial e_\lambda$,
- a strict morphism of stratified spaces

$$c_{\lambda_0,\lambda_1,\lambda_2} : P_{\lambda_1,\lambda_2} \times P_{\lambda_0,\lambda_1} \longrightarrow P_{\lambda_0,\lambda_2}$$

  for each sequence $\overline{e_{\lambda_0}} \subset \overline{e_{\lambda_1}} \subset \overline{e_{\lambda_2}}$

satisfying the following conditions:

(1) The restriction of $b_{\mu,\lambda}$ to $P_{\mu,\lambda} \times \mathrm{Int}(D_\mu)$ is a homeomorphism onto its image.
(2) The following three types of diagrams are commutative:

$$\begin{array}{ccc} D_\lambda & \xrightarrow{\varphi_\lambda} & X \\ {\scriptstyle b_{\mu,\lambda}}\uparrow & & \uparrow{\scriptstyle \varphi_\mu} \\ P_{\mu,\lambda} \times D_\mu & \xrightarrow{\mathrm{pr}_2} & D_\mu. \end{array}$$

$$\begin{array}{ccc} P_{\lambda_1,\lambda_2} \times P_{\lambda_0,\lambda_1} \times D_{\lambda_0} & \xrightarrow{1\times b_{\lambda_0,\lambda_1}} & P_{\lambda_1,\lambda_2} \times D_{\lambda_1} \\ \downarrow{\scriptstyle c_{\lambda_0,\lambda_1,\lambda_2}\times 1} & & \downarrow{\scriptstyle b_{\lambda_1,\lambda_2}} \\ P_{\lambda_0,\lambda_2} \times D_{\lambda_0} & \xrightarrow[b_{\lambda_0,\lambda_2}]{} & D_{\lambda_2} \end{array}$$

$$\begin{array}{ccc} P_{\lambda_2,\lambda_3} \times P_{\lambda_1,\lambda_2} \times P_{\lambda_0,\lambda_1} & \xrightarrow{c\times 1} & P_{\lambda_1,\lambda_3} \times P_{\lambda_0,\lambda_1} \\ \downarrow{\scriptstyle 1\times c} & & \downarrow{\scriptstyle c} \\ P_{\lambda_2,\lambda_3} \times P_{\lambda_0,\lambda_2} & \xrightarrow[c]{} & P_{\lambda_0,\lambda_3}. \end{array}$$

(3) We have

$$\partial D_\lambda = \bigcup_{e_\mu \subset \partial e_\lambda} b_{\mu,\lambda}(P_{\mu,\lambda} \times \operatorname{Int}(D_\mu))$$

as a stratified space.

The space $P_{\mu,\lambda}$ is called the *parameter space* for the inclusion $e_\mu \subset \overline{e_\lambda}$. When $\mu = \lambda$, we define $P_{\lambda,\lambda}$ to be a single point. A stellar stratified space equipped with a cylindrical structure is called a *cylindrically normal stellar stratified space.*

When the map $b_{\lambda,\mu}$ is an embedding for each pair $e_\mu \subset \overline{e_\lambda}$, the stratification is said to be *strictly cylindrical.*

When X is a stellar stratified space, we require that the normal stratification on each ∂D_λ is a coarsening (Definition 2.15) of the stratification in the definition of a stellar stratified space.

Remark 4.2: The author first intended to call such a structure a "locally product-like" or "locally trivial" cellular stratification. But it turns out the term "locally trivial stratification" is already used in [21] in a different sense.

We require morphisms between cylindrically normal cellular or stellar stratified spaces to preserve cylindrical structures.

Definition 4.3: Let (X, π_X, Φ_X) and (Y, π_Y, Φ_Y) be cylindrically normal cellular stratified spaces with cylindrical structures given by $\{b^X_{\mu,\lambda} : P^X_{\mu,\lambda} \times D_\mu \to D_\lambda\}$ and $\{b^Y_{\alpha,\beta} : P^Y_{\alpha,\beta} \times D_\alpha \to D_\beta\}$ respectively.

A *morphism of cylindrically normal cellular stratified spaces* from (X, π_X, Φ_X) to (Y, π_Y, Φ_Y) is a morphism of stellar stratified spaces

$$\boldsymbol{f} = (f, \underline{f}) : (X, \pi_X, \Phi_X) \longrightarrow (Y, \pi_Y, \Phi_Y)$$

together with maps $f_{\mu,\lambda} : P^X_{\mu,\lambda} \to P^Y_{\underline{f}(\mu),\underline{f}(\lambda)}$ making the diagram commutative:

$$\begin{array}{ccc} P^X_{\mu,\lambda} \times D_\mu & \xrightarrow{f_{\mu,\lambda}\times f_\mu} & P^Y_{\underline{f}(\mu),\underline{f}(\lambda)} \times D_{\underline{f}(\mu)} \\ {\scriptstyle b^X_{\mu,\lambda}}\big\downarrow & & \big\downarrow{\scriptstyle b^Y_{\underline{f}(\mu),\underline{f}(\lambda)}} \\ D_\lambda & \xrightarrow[f_\lambda]{} & D_{\underline{f}(\lambda)}. \end{array}$$

Morphisms of cylindrically normal stellar stratified spaces are defined analogously.

The categories of cylindrically normal cellular and stellar stratified spaces are denoted by $\mathbf{CSSpaces}^{\mathrm{cyl}}$ and $\mathbf{SSSpaces}^{\mathrm{cyl}}$ respectively.

For cylindrically normal cellular/stellar stratified spaces, the relative-compactness of closures of cells can be easily verified.

Lemma 4.4: *Let X be a cylindrically normal cellular (stellar) stratified space with parameter spaces $\{P_{\mu,\lambda}\}_{\mu\le\lambda}$. A cell $\varphi_\lambda : D_\lambda \to \overline{e_\lambda}$ is relatively compact if and only if $P_{\mu,\lambda}$ is compact for each $\mu \le \lambda$.*

Proof: For $y \in \partial e_\lambda$, there exists $\mu \le \lambda$ with $y \in e_\mu \subset \partial e_\lambda$. In the commutative diagram

$$\begin{array}{ccccc} D_\lambda & & \xrightarrow{\quad\varphi_\lambda\quad} & & \overline{e_\lambda} \\ {\scriptstyle b_{\mu,\lambda}|_{P_{\mu,\lambda}\times \mathrm{Int}(D_\mu)}}\big\uparrow & & & & \big\uparrow \\ P_{\mu,\lambda} \times \mathrm{Int}(D_\mu) & \longrightarrow & \mathrm{Int}(D_\mu) & \xrightarrow[\varphi_\mu|_{\mathrm{Int}(D_\mu)}]{} & e_\mu, \end{array}$$

the restriction $b_{\mu,\lambda}|_{P_{\mu,\lambda}\times\mathrm{Int}(D_\mu)}$ is an embedding and ∂D_λ is covered by the disjoint union of such images. Thus we have $\varphi_\lambda^{-1}(y) \cong P_{\mu,\lambda} \times \{y\}$ and the result follows from Corollary A.15 and the assumption. $\square$

Corollary 4.5: *Let X be a CW cylindrically normal cellular (stellar) stratified space. If all parameter spaces are compact, X is paracompact.*

4.2. *Polyhedral cellular stratified spaces*

Although total normality and cylindrical normality play central roles in our study of cellular and stellar stratified spaces, it is not easy to prove that a given cellular stratification is totally normal or cylindrically normal.

In order to address this problem, we found polyhedral complexes and PL maps are useful. It turns out that PL maps also play an important role in our proof of Theorem 5.18. In this section, we generalize polyhedral complexes and introduce the notion of polyhedral cellular stratified spaces.

Let us first recall the definition of polyhedral complexes. See §2.2.4 of Kozlov's book [28] for details.

Definition 4.6: A *Euclidean polyhedral complex* is a subspace K of $\mathbb{R}^N$ for some N equipped with a finite family of maps

$$\{\varphi_i : F_i \longrightarrow K \mid i = 1, \ldots, n\}$$

satisfying the conditions that

(1) each F_i is a convex polytope;
(2) each φ_i is an affine equivalence onto its image;
(3) $K = \bigcup_{i=1}^{n} \varphi_i(\mathrm{Int}(F_i))$, where $\mathrm{Int}(F_i)$ is the relative interior, i.e. the interior in its affine hull, of F_i;
(4) for $i \neq j$, $\varphi_i(F_i) \cap \varphi_j(F_j)$ is a proper face of $\varphi_i(F_i)$ and $\varphi_j(F_j)$.

The polytopes F_i's are called *generating polytopes.*

Obviously a polyhedral complex is a regular cell complex. By replacing affine cell structure maps φ_i by continuous maps, Kozlov defined a more general kind of polyhedral complexes in his book. The requirement of being a subspace of $\mathbb{R}^N$ can be removed if we assume cylindrical normality. We may also remove the condition that all cells are closed. Here is our definition of a "polyhedral" structure.

Definition 4.7: A *polyhedral stellar stratified space* consists of

- a CW cylindrically normal stellar stratified space X,
- a family of Euclidean polyhedral complexes $\widetilde{F}_\lambda$ indexed by $\lambda \in P(X)$,
- a family of homeomorphisms $\alpha_\lambda : \widetilde{F}_\lambda \to \overline{D_\lambda}$ indexed by $\lambda \in P(X)$, where $\overline{D_\lambda}$ is the closure of the domain stellar cell D_λ for e_λ in a disk containing D_λ,

satisfying the following conditions:

(1) For each cell e_λ, $\alpha_\lambda : \widetilde{F}_\lambda \to \overline{D_\lambda}$ is a subdivision of a stratified space, where the stratification on $\overline{D_\lambda}$ is defined by the cylindrical structure.
(2) For each pair $e_\mu < e_\lambda$, the parameter space $P_{\mu,\lambda}$ is a locally cone-like space (Definition B.22) and the composition

$$P_{\mu,\lambda} \times F_\mu \overset{1\times\alpha_\mu}{\longrightarrow} P_{\mu,\lambda} \times D_\mu \overset{b_{\mu,\lambda}}{\longrightarrow} D_\lambda \overset{\alpha_\lambda^{-1}}{\longrightarrow} F_\lambda$$

is a PL map (Definition B.30), where $F_\lambda = \alpha_\lambda^{-1}(D_\lambda)$.

Each α_λ is called a *polyhedral replacement* of the cell structure map of e_λ. The collection $A = \{\alpha_\lambda\}_{\lambda\in P(X)}$ is called a *polyhedral structure* on X.

Remark 4.8: When X is cellular, $\overline{D_\lambda} = D^{\dim e_\lambda}$.

Remark 4.9: An analogous structure for cell complexes is introduced by A. Kirillov, Jr. in [15] as PLCW complexes for a different purpose. M_κ-polyhedral complexes in the book [20] by Bridson and Haefliger are also closely related.

Definition 4.10: A *morphism of polyhedral stellar stratified spaces* from (X,π,Φ,A) to (X',π',Φ',A') consists of a morphism of stellar stratified spaces

$$(\boldsymbol{f},\{f_\lambda\}) : (X,\pi,\Phi) \longrightarrow (X',\pi',\Phi')$$

and a family of PL maps $\tilde{f}_\lambda : \widetilde{F}_\lambda \to \widetilde{F}'_{\underline{f}(\lambda)}$ for $\lambda \in P(X)$ that are compatible with polyhedral structures.

Any polyhedral complex is polyhedral. More generally, we have the following criterion:

Lemma 4.11: *Let X be a subspace of $\mathbb{R}^N$ equipped with a structure of a cylindrically normal CW stellar stratified space whose parameter spaces $P_{\mu,\lambda}$ are locally cone-like spaces. Suppose, for each $e_\lambda \in P(X)$, there exists a polyhedral complex $\widetilde{F}_\lambda$ and a homeomorphism $\alpha_\lambda : \widetilde{F}_\lambda \to D^{\dim e_\lambda}$ such that the composition*

$$P_{\mu,\lambda} \times F_\mu \overset{1\times\alpha_\mu}{\longrightarrow} P_{\mu,\lambda} \times D_\mu \overset{b_{\mu,\lambda}}{\longrightarrow} D_\lambda \overset{\varphi_\lambda}{\longrightarrow} X \hookrightarrow \mathbb{R}^N$$

is a PL map, where $F_\lambda = \alpha_\lambda^{-1}(D_\lambda)$. Suppose further that $\alpha_\lambda : \widetilde{F}_\lambda \to D^{\dim e_\lambda}$ is a cellular subdivision, where the cell decomposition on $D^{\dim e_\lambda}$ is the one in the definition of a cylindrical structure. Then the collection $\{\alpha_\lambda\}_{\lambda\in P(X)}$ defines a polyhedral structure on X.

Proof: For each pair $e_\mu < e_\lambda$, define $\tilde{b}_{\mu,\lambda} : P_{\mu,\lambda} \times F_\mu \to F_\lambda$ to be the composition

$$\tilde{b}_{\mu,\lambda} : P_{\mu,\lambda} \times F_\mu \overset{1\times\alpha_\mu}{\longrightarrow} P_{\mu,\lambda} \times D_\mu \overset{b_{\mu,\lambda}}{\longrightarrow} D_\lambda \overset{\alpha_\lambda^{-1}}{\longrightarrow} F_\lambda.$$

Note that, when $\mu = \lambda$, $b_{\mu,\lambda}$ is the identity map. Thus the top horizontal composition in the following diagram is a PL map:

$$\begin{array}{ccccccccc}
F_\lambda & \xrightarrow{\alpha_\lambda} & D_\lambda & \xrightarrow{\varphi_\lambda} & X & & \hookrightarrow & & \mathbb{R}^N \\
\uparrow{\scriptstyle \tilde{b}_{\mu,\lambda}} & & & & & & & & \| \\
P_{\mu,\lambda} \times F_\mu & \xrightarrow[1\times\alpha_\mu]{} & P_{\mu,\lambda} \times D_\mu & \xrightarrow[b_{\mu,\lambda}]{} & D_\lambda & \xrightarrow[\varphi_\lambda]{} & X & \hookrightarrow & \mathbb{R}^N.
\end{array}$$

The bottom horizontal composition is also a PL map by assumption and $\tilde{b}_{\mu,\lambda}$ is an embedding when restricted to $P_{\mu,\lambda} \times \alpha_\mu^{-1}(\mathrm{Int}D^{\dim e_\mu})$. Thus $\tilde{b}_{\mu,\lambda}$ is also PL by Lemma B.32. □

Example 4.12: Consider the minimal cell decomposition $S^2 = e^0 \cup e^2$. We have an embedding $f : S^2 \to \mathbb{R}^3$ whose image is the boundary $\partial\Delta^3$ of the standard 3-simplex and $f(e^0)$ is a vertex of $\partial\Delta^3$. This is a cylindrically normal cellular stratified space by Example 4.24. Let P be a 2-dimensional polyhedral complex in $\mathbb{R}^2$ described in Figure 18. By collapsing the outer

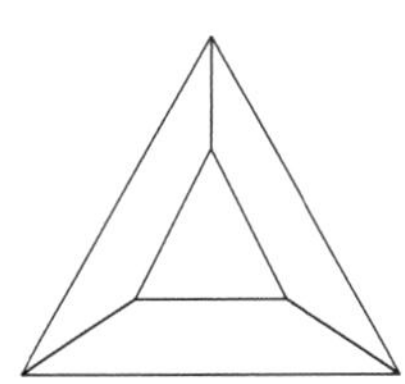

Fig. 18. A polyhedral structure on S^2

triangle, we obtain a map

$$\psi : P \longrightarrow \partial\Delta^3$$

whose restriction to the interior is a homeomorphism. Let $\alpha : P \to D^2$ be a homeomorphism given by a radial expansion. Then maps f and φ can be chosen in such a way they make the following diagram commutative

$$\begin{array}{ccccc}
D^2 & \xrightarrow{\varphi} & S^2 & & \\
\uparrow{\scriptstyle \alpha} & & \downarrow{\scriptstyle f} & & \\
P & \xrightarrow[\psi]{} & \partial\Delta^3 & \hookrightarrow & \mathbb{R}^3.
\end{array}$$

By Lemma 4.11, we obtain a structure of a polyhedral cellular stratified space on S^2 or $\partial\Delta^3$.

Definition 4.13: A cellular stratified space satisfying the assumption of Lemma 4.11 is called a *Euclidean polyhedral cellular stratified space.*

Totally normal cellular stratified spaces form an important class of polyhedral cellular stratified spaces.

Lemma 4.14: *Any CW totally normal cellular stratified space has a polyhedral structure.*

Proof: Let X be a CW totally normal cellular stratified space. By definition, for each cell $\varphi_\lambda : D_\lambda \to \overline{e}_\lambda$ in X, there exists a regular cell decomposition of $D^{\dim e_\lambda}$ containing D_λ as a cellular stratified subspace. Since the barycentric subdivision of a regular cell complex has a structure of a simplicial complex, $D^{\dim e_\lambda}$ can be embedded in a Euclidean space as a finite simplicial complex $\widetilde{F}_\lambda$. By induction on dimensions of cells, we may choose homeomorphisms $\{\alpha_\lambda : \widetilde{F}_\lambda \to D^{\dim e_\lambda}\}_{\lambda \in P(X)}$ in such a way that the composition

$$F_\mu \xrightarrow{\alpha_\mu} D_\mu \xrightarrow{b} D_\lambda \xrightarrow{\alpha_\lambda^{-1}} F_\lambda$$

is a PL map for each lift $b : D_\mu \to D_\lambda$ of the cell structure map of e_μ for each pair $e_\mu < e_\lambda$. Thus we obtain a polyhedral structure. □

Conversely, the polyhedrality condition provides us with a useful criterion for a stratification to be totally normal. The following fact first appeared in [14].

Proposition 4.15: *Let X be a normal CW complex embedded in $\mathbb{R}^N$ for some N. Suppose for each cell $e \subset X$ with cell structure map φ, there exists a Euclidean polyhedral complex F and a homeomorphism $\alpha : F \to D^{\dim e}$ such that the composition*

$$F \xrightarrow{\alpha} D^{\dim e} \xrightarrow{\varphi} X \subset \mathbb{R}^N$$

is a PL map. Then any regular cellular stratified subspace of X is polyhedral.

Proof: Let A be a regular cellular stratified subspace of X. It suffices to find polyhedral replacements of cell structures for A. For an n-cell e in A, let $\varphi : D^n \to X$ be the cell structure of e in X. The cell structure for e in A is, by definition, given by

$$\varphi_A = \varphi|_{D_A} : D_A = \varphi^{-1}(\overline{e} \cap A) \longrightarrow A.$$

By assumption, there exists a polyhedral complex P and a homeomorphism $\alpha : P \to D^n$ with $\varphi \circ \alpha : P \to X \hookrightarrow \mathbb{R}^N$ a PL map. □

Example 4.16: The product cell decomposition of $(S^n)^k$ induced by the minimal cell decomposition on S^n is polyhedral. In [14], a subdivision of the product cell decomposition containing $\mathrm{Conf}_k(S^n)$ as a stratified subspace was defined by using the stratification on $(\mathbb{R}^n)^\ell$ associated with the braid arrangement $\mathcal{A}_{\ell-1}$ for $1 \leq \ell \leq k$.

The author expected that the induced stratification on $\mathrm{Conf}_k(S^n)$ to be totally normal. It turned out that the stratification is much more complicated than the author imagined.

We propose the following strategy to prove a given cellular stratification on X is cylindrically normal or totally normal:

(1) Embed X into a cylindrically normal or totally normal cell complex $\widetilde{X}$.
(2) Find an appropriate subdivision of $\widetilde{X}$ which includes X as a stratified subspace.

In fact, it is easy to prove that, if X is polyhedral, X can be embedded in a cell complex as a stratified subspace.

Lemma 4.17: *Let X be a polyhedral cellular stratified space. Given a pair of cells $e_\mu \subset \partial e_\lambda$, the structure map*

$$b_{\mu,\lambda} : P_{\mu,\lambda} \times D_\mu \longrightarrow D_\lambda$$

has a unique extension to the whole disks

$$\overline{b_{\mu,\lambda}} : P_{\mu,\lambda} \times D^{\dim e_\mu} \longrightarrow D^{\dim e_\lambda}.$$

Proof: By Lemma B.32. □

These maps allow us to construct a canonical closure of any polyhedral cellular stratified space.

Definition 4.18: Let X be a polyhedral cellular stratified space with cells $\{e_\lambda\}_{\lambda \in P(X)}$. Define a cell complex $U(X)$ by

$$U(X) = \left(\coprod_{\lambda \in P(X)} D^{\dim e_\lambda} \right) \Big/ \sim$$

where the relation $\sim$ is the equivalence relation generated by the following relation: for $x \in D^{\dim e_\lambda}$ and $y \in D^{\dim \mu}$, $x \sim y$ if $e_\mu \subset \partial e_\lambda$ and and there exists $z \in P_{\mu,\lambda}$ such that $\overline{b_{\mu,\lambda}}(z,y) = x$.

There is a canonical inclusion

$$i : X \hookrightarrow U(X).$$

This space $U(X)$ is called the *cellular closure* of X.

By definition, we have the following:

Lemma 4.19: *When X is an polyhedral cellular stratified space, $U(X)$ is a cylindrically normal CW complex containing X as a cellular stratified subspace.*

Example 4.20: In the case of the punctured torus in Example 3.15, $U(X)$ is obtained by gluing parallel edges in I^2 and is homeomorphic to T^2. In other words, $U(X)$ is obtained by closing the hole in X.

The existence of cellular closure implies that cell structures of a polyhedral cellular stratified space are bi-quotient. (See Definition A.1.)

Corollary 4.21: *Let X be a polyhedral cellular stratified space. Then any cell structure $\varphi_\lambda : D_\lambda \to \overline{e_\lambda}$ is bi-quotient.*

Proof: For an n-cell $\varphi_\lambda : D_\lambda \to \overline{e_\lambda}$ in X, let $\widetilde{\varphi_\lambda} : D^n \to U(X)$ be the extension. Since $\widetilde{\varphi_\lambda}$ is proper, it is bi-quotient and hence is hereditarily quotient. By Lemma A.5, φ_λ is also hereditarily quotient.

For $y \in \overline{e_\lambda} \subset X$, the fiber $\varphi_\lambda^{-1}(y)$ can be identified with one of the parameter spaces by the proof of Lemma 4.4, which is a cellular stratified subspace of a regular cell decomposition of ∂D^n. Thus the boundary $\partial\varphi_\lambda^{-1}(y)$ is compact. The result follows from (3) in Lemma A.8. □

Corollary 4.22: *Any polyhedral cellular stratified space is paracompact.*

4.3. *Examples of cylindrically normal cellular stratified spaces*

Here is a collection of examples and nonexamples of cylindrically normal cellular stratified spaces.

Example 4.23: A stellar stratified space X is totally normal if and only if it is strictly cylindrically normal and each parameter space $P_{\mu,\lambda}$ is a finite set (with the discrete topology).

Consider the cell decomposition of D^3 in Example 3.17. It is easily seen to be cylindrically normal with finite parameter spaces. However, it is not strictly cylindrical, as we have seen in Example 3.17. In other words, a cylindrically normal cell complex with finite parameter spaces is a PLCW complex if it satisfies the PL requirement in the definition of PLCW complexes.

Example 4.24: Consider the minimal cell decomposition of $S^2 = \mathbb{C}\mathrm{P}^1$ in Example 3.19. The trivial stratification on ∂D^2 and the canonical inclusion

$$b_{0,2} : S^1 \times D^0 \longrightarrow \partial D^2 \subset D^2$$

define a cylindrical structure on S^2 with $P_{0,2} = S^1$, for we have a commutative diagram

$$\begin{array}{ccc} D^2 & \xrightarrow{\varphi_2} & \mathbb{C}\mathrm{P}^1 \\ {\scriptstyle b_{0,2}} \uparrow & & \uparrow {\scriptstyle \varphi_0} \\ S^1 \times D^0 & \xrightarrow[\mathrm{pr}_2]{} & D^0. \end{array}$$

We may also consider S^2 as a stellar stratified space by choosing a regular cell decomposition of ∂D^2. For example, we may use the minimal regular cell decomposition of ∂D^2.

Then we have embeddings

$$\begin{aligned} b_{0,+} &: D^0 \times D^0 \longrightarrow \partial D^2 \\ b_{0,-} &: D^0 \times D^0 \longrightarrow \partial D^2 \\ b_{1,+} &: D^0 \times D^1 \longrightarrow \partial D^2 \\ b_{1,-} &: D^0 \times D^1 \longrightarrow \partial D^2 \end{aligned}$$

corresponding to cells $e^0_+ \cup e^0_- \cup e^1_+ \cup e^1_-$ in ∂D^2. And we obtain a cylindrical structure as a stellar stratified space.

Example 4.25: Let us extend the cylindrically normal cell decomposition on $S^2 = \mathbb{C}\mathrm{P}^1$ to $\mathbb{C}\mathrm{P}^2$. The minimal cell decomposition of $\mathbb{C}\mathrm{P}^2$ is given by

$$\mathbb{C}\mathrm{P}^2 = S^2 \cup e^4 = e^0 \cup e^2 \cup e^4.$$

Consider the cell structure map of the 4-cell

$$\varphi_4 = \tilde{\eta} : D^4 \longrightarrow \mathbb{C}\mathrm{P}^2$$

whose restriction to the boundary is the Hopf map

$$\eta : S^3 \longrightarrow S^2.$$

This is a fiber bundle with fiber S^1 and thus the cell decomposition $S^2 = e^0 \cup e^2$ induces a decomposition

$$S^3 \cong e^0 \times S^1 \cup e^2 \times S^1,$$

as we have seen in Example 2.17.

Let

$$\begin{aligned} \varphi_0 &: D^0 \to \overline{e^0} \subset S^2 \\ \varphi_2 &: D^2 \to \overline{e^2} \subset S^2 \end{aligned}$$

be the cell structure maps of e^0 and e^2, respectively. We have a trivialization

$$t : \varphi_2^*(S^3) \xrightarrow{\cong} D^2 \times S^1.$$

Let $b_{2,4} : S^1 \times D^2 \to S^3 = \partial D^4$ be the composition

$$S^1 \times D^2 \xrightarrow{t^{-1}} \varphi_2^*(S^3) \xrightarrow{\tilde{\varphi}_2} S^3,$$

then we have the following commutative diagram

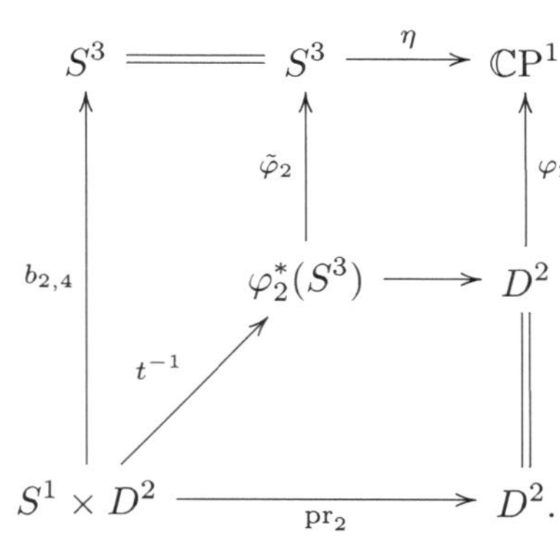

Let $b_{0,4} : S^1 \times D^0 \to S^3$ be the inclusion of the fiber over e^0. Then we have

$$\partial D^4 = b_{0,4}(S^1 \times D^0) \cup b_{2,4}(S^1 \times D^2).$$

Let $P_{0,2} = P_{2,4} = P_{0,4} = S^1$ and define

$$c_{0,2,4} : P_{2,4} \times P_{0,2} \longrightarrow P_{0,4}$$

by the multiplication of S^1.

Let us check that the above data defines a cylindrical structure on $\mathbb{C}P^2 = e^0 \cup e^2 \cup e^4$. It remains to verify the commutativity of the diagram

$$\begin{array}{ccc} P_{2,4} \times P_{0,2} \times D^0 & \xrightarrow{1\times b_{0,2}} & P_{2,4} \times D^2 \\ \downarrow{\scriptstyle c\times 1} & & \downarrow{\scriptstyle b_{2,4}} \\ P_{0,4} \times D^0 & \xrightarrow[b_{0,4}]{} & D^4. \end{array}$$

In other words, we need to show the restriction of $b_{2,4}$ to

$$b_{2,4}|_{S^1\times S^1} : S^1 \times S^1 \longrightarrow S^3 = \partial D^4$$

is given by the multiplication of S^1 followed by the inclusion of the fiber $\eta^{-1}(e^0)$. Recall that the Hopf map η is given by

$$\eta(z_1, z_2) = (2|z_1|^2 - 1, 2z_1\bar{z}_2),$$

where we regard

$$\begin{aligned} S^2 &= \{(x, z) \in \mathbb{R} \times \mathbb{C} \mid x^2 + |z|^2 = 1\} \\ S^3 &= \{(z_1, z_2) \in \mathbb{C}^2 \mid |z_1|^2 + |z_2|^2 = 1\}, \end{aligned}$$

and, on $U_+ = S^2 - \{(1, 0)\}$, the local trivialization

$$\varphi_+ : \eta^{-1}(U_+) \longrightarrow U_+ \times S^1$$

is given by

$$\varphi_+(z_1, z_2) = \left(2|z_1|^2 - 1, 2z_1\bar{z}_2, \frac{z_2}{|z_2|}\right).$$

The inverse of φ_+ is given by

$$\varphi_+^{-1}(x, z, w) = \left(\frac{zw}{2\sqrt{\frac{1-x}{2}}}, w\sqrt{\frac{1-x}{2}}\right).$$

Define

$$\mathrm{wrap} : D^2 \longrightarrow S^2$$

by

$$\mathrm{wrap}(z) = \left(2|z| - 1, 2\sqrt{|z|(1 - |z|)}\frac{z}{|z|}\right);$$

then the restriction of wrap to $\mathrm{Int}(D^2)$ is a homeomorphism onto $S^2 \setminus \{(1, 0)\}$. The map $b_{2,4}$ is defined by the composition

$$S^1 \times \mathrm{Int}(D^2) \xrightarrow{1\times\mathrm{wrap}} S^1 \times U_+ \cong U_+ \times S^1 \xrightarrow{\varphi_+^{-1}} \eta^{-1}(U_+) \hookrightarrow S^3,$$

which is given by

$$\begin{aligned} b_{2,4}(w,z) &= \varphi_+^{-1}\left(2|z|-1, 2\sqrt{|z|(1-|z|)}\frac{z}{|z|}, w\right) \\ &= \left(\frac{2\sqrt{|z|(1-|z|)}\frac{z}{|z|}w}{2\sqrt{\frac{1-(2|z|-1)}{2}}}, w\sqrt{\frac{1-(2|z|-1)}{2}}\right) \\ &= \left(\frac{zw}{\sqrt{|z|}}, w\sqrt{1-|z|}\right). \end{aligned}$$

From this calculation, we see $b_{2,4}(w,z) \to (zw, 0)$ as $|z| \to 1$.

Thus this is a cylindrically normal cellular stratification.

Example 4.26: There is an alternative way of describing the cylindrical structure in the above example. Recall that complex projective spaces are typical examples of quasitoric manifolds. Define an action of $T^n = (S^1)^n$ on $\mathbb{C}\mathrm{P}^n$ by

$$(t_1, \ldots, t_n) \cdot [z_0, \ldots, z_n] = [z_0, t_1 z_1, \ldots, t_n z_n].$$

As we have seen in Example 2.12, this action induces a stratifications on $\mathbb{C}\mathrm{P}^n$ which descends to $\mathbb{C}\mathrm{P}^n/T^n$

$$\begin{aligned} \pi_{\mathbb{C}\mathrm{P}^n} &: \mathbb{C}\mathrm{P}^n \longrightarrow I(T^n) \\ \pi_{\mathbb{C}\mathrm{P}^n/T^n} &: \mathbb{C}\mathrm{P}^n/T^n \longrightarrow I(T^n). \end{aligned}$$

The quotient space $\mathbb{C}\mathrm{P}^n/T^n$ is known to be homeomorphic to Δ^n and the stratification $\pi_{\mathbb{C}\mathrm{P}^n/T^n}$ can be identified with the stratification π_n on Δ^n in Example 2.11. This stratification, however, does not induce the minimal cell decomposition of $\mathbb{C}\mathrm{P}^n$. The other stratification $\pi_n^{\max}$ on Δ^n defined in Example 2.11 induces the minimal cell decomposition on $\mathbb{C}\mathrm{P}^n$ by the composition

$$\mathbb{C}\mathrm{P}^n \xrightarrow{p} \mathbb{C}\mathrm{P}^n/T^n \cong \Delta^n \xrightarrow{\pi_n^{\max}} [n].$$

Let us show that this cell decomposition is cylindrically normal. To this end, we first rewrite $\mathbb{C}\mathrm{P}^n$ by using the construction introduced in [25] by Davis and Januszkiewicz. Recall that a d-dimensional convex polytope is said to be *simple* if each vertex is adjacent to exactly d edges. Given a simple polytope P of dimension n and a function λ : {codimension-1 faces in P} $\to$

$\mathbb{C}^n$ satisfying certain conditions, they constructed a space $M(\lambda)$ with T^n-action. Suppose $P = \Delta^n$ and define

$$\lambda_n(C_i) = \begin{cases} (1, \ldots, 1), & i = 0 \\ (\underbrace{0, \ldots, 0}_{i-1}, 1, \underbrace{0, \ldots, 0}_{n-i}), & i = 1, \ldots, n, \end{cases}$$

where C_i is the codimension-1 face with vertices in $[n] - \{i\}$. In this case, $M(\lambda_n)$ can be described as

$$M(\lambda_n) = (T^n \times \Delta^n)/_\sim,$$

where the equivalence relation $\sim$ is generated by the following relations: let $p = (p_0, \ldots, p_n) \in \Delta^n$.

- When $p_i = 0$ for $1 \leq i \leq n$,
$$(t_1, \ldots, t_i, \ldots, t_n; p_0, \ldots, p_n) \sim (t_1, \ldots, t_i', \ldots, t_n; p_0, \ldots, p_n)$$
for any $t_i, t_i' \in S^1$.
- When $p_0 = 0$,
$$(t_1, \ldots, t_n; 0, p_1, \ldots, p_n) \sim (\omega t_1, \ldots, \omega t_n; 0, p_1, \ldots, p_n)$$
for any $\omega \in S^1$.

An explicit homeomorphism $p_n : \mathbb{C}\mathrm{P}^n \to M(\lambda_n)$ and its inverse q_n are given by

$$p_n([z_0 : \ldots : z_n]) = \begin{cases} \left[\frac{z_1/z_0}{|z_1/z_0|}, \ldots, \frac{z_n/z_0}{|z_n/z_0|}; \frac{|z_0|^2}{\sum_{i=0}^n |z_i|^2}, \ldots, \frac{|z_n|^2}{\sum_{i=0}^n |z_i|^2}\right] & z_0 \neq 0 \\ \left[\frac{z_1}{|z_1|}, \ldots, \frac{z_n}{|z_n|}; 0, \frac{|z_1|^2}{\sum_{i=0}^n |z_i|^2}, \ldots, \frac{|z_n|^2}{\sum_{i=0}^n |z_i|^2}\right] & z_0 = 0 \end{cases}$$

$$q_n([z_1, \ldots, z_n; x_0, \ldots, x_n]) = [\sqrt{x_0}, \sqrt{x_1} z_1, \ldots, \sqrt{x_n} z_n].$$

Under this identification, the minimal cell decomposition on $\mathbb{C}\mathrm{P}^n$ can be described as

$$\mathbb{C}\mathrm{P}^n \cong M(\lambda_n) = \bigcup_{i=1}^n \left(T^n/T^{n-i} \times (\Delta^i \setminus \Delta^{i-1})\right)/_\sim.$$

Regard $D^{2n} = \left\{(z_1, \ldots, z_n) \in \mathbb{C}^n \,\middle|\, \sum_{i=1}^n |z_i|^2 \leq 1\right\}$ and define a map

$$\varphi_{2n} : D^{2n} \longrightarrow M(\lambda_n)$$

by

$$\varphi_{2n}(z_1, \ldots, z_n) = \left[\frac{z_1}{|z_1|}, \ldots, \frac{z_n}{|z_n|}; 1 - \sum_{i=1}^n |z_i|^2, |z_1|^2, \ldots, |z_n|^2\right].$$

This is a cell structure map for the $2n$-cell. For $m < n$, define

$$b_{2m,2n} : S^1 \times D^{2m} \longrightarrow D^{2n}$$

by

$$b_{2m,2n}(\omega, z_1, \ldots, z_m) = \left(0, \ldots, 0, \omega\sqrt{1 - \sum_{i=1}^{m} |z_i|^2}, \omega z_1, \ldots, \omega z_m\right).$$

Then each $b_{2m,2n}|_{S^1 \times \mathrm{Int}(D^{2m})}$ is a homeomorphism onto its image and we have a stratification

$$\partial D^{2n} = S^{2n-1} = \bigcup_{m=0}^{n-1} b_{2m,2n}\left(S^1 \times \mathrm{Int}(D^{2m})\right).$$

Furthermore the diagram

$$\begin{array}{ccccc} & & D^{2n} & \xrightarrow{\varphi_{2n}} & M(\lambda_n) \\ & & \uparrow & & \uparrow \\ S^1 \times D^{2m} & \xrightarrow{b_{2m,2n}} & S^{2n-1} & & \\ \downarrow \mathrm{pr}_2 & & & & \\ D^{2m} & & \xrightarrow[\varphi_{2m}]{} & & M(\lambda_m), \end{array}$$

is commutative, where the inclusion $M(\lambda_m) \hookrightarrow M(\lambda_n)$ is given by

$$[t_1, \ldots, t_m; p_0, \ldots, p_m] \longmapsto [1, \ldots, 1, t_1, \ldots, t_m; 0, \ldots, 0, p_0, \ldots, p_m].$$

Now define $P_{2m,2n} = S^1$ for $m < n$. The group structure of S^1 defines a map

$$c_{2\ell,2m,2n} : P_{2m,2n} \times P_{2\ell,2n} \longrightarrow P_{2\ell,2n}$$

making the diagram

$$\begin{array}{ccc} S^1 \times S^1 \times D^{2\ell} & \xrightarrow{1 \times b_{2\ell,2m}} & S^1 \times D^{2m} \\ {\scriptstyle c_{2\ell,2m,2n} \times 1} \downarrow & & \downarrow {\scriptstyle b_{2m,2n}} \\ S^1 \times D^{2\ell} & \xrightarrow[b_{2\ell,2n}]{} & D^{2n} \end{array}$$

commutative and we have a cylindrical structure.

More generally, Davis and Januszkiewicz [25] proved that any quasitoric manifold M of dimension $2n$ can be expressed as $M \cong M(\lambda)$ for a simple convex polytope P of dimension n and a function λ. The right hand side is a space constructed as a quotient of $T^n \times P$ under an equivalence relation analogous to the case of $\mathbb{C}\mathrm{P}^n$. Davis and Januszkiewicz proved in §3 of their paper that there is a "perfect Morse function" on $M(\lambda)$ which induces a cell decomposition of $M(\lambda)$ or M with exactly $h_i(P)$ cells of dimension $2i$, where $(h_0(P), \ldots, h_n(P))$ is the h-vector of P. It seems very likely that the above construction of a cylindrical structure on $\mathbb{C}\mathrm{P}^n$ can be extended to quasitoric manifolds.

Example 4.27: In the same paper, Davis and Januszkiewicz introduced the notion of small covers as a real analogue of quasitoric manifolds by replacing S^1 by $\mathbb{Z}_2$. Small covers have many properties in common with quasitoric (or torus) manifolds.

For example, we have $\mathbb{R}\mathrm{P}^n/(\mathbb{Z}_2)^n \cong \Delta^n$ and the stratification $\pi_n^{\max}$ on Δ^n induces the minimal cell decomposition of $\mathbb{R}\mathrm{P}^n$. An argument analogous to the case of $\mathbb{C}\mathrm{P}^n$ can be used to prove that this stratification is totally normal.

Example 4.28: Let $X = D^3$ and consider the cell decomposition

$$X = e_1^0 \cup e_2^0 \cup e_1^1 \cup e_2^1 \cup e_1^2 \cup e_2^2 \cup e^3$$

given as follows. The interior of D^3 is the unique 3-cell and the boundary

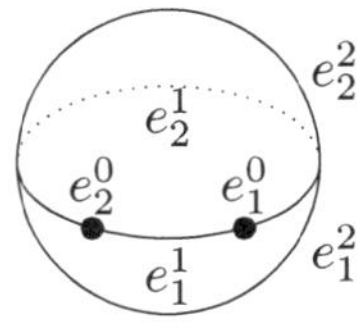

Fig. 19. A regular cell decomposition of S^2

S^2 is cut into two 2-cells by the equator, which is cut into two 1-cells by two 0-cells on it.

Consider the map between S^2 given by collapsing the shaded region in the figure below (the wedge of two 2-disks embedded in S^2) "vertically" and expanding the remaining part of S^2 continuously. Extend it to a continuous map $\varphi_3 : D^3 \to D^3$. It defines a 3-cell structure on e^3. Let $P_{(1,1),3} = D^1$. We have a continuous map

$$b : P_{(1,1),3} \times D^1 \longrightarrow \partial D^3$$

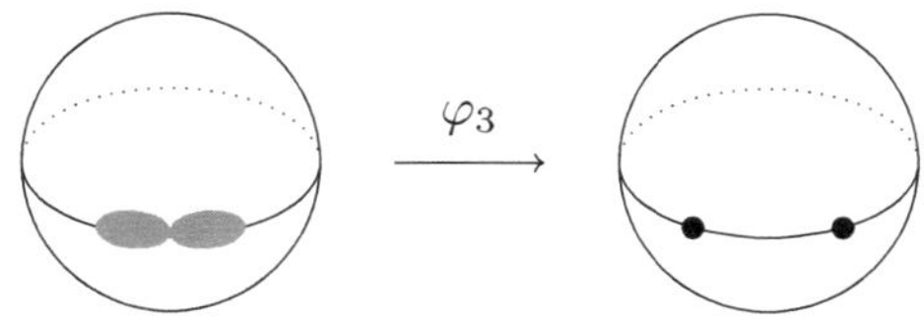

Fig. 20. A 3-cell structure which is not cylindrically normal

making the diagram

$$\begin{array}{ccc} D^3 & \xrightarrow{\varphi_3} & X \\ {\scriptstyle b}\uparrow & & \uparrow{\scriptstyle \varphi_{1,1}} \\ P_{(1,1),3} \times D^1 & \longrightarrow & D^1 \end{array}$$

commutative. However, b is not a homeomorphism when restricted to $P_{(1,1),3} \times \mathrm{Int}(D^1)$. And this is not a cylindrical structure.

In Examples 4.24 and 4.25, the restrictions of cell structure maps to the boundary spheres are fiber bundles onto their images. These facts seem to be closely related to the existence of cylindrical structures in these examples. Of course, there are many cylindrically normal cellular stratified spaces that do not have such bundle structures. We may be able to characterize cylindrically normal cellular stratified spaces by using an appropriate notion of stratified fiber bundles. See [23] for example.

Example 4.29: Let X be the subspace of the unit 3-disk D^3 obtained by removing the interior of the 3-disk of radius $\frac{1}{2}$ centered at the origin. It has a cell decomposition with two 0-cells, a 1-cell, two 2-cells, and a 3-cell depicted as follows. Cells of dimension at most 1 are regular. The cell structure maps for 2-cells are given as in Example 3.19. The restriction of the cell structure map φ_3 for the 3-cell to S^2 is given by collapsing a middle part of a sphere to a segment.

This cell decomposition is cylindrically normal, because of the "local triviality" of the middle band. However the restriction $\varphi_3|_{S^2}$ is not a fiber bundle onto its image.

Example 4.30: We have seen in Example 3.16 that the standard cell decomposition of the geometric realization of a Δ-set is totally normal. Let

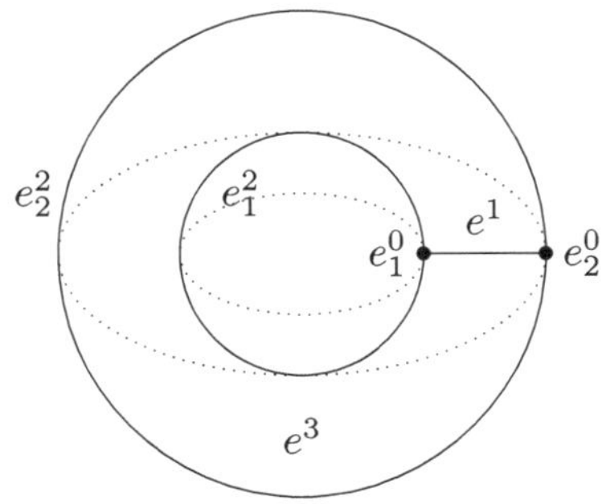

Fig. 21. A cell decomposition of thick sphere

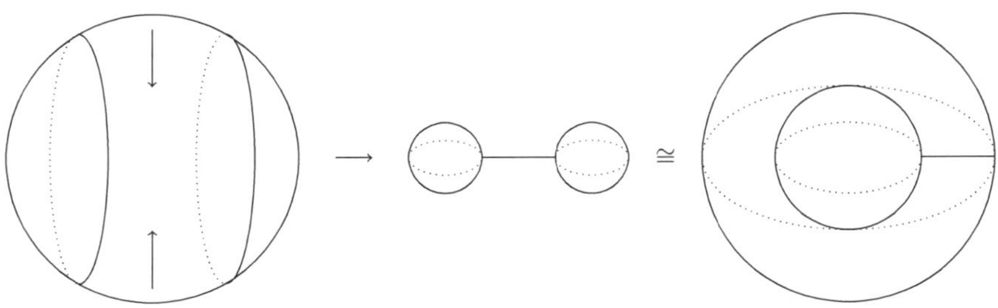

Fig. 22. A 3-cell structure on thick sphere

us consider the geometric realization of a simplicial set X. Define

$$P(X) = \coprod_{n=0}^{\infty} \left(X_n \setminus \bigcup_{i=0}^{n} s_i(X_{n-1}) \right)$$

to be the set of nondegenerate simplices. For $\sigma, \tau \in P(X)$, define $\tau \leq \sigma$ if there exists an injective morphism $u : [m] \to [n]$ with $X(u)(\sigma) = \tau$. Define

$$\pi_X : |X| \longrightarrow P(X)$$

analogously to the case of Δ-sets. Then π_X is a cellular stratification. In other words, cells in $|X|$ are in one-to-one correspondence to nondegenerate simplices.

Suppose $\tau \leq \sigma$ in $P(X)$. Then the set

$$\mathcal{P}(\tau, \sigma) = \{u : [m] \to [n] \mid u \text{ injective and } X(u)(\sigma) = \tau\}$$

is nonempty. For $u, v \in \mathcal{P}(\tau, \sigma)$, define $u \leq v$ if and only if $u(i) \leq v(i)$ for all $i \in [m]$. Let us denote the order complex of this poset by $P_{\tau,\sigma}$. This is a simplicial complex whose simplices are indexed by chains in $\mathcal{P}(\tau, \sigma)$ and can be written as

$$P_{\tau,\sigma} = \bigcup_{k=0}^{\infty} \bigcup_{\boldsymbol{u} \in N_k(\mathcal{P}(\tau,\sigma))} \Delta^k \times \{\boldsymbol{u}\}.$$

For a k-chain $\boldsymbol{u} = \{u_0 < \cdots < u_k\} \in N_k(\mathcal{P}(\tau, \sigma))$, define a map

$$\beta^{\boldsymbol{u}} : [k] \times [m] \longrightarrow [n]$$

by

$$\beta^{\boldsymbol{u}}(i, j) = u_i(j).$$

This map induces an affine map

$$b^{\boldsymbol{u}} : (\Delta^k \times \{\boldsymbol{u}\}) \times \Delta^m \longrightarrow \Delta^n.$$

These maps can be glued together to give us a map

$$b_{\tau,\sigma} : P_{\tau,\sigma} \times \Delta^m \longrightarrow \Delta^n.$$

For $\sigma_0 < \sigma_1 < \sigma_2$ in $P(X)$, the composition

$$\mathcal{P}(\sigma_1, \sigma_2) \times \mathcal{P}(\sigma_0, \sigma_1) \longrightarrow \mathcal{P}(\sigma_0, \sigma_2)$$

is a morphism of posets and induces a map

$$c_{\sigma_0,\sigma_1,\sigma_2} : P_{\sigma_1,\sigma_2} \times P_{\sigma_0,\sigma_1} \longrightarrow P_{\sigma_0,\sigma_2}.$$

It is straightforward to check that these maps satisfy the requirements of a cylindrical structure. Thus the standard cell decomposition of the geometric realization $|X|$ is cylindrically normal.

5. Topological face categories and their classifying spaces

Recall that the collection of all cells in a regular cell complex X forms a poset whose order complex is homeomorphic to X. We also have a face poset for any cellular stratified space. When X is a non-regular cell complex or a cellular stratified space, however, we cannot expect to recover the homotopy type of X from its face poset, as we will see in Example 5.3.

The aim of this section is to show that there is a canonical way to construct an acyclic topological category $C(X)$ from a cylindrically normal cellular stratified space X and that its classifying space $BC(X)$ has the same homotopy type as X under appropriate conditions. See C for basics of topological categories.

5.1. *Face categories*

There are several ways to construct a category from a cellular or stellar stratified space. A naive idea is the following:

Definition 5.1: For a cellular or a stellar stratified space (X, π, Φ) and cells $\varphi_\mu : D_\mu \to \overline{e_\mu}$ and $\varphi_\lambda : D_\lambda \to \overline{e_\lambda}$ with $e_\mu \subset \overline{e_\lambda}$, define $F(X)(e_\mu, e_\lambda)$ to be the set of all maps $b : D_\mu \to D_\lambda$ making the following diagram commutative

$$\begin{array}{ccc} D_\lambda & \xrightarrow{\varphi_\lambda} & \overline{e_\lambda} \\ {\scriptstyle b}\uparrow & & \uparrow \\ D_\mu & \xrightarrow[\varphi_\mu]{} & \overline{e_\mu}. \end{array}$$

The set $F(X)(e_\mu, e_\lambda)$ is topologized by the compact-open topology as a subspace of $\mathrm{Map}(D_\mu, D_\lambda)$. $F(X)(e_\mu, e_\lambda)$ is defined to be empty if $e_\mu \not\subset \overline{e_\lambda}$.

By defining the set of objects to be cells in X and morphisms from e_μ to e_λ to be $F(X)(e_\mu, e_\lambda)$, we obtain a topological category $F(X)$. The composition is given by the composition of maps. This topological category $F(X)$ is called the *naive face category* of X. It is also denoted by $F(X, \pi)$ or $F(X, \pi, \Phi)$.

Lemma 5.2: *The naive face category $F(X)$ is an acyclic category. When X is regular, $F(X)$ is a poset and coincides with the face poset $P(X)$. In particular, when X is a regular cell complex, our construction coincides with the classical face poset construction.*

Proof: If both $F(X)(e, e')$ and $F(X)(e', e)$ are nonempty, we have $\dim e = \dim e'$, since the existence of a morphism $e \to e'$ in $F(X)$ implies $\dim e \leq \dim e'$. The compatibility of lifts with cell structure maps implies that the only case we have a morphism is $e = e'$ and the morphism should be the identity. Thus $F(X)$ is acyclic.

When X is regular, the regularity implies that there is at most one morphism between two objects. Hence it is a poset. □

Even when X is not regular, we have the underlying poset as is defined in Definition C.6, since $F(X)$ is an acyclic category. Obviously it is isomorphic to the face poset $P(X, \pi) = \mathrm{Im}\, \pi$.

Example 5.3: Consider the minimal cell decomposition

$$\pi_n : S^n = e^0 \cup e^n \longrightarrow \{0 < n\}.$$

This is a typical non-regular cell complex. The face poset $P(S^n, \pi_n)$ is a totally ordered set of two elements and its order complex is homeomorphic to an interval $BP(S^n, \pi_n) \cong [0,1]$. The homotopy type of S^n cannot be recovered from its face poset.

On the other hand, the face category $F(S^n, \pi_n)$ has more information. It has two objects e^0 and e^n. When $n = 1$, as we have seen in Example 3.13, the set of morphisms from e^0 to e^1 is given by

$$F(S^1, \pi_1)(e^0, e^1) = \{b_1, b_{-1}\}.$$

We also have the identity morphism on each object. The resulting category is depicted in Figure 23. When $n > 1$, there are infinitely many morphisms

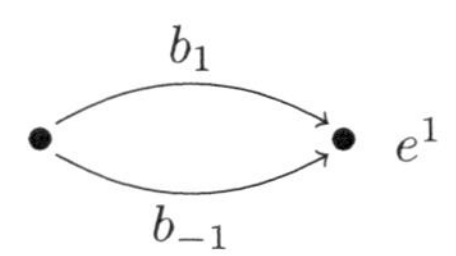

Fig. 23. The face category of $S^1 = e^0 \cup e^1$

from e^0 to e^n parametrized by ∂D^n, and we have a homeomorphism

$$F(S^n, \pi_n)(e^0, e^n) \cong S^{n-1}.$$

In the above example of S^n, the compact-open topology on the morphism space $F(S^n, \pi_n)(e^0, e^n)$ can be replaced with a more understandable topology of S^{n-1}. In general, we cannot expect such simplicity. Under the assumption of cylindrical normality, however, we may define a smaller face category.

Definition 5.4: Let X be a cylindrically normal stellar stratified space. Define a category $C(X)$ as follows: objects are cells in X. For each pair $e_\mu \subset \overline{e_\lambda}$, define

$$C(X)(e_\mu, e_\lambda) = P_{\mu,\lambda}.$$

The composition of morphisms is given by

$$c_{\lambda_0,\lambda_1,\lambda_2} : P_{\lambda_1,\lambda_2} \times P_{\lambda_0,\lambda_1} \longrightarrow P_{\lambda_0,\lambda_2}.$$

The category $C(X)$ is called the *cylindrical face category* of X.

Lemma 5.5: *For any cylindrically normal stellar stratified space X, its face category $C(X)$ is an acyclic topological category. The maps $b_{\mu,\lambda}$ induce a continuous functor*

$$b : C(X) \longrightarrow F(X),$$

which is natural with respect to morphisms of cylindrically normal stellar stratified spaces.

Furthermore the underlying poset of $C(X)$ is also $P(X)$ and the diagram

$$\begin{array}{ccc} C(X) & \xrightarrow{\quad b \quad} & F(X) \\ & \searrow \quad \swarrow & \\ & P(X) & \end{array}$$

is commutative.

Proof: The continuity of

$$b_{\mu,\lambda} : P_{\mu,\lambda} \times D_\mu \longrightarrow D_\lambda$$

implies the continuity of its adjoint

$$\mathrm{ad}(b_{\mu,\lambda}) : P_{\mu,\lambda} \longrightarrow \mathrm{Map}(D_\mu, D_\lambda),$$

which factors through $F(X)(e_\mu, e_\lambda)$. It is immediate to verify that these maps form a continuous functor

$$b : C(X) \longrightarrow F(X).$$

Morphisms of cylindrically normal stellar stratified spaces are required to be compatible with maps $b_{\mu,\lambda}$ and thus the functor b is natural with respect to morphisms of cylindrically normal stellar stratified spaces. The commutativity of the triangle is obvious from the definition. □

Example 5.6: Consider the minimal cell decomposition on $\mathbb{C}\mathrm{P}^2$

$$\mathbb{C}\mathrm{P}^2 = e^0 \cup e^2 \cup e^4.$$

It is shown in Example 4.25 that it has a cylindrical structure. The cylindrical face category $C(\mathbb{C}\mathrm{P}^2)$ has three objects, e^0, e^2, and e^4. We have seen that

$$F(\mathbb{C}\mathrm{P}^2)(e^0, e^2) = F(S^2, \pi_2)(e^0, e^2) \cong S^1 = C(\mathbb{C}\mathrm{P}^2)(e^0, e^2).$$

Since the attaching map of e^4 is the Hopf map

$$\eta : S^3 \longrightarrow \mathbb{C}\mathrm{P}^1,$$

we have

$$F(\mathbb{C}\mathrm{P}^2)(e^0, e^4) \cong \eta^{-1}(e^0) \cong S^1 = C(\mathbb{C}\mathrm{P}^2)(e^0, e^4).$$

By using the local trivialization

$$\eta^{-1}(e^2) \cong e^2 \times S^1,$$

we see that $F(\mathbb{C}\mathrm{P}^2)(e^2, e^4)$ is the set of sections of the trivial bundle

$$D^2 \times S^1 \longrightarrow D^2$$

and thus $F(\mathbb{C}\mathrm{P}^2)(e^2, e^4) = \mathrm{Map}(D^2, S^1)$. On the other hand, we have

$$C(\mathbb{C}\mathrm{P}^2)(e^2, e^4) = S^1$$

by definition. The composition

$$C(\mathbb{C}\mathrm{P}^2)(e^2, e^4) \times C(\mathbb{C}\mathrm{P}^2)(e^0, e^2) \longrightarrow C(\mathbb{C}\mathrm{P}^2)(e^0, e^4)$$

is given by the multiplication of S^1, as is shown in Example 4.25.

In general, Example 4.26 says that the face category $C(\mathbb{C}\mathrm{P}^n)$ of the minimal cell decomposition of $\mathbb{C}\mathrm{P}^n$ can be described as a "poset enriched by S^1" in the sense that, for any pair of objects e^{2k}, e^{2m} $(k < m)$, the space

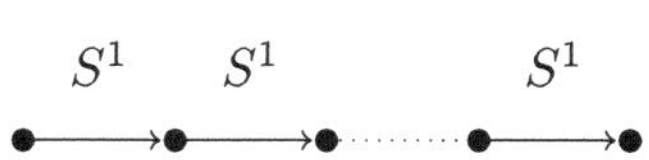

Fig. 24. The face category of $\mathbb{C}\mathrm{P}^n$

of morphisms $C(\mathbb{C}\mathrm{P}^n)(e^{2k}, e^{2m})$ is S^1 and the composition of morphisms is given by the group structure of S^1.

Recall that the order complex of the face poset of a regular cell complex X is the barycentric subdivision of X. With this fact in mind, we introduce the following notation:

Definition 5.7: Let X be a cylindrically normal stellar stratified space. Define its *barycentric subdivision* $\mathrm{Sd}(X)$ to be the classifying space of the cylindrical face category

$$\mathrm{Sd}(X) = BC(X).$$

Remark 5.8: There is a notion of barycentric subdivision $\mathrm{Sd}(C)$ of a small category C. A good reference is a paper [29] by del Hoyo. See also Noguchi's papers [30, 31]. We will show that, for a totally normal stellar stratified space X, there is an isomorphism of categories $\mathrm{Sd}(C(X)) \cong C(\mathrm{Sd}(X))$ in §7.

When X is not a regular cell complex, we usually do not have a homeomorphism between $\mathrm{Sd}(X)$ and X.

Example 5.9: Consider $X = \mathbb{R}^n$. This is a regular totally normal cellular stratification consisting of a single n-cell. The barycentric subdivision is a single point.

Example 5.10: Consider the minimal cell decomposition π_n of S^n. When $n = 1$, it is easy to see that $\mathrm{Sd}(S^1, \pi_1)$ is the cell complex in Figure 25 and is homeomorphic to S^1. Note that this complex is obtained by subdividing

Fig. 25. $\mathrm{Sd}(S^1, \pi_1)$

the 1-cell in π_1 and can be regarded as the barycentric subdivision of π_1.

When $n > 1$, $C(S^n, \pi_n)$ is a topological category with nontrivial topology on $C(S^n, \pi_n)(e^0, e^n)$. Since we have a homeomorphism

$$C(S^n, \pi_n)(e^0, e^n) \cong S^{n-1},$$

it is easy to determine $\mathrm{Sd}(S^n, \pi_n)$ and we have

$$\mathrm{Sd}(S^n, \pi_n) = BC(S^n, \pi_n) \cong \Sigma(S^{n-1}) \cong S^n.$$

Again we recovered S^n.

Example 5.11: Consider $X = \mathrm{Int}D^n \cup \{(1, 0)\}$ with the obvious stratification.

This is a regular cellular stratification and $\mathrm{Sd}(X)$ is a 1-simplex $[0, 1]$. We have an embedding

$$i : [0, 1] \longrightarrow X$$

by

$$i(t) = (1 - t)(1, 0) + t(0, 0).$$

See Figure 27. Obviously $i([0, 1])$ is a strong deformation retract of X.

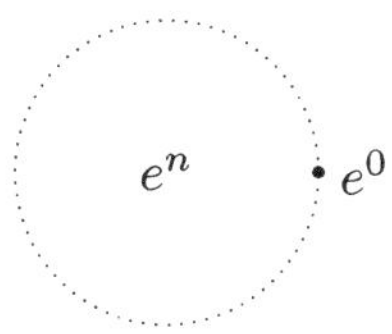

Fig. 26. $\mathrm{Int}D^n \cup \{(1,0)\}$

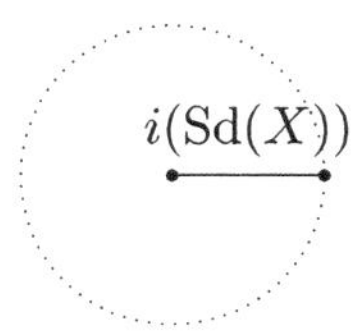

Fig. 27. $i(\mathrm{Sd}(\mathrm{Int}D^2 \cup \{(1,0)\}))$

Example 5.12: Consider the punctured torus in Example 3.15. There is a totally normal cellular stratification on $X = S^1 \times S^1 \setminus e^0 \times e^0$ induced from the product cell decomposition π_1^2

$$S^1 \times S^1 = e^0 \times e^0 \cup e^0 \times e^1 \cup e^1 \times e^0 \cup e^1 \times e^1.$$

Let

$$\varphi_{1,1} : D_{1,1} = [-1,1]^2 \setminus \{(-1,-1),(-1,1),(1,-1),(1,1)\} \longrightarrow X$$

be the cell structure map of the 2-cell in X and

$$\varphi_{0,1} : D_{0,1} = (-1,1) \longrightarrow X$$
$$\varphi_{1,0} : D_{1,0} = (-1,1) \longrightarrow X$$

be the cell structure maps for 1-cells.

As we have seen in Example 3.15, there are two ways to lift each cell structure map of a 1-cell and these four lifts cover $\partial D_{1,1}$.

$$\partial D_{1,1} = b'_{0,1}(D_{0,1}) \cup b''_{0,1}(D_{0,1}) \cup b'_{1,0}(D_{1,0}) \cup b''_{1,0}(D_{1,0}).$$

The cylindrical face category $C(X)$ consists of three objects $e^0 \times e^1$, $e^1 \times e^0$, $e^1 \times e^1$. Nontrivial morphisms are

$$C(X)(e^0 \times e^1, e^1 \times e^1) = \{b'_{0,1}, b''_{0,1}\},$$
$$C(X)(e^1 \times e^0, e^1 \times e^1) = \{b'_{1,0}, b''_{1,0}\}.$$

By allowing multiple edges, we can draw a "Hasse diagram" of this acyclic category as in Figure 28.

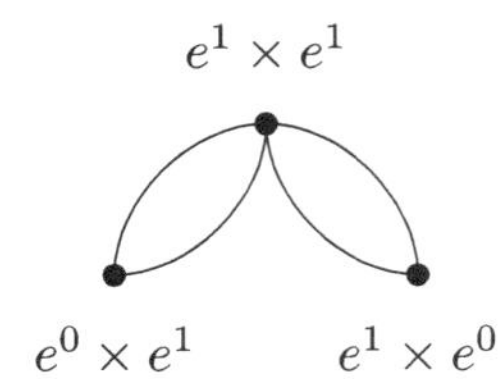

Fig. 28. The barycentric subdivision of the punctured torus

Obviously, the classifying space of this category is the wedge of two circles,

$$\mathrm{Sd}(X) = BC(X) = S^1 \vee S^1.$$

It is easy to find an embedding of $\mathrm{Sd}(X)$ into X by using lifts of cell structure maps. The images of 0 under the four maps $b'_{0,1}$, $b''_{0,1}$, $b'_{1,0}$, $b''_{1,0}$ constitute four points in $\partial D_{1,1}$ that are mapped to a single point under $\varphi_{1,1}$. By connecting each of these four points and $(0,0)$ in $D_{1,1}$ by a segment, respectively, we obtain a 1-dimensional stratified space $\widetilde{K}$ in $D_{1,1}$. See Figure 29.

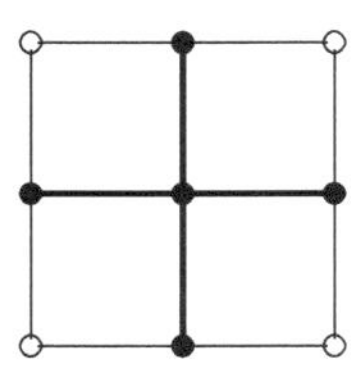

Fig. 29. A thin subcomplex $\widetilde{K}$ in $D_{1,1}$

The complex $\widetilde{K}$ in Figure 29 corresponds to cells in $\mathrm{Sd}(X)$ and we have an embedding

$$\mathrm{Sd}(X) \hookrightarrow X.$$

Figure 29 can be also used to construct a deformation retraction of X onto $\mathrm{Sd}(X)$.

The above examples show that, when a cellular stratification π is totally normal, or more generally, cylindrically normal, the barycentric subdivision $\mathrm{Sd}(X, \pi)$ is closely related to X.

The work of Cohen, Jones, and Segal [16] suggests that one should analyze the nerve of the face category of a cylindrically normal stellar stratified space by using the underlying poset functor

$$\pi : C(X,\pi) \longrightarrow P(X,\pi).$$

The following easily verifiable fact will be used in the next section.

Lemma 5.13: *For a cylindrically normal stellar stratified space (X,π), consider the induced morphism of simplicial sets*

$$N(\pi) : N(C(X,\pi)) \longrightarrow N(P(X,\pi)).$$

Here we forget the topology on $C(X,\pi)$ temporarily. For each n-chain in the face poset $\boldsymbol{e} = (e_{\lambda_0}, \ldots, e_{\lambda_n}) \in N_n(P(X,\pi))$, we have

$$N(\pi)_n^{-1}(\boldsymbol{e}) = P_{\lambda_{n-1},\lambda_n} \times \cdots \times P_{\lambda_0,\lambda_1}.$$

Consequently the space of n-chains has the following decomposition

$$N_n(C(X,\pi)) = \coprod_{\boldsymbol{e}\in N_n(P(X,\pi))} \{\boldsymbol{e}\} \times P_{\lambda_{n-1},\lambda_n} \times \cdots \times P_{\lambda_0,\lambda_1}.$$

The space of nondegenerate n-chains is, therefore, given by

$$\overline{N}_n(C(X,\pi)) = \coprod_{\boldsymbol{e}\in \overline{N}_n(P(X,\pi))} \{\boldsymbol{e}\} \times P_{\lambda_{n-1},\lambda_n} \times \cdots \times P_{\lambda_0,\lambda_1}.$$

By using the face category, we may regard a cylindrically normal stellar stratified space as a functor.

Definition 5.14: For a cylindrically normal stellar stratified space X, define a functor

$$D^X : C(X) \longrightarrow \textbf{Spaces}$$

by assigning the domain D_λ to each cell $\varphi_\lambda : D_\lambda \to \overline{e_\lambda}$. For a morphism $p \in C(X)(e_\mu, e_\lambda) = P_{\mu,\lambda}$, define $D^X(p) = b_{\mu,\lambda}(p) \in \mathrm{Map}(D_\mu, D_\lambda)$.

The following is an extension of Proposition 2.47 of [1]. See Definition C.10 for the definitions of the continuity of a functor to **Spaces** and its colimit.

Proposition 5.15: *For a CW cylindrically normal stellar stratified space X, the functor D^X is a continuous functor and we have a natural homeomorphism*

$$\operatorname*{colim}_{C(X)} D^X \xrightarrow{\cong} X.$$

Proof: The continuity of the functor D^X is obvious from the definition.

By definition, $\operatorname{colim}_{C(X)} D^X$ is a quotient space of $\operatorname{tot}(D^X) = D(X)$. Let $\sim_c$ be the defining equivalence relation so that $\operatorname{colim}_{C(X)} D^X = D(X)/_{\sim_c}$. On the other hand, by Lemma 2.56, we have a description of X as a quotient of $D(X)$

$$X \cong D(X)/_{\sim_\Phi},$$

where the relation $\sim_\Phi$ is defined by $x \sim_\Phi y$ if and only if $\varphi_\mu(x) = \varphi_\lambda(y)$ for $x \in D_\mu$ and $y \in D_\lambda$.

It remains to verify that $\sim_c$ is identical to $\sim_\Phi$. The proof is essentially the same as that of Proposition 2.47 in [1]. Details are omitted. □

5.2. *Barycentric subdivisions of cellular stratified spaces*

As we can see from the examples in Section 5.1, we often have an embedding of the barycentric subdivision $\operatorname{Sd}(X)$ in X, even if X is neither a cell complex nor regular. In this section, we first show that such an embedding always exists for any cylindrically normal cellular stratified space, and then we show that it often admits a strong deformation retraction.

In order to describe our embedding, we use the language of Δ-spaces. See Definition B.14.

Theorem 5.16: *Let (X, π) be a cylindrically normal CW stellar stratified space. Then there exists an embedding*

$$i_X : \operatorname{Sd}(X, \pi) \hookrightarrow X,$$

which is natural with respect to strict morphisms of cylindrically normal stellar stratified spaces. Furthermore, when all cells in X are closed, i_X is a homeomorphism onto X.

The construction of i_X and the proof of the fact that i_X is an embedding are essentially the same as those of Proposition 2.51 in [1]. We record them for the convenience of the reader.

Definition 5.17: We construct i_X as follows. Since the face category $C(X, \pi) = C(X)$ is acyclic, $\operatorname{Sd}(X) = BC(X) = \|\overline{N}(C(X))\|$. Thus it suffices to construct a series of maps

$$i_n : \overline{N}_n(C(X)) \times \Delta^n \longrightarrow X$$

making the following diagram commutative for all i

$$\begin{array}{ccc}
\overline{N}_n(C(X,\pi)) \times \Delta^{n-1} & \xrightarrow{1\times d^i} & \overline{N}_n(C(X,\pi)) \times \Delta^{n} \\
{\scriptstyle d_i\times 1}\downarrow & & \downarrow{\scriptstyle i_n} \\
\overline{N}_{n-1}(C(X,\pi)) \times \Delta^{n-1} & \xrightarrow[i_{n-1}]{} & X.
\end{array}$$

Let us construct i_n by induction on n. The set $\overline{N}_0(C(X))$ is the set of cells for π. For each cell e_λ with cell structure map $\varphi_\lambda : D_\lambda \to \overline{e_\lambda} \subset X$, define

$$i_0(e_\lambda) = \varphi_\lambda(0)$$

and we obtain a map

$$i_0 : \overline{N}_0(C(X)) \cong \overline{N}_0(C(X)) \times \Delta^0 \hookrightarrow X.$$

Note that this map makes the diagram

$$\begin{array}{ccc}
\overline{N}_0(C(X,\pi)) \times \Delta^0 & \xrightarrow{i_0} & X \\
{\scriptstyle z_0}\downarrow & \nearrow_{\Phi} & \\
D(X) & &
\end{array}$$

commutative, where z_0 is induced by the inclusion of Δ^0 to the center of each disk and $D(X)$ and Φ are defined in Lemma 2.56.

Suppose that we have constructed a map

$$i_{k-1} : \overline{N}_{k-1}(C(X)) \times \Delta^{k-1} \longrightarrow X$$

satisfying the above compatibility conditions and that the restriction of i_{k-1} to $\overline{N}_{k-1}(C(X)) \times \mathrm{Int}(\Delta^{k-1})$ is an embedding. Suppose, further, that there exists a map

$$z_{k-1} : \overline{N}_{k-1}(C(X)) \times \Delta^{k-1} \longrightarrow D(X)$$

making the diagram

$$\begin{array}{ccc}
\overline{N}_{k-1}(C(X)) \times \Delta^{k-1} & \xrightarrow{i_{k-1}} & X \\
{\scriptstyle z_{k-1}}\downarrow & \nearrow_{\Phi} & \\
D(X) & &
\end{array}$$

commutative. We construct an embedding

$$z_k : \overline{N}_k(C(X)) \times \Delta^k \longrightarrow D(X)$$

satisfying the compatibility conditions corresponding to those of i_k and define i_k to be $\Phi \circ z_k$.

Under the decomposition in Lemma 5.13

$$\overline{N}_k(C(X)) = \coprod_{\boldsymbol{e}\in\overline{N}_k(P(X))} \{\boldsymbol{e}\} \times \overline{N}(\pi)_k^{-1}(\boldsymbol{e}),$$

it suffices to construct a map

$$z_{\boldsymbol{e}} : \overline{N}(\pi)_k^{-1}(\boldsymbol{e}) \times \Delta^k \longrightarrow D_{\lambda_k}$$

for each nodegenerate k-chain $\boldsymbol{e} : e_{\lambda_0} < \cdots < e_{\lambda_k}$ in $P(X)$. The maps $\{z_{\boldsymbol{e}}\}$ should satisfy the following conditions:

(1) For each $0 \leq j < k$, the following diagram is commutative:

$$\begin{array}{ccc}
\overline{N}(\pi)_k^{-1}(\boldsymbol{e}) \times \Delta^k & \xrightarrow{z_{\boldsymbol{e}}} & D_{\lambda_k} \\
\uparrow{\scriptstyle 1\times d^j} & & \uparrow{\scriptstyle z_{d_j(\boldsymbol{e})}} \\
\overline{N}(\pi)_k^{-1}(\boldsymbol{e}) \times \Delta^{k-1} & \xrightarrow[d_j\times 1]{} & \overline{N}(\pi)_{k-1}^{-1}(d_j(\boldsymbol{e})) \times \Delta^{k-1}.
\end{array} \tag{5.1}$$

(2) When $j = k$, the following diagram is commutative:

$$\begin{array}{ccc}
\overline{N}(\pi)_k^{-1}(\boldsymbol{e}) \times \Delta^k & \xrightarrow{z_{\boldsymbol{e}}} & D_{\lambda_k} \\
\uparrow{\scriptstyle 1\times d^k} & & \\
\overline{N}(\pi)_k^{-1}(\boldsymbol{e}) \times \Delta^{k-1} & & \uparrow{\scriptstyle b_{\lambda_{k-1},\lambda_k}} \\
\| & & \\
P_{\lambda_{k-1},\lambda_k} \times \overline{N}(\pi)_{k-1}^{-1}(d_k(\boldsymbol{e})) \times \Delta^{k-1} & \xrightarrow[1\times z_{d_k(\boldsymbol{e})}]{} & P_{\lambda_{k-1},\lambda_k} \times D_{\lambda_{k-1}}.
\end{array} \tag{5.2}$$

By the inductive assumption, we have an embedding

$$z_{d_k(\boldsymbol{e})} : \overline{N}(\pi)_{k-1}^{-1}(d_k(\boldsymbol{e})) \times \Delta^{k-1} \longrightarrow D_{\lambda_{k-1}}$$

corresponding to the $(k-1)$-chain $d_k(\boldsymbol{e}) : e_{\lambda_0} < \cdots < e_{\lambda_{k-1}}$. Note that

$$\overline{N}(\pi)_k^{-1}(\boldsymbol{e}) = P_{\lambda_{k-1},\lambda_k} \times \overline{N}(\pi)_{k-1}^{-1}(d_k(\boldsymbol{e})).$$

Compose with $b_{\lambda_{k-1},\lambda_k}$ and we obtain a map

$$\begin{aligned}\overline{N}(\pi)_k^{-1}(\boldsymbol{e}) \times \Delta^{k-1} &= P_{\lambda_{k-1},\lambda_k} \times \overline{N}(\pi)_{k-1}^{-1}(d_k(\boldsymbol{e})) \times \Delta^{k-1} \\ &\stackrel{1\times z_{d_k(\boldsymbol{e})}}{\longrightarrow} P_{\lambda_{k-1},\lambda_k} \times D_{\lambda_{k-1}} \\ &\stackrel{b_{\lambda_{k-1},\lambda_k}}{\longrightarrow} \partial D_{\lambda_k}.\end{aligned}$$

Since D_{λ_k} is an aster, it can be extended to an embedding

$$z_{\boldsymbol{e}} : \overline{N}(\pi)_k^{-1}(\boldsymbol{e}) \times \Delta^k = \overline{N}(\pi)_k^{-1}(\boldsymbol{e}) \times \Delta^{k-1} * \boldsymbol{v}_k \longrightarrow \partial D_{\lambda_k} * 0 \subset D_{\lambda_k}$$

by

$$z_{\boldsymbol{e}}(\boldsymbol{p}, (1-t)\boldsymbol{s} + t\boldsymbol{v}_k) = (1-t)z_{d_k(\boldsymbol{e})}(\boldsymbol{p}, \boldsymbol{s}) + t \cdot 0 = (1-t)z_{d_k(\boldsymbol{e})}(\boldsymbol{p}, \boldsymbol{s}),$$

where $\boldsymbol{v}_k = (0, \ldots, 0, 1)$ is the last vertex in Δ^k. Recall that the join operation $*$ used in these notes is defined by connecting points by line segments (Definition B.20) and is not the same as the join operation used in algebraic topology, for example in Milnor's paper [32].

By definition, $z_{\boldsymbol{e}}$ makes the diagram (5.2) commutative. Let us verify that $z_{\boldsymbol{e}}$ also makes the diagram (5.1) commutative for $0 \leq j < k$. By the inductive assumption, the following diagram is commutative:

$$\begin{array}{ccccc}
\overline{N}(\pi)_{k-1}^{-1}(d_k(\boldsymbol{e})) \times \Delta^{k-2} & & \xrightarrow{d_j \times 1} & & \overline{N}(\pi)_{k-2}^{-1}(d_j d_k(\boldsymbol{e})) \times \Delta^{k-2} \\
\downarrow {\scriptstyle 1\times d^j} & & \swarrow {\scriptstyle z_{d_j d_k(\boldsymbol{e})}} & & \| \\
\overline{N}(\pi)_{k-1}^{-1}(d_k(\boldsymbol{e})) \times \Delta^{k-1} & \xrightarrow{z_{d_k(\boldsymbol{e})}} & D_{\lambda_{k-1}} & \xleftarrow[z_{d_{k-1} d_j(\boldsymbol{e})}]{} & \overline{N}(\pi)_{k-2}^{-1}(d_{k-1} d_j(\boldsymbol{e})) \times \Delta^{k-2}
\end{array}$$

Under the identification $\Delta^k = \Delta^{k-1} * \boldsymbol{v}_k$, $d^j : \Delta^{k-1} \to \Delta^k$ can be identified with the composition

$$\Delta^{k-2} * \boldsymbol{v}_{k-1} \cong \Delta^{k-2} * \boldsymbol{v}_k \stackrel{d^j * 1_{\boldsymbol{v}_k}}{\longrightarrow} \Delta^{k-1} * \boldsymbol{v}_k.$$

On the other hand, since $j < k$, we have

$$\begin{aligned}\overline{N}(\pi)_k(\boldsymbol{e}) &= P_{\lambda_{k-1},\lambda_k} \times \overline{N}(\pi)_{k-1}^{-1}(d_k(\boldsymbol{e})) \\ \overline{N}(\pi)_{k-1}(d_j(\boldsymbol{e})) &= P_{\lambda_{k-1},\lambda_k} \times \overline{N}(\pi)_{k-2}^{-1}(d_{k-1}d_j(\boldsymbol{e})),\end{aligned}$$

and the face operator $d_j : \overline{N}(\pi)_k(\boldsymbol{e}) \to \overline{N}(\pi)_{k-1}(d_j(\boldsymbol{e}))$ coincides with the map

$$\begin{aligned}P_{\lambda_{k-1},\lambda_k} \times \overline{N}(\pi)_{k-1}^{-1}(d_k(\boldsymbol{e})) &\stackrel{1\times d_j}{\longrightarrow} P_{\lambda_{k-1},\lambda_k} \times \overline{N}(\pi)_{k-1}^{-1}(d_j d_k(\boldsymbol{e})) \\ &= P_{\lambda_{k-1},\lambda_k} \times \overline{N}(\pi)_{k-1}^{-1}(d_{k-1}d_j(\boldsymbol{e})).\end{aligned}$$

Thus we obtain the commutative diagram

$$\begin{array}{ccc}
\overline{N}(\pi)_k^{-1}(\boldsymbol{e}) \times \Delta^{k-1} & \xrightarrow{d_j \times 1} & \overline{N}(\pi)_{k-1}^{-1}(d_j(\boldsymbol{e})) \times \Delta^{k-1} \\
\| & & \| \\
P_{\lambda_{k-1},\lambda_k} \times \overline{N}(\pi)_{k-1}^{-1}(d_k(\boldsymbol{e})) \times \Delta^{k-1} & \xrightarrow{1 \times d_j \times 1} & P_{\lambda_{k-1},\lambda_k} \times \overline{N}(\pi)_{k-2}^{-1}(d_{k-1}d_j(\boldsymbol{e})) \times \Delta^{k-1} \\
\downarrow {\scriptstyle 1 \times 1 \times d^j} & & \downarrow {\scriptstyle 1 \times \tilde{z}_{d_{k-1}d_j(\boldsymbol{e})}} \\
P_{\lambda_{k-1},\lambda_k} \times \overline{N}(\pi)_{k-1}^{-1}(d_k(\boldsymbol{e})) \times \Delta^{k} & \xrightarrow{1 \times \tilde{z}_{d_k(\boldsymbol{e})}} & P_{\lambda_{k-1},\lambda_k} \times ((D_{\lambda_{k-1}} \times \{0\}) * (0,1))
\end{array}$$

where we embed $D_{\lambda_{k-1}}$ into $D_{\lambda_{k-1}} \times \mathbb{R}$ as $D_{\lambda_{k-1}} \times \{0\}$ and $\tilde{z}_{d_k(\boldsymbol{e})}$ is defined by

$$\tilde{z}_{d_k(\boldsymbol{e})}(\boldsymbol{p}, (1-t)\boldsymbol{s} + t\boldsymbol{v}_k) = (1-t)(z_{d_k(\boldsymbol{e})}(\boldsymbol{p}, \boldsymbol{s}), 0) + t(0,1).$$

The map $\tilde{z}_{d_{k-1}d_j(\boldsymbol{e})}$ is defined analogously. We also have an extension of $b_{\lambda_{k-1},\lambda_k}$

$$\tilde{b}_{\lambda_{k-1},\lambda_k} : P_{\lambda_{k-1},\lambda_k} \times ((D_{\lambda_{k-1}} \times \{0\}) * (0,1)) \longrightarrow D_{\lambda_k}.$$

By definition, we have the commutative diagram

$$\begin{array}{ccc}
P_{\lambda_{k-1},\lambda_k} \times N(\pi)_{k-2}^{-1}(d_{k-1}d_j(\boldsymbol{e})) \times \Delta^{k-1} & = & N(\pi)_{k-1}^{-1}(d_j(\boldsymbol{e})) \times \Delta^{k-1} \\
\downarrow {\scriptstyle 1 \times \tilde{z}_{d_{k-1}d_j(\boldsymbol{e})}} & & \downarrow {\scriptstyle z_{d_j(\boldsymbol{e})}} \\
P_{\lambda_{k-1},\lambda_k} \times ((D_{\lambda_{k-1}} \times \{0\}) * (0,1)) & \xrightarrow{\tilde{b}_{\lambda_{k-1},\lambda_k}} & D_{\lambda_k} \\
\uparrow {\scriptstyle 1 \times \tilde{z}_{d_k(\boldsymbol{e})}} & & \uparrow {\scriptstyle z_{\boldsymbol{e}}} \\
P_{\lambda_{k-1},\lambda_k} \times N(\pi)_{k-1}^{-1}(d_k(\boldsymbol{e})) \times \Delta^{k} & = & N(\pi)_{k}^{-1}(\boldsymbol{e}) \times \Delta^{k}.
\end{array}$$

This completes the proof of the commutativity of the diagram (5.1) and we obtain a map

$$z_k : \overline{N}_k(C(X)) \times \Delta^k \longrightarrow D(X)$$

for all k. By composing with $\Phi : D(X) \to X$, we obtain a family of continuous maps

$$\{i_k : \overline{N}_k(C(X)) \times \Delta^k \longrightarrow X\}_{k \geq 0}$$

that are compatible with d_i, which induces a continuous map

$$i_X : \mathrm{Sd}(X) = \left\|\overline{N}(C(X))\right\| \longrightarrow X.$$

The next step is to prove that i_X is an embedding.

Proof of Theorem 5.16: Let us show that $i_X : BC(X) \to \mathrm{Im}(\iota_X)$ is a bijective closed map, hence is a homeomorphism. The bijectivity is obvious from the construction. In order to show that i_X is a closed map, consider the diagram

$$\begin{array}{ccc}
\displaystyle\coprod_{n,\boldsymbol{e}\in\overline{N}_n(P(X))} \{\boldsymbol{e}\} \times \overline{N}(\pi)_n^{-1}(\boldsymbol{e}) \times \Delta^n & \xrightarrow{\coprod_{\boldsymbol{e}} z_{\boldsymbol{e}}} & D(X) \\
\Big\downarrow{\scriptstyle p} & & \Big\downarrow{\scriptstyle \Phi} \\
\|\overline{N}(C(X))\| & \xrightarrow[i_X]{} & X
\end{array}$$

whose vertical arrows are quotient maps. For simplicity, let us denote $z = \coprod_{\boldsymbol{e}} z_{\boldsymbol{e}}$. For $A \subset \mathrm{Sd}(X)$, we show that

$$\Phi^{-1}(i_X(A)) = z(p^{-1}(A)), \tag{5.3}$$

which immediately implies that i_X is a closed map onto its image, since Φ is a quotient map and z is an embedding.

The commutativity of the above diagram implies

$$\Phi^{-1}(i_X(A)) \supset z(p^{-1}(A)).$$

On the other hand, suppose $x \in \Phi^{-1}(i_X(A))$. Then there exist $a \in A$ and $\lambda \in \Lambda$ with $\Phi(x) = i_X(a)$ and $x \in D_\lambda$. Write $a = p(\boldsymbol{b}, \boldsymbol{t}))$ for $\boldsymbol{b} \in \overline{N}(\pi)_k^{-1}(\boldsymbol{e})$ and $\boldsymbol{t} \in \Delta^k$.

If $x \in \mathrm{Int}(D_\lambda)$, we have

$$\varphi_\lambda(x) = i_X(a) = i_X(p(\boldsymbol{b}, \boldsymbol{t})) = \Phi(z_{\boldsymbol{e}}(\boldsymbol{b}, \boldsymbol{t})).$$

Since $x \in \mathrm{Int}(D_\lambda)$, $\boldsymbol{t}$ is of the form $\boldsymbol{t} = (1-t)\boldsymbol{s} + t\boldsymbol{v}_k$ for some $0 < t < 1$ and $\boldsymbol{s} \in \Delta^{k-1}$. This implies that $z_{\boldsymbol{e}}(\boldsymbol{b}, \boldsymbol{t}) \in \mathrm{Int}(D_\lambda)$ and $\varphi_\lambda(x) = \varphi_\lambda(z_{\boldsymbol{e}}(\boldsymbol{b}, \boldsymbol{t}))$. Thus $x = z_{\boldsymbol{e}}(\boldsymbol{b}, \boldsymbol{t})$ with $p(\boldsymbol{b}, \boldsymbol{t}) = a \in A$. In other words, $x \in z(p^{-1}(A))$.

Suppose $x \in \partial D_\lambda$ and let e_μ be a cell containing $\varphi_\lambda(x)$. By the cylindrical normality, ∂D_λ is a stratified space and

$$b_{\mu,\lambda} : P_{\mu,\lambda} \times D_\mu \longrightarrow \partial D_\lambda$$

is a strict morphism of stratified space. Choose $(b, y) \in P_{\mu,\lambda} \times D_\mu$ with $b_{\mu,\lambda}(b, y) = x$. Then $\varphi_\mu(y) = \varphi_\lambda(x) = \Phi(x) \in i_X(A)$. Since $y \in \mathrm{Int}(D_\mu)$, the above argument implies that there exists $(\boldsymbol{b}', \boldsymbol{t}') \in p^{-1}(A)$ with $y =$

$z_{\pi(\boldsymbol{b}')}(\boldsymbol{b}',\boldsymbol{t}')$. Define $\boldsymbol{b}=(b,\boldsymbol{b}')$ and $\boldsymbol{t}=d^k(\boldsymbol{t}')$. Then the defining relation of the geometric realization of a Δ-space implies that

$$p(\boldsymbol{b},\boldsymbol{t})=[\boldsymbol{b},d^k(\boldsymbol{t}')]=[d_k(\boldsymbol{b}),\boldsymbol{t}']=[\boldsymbol{b}',\boldsymbol{t}']=p(\boldsymbol{b}',\boldsymbol{t}')\in A.$$

On the other hand, the commutativity of (5.2) implies that

$$x=b_{\mu,\lambda}(b,y)=b_{\mu,\lambda}(b,z_{p(\boldsymbol{b}')}(\boldsymbol{b}',\boldsymbol{t}'))=z_{p(b,\boldsymbol{b}')}(p(b,\boldsymbol{b}),d^k(\boldsymbol{t}'))=z_{p(\boldsymbol{b})}(\boldsymbol{b},\boldsymbol{t}).$$

And we have $x\in z(p^{-1}(A))$. This completes the proof of (5.3).

Finally when all cells are closed, each z_k is surjective and thus i_X is a homeomorphism. □

The next task is to show that the image of i_X is a strong deformation retract of X under a suitable condition. The idea of proof is essentially the same as Theorem 5.16. We construct deformation retractions on each cell and glue them together. In order to define cell-wise deformation retraction, we assume the polyhedrality.

Theorem 5.18: *For a polyhedral stellar stratified space X, the image of embedding $i_X:\mathrm{Sd}(X,\pi)\to X$ is a strong deformation retract of X. The deformation retraction can be taken to be natural with respect to strict morphisms of polyhedral stellar stratified spaces.*

Corollary 5.19: *For a totally normal cellular stratified space (X,π), the map i_X embeds $\mathrm{Sd}(X,\pi)$ into X as a strong deformation retract.*

The following fact is essential.

Lemma 5.20: *Let π be a regular cell decomposition of S^{n-1} and $L\subset S^{n-1}$ be a stratified subspace. Let $\tilde{\pi}$ be the cellular stratification on $K=\mathrm{Int}D^n\cup L$ obtained by adding $\mathrm{Int}D^n$ as an n-cell. Then there is a deformation retraction H of K to $i_K(\mathrm{Sd}(K,\tilde{\pi}))$. Furthermore, if a deformation retraction h of L onto $i_L(\mathrm{Sd}(L))$ is given, H can be taken to be an extension of h.*

This lemma can be proved by using good old simplicial topology. It was first proved in a joint work with Basabe, González, and Rudyak [14] but removed from the published version [5]. The proof is now contained in [1] as Appendix A.

Now we are ready to prove Theorem 5.18.

Proof of Theorem 5.18: Let us show the embedding i_X constructed in the proof of Theorem 5.16 has a homotopy inverse.

Since X is polyhedral, each cell $\varphi_\lambda : D_\lambda \to \overline{e_\lambda}$ has a polyhedral replacement $\alpha_\lambda : \widetilde{F}_\lambda \to D^{\dim e_\lambda}$. For simplicity, we identify D_λ with $\alpha_\lambda^{-1}(D_\lambda)$. Thus D_λ is a stratified subspace of a polyhedral complex and $b_{\mu,\lambda} : P_{\mu,\lambda} \times D_\mu \to D_\lambda$ is a PL map.

We construct, by induction on k, a PL homotopy

$$H_\lambda : D_\lambda \times [0,1] \longrightarrow D_\lambda$$

for each k-cell e_λ satisfying the following conditions:

(1) It is a strong deformation retraction of D_λ onto $i_{D_\lambda}(\mathrm{Sd}(D_\lambda))$, where the stratification on D_λ is given by adding $\mathrm{Int}(D_\lambda)$ as a unique k cell to the stratification on ∂D_λ.

(2) The diagram

$$\begin{array}{ccc} D_\lambda \times [0,1] & \xrightarrow{H_\lambda} & D_\lambda \\ {\scriptstyle b_{\mu,\lambda}\times 1}\uparrow & & \uparrow{\scriptstyle b_{\mu,\lambda}} \\ P_{\mu,\lambda} \times D_\mu \times [0,1] & \xrightarrow[1\times H_\mu]{} & P_{\mu,\lambda} \times D_\mu. \end{array} \tag{5.4}$$

is commutative for any pair $e_\mu \subset \overline{e_\lambda}$.

When $k=0$, the homotopy is the canonical projection. Suppose we have constructed H_μ for all i-cells e_μ with $i \leq k-1$. We would like to extend them to k-cells. For a k-cell e_λ with stellar structure map $\varphi_\lambda : D_\lambda \to X$ and a cell e_μ with $e_\mu \subset \overline{e_\lambda}$, consider the diagram

$$\begin{array}{ccc} P_{\mu,\lambda} \times D_\mu \times [0,1] & \xrightarrow{1\times H_\mu} & P_{\mu,\lambda} \times D_\mu \\ {\scriptstyle b_{\mu,\lambda}\times 1}\downarrow & & \downarrow{\scriptstyle b_{\mu,\lambda}} \\ b_{\mu,\lambda}(P_{\mu,\lambda} \times D_\mu) \times [0,1] & \dashrightarrow & b_{\mu,\lambda}(P_{\mu,\lambda} \times D_\mu). \end{array}$$

Since $b_{\mu,\lambda}$ is a homeomorphism when restricted to $P_{\mu,\lambda} \times \mathrm{Int}(D_\mu)$ and $b_{\mu,\lambda}$ and H_μ are assumed to be PL, we have the dotted arrow making the diagram commutative by Lemma B.32.

The decomposition

$$\partial D_\lambda = \bigcup_{e_\mu \subset \overline{e_\lambda}} b_{\mu,\lambda}(P_{\mu,\lambda} \times D_\mu)$$

allows us to glue these homotopies together. Thus we obtain a homotopy

$$H_\lambda : \partial D_\lambda \times [0,1] \longrightarrow \partial D_\lambda.$$

By Lemma 5.20, this homotopy can be extended to a strong deformation retraction

$$H_\lambda : D_\lambda \times [0,1] \longrightarrow D_\lambda$$

of D_λ onto $i_{D_\lambda}(\mathrm{Sd}(D_\lambda))$. This completes the inductive step.

The second condition above (5.4) and the CW condition imply that these deformation retractions can be assembled together to give a strong deformation retraction

$$H : X \times [0,1] \longrightarrow X$$

of X onto $i_X(\mathrm{Sd}(X))$. □

The construction of the higher order Salvetti complex in [7, 8] can be regarded as a corollary to Theorem 5.18.

Example 5.21: Consider the stratification on $\mathbb{R}^n \otimes \mathbb{R}^\ell$

$$\pi_{\mathcal{A}\otimes\mathbb{R}^\ell} : \mathbb{R}^n \otimes \mathbb{R}^\ell \longrightarrow \mathrm{Map}(L, S_\ell)$$

in Example 2.10. As we have seen in Example 2.42, this is a normal cellular stratification. It is also regular and polyhedral.

It contains

$$\mathrm{Lk}(\mathcal{A}\otimes\mathbb{R}^\ell) = \bigcup_{i=1}^{k} H_i \otimes \mathbb{R}^\ell$$

as a cellular stratified subspace. The complement

$$M(\mathcal{A}\otimes\mathbb{R}^\ell) = \mathbb{R}^n \otimes \mathbb{R}^\ell \setminus \mathrm{Lk}(\mathcal{A}\otimes\mathbb{R}^\ell)$$

is also a cellular stratified subspace. Since $\pi_{\mathcal{A}\otimes\mathbb{R}^\ell}$ is regular, we have

$$C(M(\mathcal{A}\otimes\mathbb{R}^\ell)) = P(M(\mathcal{A}\otimes\mathbb{R}^\ell)) = P(\mathcal{A}\otimes\mathbb{R}^\ell) \setminus P(\mathrm{Lk}(\mathcal{A}\otimes\mathbb{R}^\ell)).$$

By Theorem 5.18, $BC(M(\mathcal{A}\otimes\mathbb{R}^\ell))$ can be embedded in the complement $M(\mathcal{A}\otimes\mathbb{R}^\ell)$ as a strong deformation retract. This simplicial complex $BC(M(\mathcal{A}\otimes\mathbb{R}^\ell))$ is nothing but the higher order Salvetti complex in [7, 8].

The next example is totally normal but not regular.

Example 5.22: By Example 3.14, a graph X can be regarded as a totally normal cellular stratified space. As is described in [1], we may define a totally normal cellular stratification on the k-fold product X^k by using the braid arrangements, which can be restricted to a totally normal cellular stratification on $\mathrm{Conf}_k(X)$. By Corollary 5.19, $BC(\mathrm{Conf}_k(X))$ can be embedded in $\mathrm{Conf}_k(X)$ as a Σ_k-equivariant strong deformation retract.

$BC(\mathrm{Conf}_k(X))$ is a regular CW complex model for $\mathrm{Conf}_k(X)$. This is better than Abrams' model in the sense that it works for all graphs.

The cellular stratification on $\mathbb{C}\mathrm{P}^n$ in Example 4.26 is cylindrically normal and Theorem 5.16 applies.

Example 5.23: Consider the cylindrically normal cellular stratification on $\mathbb{C}\mathrm{P}^n$ in Example 4.25 and Example 4.26. By Theorem 5.16, $BC(\mathbb{C}\mathrm{P}^n)$ is homeomorphic to $\mathbb{C}\mathrm{P}^n$. Let us compare the description of $BC(\mathbb{C}\mathrm{P}^n)$ with the one in Example 4.26 as the Davis-Januszkiewicz construction $M(\lambda_n)$.

As we have seen in Example 5.6, the cylindrical face category $C(\mathbb{C}\mathrm{P}^n)$ can be obtained from the underlying poset $[n]$ of $C(\mathbb{C}\mathrm{P}^n)$ by replacing the set of morphisms by S^1. Compositions of morphisms are given by the group structure of S^1.

By Lemma 5.13, we have

$$\overline{N}_k(C(\mathbb{C}\mathrm{P}^n)) = \coprod_{\boldsymbol{e}\in\overline{N}_k([n])} \{\boldsymbol{e}\} \times (S^1)^k$$

and the classifying space of $C(\mathbb{C}\mathrm{P}^n)$ can be described in the form

$$\begin{aligned} BC(\mathbb{C}\mathrm{P}^n) &= \left\|\overline{N}(C(\mathbb{C}\mathrm{P}^n))\right\| \\ &= \left(\coprod_k \coprod_{\boldsymbol{e}\in\overline{N}_k([n])} \{\boldsymbol{e}\} \times (S^1)^k \times \Delta^k\right)\Big/_{\sim}. \end{aligned}$$

Note that a nondegenerate k-chain $\boldsymbol{e}$ in the poset $[n]$ can be described by a strictly increasing sequence of nonnegative integers $\boldsymbol{e} = (i_0, \ldots, i_k)$ with $i_k \leq n$. For an element $(\boldsymbol{e}; t_1, \ldots, t_k; p_0, \ldots, p_k)$ in $\overline{N}_k(C(\mathbb{C}\mathrm{P}^n))$, let $\tilde{\boldsymbol{t}}$ be an element of $(S^1)^n$ obtained from $(t_1, \ldots, t_k)$ by inserting 1 in such a way that each t_j is placed between i_{j-1}-th and i_j-th positions. Then there exists a face operator $d_I : N_n(C(\mathbb{C}\mathrm{P}^n)) \to N_k(C(\mathbb{C}\mathrm{P}^n))$ such that

$$(\boldsymbol{e}; t_1, \ldots, t_k) = (d_I(0, 1, \ldots, n), d_I(\tilde{\boldsymbol{t}})).$$

We then have

$$(\boldsymbol{e}; t_1, \ldots, t_k; p_0, \ldots, p_k) \sim (0, \ldots, n; \tilde{\boldsymbol{t}}; d^I(p_0, \ldots, p_k)).$$

Note that $d^I(p_0, \ldots, p_k)$ is obtained from $(p_0, \ldots, p_k)$ by inserting 0 in appropriate coordinates. Thus any point in $BC(\mathbb{C}\mathrm{P}^n)$ can be represented by a point in $(S^1)^n \times \Delta^n$ and $BC(\mathbb{C}\mathrm{P}^n)$ can be written as

$$BC(\mathbb{C}\mathrm{P}^n) = (S^1)^n \times \Delta^n/_\sim. \tag{5.5}$$

The relation here is not exactly the same as the defining relation of $M(\lambda_n)$.

Define a map

$$s_n : M(\lambda_n) \longrightarrow BC(\mathbb{C}\mathrm{P}^n)$$

by

$$s_n([t_1, \ldots, t_n; p_0, \ldots, p_n]) = [t_1, t_1^{-1}t_2, \ldots, t_{n-1}^{-1}t_n; p_0, \ldots, p_n].$$

It is left to the reader to verify that s_n is a well-defined homeomorphism making the diagram

$$\begin{array}{ccccc} M(\lambda_n) & & \xrightarrow{\quad s_n \quad} & & BC(\mathbb{C}\mathrm{P}^n) \\ \downarrow & & & & \downarrow{\scriptstyle B\pi_{C(\mathbb{C}\mathrm{P}^n)}} \\ \Delta^n & = & B([n]) & = & BP(\mathbb{C}\mathrm{P}^n) \end{array}$$

commutative, where $\pi_{C(\mathbb{C}\mathrm{P}^n)} : C(CP^n) \to P(\mathbb{C}\mathrm{P}^n)$ is the canonical projection onto the underlying poset (Definition C.6). Thus we see that $BC(\mathbb{C}\mathrm{P}^n)$ coincides with $M(\lambda_n)$ up to a homeomorphism.

6. Basic constructions on cellular stratified spaces

We study the following operations on cellular and stellar stratified spaces in this section:

- taking stratified subspaces, cellular stratified subspaces, and stellar stratified subspaces,
- taking products of cellular stratified spaces,
- taking subdivisions of cellular and stellar stratified spaces.

6.1. *Stratified subspaces*

In this section, we consider the problem of restricting cellular stratifications to subspaces. Obviously the category of stratified spaces is closed under taking complements of stratified subspaces.

Lemma 6.1: *Let X be a stratified space and A be a stratified subspace. Then the complement $X \setminus A$ is also a stratified subspace of X.*

This is one of the most useful facts when we work with stratifications on configuration spaces and complements of arrangements. On the other hand, when X is a CW complex and A is a subcomplex, A is always closed. This is no longer true for cellular stratified spaces. A typical example is the case of the complement $M(\mathcal{A} \otimes \mathbb{R}^\ell) = \mathbb{R}^n \otimes \mathbb{R}^\ell \setminus \mathrm{Lk}(\mathcal{A} \otimes \mathbb{R}^\ell)$ of an arrangement in Example 5.21.

Let us consider cell structures on subspaces. Suppose X is a cellular stratified space and A is a stratified subspace. In order to incorporate cell structures, we need to specify a cell structure for each cell contained in A.

Definition 6.2: Let X be a topological space and A be a subspace. For a hereditarily quotient (Definition A.2) n-cell structure $\varphi : D \to \overline{e} \subset X$ with $e \subset A$, define an n-cell structure on e in A to be $(D_A, \varphi|_{D_A})$ where $D_A = \varphi^{-1}(\overline{e} \cap A)$.

The hereditarily-quotient assumption guarantees the restriction is a quotient map by Lemma A.5.

Lemma 6.3: *Let X be a topological space and A be a subset of X. If (D, φ) is a hereditarily quotient n-cell structure on $e \subset X$ and $e \subset A$, then $(D_A, \varphi|_{D_A})$ defined above is an n-cell structure on e in A.*

Definition 6.4: Let (X, π) be a cellular stratified space. A stratified subspace $(A, \pi|_A)$ of X is said to be a *cellular stratified subspace*, provided cell structures on cells in A are given as indicated in Definition 6.2. When the inclusion $A \hookrightarrow X$ is a strict morphism, A is said to be a *strict cellular stratified subspace*.

We need to take cell decompositions of domains of cells into account for stellar stratified subspaces.

Lemma 6.5: *Let X be a topological space, A a subspace, and $\varphi : D \to \overline{e}$ a stellar n-cell of X with $e \subset A$. When $D_A = \varphi^{-1}(\overline{e} \cap A)$ is a strict stratified subspace of D and φ is hereditarily quotient, the restriction $\varphi|_{D_A} : D_A \to A$ defines a stellar n-cell structure on e in A.*

Definition 6.6: Let (X, π) be a stellar stratified space and A a subspace of X. If the assumption of Lemma 6.5 is satisfied for each cell in A, $(A, \pi|_A)$ is said to be a *stellar stratified subspace* of X.

Remark 6.7: Let X be a cellular or stellar stratified space and A be a cellular or stellar stratified subspace. Then A is a strict stratified subspace of X.

The following fact can be regarded as a generalization of the fact that any subcomplex of a regular cell complex is regular.

Proposition 6.8: *Any stellar stratified subspace A of a cylindrically normal stellar stratified space X is cylindrically normal. Furthermore, the parameter space for a pair of cells $e_\mu < e_\lambda$ in A can be identified with the parameter space for the same pair when regarded as cells in X.*

Proof: Let $e_\mu < e_\lambda$ be a pair of cells in A. Let $\varphi_\mu : D_\mu \to X$ and $\varphi_\lambda : D_\lambda \to X$ be the stellar structures for these cells in X. When regarded as cells in A, their stellar structures are denoted by

$$\begin{aligned} \varphi_{A,\mu} &: D_{A,\mu} \longrightarrow A \\ \varphi_{A,\lambda} &: D_{A,\lambda} \longrightarrow A, \end{aligned}$$

respectively.

We need to show that the structure map

$$b_{\mu,\lambda} : P_{\mu,\lambda} \times D_\mu \longrightarrow D_\lambda$$

of the cylindrical structure for the pair in X can be restricted to

$$b_{\mu,\lambda}|_{P_{\mu,\lambda}\times D_{A,\mu}} : P_{\mu,\lambda} \times D_{A,\mu} \longrightarrow D_{A,\lambda}.$$

This can be verified by the commutativity of the diagram

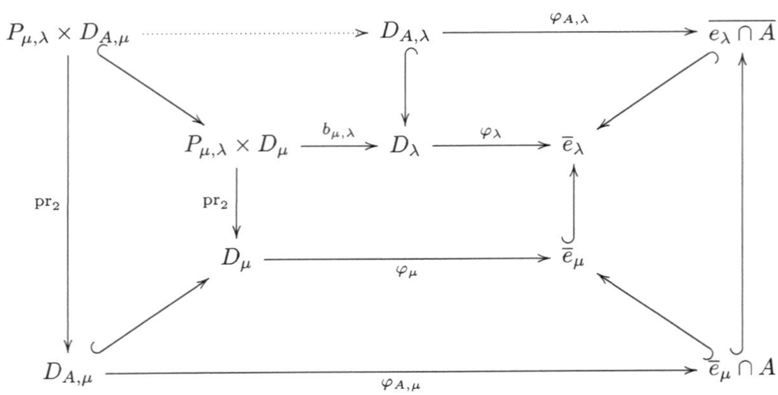

and the definition of stellar structures on A. $\square$

Example 6.9: Consider the stratification in Example 5.21. The link $\mathrm{Lk}(\mathcal{A} \otimes \mathbb{R}^\ell)$ and the complement $M(\mathcal{A} \otimes \mathbb{R}^\ell)$ are both cellular stratified subspaces of $(\mathbb{R}^n \otimes \mathbb{R}^\ell, \pi_{\mathcal{A}\otimes\mathbb{R}^\ell})$, which is regular, hence cylindrically normal.

Thanks to Corollary 4.21 and Lemma A.7, cell structures in a polyhedral stellar stratified space X are hereditarily quotient. Thus any stellar stratified subspace A inherits a cylindrically normal structure with structure maps satisfying the PL conditions in the definition of polyhedral cellular stratification. The problem is the CW condition. It is easy to see that the closure finiteness condition can be restricted freely. The question is when a stratified subspace of a CW stratified subspace inherits the weak topology.

Lemma 6.10: *A closed or an open stratified subspace A of a CW stratified space X is CW.*

Proof: This follows from the corresponding property of weak topology. □

By Lemma 4.19, any polyhedral cellular stratified space X can be embedded in a CW complex $U(X)$. In general, however, a cellular stratified subspace A of X is neither closed nor open in $U(X)$ and it is not easy to verify the weak topology condition. One of the practical conditions is the locally finiteness. The CW condition is guaranteed by Proposition 2.22.

Proposition 6.11: *Let X be a polyhedral cellular stratified space. Any locally finite cellular stratified subspace A is polyhedral.*

We end this section by an example which shows another difference between cellular stratified subspaces and subcomplexes. In the case of CW complexes, the colimit of an increasing sequence of finite subcomplexes

$$X_0 \subset X_1 \subset \cdots \subset \operatorname*{colim}_n X$$

is automatically a CW complex. This is not true for cellular stratified spaces.

Example 6.12: Consider the space

$$X = \left\{(x,y) \in \mathbb{R}^2 \,\middle|\, y > 0\right\} \cup \mathbb{Z} \times \{0\}.$$

The homeomorphism

$$p : D^2 \setminus \{(0,1)\} \longrightarrow \left\{(x,y) \in \mathbb{R}^2 \,\middle|\, y \geq 0\right\}$$

given by extending the stereographic projection $S^1 \setminus \{(0,1)\} \to \mathbb{R}$,

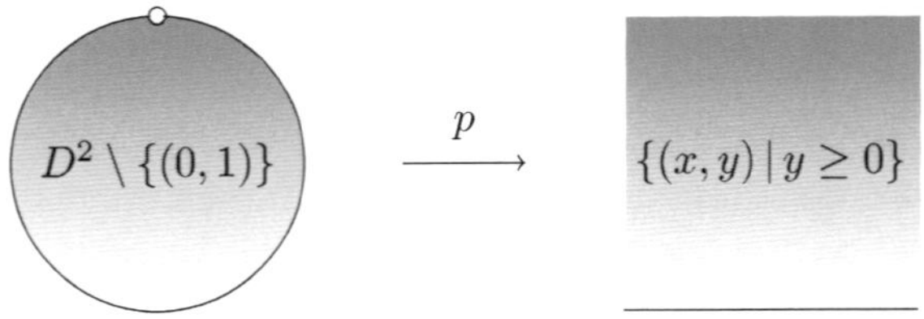

defines a 2-cell structure on

$$e^2 = \{(x,y) \in \mathbb{R}^2 \,|\, y > 0\} \subset X$$

by restricting p to $D = p^{-1}(X)$.

Each $\{(n,0)\} \subset \mathbb{Z} \times \{0\}$ can be regarded as a 0-cell e_n^0. And we have a cellular stratification on X

$$X = \left(\bigcup_{n \in \mathbb{Z}} e_n^0 \right) \cup e^2.$$

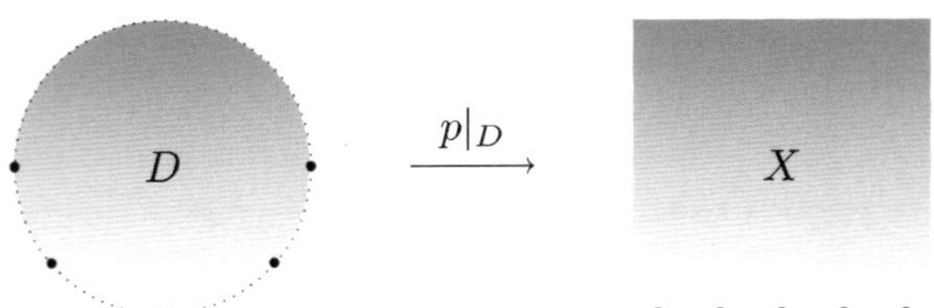

X is a colimit of

$$X_n = \{(x,y) \in \mathbb{R}^2 \,|\, y > 0\} \cup \{i \in \mathbb{Z} \,|\, |i| \leq n\} \times \{0\}.$$

Each X_n is a finite stratified subspace of X, hence is CW. But X is not CW.

6.2. *Products*

In this section, we study products of stratifications, cell structures and cellular stratified spaces, and deduce a couple of conditions under which we may take products.

It is not difficult to define a stratification on the product of two stratified spaces, as we have seen in Lemma 2.14. We have to be careful, however, when we take products of cellular stratified spaces. Even in the category of CW complexes, there is a well-known difficulty in taking products. Given CW complexes X and Y, we need to impose the local-finiteness on X or Y or to redefine a topology on $X \times Y$ in order to make $X \times Y$ into a CW complex.

In the case of cellular stratified spaces, we have another difficulty because of our requirement on cell structures. The product of two quotient maps may not be a quotient map.

Definition 6.13: Let (X, π_X, Φ_X) and (Y, π_Y, Φ_Y) be cellular stratified spaces and consider the product stratification

$$\pi_X \times \pi_Y : X \times Y \longrightarrow P(X) \times P(Y)$$

on $X \times Y$ in Lemma 2.14. For a pair of cells $e_\lambda \subset X$ and $e_\mu \subset Y$, consider the composition

$$\varphi_{\lambda,\mu} : D \cong D_\lambda \times D_\mu \stackrel{\varphi_\lambda \times \varphi_\mu}{\longrightarrow} \overline{e_\lambda} \times \overline{e_\mu} = \overline{e_\lambda \times e_\mu} \subset X \times Y,$$

where D is the subspace of $D^{\dim e_\lambda + \dim e_\mu}$ defined by pulling back $D_\lambda \times D_\mu$ via the standard homeomorphism

$$D^{\dim e_\lambda + \dim e_\mu} \cong D^{\dim e_\lambda} \times D^{\dim e_\mu}.$$

If $\varphi_{\lambda,\mu}$ is a quotient map for each pair of cells, the resulting cellular stratification is called the *product cellular stratification* on $X \times Y$.

The above definition is incomplete. Unless we have a general criterion for $\varphi_{\lambda,\mu}$ to be a quotient map, this definition is useless.

By Lemma A.7 and Proposition A.13, we can take products of bi-quotient cell structures. The question is when a cell structure is a bi-quotient map. By Corollary A.15 and Corollary 4.21, we obtain the following practical conditions under the assumption of cylindrical normality.

Proposition 6.14: *Let X and Y be cylindrically normal cellular stratified spaces. If they satisfy one of the following conditions:*

(1) All parameter spaces are compact;
(2) Polyhedral,

then any product $e_\lambda \times e_\mu$ has the product cell structure for $\lambda \in P(X)$ and $\mu \in P(Y)$.

Proof: If X or Y is polyhedral, all cell structures are bi-quotient by Corollary 4.21. By Lemma 4.4, each fiber of a cell structure in a cylindrically normal cellular stratified space can be identified with a parameter space. Thus, when all parameter spaces are compact, the cell structures are bi-quotient by Corollary A.15. □

If X and Y satisfy one of the above conditions, the product $X \times Y$ has a structure of cellular stratified space. It is reasonable to expect that $X \times Y$ is again cylindrically normal.

Theorem 6.15: *Let X and Y be cylindrically normal cellular stratified spaces satisfying one of the conditions in Proposition 6.14. Then the product $X \times Y$ is a cylindrically normal cellular stratified space.*

Proof: Let $\{\varphi_\lambda^X : D_\lambda \to \overline{e_\lambda}\}$ and $\{\varphi_\mu^Y : D_\mu \to \overline{e_\mu}\}$ be cell structures on X and Y, respectively. Cylindrical structures on X and Y are denoted by $\{b_{\lambda,\lambda'}^X : P_{\lambda,\lambda'}^X \times D_\lambda \to D_{\lambda'}\}$ and $\{b_{\mu,\mu'}^Y : P_{\mu,\mu'}^Y \times D_\mu \to D_{\mu'}\}$, respectively.

For a pair of cell structures $\varphi_\lambda^X : D_\lambda \to X$ and $\varphi_\mu^Y : D_\mu \to Y$, consider the product cell structure

$$\varphi_{\lambda,\mu} : D_{\lambda,\mu} \cong D_\lambda \times D_\mu \overset{\varphi_\lambda^X \times \varphi_\mu^Y}{\longrightarrow} \overline{e_\lambda} \times \overline{e_\mu}.$$

For $(\lambda, \mu) \leq (\lambda', \mu')$, define

$$P_{(\lambda,\mu),(\lambda',\mu')} = P_{\lambda,\lambda'}^X \times P_{\mu,\mu'}^Y.$$

When $(\lambda, \mu) < (\lambda', \mu')$, we have either $\lambda < \lambda'$ or $\mu < \mu'$ and thus the image of the composition

$$\begin{aligned} P_{(\lambda,\mu),(\lambda',\mu')} \times D_{\lambda,\mu} &\cong P_{\lambda,\lambda'}^X \times P_{\mu,\mu'}^Y \times D_\lambda \times D_\mu \\ &\overset{\cong}{\longrightarrow} P_{\lambda,\lambda'}^X \times D_\lambda \times P_{\mu,\mu'}^Y \times D_\mu \\ &\overset{b_{\lambda,\lambda'}^X \times b_{\mu,\mu'}^Y}{\longrightarrow} D_{\lambda'} \times D_{\mu'} \\ &\cong D_{\lambda',\mu'} \end{aligned}$$

lies in $\partial D_{\lambda,\mu} \cong (D_\lambda \times \partial D_\mu) \cup (\partial D_\lambda \times D_\mu)$. And we obtain a map

$$b_{(\lambda,\mu),(\lambda',\mu')} : P_{(\lambda,\mu),(\lambda',\mu')} \times D_{\lambda,\mu} \longrightarrow \partial D_{\lambda',\mu'}.$$

The composition operations

$$P_{(\lambda_1,\mu_1),(\lambda_2,\mu_2)} \times P_{(\lambda_0,\mu_0),(\lambda_1,\mu_1)} \longrightarrow P_{(\lambda_0,\mu_0),(\lambda_2,\mu_2)}$$

are defined in an obvious way.

It is straightforward to verify that these maps define a cylindrical structure on $X \times Y$ under the product stratification and the product cell structures. □

Let us consider the CW conditions on products next. The closure finiteness condition is automatic.

Lemma 6.16: *If X and Y are stratified spaces satisfying the closure finiteness condition, then so is $X \times Y$.*

As is the case of CW complexes [33], the product of two CW stratifications may not satisfy the weak topology condition. In the case of CW complexes, the local finiteness of X implies that $X \times Y$ is CW for any CW complex Y. The author does not know if an analogous fact holds for CW stratified spaces in general. The following obvious fact is still useful in many cases.

Lemma 6.17: *Let X and Y be CW stratified spaces. If both X and Y are locally finite, then $X \times Y$ is a CW stratified space.*

Proof: The product of locally finite stratified spaces is again locally finite. The result follows from Proposition 2.22. □

Thus, by Theorem 6.15, the product $X \times Y$ of locally finite cylindrically normal cellular stratified spaces X and Y is a CW cylindrically normal cellular stratified space. When X and Y are polyhedral, it is easy to verify that $X \times Y$ inherits a polyhedral structure.

Corollary 6.18: *Let X and Y be polyhedral cellular stratified spaces. Suppose both X and Y are locally finite. Then the product $X \times Y$ is a polyhedral cellular stratified space.*

Thus we may freely take finite products of Euclidean polyhedral cellular stratified spaces.

Corollary 6.19: *For any Euclidean polyhedral cellular stratified spaces X and Y, the product $X \times Y$ is polyhedral.*

Proof: By Lemma 2.24, X and Y are locally finite. □

In Proposition 6.15 and Definition 6.13, we made an implicit choice of a homeomorphism

$$D^{m+n} \cong D^m \times D^n.$$

This procedure can be avoided by using cubes as domains of cell structures, in which case it is a reasonable idea to require the structure maps to be compatible with cubical structures. Thus we introduce the following variant of polyhedral cellular stratified spaces:

Definition 6.20: Let X be a CW cellular stratified space. We consider $I^n = (\Delta^1)^n$ as a stratified space (regular cell complex) under the product stratification of the simplicial stratification on Δ^1. A *cubical structure* on X consists of

- a cylindrical structure $(\{b_{\mu,\lambda} : P_{\mu,\lambda} \times D_\mu \to \partial D_\lambda\}, \{c_{\lambda_0,\lambda_1,\lambda_2} : P_{\lambda_1,\lambda_2} \times P_{\lambda_0,\lambda_1} \to P_{\lambda_0,\lambda_2}\})$,
- a stratified subspace Q_λ of $I^{\dim e_\lambda}$ for each $\lambda \in P(X)$ under a suitable regular cellular subdivision of $I^{\dim e_\lambda}$,
- a homeomorphism $\alpha_\lambda : Q_\lambda \to D_\lambda$ for each $\lambda \in P(X)$,

satisfying the following conditions:

(1) Each $P_{\mu,\lambda}$ is a stratified subspace of $I^{\dim e_\lambda - \dim e_\mu}$.
(2) The composition
$$\tilde{b}_{\mu,\lambda} : P_{\mu,\lambda} \times Q_\mu \overset{1_{P_{\mu,\lambda}} \times \alpha_\mu}{\longrightarrow} P_{\mu,\lambda} \times D_\mu \overset{b_{\mu,\lambda}}{\longrightarrow} \partial D_\lambda \overset{\alpha_\lambda^{-1}}{\longrightarrow} Q_\lambda$$
is a strict morphism of stratified spaces.
(3) The map $\tilde{b}_{\mu,\lambda}$ is an affine embedding onto its image when restricted to each face.

The family of maps $\{\alpha_\lambda : Q_\lambda \to D_\lambda\}_{\lambda \in P(X)}$ is also called a *cubical structure.*

A cylindrically normal CW cellular stratified space equipped with a cubical structure is called a *cubically normal cellular stratified space.*

Example 6.21: The minimal cell decomposition of S^n is cubically normal. The radial expansion
$$\alpha_n : I^n \longrightarrow D^n$$
defines a cubical structure. The parameter space between the 0-cell and the n-cell is ∂I^n and is a stratified subspace of I^n.

Example 6.22: Recall that any 1-dimensional cellular stratified space is totally normal, hence is cylindrically normal. They are cubically normal for an obvious reason, if they are CW.

Example 6.23: Recall that $\mathbb{R}\mathrm{P}^2$ can be obtained by gluing the edges of I^2 as follows:

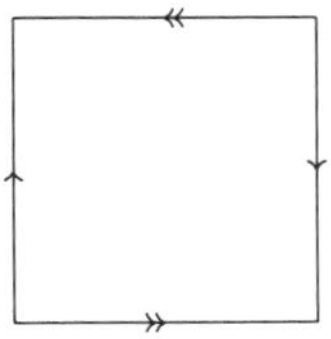

This can be regarded as a description of a cell decomposition of $\mathbb{R}\mathrm{P}^2$, consisting of two 0-cells e_1^0, e_2^0, two 1-cells e_1^1, e_2^1 and a 2-cell e^2. This is obviously cubically normal.

Remark 6.24: When the domains of cells are not globular (Definition 2.25), the situation is more complicated and we do not discuss the product structure here.

6.3. *Subdivisions of cells*

We defined the notion of cellular stratified subspace in Definition 6.4. It often happens that we need to subdivide cells before we take a stratified subspace. For example, the complement of an arrangement $M(\mathcal{A} \otimes \mathbb{R}^\ell)$ in Example 5.21 was defined first by taking a subdivision of the trivial stratification on $\mathbb{R}^n \otimes \mathbb{R}^\ell$ and then by taking the complement of $\mathrm{Lk}(\mathcal{A} \otimes \mathbb{R}^\ell)$.

We have already defined subdivisions of stratified spaces in Definition 2.15. We impose the following "regularity condition" on the definition of subdivisions of cell structures.

Definition 6.25: Let (π, Φ) be a cellular stratification on X. A *cellular subdivision* of (π, Φ) consists of:

- a subdivision of stratified spaces
$$\boldsymbol{s} = (1_X, \underline{s}) : (X, \pi') \longrightarrow (X, \pi);$$
- a regular cellular stratification $(\pi_\lambda, \Phi_\lambda)$ on the domain D_λ of each cell e_λ in (π, Φ) containing $\mathrm{Int}(D_\lambda)$ as a strict stratified subspace,

satisfying the following conditions:

(1) For each $\lambda \in P(X, \pi)$, the cell structure
$$\varphi_\lambda : (D_\lambda, \pi_\lambda) \longrightarrow (X, \pi')$$
of e_λ is a strict morphism of stratified spaces.

(2) The maps
$$P(\varphi_\lambda) : P(\mathrm{Int}(D_\lambda)) \longrightarrow P(X, \pi)$$
induced by the cell structures $\{\varphi_\lambda\}$ give rise to a bijection
$$P(\Phi) = \coprod_{\lambda \in P(X,\pi)} P(\varphi) : \coprod_{\lambda \in P(X,\pi)} P(\mathrm{Int}(D_\lambda), \pi_\lambda) \longrightarrow P(X, \pi').$$

Remark 6.26: The morphism $\boldsymbol{s}$ induces a surjective map

$$P(\boldsymbol{s}) : P(X, \pi') \longrightarrow P(X, \pi),$$

which gives rise to a decomposition

$$P(X, \pi') = \coprod_{\lambda \in P(X,\pi)} P(\boldsymbol{s})^{-1}(\lambda)$$

of sets. Since $\varphi_\lambda : (D_\lambda, \pi_\lambda) \to (X, \pi')$ is a strict morphism, it induces an isomorphism of posets

$$P(\varphi_\lambda) : P(\mathrm{Int}(D_\lambda), \pi_\lambda) \xrightarrow{\cong} P(\boldsymbol{s})^{-1}(\lambda).$$

In other words, the restriction of the bijection $P(\Phi)$ in the second condition, to each $P(\mathrm{Int}(D_\lambda), \pi_\lambda)$, is an embedding of posets, although $P(\Phi)$ itself is rarely an isomorphism of posets.

Remark 6.27: Cellular subdivisions of stellar stratified spaces can be defined in a similar way. We may also define stellar subdivisions of cellular or stellar stratified spaces. We do not pursue such generalizations in these notes.

It is easy to verify that the category of cellular stratified spaces is closed under cellular subdivisions, when all cell structures are hereditarily quotient.

Lemma 6.28: *Let (X, π, Φ) be a cellular stratified space whose cell structures are hereditarily quotient. Then any cellular subdivision of (X, π, Φ) defines a structure of a cellular stratified space on X, under which, for $\lambda \in P(X, \pi)$ and $\lambda' \in P(\mathrm{Int}(D_\lambda), \pi_\lambda)$, the composition*

$$D_{\lambda'} \xrightarrow{s_{\lambda'}} D_\lambda \xrightarrow{\varphi_\lambda} X$$

is the cell structure of the cell $e_{\lambda'}$ in $P(X, \pi')$, where $s_{\lambda'} : D_{\lambda'} \to D_\lambda$ is the cell structure of the cell in $P(D_\lambda, \pi_\lambda)$ indexed by λ'.

Proof: By assumption, each cell structure φ_λ is hereditarily quotient and hence the composition $\varphi_\lambda \circ s_{\lambda'} : D_{\lambda'} \to \overline{e_{\lambda'}}$ is a quotient map. Each new stratum is connected, since $e_{\lambda'}$ is connected. It is also locally closed, since $\varphi_\lambda|_{\mathrm{Int}(D_\lambda)}$ is a homeomorphism onto its image. Other conditions can be verified immediately. □

Remark 6.29: The reader might want to define a subdivision of a cellular stratified space (X, π, Φ) as a morphism

$$\boldsymbol{s} = (1_X, \underline{s}, \{s_{\lambda'}\}_{\lambda' \in P(X,\pi')}) : (X, \pi', \Phi') \longrightarrow (X, \pi, \Phi)$$

of cellular stratified spaces satisfying the condition that, for each cell e_λ in (π, Φ), the stratification of the interior of the domain D_λ is indexed by $\underline{s}^{-1}(\lambda)$

$$\operatorname{Int}(D_\lambda) = \bigcup_{\lambda' \in \underline{s}^{-1}(\lambda)} s_{\lambda'}(\operatorname{Int}(D_{\lambda'})).$$

However, we also need to specify the behavior of each cell structure on the boundary ∂D_λ.

Example 6.30: Let $\mathcal{A} = \{H_1, \dots, H_k\}$ be a hyperplane arrangement in $\mathbb{R}^n$ defined by affine 1-forms $L = \{\ell_1, \dots, \ell_k\}$. The stratification $\pi_{\mathcal{A}\otimes\mathbb{R}^\ell}$ on $\mathbb{R}^n \otimes \mathbb{R}^\ell$ in Example 5.21 is one of the coarsest cellular stratifications containing $\operatorname{Lk}(\mathcal{A} \otimes \mathbb{R}^\ell) = \bigcup_{i=1}^k H_k \otimes \mathbb{R}^\ell$ as a stratified subspace.

This efficiency is achieved by sacrificing symmetry. The stratification $\pi_{\mathcal{A}\otimes\mathbb{R}^\ell}$ is not compatible with the action of the symmetric group Σ_ℓ on $\mathbb{R}^\ell$.

As is stated by Björner and Ziegler in [7], we may subdivide $\pi_{\mathcal{A}\otimes\mathbb{R}^\ell}$ by using the product sign vector S_1^ℓ. Define

$$\pi^s_{\mathcal{A}\otimes\mathbb{R}^\ell} : \mathbb{R}^n \otimes \mathbb{R}^\ell \longrightarrow \operatorname{Map}(L, S_1^\ell)$$

by

$$\pi^s_{\mathcal{A}\otimes\mathbb{R}^\ell}(x_1, \dots, x_\ell)(\ell_i) = (\operatorname{sign}(\ell_i(x_1)), \dots, \operatorname{sign}(\ell_i(x_\ell))).$$

Define a map $c : S_1^\ell \to S_\ell$ of posets by

$$c(\varepsilon_1, \dots, \varepsilon_\ell) = \varepsilon_i e_i,$$

where $i = \max\{j \mid \varepsilon_j \neq 0\}$. This is surjective and the diagram

$$\begin{array}{ccc}
\mathbb{R}^n \otimes \mathbb{R}^\ell & = \!= \!= \!= & \mathbb{R}^n \otimes \mathbb{R}^\ell \\
\Big\downarrow {\scriptstyle \pi^s_{\mathcal{A}\otimes\mathbb{R}^\ell}} & & \Big\downarrow {\scriptstyle \pi_{\mathcal{A}\otimes\mathbb{R}^\ell}} \\
\operatorname{Map}(L, S_1^\ell) & \xrightarrow[\operatorname{Map}(L,c)]{} & \operatorname{Map}(L, S_\ell)
\end{array}$$

is commutative. Thus $\pi^s_{\mathcal{A}\otimes\mathbb{R}^\ell}$ is a subdivision of the stratification $\pi_{\mathcal{A}\otimes\mathbb{R}^\ell}$. The regularity of the cellular stratification $\pi_{\mathcal{A}\otimes\mathbb{R}^\ell}$ implies that this is a cellular subdivision. Note that $(\mathbb{R}^n \otimes \mathbb{R}^\ell, \pi^s_{\mathcal{A}\otimes\mathbb{R}^\ell})$ is a Σ_ℓ-cellular stratified space.

Example 6.31: Let $\pi_{\min}$ be the minimal cell decomposition of S^2. By dividing the 2-cell by the equator, we obtain a subdivision π of $\pi_{\min}$ as a stratified space. By dividing the domain of the cell structure of the 2-cell in

$\pi_{\min}$, this subdivision can be made into a subdivision of cellular stratified spaces, as is depicted in Figure 30.

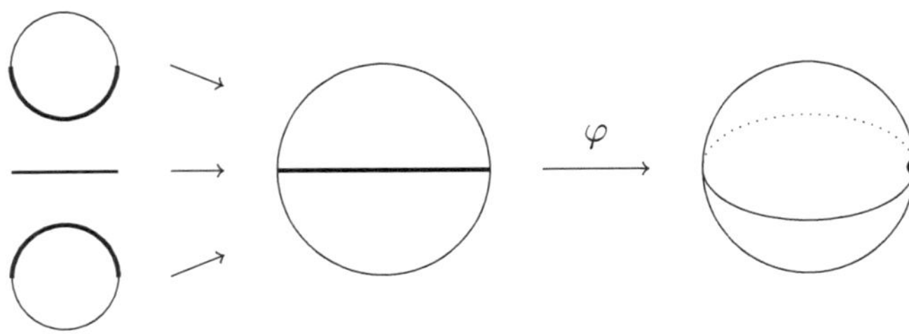

Fig. 30. A subdivision of the minimal cell decomposition of S^2.

There is another way to make the stratification π into a cellular stratified space. Choose a small disk D in D^2 touching ∂D^2 at a point p. The small

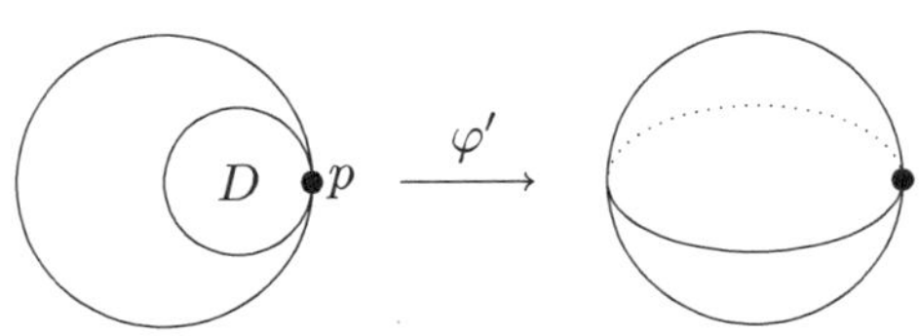

Fig. 31. Not a subdivision of $\pi_{\min}$.

disk D is mapped homeomorphically onto the lower hemisphere via φ'. There is a map

$$\psi : D^2 \longrightarrow D^2 \setminus \mathrm{Int}(D)$$

such that the composition $\varphi \circ \psi$ is a cell structure for the upper hemisphere. For example, such a map can be defined by identifying the two points in Figure 32 to p. The morphism of stratified spaces $\pi \to \pi_{\min}$ also becomes a morphism of cellular stratified spaces with this cellular stratification on π, but it is not a subdivision of cellular stratified spaces, since the above stratification on the domain D^2 is not a regular cellular stratification.

We often define a cellular subdivision of a cellular stratified space (X, π, Φ) by defining subdivisions of domains of cell structures. The problem is of course the compatibility.

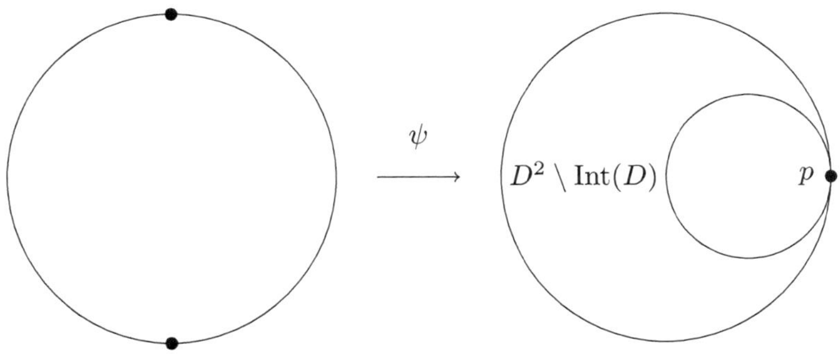

Fig. 32. A definition of ψ.

Lemma 6.32: *Let (X, π, Φ) be a cellular stratified space whose cell structures are hereditarily quotient. Suppose that a regular cellular stratification π_λ on each domain D_λ is given. Define a stratification π' on X by*

$$X = \bigcup_{\lambda \in P(X,\pi)} \bigcup_{\lambda' \in P(\mathrm{Int}(D_\lambda))} \varphi_\lambda(e_{\lambda'}).$$

Suppose further that the following conditions are satisfied:

(1) *For each $\lambda \in P(X)$, $\mathrm{Int}(D_\lambda)$ is a strict stratified subspace of D_λ.*
(2) *For $\lambda' \in P(D_\lambda)$ and , $\mu' \in P(D_\mu)$,*

$$\varphi_\lambda \circ s_{\lambda'}\left(\mathrm{Int}(D_{\lambda'})\right) \cap \varphi_\mu \circ s_{\mu'}\left(\mathrm{Int}(D_{\mu'})\right) \neq \emptyset$$

implies

$$\varphi_\lambda \circ s_{\lambda'}\left(\mathrm{Int}(D_{\lambda'})\right) = \varphi_\mu \circ s_{\mu'}\left(\mathrm{Int}(D_{\mu'})\right).$$

For $\lambda' \in P(\mathrm{Int}(D_\lambda))$, define a map $\psi_{\lambda'} : D_{\lambda'} \to X$ by the composition

$$D_{\lambda'} \xrightarrow{s_{\lambda'}} D_\lambda \xrightarrow{\varphi_\lambda} X,$$

where $s_{\lambda'}$ is the cell structure of $e_{\lambda'}$. Then these structures define a cellular stratification on X with $\{\psi_{\lambda'}\}$ cell structure maps.

Proof: It suffices to verify that φ_λ is a strict morphism of cellular stratified spaces from (D_λ, π_λ) to (X, π').

By assumption,

$$\varphi_\lambda|_{\mathrm{Int}(D_\lambda)} : \mathrm{Int}(D_\lambda) \longrightarrow e_\lambda$$

is an isomorphism of stratified spaces. We need to show that φ_λ is a strict morphism of stratified spaces on ∂D_λ. For a cell $e_{\lambda'} \subset \partial D_\lambda$, $\varphi(e_\lambda)$ is contained in $X \setminus e_\lambda$ and there exist $\mu \in P(X)$ and $\mu' \in P(\mathrm{Int}(D_\mu))$ such that

$$\varphi_\lambda \circ s_{\lambda'}(\mathrm{Int}(D_{\lambda'})) \cap \varphi_\mu \circ s_{\mu'}(\mathrm{Int}(D_{\mu'})) \neq \emptyset.$$

By assumption,

$$\varphi_\lambda \circ s_{\lambda'}(\mathrm{Int}(D_{\lambda'})) = \varphi_\mu \circ s_{\mu'}(\mathrm{Int}(D_{\mu'}))$$

and thus $\varphi_\lambda : (D_\lambda, \pi_\lambda) \to (X, \pi')$ is a strict morphism of stratified spaces. □

Definition 6.33: We say that the above cellular stratification is *induced* by the family $\{\pi_\lambda\}_{\lambda \in P(X)}$ of cellular stratifications on the domains of cells.

Example 6.34: Consider the cellular stratification $A = e^1 \cup e^2$ of an open annulus $A = S^1 \times (0,1)$ in Figure 33.

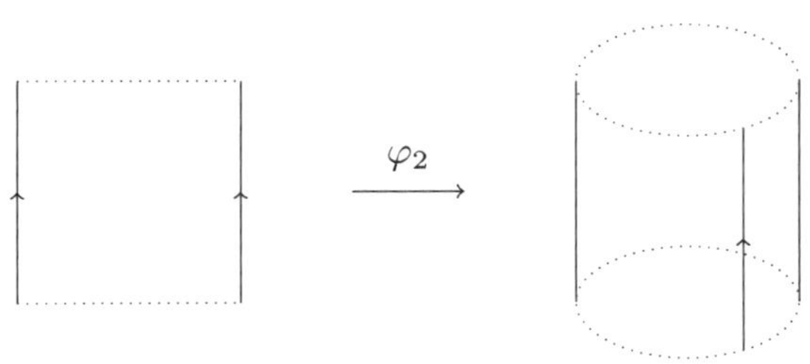

Fig. 33. A cellular stratification on an open annulus.

We use $(0,1)$ and $(0,1) \times [0,1]$ as the domains of the cell structures for the 1-cell and the 2-cell, respectively. The subdivisions of $(0,1)$ in the middle and of $(0,1) \times [0,1]$ by horizontal cut in the middle induce a subdivision of this stratification.

The following "tilted subdivision" of $(0,1) \times [0,1]$, however, does not induce a subdivision of the annulus, since it does not satisfy the second condition of cellular subdivision.

By subdividing the boundary further, we obtain an induced subdivision, as is shown in the right-hand figure.

For totally normal cellular stratified spaces, we require the following conditions.

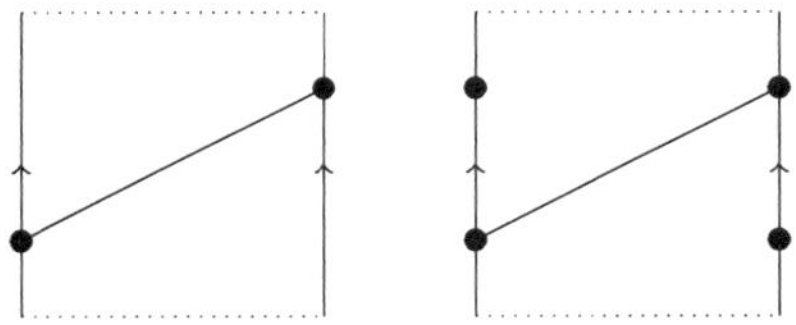

Fig. 34. A tilted subdivision of an open annulus.

Definition 6.35: Let (X, π, Φ) be a totally normal cellular stratified space. We say a cellular subdivision (X, π', Φ') is a *subdivision* of (X, π, Φ) as a totally normal cellular stratified space if the following conditions are satisfied:

(1) For each $\lambda' \in P(X, \pi')$, there exists a structure of regular cell complex on $\partial D^{\dim e_{\lambda'}}$ containing $\partial D_{\lambda'}$ as a strict stratified subspace.
(2) For each $b \in C(X, \pi)(e_\mu, e_\lambda)$, the associated map on domains of cells is a strict morphism $b : (D_\mu, \pi_\mu) \to (D_\lambda, \pi_\lambda)$ of cellular stratified spaces.

Proposition 6.36: *A cellular subdivision of a totally normal cellular stratified space satisfying the conditions in Definition 6.35 is totally normal.*

Remark 6.37: An analogous statement was stated as Proposition 2.45 in [1] without the first condition of Definition 6.35. It was pointed out by Priyavrat Deshpande that this condition was missing. He also pointed out typos and errors in the proof, that are corrected in the following proof.

Proof: Let (X, π, Φ) be a totally normal cellular stratified space and $s : P(X, \pi') \to P(X, \pi)$ be a cellular subdivision of (X, π) satisfying the conditions in Definition 6.35. Its cell structure is denoted by Φ'.

We need to verify that, for each cell $e_{\lambda'}$ in (X, π', Φ') and a cell e' in $\partial D_{\lambda'}$, there exists a cell $e_{\mu'}$ in (X, π', Φ') and a map $b' : D_{\mu'} \to D_{\lambda'}$ making the diagram

$$\begin{array}{ccccc} D_{\lambda'} & \xrightarrow{\varphi'_{\lambda'}} & \overline{e_{\lambda'}} & \hookrightarrow & X \\ \uparrow{\scriptstyle b'} & & & \nearrow & \\ D_{\mu'} & \xrightarrow[\varphi'_{\mu'}]{} & \overline{e_{\mu'}} & & \end{array}$$

commutative and satisfying $b'(\mathrm{Int}(D_{\mu'})) = e'$.

Let $\lambda = s(\lambda')$. By the definition of cellular subdivision, D_λ has a structure of regular cellular stratified space $(D_\lambda, \pi_\lambda, \Phi_\lambda)$ under which e' is a cell.

Under the identification

$$P(\mathrm{Int}(D_\lambda), \pi_\lambda) \cong P(\boldsymbol{s})^{-1}(\lambda),$$

there exists a cell in $(\mathrm{Int}(D_\lambda), \pi_\lambda)$ corresponding to $e_{\lambda'}$. Let $s_{\lambda'} : D_{\lambda'} \to D_\lambda$ be its characteristic map. Then the characteristic map for $e_{\lambda'}$ in (X, π', Φ') is given by the composition $\varphi_\lambda \circ s_{\lambda'}$.

On the other hand, there exists a cell e in ∂D_λ, before subdivision, containing $s_{\lambda'}(e')$ in its interior. By the total normality of (X, π, Φ), there exists a cell e_μ in (X, π, Φ) and a morphism $b : e_\mu \to e_\lambda$ in $C(X, \pi)$ with $b(\mathrm{Int}(D_\mu)) = e$ and $\varphi_\lambda \circ b = \varphi_\mu$. Since b is a strict morphism of stratified spaces and $s_{\lambda'}(e')$ is contained in the interior of e, there exists a unique cell e'' in $(D_\mu, \pi_\mu, \Phi_\mu)$ such that $e' = b(e'')$. φ_μ is also a strict morphism of stratified spaces and thus $\varphi_\mu(e'')$ is a cell in (X, π', Φ'). Let us denote this cell by $e_{\mu'}$. Then the characteristic map for $e_{\mu'}$ is given by the composition $\varphi_\mu \circ s_{\mu'}$, where $s_{\mu'} : D_{\mu'} \to D_\mu$ is the characteristic map for e''.

When $\mu = \lambda$, both e and e' are cells in $P(D_\lambda, \pi_\lambda)$ and b is the identity map. Hence the regularity of $(D_\lambda, \pi_\lambda, \Phi_\lambda)$ implies the existence of a unique map $b' : D_{\mu'} \to D_{\lambda'}$ satisfying the required conditions.

Suppose $\mu < \lambda$. The relation among these cells is depicted in Figure 35. The image of the composition $b \circ s_{\mu'}$ is contained in the image of $s_{\lambda'}$, i.e. $\overline{e}$. The regularity of $(D_\lambda, \pi_\lambda, \Phi_\lambda)$ and the fact that b is an embedding imply that the above diagram can be completed by a map b'. □

Remark 6.38: A subdivision $(\boldsymbol{s}, \{(\pi_\lambda, \Phi_\lambda)\})$ of a totally normal cellular stratified space (X, π, Φ) gives rise to a functor

$$P(\boldsymbol{s}) : C(X, \pi) \longrightarrow \mathbf{Posets}$$

in a canonical way. On objects it is defined by $P(\boldsymbol{s})(\lambda) = P(D_\lambda, \pi_\lambda)$. The second condition in Definition 6.35 guarantees that this extends to a functor.

In general, for a functor $F : C \to \mathbf{Cats}$ from a small category to the category **Cats** of small categories, there is a construction of a single category $\mathrm{Gr}(F)$, called the Grothendieck construction of F. A definition can be found in Thomason's paper [34], for example.

It is easy to verify that, in the case of a subdivision of a totally normal cellular stratified space, we have an isomorphism categories

$$\mathrm{Gr}(P(\boldsymbol{s})) \cong C(X, \pi').$$

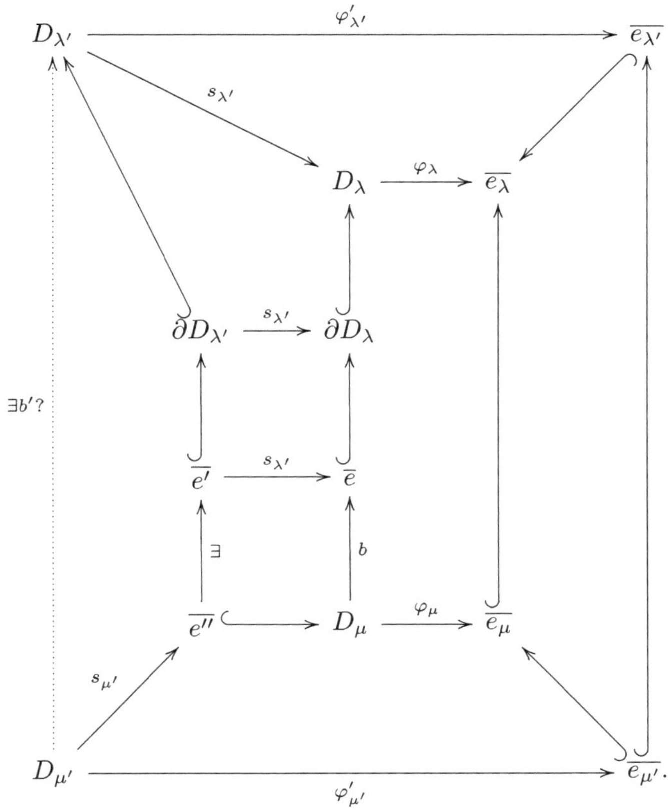

Fig. 35. Cells in the proof of Proposition 6.36

The next step would be to study subidivisions of cylindrical and polyhedral structures but it is not easy to describe appropriate conditions on parameter spaces.

7. Duality

In [6], Salvetti first constructed a simplicial complex $\mathrm{Sd}(M(\mathcal{A}\otimes\mathbb{C})) = BC(M(\mathcal{A}\otimes\mathbb{C}))$ modelling the homotopy type of the complement of the complexification of a real hyperplane arrangement $\mathcal{A}$ and then defined a structure of regular cell complex by gluing simplices. This process reduces the number of cells and allows us to relate the combinatorics of the arrangement $\mathcal{A}$ and the topology of Salvetti's model $\mathrm{Sd}(M(\mathcal{A}\otimes\mathbb{C}))$. The process is closely related to the classical concept of duality in PL topology.

In this section, we introduce an analogous process for polyhedral stellar stratified spaces. With stellar stratifications, we are able to extend both Salvetti's construction and the duality in PL topology.

7.1. *A canonical cellular stratification on the barycentric subdivision*

Let us first study stratifications on the barycentric subdivision $\mathrm{Sd}(X)$ of a cylindrically normal stellar stratified space X.

In the case of totally normal stellar stratified spaces, Lemma 5.5 and Lemma C.16 imply the following:

Proposition 7.1: *For a totally normal stellar stratified space X, $\mathrm{Sd}(X)$ has a structure of regular cell complex.*

Let us extend this structure to a cellular stratification on $\mathrm{Sd}(X)$ for a cylindrically normal stellar stratified space X.

Given a cylindrically normal stellar stratified space X, we have an acyclic topological category $C(X)$. By forgetting the topology, Lemma C.16 and Lemma 3.16 give us a map

$$\pi_{\mathrm{Sd}(X)} : \mathrm{Sd}(X) = BC(X) = \|\overline{N}(C(X))\| \stackrel{\pi_{\overline{N}(C(X))}}{\longrightarrow} \coprod_k \overline{N}_k(C(X)).$$

As we will see in Lemma 7.33, the set $\coprod_k \overline{N}_k(C(X))$ can be identified with the set of objects of the barycentric subdivision of the acyclic category $C(X)$ and has a structure of poset. The partial order is defined as follows:

Lemma 7.2: *For $\boldsymbol{b} \in \overline{N}_k(C(X))$ and $\boldsymbol{b}' \in \overline{N}_\ell(C(X))$, regarded as functors $\boldsymbol{b} : [k] \to C(X)$ and $\boldsymbol{b}' : [\ell] \to C(X)$, define $\boldsymbol{b} \leq \boldsymbol{b}'$ if and only if there exists a morphism $\varphi : [k] \to [\ell]$ in Δ_{inj} such that $\boldsymbol{b}' \circ \varphi = \boldsymbol{b}$. Then the relation $\leq$ is a partial order.*

Proof: See Lemma 7.33. □

Let us verify that the map $\pi_{\mathrm{Sd}(X)}$ defines a cellular stratification on $\mathrm{Sd}(X)$.

Proposition 7.3: *For a cylindrically normal stellar stratified space X, $\pi_{\mathrm{Sd}(X)}$ defines a cellular stratification.*

Proof: Let X be a cylindrically normal stellar stratified space. For each nondegenerate k-chain $\boldsymbol{b} \in \overline{N}_k(C(X))$, $\pi_{\mathrm{Sd}(X)}^{-1}(\boldsymbol{b})$ is homeomorphic to

$\mathrm{Int}(\Delta^k)$. The closure of $\pi_{\mathrm{Sd}(X)}^{-1}(\boldsymbol{b})$ in $\mathrm{Sd}(X)$ is an identification space of Δ^n. Since the identification is defined only on the boundary, $\pi_{\mathrm{Sd}(X)}^{-1}(\boldsymbol{b})$ is open in its closure and is locally closed. For nondegenerate chains $\boldsymbol{b}$ and $\boldsymbol{b}'$, $\pi_{\mathrm{Sd}(X)}^{-1}(\boldsymbol{b}) \subset \overline{\pi_{\mathrm{Sd}(X)}^{-1}(\boldsymbol{b}')}$ if and only if $\pi_{\mathrm{Sd}(X)}^{-1}(\boldsymbol{b})$ is included in $\pi_{\mathrm{Sd}(X)}^{-1}(\boldsymbol{b}')$ as a face. In other words, this holds if and only if there exists a sequence of face operators mapping $\boldsymbol{b}'$ to $\boldsymbol{b}$, which is equivalent to saying $\boldsymbol{b} \leq \boldsymbol{b}'$. Thus $\pi_{\mathrm{Sd}(X)}$ is a stratification in the sense of Definition 2.2.

Under the standard homeomorphism $\Delta^k \cong D^k$, we obtain a continuous map

$$\varphi_{\boldsymbol{b}} : D^k \cong \{\boldsymbol{b}\} \times \Delta^k \hookrightarrow \coprod_k \overline{N}_k(C(X)) \times \Delta^k \longrightarrow \|\overline{N}(C(X))\| \cong \mathrm{Sd}(X).$$

By the compactness of D^k, $\varphi_{\boldsymbol{b}}$ is a quotient map onto its image, which is the closure of $\pi_{\mathrm{Sd}(X)}^{-1}(\boldsymbol{b})$. The boundary of $\pi_{\mathrm{Sd}(X)}^{-1}(\boldsymbol{b})$ consists of points in $\mathrm{Sd}(X)$ of the form $[\boldsymbol{b}, t]$ with $t \in \partial\Delta^k$. Under the defining relation, such a point is equivalent to a point in $\overline{N}_\ell(X) \times \Delta^\ell$ for $\ell < k$. Thus we obtain a cellular stratification. □

Note that this cellular stratification is rarely a CW stratification, when parameter spaces have non-discrete topology.

Example 7.4: Consider the minimal cell decomposition $\pi_n : S^n = e^0 \cup e^n$ of the n-sphere. We have

$$\begin{aligned} \overline{N}_0(C(S^n)) &= C(S^n)_0 = \{e^0, e^n\} \\ \overline{N}_1(C(S^n)) &= C(S^n)(e^0, e^n) = S^{n-1} \\ \coprod_{k\geq 2} \overline{N}_k(C(S^n)) &= \emptyset. \end{aligned}$$

Thus $\mathrm{Sd}(S^n; \pi_n)$ has only 0-cells and 1-cells. There are two 0-cells $v(e_0)$ and $v(e^n)$ corresponding to e^0 and e^n. There are infinitely many 1-cells parametrized by the equator S^{n-1}. For $x \in S^{n-1}$, the 1-cell e^1_x corresponding to x is given by the great half circle from the south pole to the north pole through x under the identification $\mathrm{Sd}(S^n; \pi_n) \cong S^n$.

7.2. *Stars*

As Example 7.4 shows, the cellular stratification on $\mathrm{Sd}(X)$ defined in the previous section is not very useful. However, Example 7.4 also suggests that by gluing cells in $\mathrm{Sd}(X)$ together, we may construct a good cellular

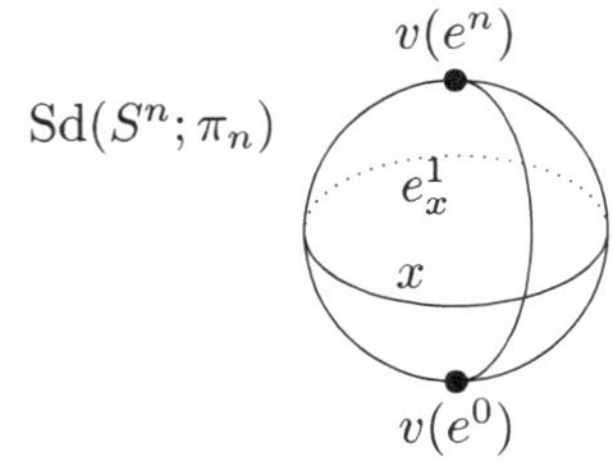

Fig. 36. The canonical cellular stratification on $\mathrm{Sd}(S^n, \pi_n)$.

stratification. In the classical PL topology, such a construction is called a star. See Definition B.18 for the classical definition.

Definition 7.5: Let X be a stellar stratified space. For $x \in X$, define

$$\mathrm{St}(x;X) = \bigcup_{x\in\overline{e_\lambda}} e_\lambda.$$

For a subset $A \subset X$, define

$$\mathrm{St}(A;X) = \bigcup_{x\in A} \mathrm{St}(x;X).$$

The stratified subspace $\mathrm{St}(A;X)$ is called the *star* of A in X.

Example 7.6: In the cellular stratification on $\mathrm{Sd}(S^n;\pi_n)$ in Example 7.4, the star $\mathrm{St}(v(e^n);\mathrm{Sd}(S^n;\pi_n))$ coincides with $S^n \setminus \{v(e^0)\}$ and we recover the original cellular stratification on S^n as

$$\mathrm{Sd}(S^n) = v(e^0) \cup \mathrm{St}(v(e^n);\mathrm{Sd}(S^n;\pi_n)).$$

Note that we also have another cellular stratification

$$\mathrm{Sd}(S^n) = v(e^n) \cup \mathrm{St}(v(e^0);\mathrm{Sd}(S^n;\pi_n)),$$

which can be regarded as a dual to the original cellular stratification.

We introduce analogous constructions for acyclic categories. Recall that, given an object x in a small category C, the comma category $x \downarrow C$ is the category whose objects are morphisms u in C with $s(u) = x$. Morphisms are morphisms in C making the obvious triangle commutative.

Definition 7.7: Let C be an acyclic topological category and x an object of C. The nondegenerate nerve $\overline{N}(x \downarrow C)$ of the comma category $x \downarrow C$ is denoted by $\mathrm{St}_{\geq x}(C)$ and is called the *upper star* of x in C.

The full subcategory of $x \downarrow C$ consisting of $(x \downarrow C)_0 \setminus \{1_x\}$ is denoted by $C_{>x}$. The nondegenerate nerve $\overline{N}(C_{>x})$ is denoted by $\mathrm{Lk}_{>x}(C)$ and is called the *upper link* of x in C.

The functor induced by the target map in C is denoted by

$$t_x : C_{>x} \subset x \downarrow C \longrightarrow C.$$

The induced map of Δ-spaces is also denoted by

$$t_x : \mathrm{Lk}_{>x}(C) \subset \mathrm{St}_{\geq x}(C) \longrightarrow \overline{N}(C).$$

Dually the nondegenerate nerves of the comma category $C \downarrow x$ and of its full subcategory $C_{<x}$ consisting of $(C \downarrow x)_0 \setminus \{1_x\}$ are denoted by $\mathrm{St}_{\leq x}(C)$ and $\mathrm{Lk}_{<x}(C)$ and called the *lower star* and *lower link* of x in C, respectively. We also have a functor and a map

$$\begin{aligned} s_x &: C_{<x} \longrightarrow C \\ s_x &: \mathrm{Lk}_{<x}(C) \longrightarrow \overline{N}(C) \end{aligned}$$

induced by the source map.

Remark 7.8: The notation $C_{>x}$ and its definition is borrowed from Kozlov's book [28]. Note that $\mathrm{Lk}_{>x}(C)$ is different from the usual link of x in $\overline{N}(C)$ in general.

Lemma 7.9: *For an acyclic topological category C and an object $x \in C_0$, we have the following identification*

$$\mathrm{Lk}_{>x}(C)_k \cong \begin{cases} \coprod_{x \neq y} C(x,y), & k = 0 \\ \left\{ \boldsymbol{u} \in \overline{N}_{k+1}(C) \,\middle|\, s(u) = x \right\}, & k > 0, \end{cases}$$

under which the face operators

$$d_i^{\mathrm{Lk}} : \mathrm{Lk}_{>x}(C)_k \longrightarrow \mathrm{Lk}_{>x}(C)_{k-1}$$

are identified as $d_i^{\mathrm{Lk}} = d_{i+1}$, where d_{i+1} is the $(i+1)$-st face operator in $\overline{N}(C)$.

Example 7.10: Consider the poset $[2] = \{0 < 1 < 2\}$ regarded as a category $0 \to 1 \to 2$. The category $[2]_{>1}$ has a unique object $1 \to 2$ and no nontrivial morphism. Thus $\mathrm{Lk}_{>1}([2])$ is a single point. Under the map $t_1 : \mathrm{Lk}_{>1}([2]) \to \overline{N}([2]) \cong \Delta^2$, $\mathrm{Lk}_{>1}([2])$ can be identified with the vertex 2 in Δ^2. On the other hand, the usual link of 1 in Δ^2 is the 1-simplex spanned by vertices 0 and 2.

Example 7.11: Consider the face category $C(S^1;\pi_1)$ of the minimal cell decomposition of S^1. $C(S^1;\pi_1)_{>e^0}$ consists of two objects $(C(S^1;\pi_1)_{>e^0})_0 = C(S^1;\pi_1)(e^0,e^1) = \{b_-,b_+\}$ and no nontrivial morphisms. Thus $\mathrm{St}_{\geq e^0}(C(S^1;\pi_1))$ is the cell complex $[-1,1] = \{-1\}\cup(-1,0)\cup\{0\}\cup(0,1)\cup\{1\}$ and $\mathrm{Lk}_{>e^0}(C(S^1;\pi_1))$ is S^0. The map t_{e^0} maps the boundary $\partial[-1,1] = S^0$ to $v(e^2)$ in $\mathrm{Sd}(S^1)$ and defines a 1-cell structure.

Note that, the comma category $x \downarrow C$ has an initial object 1_x.

Lemma 7.12: *For an acyclic topological category C and an object $x \in C_0$, we have a homeomorphism*

$$\|\mathrm{St}_{\geq x}(C)\| \cong \{1_x\} * \|\mathrm{Lk}_{>x}(C)\|.$$

Proof: Define a map $h_x : \|\mathrm{St}_{\geq x}(C)\| \to \{1_x\} * \|\mathrm{Lk}_{>x}(C)\|$ as follows. For $[(\boldsymbol{u},\boldsymbol{t})] \in \|\mathrm{St}_{\geq x}(C)\| = \|\overline{N}(x\downarrow C)\|$, choose a representative $(\boldsymbol{u},\boldsymbol{t}) \in \overline{N}_k(x\downarrow C)\times\Delta^k$. Here we regard $\boldsymbol{u}$ as a sequence of composable $k+1$ morphisms in C starting from x;

$$\boldsymbol{u} : x \xrightarrow{u_0} x_0 \xrightarrow{u_1} x_1 \xrightarrow{u_2} \cdots \xrightarrow{u_k} x_k$$

with u_1, ..., u_k non-identity morphisms. When u_0 is not the identity morphism, $\boldsymbol{u}$ belongs to $\mathrm{Lk}_{>x}(C)_k$ and $[(\boldsymbol{u},\boldsymbol{t})]$ can be regarded as an element of $\|\mathrm{Lk}_{>x}(C)\|$. Define

$$h_x([\boldsymbol{u},\boldsymbol{t}]) = 01_x + 1[(\boldsymbol{u},\boldsymbol{t})].$$

When $u_0 = 1_x$, write $\boldsymbol{t} = t_0 0 + (1-t_0)\boldsymbol{t}'$ under the identification $\Delta^k \cong \{0\} * \Delta^{k-1}$ and define

$$h_x([\boldsymbol{u},\boldsymbol{t}]) = t_0 1_x + (1-t_0)[(\boldsymbol{u}',\boldsymbol{t}')],$$

where

$$\boldsymbol{u}' : x \xrightarrow{u_1} x_1 \xrightarrow{u_2} \cdots \xrightarrow{u_k} x_k$$

is the $(k-1)$-chain obtained from $\boldsymbol{u}$ by removing u_0. Since $\boldsymbol{u}$ is a nondegenerate chain, u_1 is not the identity morphism and $\boldsymbol{u}' \in \mathrm{Lk}_{>x}(C)_{k-1}$.

Since the set of objects C_0 has the discrete topology, the decomposition

$$\begin{aligned}
&\overline{N}_k(x \downarrow C) \\
&= \left(\coprod_{x<x_1<\cdots<x_k} C(x,x) \times C(x,x_1) \times \cdots \times C(x_{k-1},x_k) \right) \\
&\qquad \amalg \left(\coprod_{x<x_0<\cdots<x_k} C(x,x_0) \times \cdots \times C(x_{k-1},x_k) \right) \\
&= (\{1_x\} \times \mathrm{Lk}_{>x}(x \downarrow C)) \amalg \mathrm{Lk}_{>x}(x \downarrow C)
\end{aligned}$$

is a decomposition of topological spaces. The map h_x is continuous on each component of $\overline{N}_k(x \downarrow C) \times \Delta^k$ and defines a continuous map

$$h_x : \|\mathrm{St}_{\geq x}(C)\| \longrightarrow \{1_x\} * \|\mathrm{Lk}_{>x}(C)\|.$$

It is easy to define an inverse to h_x, and thus h_x is a homeomorphism. □

7.3. *Canonical stellar stratifications on the barycentric subdivision*

Let us now go back to the discussion of stellar stratifications on $\mathrm{Sd}(X)$.

Definition 7.13: Let X be a cylindrically normal stellar stratified space. Define a map

$$\pi_{X^{\mathrm{op}}} : \mathrm{Sd}(X) \longrightarrow P(X)$$

by the composition

$$\mathrm{Sd}(X) \stackrel{\pi_{\mathrm{Sd}(X)}}{\longrightarrow} \coprod_k \overline{N}_k(C(X)) \stackrel{s}{\longrightarrow} C(X)_0 = P(X),$$

where s is the map induced by the source map in the category $C(X)$ and is also called the *source map*.

By definition (Definition 2.15), $\pi_{\mathrm{Sd}(X)}$ is a subdivision of $\pi_{X^{\mathrm{op}}}$ if $\pi_{X^{\mathrm{op}}}$ defines a stratification on $\mathrm{Sd}(X)$. Each inverse image $\pi_{X^{\mathrm{op}}}^{-1}(\lambda)$ can be described by using the upper stars.

Definition 7.14: For a cell e_λ in a cylindrically normal stellar stratified space X, define

$$D_\lambda^{\mathrm{op}} = \|\mathrm{St}_{\geq e_\lambda}(C(X))\|.$$

We also denote

$$D_\lambda^{\mathrm{op},\circ} = \|\mathrm{St}_{\geq e_\lambda}(C(X))\| \setminus \|\mathrm{Lk}_{>e_\lambda}(C(X))\|,$$

where $\|\mathrm{Lk}_{>e_\lambda}(C(X))\|$ is regarded as the bottom subspace of the cone under the identification

$$\|\mathrm{St}_{\geq e_\lambda}(C(X))\| \cong \{1_{e_\lambda}\} * \|\mathrm{Lk}_{>e_\lambda}(C(X))\|$$

in Lemma 7.12.

Lemma 7.15: *Let X be a cylindrically normal stellar stratified space. In the stratification $\pi_{X^{\mathrm{op}}}$, each stratum is given by the image of $D_\lambda^{\mathrm{op},\circ}$ under the map $t_\lambda = t_{e_\lambda} : \|\mathrm{St}_{\geq e_\lambda}(C(X))\| \to \mathrm{Sd}(X)$ defined in Definition 7.7. Namely*

$$\pi_{X^{\mathrm{op}}}^{-1}(\lambda) = t_\lambda\left(D_\lambda^{\mathrm{op},\circ}\right).$$

Hence the closure of each stratum is given by

$$\overline{\pi_{X^{\mathrm{op}}}^{-1}(\lambda)} = t_\lambda\left(D_\lambda^{\mathrm{op}}\right).$$

Proof: Elements in $D_\lambda^{\mathrm{op},\circ}$ are those which are represented by $(\boldsymbol{u}, t_0 0 + (1-t_0)\boldsymbol{t}')$, where $t_0 < 1$ and $\boldsymbol{u}$ begins with the identity morphism 1_{e_λ}. Therefore $s(t_\lambda(\boldsymbol{u})) = e_\lambda$ and $t_\lambda\left(D_\lambda^{\mathrm{op},\circ}\right) \subset \pi_{X^{\mathrm{op}}}^{-1}(\lambda)$. Conversely, by choosing a representative $(\boldsymbol{u}, \boldsymbol{t})$ of $[(\boldsymbol{u}, \boldsymbol{t})] \in \pi_{X^{\mathrm{op}}}^{-1}(\lambda)$ with $\boldsymbol{t} \in \mathrm{Int}(\Delta^k)$, we see $\pi_{X^{\mathrm{op}}}^{-1}(\lambda) \subset t_\lambda\left(D_\lambda^{\mathrm{op},\circ}\right)$.

Let $p : \coprod_k \overline{N}_k(C(X)) \times \Delta^k \to \mathrm{Sd}(X)$ be the projection. Then the topology on $\mathrm{Sd}(X)$ is the weak topology defined by the covering

$$\left\{p(C(X)(e_{\lambda_{k-1}}, e_{\lambda_k}) \times \cdots \times C(X)(e_{\lambda_0}, e_{\lambda_1}) \times \Delta^k)\right\}.$$

Thus the closure of

$$t_\lambda\left(D_\lambda^{\mathrm{op},\circ}\right) = p\left(\coprod_k \coprod_{\lambda<\lambda_0<\cdots<\lambda_k} C(X)(e_{\lambda_{k-1}}, e_{\lambda_k}) \times \cdots \times C(X)(e_\lambda, e_{\lambda_0}) \times (\Delta^k \setminus d^0(\Delta^{k-1}))\right)$$

is given by adding $C(X)(e_\lambda, e_{\lambda_1}) \times \cdots \times C(X)(e_{\lambda_{k-1}}, e_{\lambda_k}) \times d^0(\Delta^{k-1})$ and we have

$$\overline{\pi_{X^{\mathrm{op}}}^{-1}(\lambda)} = \overline{t_\lambda\left(D_\lambda^{\mathrm{op},\circ}\right)} = t_\lambda\left(D_\lambda^{\mathrm{op}}\right).$$

□

Corollary 7.16: *For a cylindrically normal stellar stratified space X, $\pi_{X^{\mathrm{op}}}$ is a stratification whose face poset is the opposite $P(X)^{\mathrm{op}}$ of $P(X)$.*

Proof: The fact that $\pi_{X^{\mathrm{op}}}^{-1}(\lambda)$ is locally closed for each $\lambda \in P(X)$ follows from the description in Lemma 7.15. It also says that $\pi_{X^{\mathrm{op}}}^{-1}(\lambda)$ is connected, since it contains the barycenter $v(e_\lambda)$ of e_λ.

The compatibility with the partial order in $P(X)^{\mathrm{op}}$ also follows from the description of the boundary in Lemma 7.15. □

Definition 7.17: The stratification $\pi_{X^{\mathrm{op}}}$ is called the *stellar dual* of π_X.

Thus when $\|\mathrm{Lk}_{>e_\lambda}(C(X))\|$ can be embedded in a sphere S^{N-1} in such a way that the closure $\overline{\|\mathrm{Lk}_{>e_\lambda}(C(X))\|}$ is a finite cell complex containing $\|\mathrm{Lk}_{>e_\lambda}(C(X))\|$ as a strict cellular stratified subspace, $\|\mathrm{St}_{\geq e_\lambda}(C(X))\|$ can be regarded as an aster in D^N.

Proposition 7.18: *Let X be a finite polyhedral relatively compact cellular stratified space. Then* $\mathrm{Sd}(X)$ *has a structure of stellar stratified space whose underlying stratification is* $\pi_{X^{\mathrm{op}}}$ *and the face poset is* $P(\mathrm{Sd}(X), \pi_{X^{\mathrm{op}}}) = P(X, \pi_X)^{\mathrm{op}}$.

Proof: For $\lambda \in P(X)$, consider the upper star $\mathrm{St}_{\geq e_\lambda}(C(X))$ and the upper link $\mathrm{Lk}_{>e_\lambda}(C(X))$ of λ in $C(X)$. Since compact locally cone-like spaces can be expressed as a union of a finite number of simplices (Theorem 2.11 in [35]), each parameter space has a structure of a finite polyhedral complex. And the comma category $e_\lambda \downarrow C(X)$ is a cellular category whose morphism spaces are finite cell complexes. By Lemma C.17 and the finiteness assumption, $\|\mathrm{St}_{\geq e_\lambda}(C(X))\| = B(e_\lambda \downarrow C(X))$ is a finite cell complex and $\|\mathrm{Lk}_{>e_\lambda}(C(X))\|$ is a subcomplex. Choose an embedding $\|\mathrm{Lk}_{>e_\lambda}(C(X))\| \hookrightarrow S^{N-1}$. Then $D_\lambda^{\mathrm{op}} = \{1_{e_\lambda}\} * \|\mathrm{Lk}_{>e_\lambda}(C(X))\|$ is embedded in D^N.

By definition,

$$t_{e_\lambda} : D_\lambda^{\mathrm{op}} \longrightarrow \overline{\pi_{X^{\mathrm{op}}}^{-1}(\lambda)} \subset \mathrm{Sd}(X)$$

is a quotient map. The fact that t_{e_λ} is a homeomorphism onto $\pi_{X^{\mathrm{op}}}^{-1}(\lambda)$ when restricted to $\mathrm{Int}\,(D_\lambda^{\mathrm{op}}) = D_\lambda^{\mathrm{op},\circ}$ follows easily from the description of elements in $D_\lambda^{\mathrm{op},\circ}$ in the proof of Lemma 7.15. □

Remark 7.19: The three assumptions, i.e. local-polyhedrality, finiteness, and relative-compactness, are imposed only for the purpose of the existence of an embedding of $\|\mathrm{Lk}_{>e_\lambda}(C(X))\|$ in a sphere. If we relax the definition of a stellar cell $\varphi_\lambda : D_\lambda \to \overline{e_\lambda}$ by dropping the embeddability of the domain D_λ in a disk, we do not need to require these conditions.

The next problem is to define a cylindrical structure for the stellar dual $(\mathrm{Sd}(X), \pi_{X^{\mathrm{op}}})$.

Theorem 7.20: *Let X be a finite polyhedral stellar stratified space. Suppose all parameter spaces $P_{\mu,\lambda}$ are compact. For $\lambda \leq^{\mathrm{op}} \mu$ in $P(\mathrm{Sd}(X), \pi_{X^{\mathrm{op}}}) = P(X, \pi_X)^{\mathrm{op}}$, define $P^{\mathrm{op}}_{\lambda,\mu} = P_{\mu,\lambda}$. Then the stellar structure in Proposition 7.18 and parameter spaces $\{P^{\mathrm{op}}_{\lambda,\mu}\}$ make $(\mathrm{Sd}(X), \pi_{X^{\mathrm{op}}})$ into a polyhedral stellar stratified space.*

Proof: It remains to construct PL structure maps

$$b^{\mathrm{op}}_{\lambda,\mu} : P^{\mathrm{op}}_{\lambda,\mu} \times D^{\mathrm{op}}_{\lambda} \longrightarrow D^{\mathrm{op}}_{\mu}$$
$$\circ^{\mathrm{op}} : P^{\mathrm{op}}_{\lambda_1,\lambda_0} \times P^{\mathrm{op}}_{\lambda_2,\lambda_1} \longrightarrow P^{\mathrm{op}}_{\lambda_2,\lambda_0}$$

for $\lambda \leq^{\mathrm{op}} \mu$ and $\lambda_2 \leq^{\mathrm{op}} \lambda_1 \leq^{\mathrm{op}} \lambda_0^{\mathrm{op}}$.

The composition map $\circ^{\mathrm{op}}$ is obviously given by the composition in X under the identification

$$P^{\mathrm{op}}_{\lambda_1,\lambda_0} \times P^{\mathrm{op}}_{\lambda_2,\lambda_1} = P_{\lambda_0,\lambda_1} \times P_{\lambda_1,\lambda_2} \cong P_{\lambda_1,\lambda_2} \times P_{\lambda_0,\lambda_1}.$$

The action $b^{\mathrm{op}}_{\lambda,\mu}$ of the parameter space is given by the following composition

$$\begin{aligned} P^{\mathrm{op}}_{\lambda,\mu} \times D^{\mathrm{op}}_{\lambda} &= P_{\mu,\lambda} \times B(e_\lambda \downarrow C(X)) \\ &= C(X)(e_\mu, e_\lambda) \times B(e_\lambda \downarrow C(X)) \\ &\cong B(e_\lambda \downarrow C(X)) \times C(X)(e_\mu, e_\lambda) \\ &\longrightarrow B(e_\mu \downarrow C(X)) = D^{\mathrm{op}}_{\mu}, \end{aligned}$$

where the last arrow is given by the composition in $C(X)$.

The compatibilities of $b_{\mu,\lambda}$ and $\circ$ in X implies that $b^{\mathrm{op}}_{\lambda,\mu}$ and $\circ^{\mathrm{op}}$ satisfy the conditions for cylindrical structure. Obviously this is polyhedral. □

Definition 7.21: The stratified space $(\mathrm{Sd}(X), \pi_{X^{\mathrm{op}}})$ equipped with the stellar and polyhedral structures defined in Theorem 7.20 is called the *stellar dual of X* and is denoted by X^{op}.

The following fundamental example shows that cells in X^{op} are usually stellar and not cellular.

Example 7.22: Consider the standard regular cellular stratification on Δ^2 as a simplicial complex. The barycentric subdivision $\mathrm{Sd}(\Delta^2)$ is a simplicial complex depicted in Figure 37

The stellar stratification on $\mathrm{Sd}(\Delta^2)$, i.e. $(\Delta^2)^{\mathrm{op}}$ is given by Figure 38. $(\Delta^2)^{\mathrm{op}}$ consists of one stellar 0-cell, three stellar 1-cells, and three stellar 2-cells. The stellar structure for the 2-cell at the left bottom in $D(\Delta^2)$ is indicated in Figure 38 as t_0.

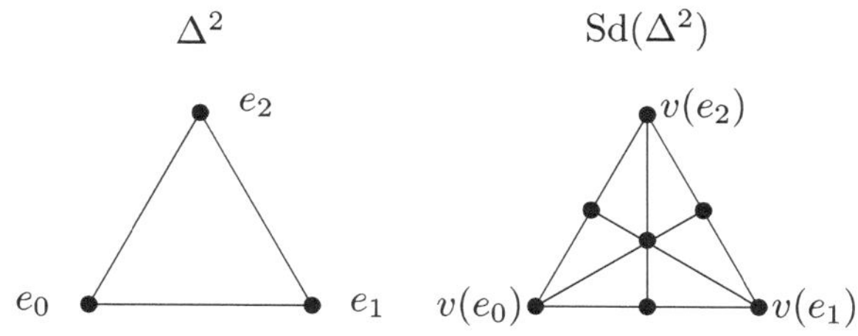

Fig. 37. Δ^2 and its barycentric subdivision.

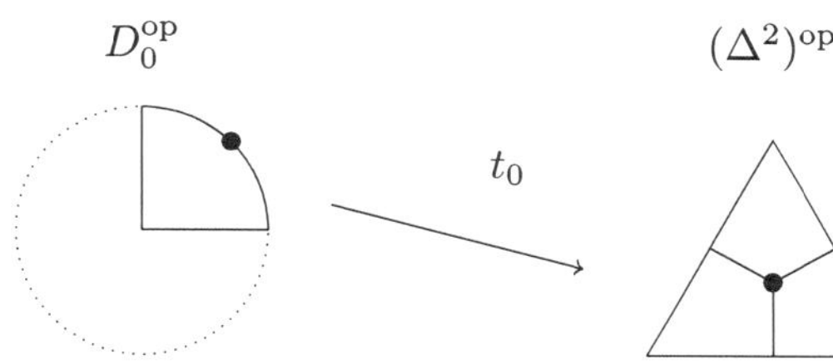

Fig. 38. The stellar structures on $(\Delta^2)^{\mathrm{op}}$.

Note that $\|\mathrm{Lk}_{>e_0}(C(\Delta^2))\|$ consists of two 1-simplices. It is embedded in S^1. The domain for this stellar structure map is the circular sector in the left. The middle point in the arc of the circular sector in D_0^{op} is mapped to the barycenter of Δ^2 and two radii in D_0^{op} are mapped to the half edges touching the lower left vertex of Δ^2.

The barycentric subdivision of $(\Delta^2)^{\mathrm{op}}$ can be easily seen to be isomorphic to $\mathrm{Sd}(\Delta^2)$ as simplicial complexes and we have $((\Delta^2)^{\mathrm{op}})^{\mathrm{op}} \cong \Delta^2$ as simplicial complexes. For example, the 2-cell corresponding to the unique

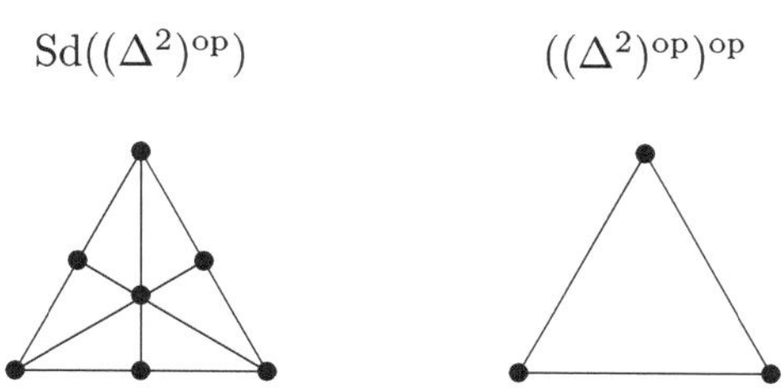

Fig. 39. The dual of the dual of Δ^2.

0-cell in $(\Delta^2)^{\mathrm{op}}$ is the whole triangle.

The next example shows that, when X contains non-closed cells, the process of taking the double dual $((X)^{\mathrm{op}})^{\mathrm{op}}$ slims down X while retaining the stratification.

Example 7.23: Consider the 1-dimensional stellar stratified space Y in Figure 40. It consists of a 0-cell e^0 and a stellar 1-cell e^1 whose domain D_1

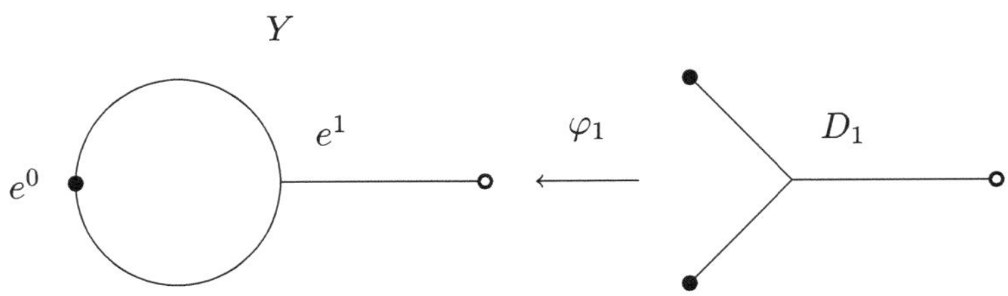

Fig. 40. A 1-dimensional stellar stratified space Y.

is a graph of the shape of Y with one vertex removed. The barycentric subdivision $\mathrm{Sd}(Y)$ is the minimal regular cell decomposition of S^1, as is shown in Figure 41. Both Y^{op} and $(Y^{\mathrm{op}})^{\mathrm{op}}$ are the minimal cell decomposition of S^1. Note that the embedding i_Y in Theorem 5.16 embeds $(Y^{\mathrm{op}})^{\mathrm{op}}$ in Y as

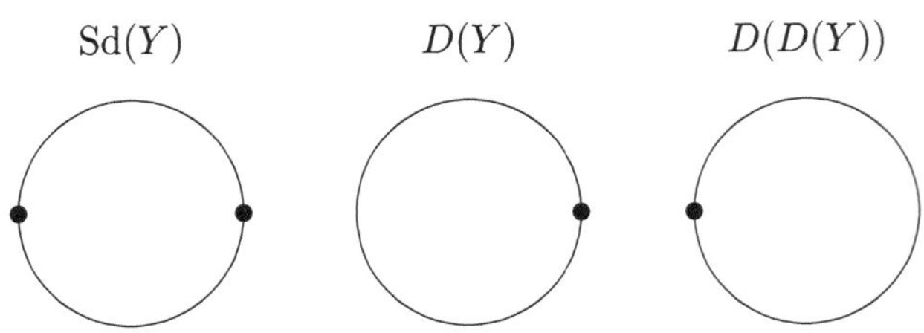

Fig. 41. The stellar dual of Y.

a stellar stratified space.

As the above example suggests, the embedding i_X in Theorem 5.16 is an embedding of stellar stratified spaces if the domain is regarded as $(X^{\mathrm{op}})^{\mathrm{op}}$. Furthermore, when all cells are closed, we can always recover X from this stellar structure.

Note that we may define this stellar stratification on $\mathrm{Sd}(X)$ directly without using $(-)^{\mathrm{op}}$.

Definition 7.24: Let X be a cylindrically normal stellar stratified space. For $\lambda \in P(X)$, define

$$\begin{aligned} D_{\lambda}^{\mathrm{Sal}(X)} &= \|\mathrm{St}_{\leq e_{\lambda}}(C(X))\| \\ D_{\lambda}^{\mathrm{Sal}(X),\circ} &= \|\mathrm{St}_{\leq e_{\lambda}}(C(X))\| \setminus \|\mathrm{Lk}_{< e_{\lambda}}(C(X))\|. \end{aligned}$$

The following fact is dual to Lemma 7.15. The proof is omitted.

Lemma 7.25: *Let X be a cylindrically normal stellar stratified space. Define a map*

$$\pi_{\mathrm{Sal}(X)} : \mathrm{Sd}(X) \longrightarrow P(X)$$

by the composition

$$\mathrm{Sd}(X) \stackrel{\pi_{\mathrm{Sd}(X)}}{\longrightarrow} \coprod_k \overline{N}_k(C(X)) \stackrel{t}{\longrightarrow} C(X)_0 = P(X),$$

where t is the target map. Then $\pi_{\mathrm{Sal}(X)}$ is a stratification whose strata and their closures are given by

$$\begin{aligned} \pi_{\mathrm{Sal}(X)}^{-1}(\lambda) &= s_{\lambda}\left(D_{\lambda}^{\mathrm{Sal}(X),\circ}\right) \\ \overline{\pi_{\mathrm{Sal}(X)}^{-1}(\lambda)} &= s_{\lambda}\left(D_{\lambda}^{\mathrm{Sal}(X)}\right), \end{aligned}$$

where $s_{\lambda} = s_{e_{\lambda}}$ is the map defined in Definition 7.7.

Definition 7.26: The stellar stratified space $(\mathrm{Sd}(X), \pi_{\mathrm{Sal}})$ defined above is called the *Salvetti complex* of X and is denoted by $\mathrm{Sal}(X)$.

Remark 7.27: When the three assumptions in Proposition 7.18 are satisfied, we have $\mathrm{Sal}(X) = (X^{\mathrm{op}})^{\mathrm{op}}$ as stellar stratified spaces.

Theorem 7.28: *Let (X, π_X) be a cylindrically normal stellar stratified space. Then $\mathrm{Sal}(X)$ has a structure of cylindrically normal stellar stratified space. When X is relatively compact, the embedding $i_X : \mathrm{Sal}(X) \hookrightarrow X$ is an embedding of cylindrically normal stellar stratified spaces. When all cells in X are closed, i_X is an isomorphism of cylindrically normal stellar stratified spaces. When X is a finite polyhedral stellar stratified space satisfying the assumptions of Proposition 7.18, we have $\mathrm{Sal}(X) = (X^{\mathrm{op}})^{\mathrm{op}}$ as stellar stratified spaces.*

Definition 7.29: When we regard $\mathrm{Sd}(X)$ as $\mathrm{Sal}(X)$, the embedding i_X is denoted by $i_{\mathrm{Sal}(X)} : \mathrm{Sal}(X) \hookrightarrow X$.

In order to prove Theorem 7.28, we use the following reformulation of the construction of i_X.

Lemma 7.30: *Let X be a CW cylindrically normal stellar stratified space X. For each stellar cell e_λ, there exists an embedding*

$$z_\lambda : D_\lambda^{\mathrm{Sal}(X)} \longrightarrow D_\lambda$$

making the diagram

$$\begin{array}{ccc} P_{\mu,\lambda} \times D_\mu^{\mathrm{Sal}(X)} & \xrightarrow{b_{\mu,\lambda}^{\mathrm{Sal}(X)}} & D_\lambda^{\mathrm{Sal}(X)} \\ {\scriptstyle 1\times z_\mu}\downarrow & & \downarrow{\scriptstyle z_\lambda} \\ P_{\mu,\lambda} \times D_\mu & \xrightarrow[b_{\mu,\lambda}]{} & D_\lambda \end{array} \tag{7.1}$$

commutative, where $b_{\mu,\lambda}^{\mathrm{Sal}(X)}$ is defined by the composition

$$\begin{aligned} P_{\mu,\lambda} \times D_\mu^{\mathrm{Sal}(X)} &= P_{\mu,\lambda} \times B(C(X) \downarrow e_\mu) \\ &= C(X)(e_\mu, e_\lambda) \times B(C(X) \downarrow e_\mu) \\ &\longrightarrow B(C(X) \downarrow e_\lambda). \end{aligned}$$

Here the last map is induced by the composition in $C(X)$.

These maps $\{z_\lambda\}$ can be glued together to define an embedding $\mathrm{Sal}(X) \hookrightarrow X$, which can be identified with i_X in Theorem 5.16.

Proof: Note that we have the following decomposition

$$\overline{N}_k(C(X)) = \coprod_{\lambda \in P(X)} \overline{N}_k(C(X) \downarrow e_\lambda).$$

The restrictions of the maps $z_k : \overline{N}_k(C(X)) \times \Delta^k \to D(X)$ to $\overline{N}_k(C(X) \downarrow e_\lambda)$ define a map

$$z_\lambda : \|\overline{N}(C(X) \downarrow e_\lambda)\| = D_\lambda^{\mathrm{op}} \longrightarrow D_\lambda.$$

The fact that these maps make the diagram (7.1) commutative can be verified by investigating the construction of z_k and is omitted. □

Proof of Theorem 7.28: Lemma 7.30 says that i_X is an embeddding of stellar stratified spaces.

When all cells in X are closed, $i_X : \mathrm{Sd}(X) \to X$ is a homeomorphism. The above argument implies that this map defines an isomorphism $i_{\mathrm{Sal}(X)} : \mathrm{Sal}(X) \xrightarrow{\cong} X$ of cylindrically normal stellar stratified spaces.

Suppose that X satisfies the assumption of Proposition 7.18. By Theorem 7.20, we have an isomorphism of categories $C(X^{\mathrm{op}}) \cong C(X)^{\mathrm{op}}$. Thus

$$D_\lambda^{(X^{\mathrm{op}})^{\mathrm{op}}} = B(e_\lambda \downarrow C(X^{\mathrm{op}})) \cong B(e_\lambda \downarrow C(X)^{\mathrm{op}}) \cong B(C(X) \downarrow e_\lambda)$$

and the stellar structure map $t_\lambda : D_\lambda^{(X^{\mathrm{op}})^{\mathrm{op}}} \to \mathrm{Sd}(X^{\mathrm{op}})$ can be identified with $s_\lambda : D_\lambda^{(\mathrm{Sal}(X))} \to \mathrm{Sd}(X)$ under the identification $\mathrm{Sd}(D(X)) \cong B(C(X)^{\mathrm{op}}) \cong B(C(X)) = \mathrm{Sd}(X)$. And we have an identification $D(D(X)) \cong \mathrm{Sal}(X)$. □

The following example justifies the name for $\mathrm{Sal}(X)$.

Example 7.31: For a real hyperplane arrangement $\mathcal{A}$, the structure of a regular cell complex on the Salvetti complex $\mathrm{Sal}(\mathcal{A})$ for the complexification of $\mathcal{A}$ defined in [6] is nothing but $D(D(M(\mathcal{A} \otimes \mathbb{C})))$. For example, in the case of the arrangement $\mathcal{A} = \{\{0\}\}$ in $\mathbb{R}$, the stratification on $\mathbb{R}$ is

$$\mathbb{R} = \{0\} \cup (-\infty, 0) \cup (0, \infty)$$

and the associated stratification on the complexification is

$$\begin{aligned}\mathbb{C} = \{(0,0)\} &\cup (-\infty, 0) \times \{0\} \cup (0, \infty) \times \{0\} \cup \{x + iy \in \mathbb{C} \mid y > 0\} \\ &\cup \{x + iy \in \mathbb{C} \mid y < 0\}.\end{aligned}$$

Then $\mathrm{Sd}(M(\mathcal{A} \otimes \mathbb{C}))$, $M(\mathcal{A} \otimes \mathbb{C})^{\mathrm{op}}$, and $(M(\mathcal{A} \otimes \mathbb{C})^{\mathrm{op}})^{\mathrm{op}}$ are given in Figure 42.

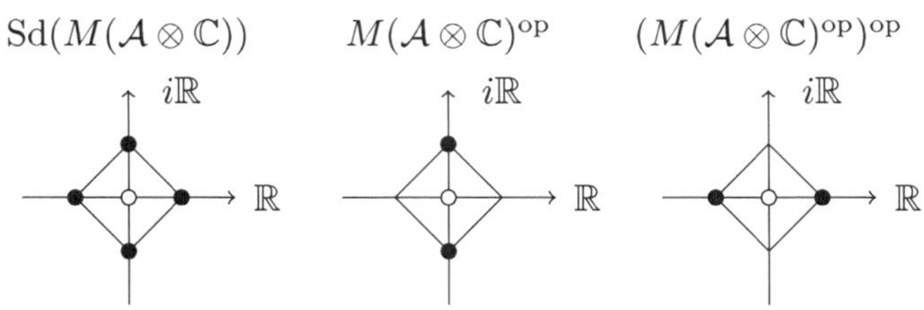

Fig. 42. Stellar duals for $M(\mathcal{A} \otimes \mathbb{C})$

7.4. *The barycentric subdivision of face categories*

We conclude these notes by proving that the barycentric subdivision of a totally normal cellular stratified space corresponds to the barycentric subdivision of the face category.

Let us first recall the definition of the barycentric subdivision of a small category. We use a definition in Noguchi's papers [30, 31]. See also the paper [29] by del Hoyo.

Definition 7.32: For a small category C, the *barycentric subdivision* $\mathrm{Sd}(C)$ is a small category defined by

$$\mathrm{Sd}(C)_0 = \coprod_n \overline{N}_n(C),$$
$$\mathrm{Sd}(C)(f,g) = \{\varphi : [m] \to [n] \,|\, g \circ \varphi = f\} /_\sim$$

for $f : [m] \to C$ and $g : [n] \to C$, where $\sim$ is the equivalence relation generated by the following relation: for functors $\varphi, \psi : [m] \to [n]$ with $g \circ \varphi = f$ and $g \circ \psi = f$, $\varphi \sim \psi$ if and only if the morphism $g\,(\min\{\varphi(i), \psi(i)\} \leq \max\{\varphi(i), \psi(i)\})$ in $C(g(\min\{\varphi(i), \psi(i)\}), g(\max\{\varphi(i), \psi(i)\}))$ is an identity morphism in C for any i in $[m]$.

The description can be simplified for acyclic categories as follows:

Lemma 7.33: *Let C be an acyclic small category. For $f, g \in \mathrm{Sd}(C)_0$, the set of morphisms $\mathrm{Sd}(C)(f,g)$ consists of a single point, if there exists φ with $f = g \circ \varphi$, and an empty set otherwise.*

Therefore $\mathrm{Sd}(C)$ is a poset.

Proof: Since C is acyclic, $C(x,x) = \{1_x\}$ for any objects $x \in C_0$. This implies that for $\varphi, \psi : [m] \to [n]$ with $f = g \circ \varphi = g \circ \psi$, $\varphi \sim \psi$ if and only if $g(\varphi(i)) = g(\psi(i))$ for all $i \in [m]$. In other words, all elements in $\{\varphi : [m] \to [n] \,|\, g \circ \varphi = f\}$ are equivalent to each other. Hence $\mathrm{Sd}(C)(f,g)$ is a single point if the above set is nonempty. □

In order to compare $C(\mathrm{Sd}(X))$ and $\mathrm{Sd}(C(X))$ for a totally normal stellar stratified space X, we need to understand the cellular stratification on $\mathrm{Sd}(X)$. By Corollary 7.1, we know that $\mathrm{Sd}(X)$ is a totally normal cell complex when X is a totally normal stellar stratified space. Cells are parametrized by elements in $\overline{N}(C(X))$. Let us denote the cell corresponding to $\boldsymbol{b} \in \overline{N}(C(X))$ by $e_{\boldsymbol{b}}$. Cell structure maps are given as follows:

Lemma 7.34: *For each k, fix a homeomorphism $D^k \cong \Delta^k$. Let X be a totally normal stellar stratified space. For $\boldsymbol{b} \in \overline{N}_k(C(X))$, the composition*

$$D^k \cong \Delta^k = B[k] \xrightarrow{B\boldsymbol{b}} BC(X) = \mathrm{Sd}(X)$$

defines a cell structure on the cell corresponding to $\boldsymbol{b}$, where we regard $\boldsymbol{b}$ as a functor $\boldsymbol{b} : [k] \to C(X)$.

Proof: The map $B\boldsymbol{b} : B[k] \to BC(X)$ is induced by the map $\overline{N}\boldsymbol{b} : \overline{N}([k]) \to \overline{N}(C(X))$. As we have seen in the proof of Proposition 7.3, a cell structure map on the cell corresponding to $\boldsymbol{b}$ is given by the composition

$$\Delta^k \cong \{\boldsymbol{b}\} \times \Delta^k \hookrightarrow \coprod_k \overline{N}_k(C(X)) \times \Delta^k \longrightarrow \|\overline{N}(C(X))\| \cong BC(X).$$

Since $\overline{N}_k([k])$ consists of a single point, the above composition can be identified with

$$\Delta^k \cong \overline{N}_k([k]) \times \Delta^k \xrightarrow{\overline{N}(\boldsymbol{b})\times 1} \coprod_k \overline{N}_k(C(X)) \times \Delta^k \longrightarrow \|\overline{N}(C(X))\| \cong BC(X)$$

and the result follows. □

For simplicity, we use the standard simplices Δ^k as the domains of cells in $\mathrm{Sd}(X)$. The cell structure map for $e_{\boldsymbol{b}}$ is identified with $B\boldsymbol{b}$ by Lemma 7.34.

Theorem 7.35: *For any totally normal stellar stratified space X, we have an isomorphism of categories*

$$\mathrm{Sd}(C(X)) \cong C(\mathrm{Sd}(X)).$$

Proof: By definition, objects in $\mathrm{Sd}(C(X))$ are elements of the nondegenerate nerve of $C(X)$. On the other hand, objects in $C(\mathrm{Sd}(X))$ are in one-to-one correspondence with cells in $\mathrm{Sd}(X) = BC(X)$. Under the stratification in Proposition 7.3, we obtain a bijection $C(\mathrm{Sd}(X))_0 \cong \mathrm{Sd}(C(X))_0$.

For $\boldsymbol{b} \in \overline{N}_k(C(X))$ and $\boldsymbol{b}' \in \overline{N}_m(C(X))$, we have

$$C\left(\mathrm{Sd}(X); \pi_{\mathrm{Sd}(X)}\right)(e_{\boldsymbol{b}'}, e_{\boldsymbol{b}}) = \left\{f : \Delta^m \to \Delta^k \,\middle|\, B\boldsymbol{b}' = B\boldsymbol{b} \circ f\right\}.$$

Since $B\boldsymbol{b}'|_{\mathrm{Int}(\Delta^m)}$ is injective, $f|_{\mathrm{Int}(\Delta^m)}$ is also injective. The condition $B\boldsymbol{b}' = B\boldsymbol{b} \circ f$ implies that $f|_{\mathrm{Int}(\Delta^m)}$ is a PL map and hence f is a PL map. Since $B\boldsymbol{b}|_{\mathrm{Int}(\Delta^k)}$ is injective, such a PL map is unique if it were to exist. It is given by $f = B\varphi$ for some poset map $\varphi : [m] \to [k]$.

On the other hand, by Lemma 7.33, $\mathrm{Sd}(C(X))(\boldsymbol{b}', \boldsymbol{b})$ is nonempty (and a single point set) if and only if there exists a poset map $\varphi : [m] \to [k]$ with $\boldsymbol{b}' = \boldsymbol{b} \circ \varphi$. Thus the classifying space functor $B(-)$ induces an isomorphism of categories

$$B : \mathrm{Sd}(C(X)) \longrightarrow C(\mathrm{Sd}(X)). \qquad \square$$

Remark 7.36: Note that we obtained an isomorphism of categories instead of an equivalence. Since $\mathrm{Sd}(C(X))$ is a poset, it implies that $C(\mathrm{Sd}(X))$ is also a poset. Thus the barycentric subdivision $\mathrm{Sd}(X)$ of a totally normal stellar stratified space is a regular cell complex.

Acknowledgments

This work originates in fruitful and enjoyable discussions with Jesús González on the Google Wave system in 2010. Although Google terminated the development of the system, I would like to thank Google for providing us with such a useful collaboration tool for free. I would also like to express my appreciation to the Centro di Ricerca Matematica Ennio De Giorgi, Scuola Normale Superiore di Pisa, for supporting my participation in the research program "Configuration Spaces: Geometry, Combinatorics and Topology" in 2010, during which a part of this work was done. I would also like to thank the National University of Signapore for inviting me to the workshop "Combinatorial and Toric Topology" in 2015.

The discussion on the Google Wave has been incorporated into a joint project with Ibai Basabe and Yuli Rudyak with the aim of applying cellular stratifies spaces to study the homotopy type of configuration spaces of spheres. It turns out that configuration spaces of spheres are much more complicated than we imagined and the sections covering cellular stratified spaces were removed from the published paper [5]. Nonetheless the discussions during the project with them have been essential. I would like to thank Basabe, González, and Rudyak, as well as Peter Landweber, who read earlier versions of the paper and made useful comments.

Before I began the discussion with González on Google Wave, some of the ideas had been already developed during discussions with my students. Takamitsu Jinno and Mizuki Furuse worked on Hom complexes in 2009 and configuration spaces of graphs in 2010, respectively, in their masters theses. The possibility of finding a better combinatorial model for configuration spaces was suggested by their work. Furuse's work was developed further by another student Takashi Mukouyama. Their work is contained in a paper

[1], in which a theory of totally normal cellular stratified spaces has been developed. The connection with the Cohen-Jones-Segal Morse theory, which resulted in the current definition of cylindrical structures, was discovered during the discussion with another former student, Kohei Tanaka. I am grateful to all these former students.

Mikiya Masuda pointed out my misunderstanding in examples concerning complex projective spaces in an early draft of these notes. Priyavrat Deshpande pointed out a mistake in [1] on subdivisions of totally normal cellular stratified spaces. The mistake is corrected in these notes as Proposition 6.36. I would like to thank both of them.

This work is supported by JSPS KAKENHI Grant Number 23540082 and 15K04870.

References

1. M. Furuse, T. Mukouyama, and D. Tamaki, Totally normal cellular stratified spaces and applications to the configuration space of graphs, *Topol. Methods Nonlinear Anal.* **45** (2015), 169–214. doi: 10.12775/TMNA.2015.010. URL `http://dx.doi.org/10.12775/TMNA.2015.010`.
2. D. Tamaki, *Homological methods in the unstable chromatic phenomena.* ProQuest LLC, Ann Arbor, MI, 1993. Thesis (Ph.D.)–University of Rochester.
3. D. Tamaki, A dual Rothenberg-Steenrod spectral sequence, *Topology.* **33** (1994), 631–662. doi: 10.1016/0040-9383(94)90002-7.
4. D. Tamaki. On a space realizing the v_1-torsion part of the mod p homotopy groups of S^3. In *Topology and representation theory (Evanston, IL, 1992)*, vol. 158, *Contemp. Math.*, pp. 229–267. Amer. Math. Soc., Providence, RI, 1994. doi: 10.1090/conm/158/01462.
5. I. Basabe, J. González, Y. B. Rudyak, and D. Tamaki, Higher topological complexity and its symmetrization, *Algebr. Geom. Topol.* **14** (2014), 223–244. doi: 10.2140/agt.2014.14.223. URL `http://dx.doi.org/10.2140/agt.2014.14.223`.
6. M. Salvetti, Topology of the complement of real hyperplanes in $\mathbf{C}^N$, *Invent. Math.* **88** (1987), 603–618. doi: 10.1007/BF01391833.
7. A. Björner and G. M. Ziegler, Combinatorial stratification of complex arrangements, *J. Amer. Math. Soc.* **5** (1992), 105–149. doi: 10.2307/2152753.
8. C. De Concini and M. Salvetti, Cohomology of Coxeter groups and Artin groups, *Math. Res. Lett.* **7** (2000), 213–232. doi: 10.4310/MRL.2000.v7.n2.a7.
9. A. Abrams. *Configuration Spaces and Braid Groups of Graphs.* PhD thesis, University of California at Berkeley, 2000.
10. A. Abrams, D. Gay, and V. Hower, Discretized configurations and partial partitions, *Proc. Amer. Math. Soc.* **141** (2013), 1093–1104. doi: 10.1090/S0002-9939-2012-10816-0. URL `http://dx.doi.org/10.1090/S0002-9939-2012-10816-0`.

11. W. Fulton and R. MacPherson, A compactification of configuration spaces, *Ann. of Math. (2)*. **139** (1994), 183–225.
12. S. Axelrod and I. M. Singer. Chern-Simons perturbation theory. II. In *Perspectives in mathematical physics*, Conf. Proc. Lecture Notes Math. Phys., III, pp. 17–49. Int. Press, Cambridge, MA, 1994.
13. C. De Concini and C. Procesi, Wonderful models of subspace arrangements, *Selecta Math. (N.S.)*. **1** (1995), 459–494. doi: 10.1007/BF01589496.
14. I. Basabe, J. Gonzalez, Y. B. Rudyak, and D. Tamaki. Higher topological complexity and homotopy dimension of configuration spaces on spheres, .
15. A. Kirillov Jr., On piecewise linear cell decompositions, *Algebr. Geom. Topol.* **12** (2012), 95–108. doi: 10.2140/agt.2012.12.95. URL `http://dx.doi.org/10.2140/agt.2012.12.95`.
16. R. L. Cohen, J. Jones, and G. B. Segal. Morse theory and classifying spaces, preprint.
17. R. Forman. A discrete Morse theory for cell complexes. In *Geometry, topology, & physics*, Conf. Proc. Lecture Notes Geom. Topology, IV, pp. 112–125. Int. Press, Cambridge, MA, 1995.
18. R. Forman, Morse theory for cell complexes, *Adv. Math.* **134** (1998), 90–145. doi: 10.1006/aima.1997.1650.
19. F. Kirwan, *An introduction to intersection homology theory*. vol. 187, *Pitman Research Notes in Mathematics Series*, Longman Scientific & Technical, Harlow, 1988. ISBN 0-582-02879-5.
20. M. R. Bridson and A. Haefliger, *Metric spaces of non-positive curvature*. vol. 319, *Grundlehren der Mathematischen Wissenschaften [Fundamental Principles of Mathematical Sciences]*, Springer-Verlag, Berlin, 1999. ISBN 3-540-64324-9.
21. M. J. Pflaum, *Analytic and geometric study of stratified spaces*. vol. 1768, *Lecture Notes in Mathematics*, Springer-Verlag, Berlin, 2001. ISBN 3-540-42626-4.
22. J. Schürmann, *Topology of singular spaces and constructible sheaves*. vol. 63, *Instytut Matematyczny Polskiej Akademii Nauk. Monografie Matematyczne (New Series) [Mathematics Institute of the Polish Academy of Sciences. Mathematical Monographs (New Series)]*, Birkhäuser Verlag, Basel, 2003. ISBN 3-7643-2189-X. doi: 10.1007/978-3-0348-8061-9.
23. M. Davis, Smooth G-manifolds as collections of fiber bundles, *Pacific J. Math.* **77** (1978), 315–363.
24. A. Hattori and M. Masuda, Theory of multi-fans, *Osaka J. Math.* **40** (2003), 1–68.
25. M. W. Davis and T. Januszkiewicz, Convex polytopes, Coxeter orbifolds and torus actions, *Duke Math. J.* **62** (1991), 417–451. doi: 10.1215/S0012-7094-91-06217-4.
26. M. W. Davis, Groups generated by reflections and aspherical manifolds not covered by Euclidean space, *Ann. of Math. (2)*. **117** (1983), 293–324. doi: 10.2307/2007079.
27. A. T. Lundell and S. Weingram, *Topology of CW-Complexes*. Van Nostrand Reinhold, New York, 1969.

28. D. Kozlov, *Combinatorial algebraic topology*. vol. 21, *Algorithms and Computation in Mathematics*, Springer, Berlin, 2008. ISBN 978-3-540-71961-8. doi: 10.1007/978-3-540-71962-5.
29. M. L. del Hoyo, On the subdivision of small categories, *Topology Appl.* **155** (2008), 1189–1200. doi: 10.1016/j.topol.2008.02.006. URL `http://dx.doi.org/10.1016/j.topol.2008.02.006`.
30. K. Noguchi, The Euler characteristic of acyclic categories, *Kyushu J. Math.* **65** (2011), 85–99. doi: 10.2206/kyushujm.65.85. URL `http://dx.doi.org/10.2206/kyushujm.65.85`.
31. K. Noguchi, Euler characteristics of categories and barycentric subdivision, *Münster J. Math.* **6** (2013), 85–116.
32. J. Milnor, Construction of universal bundles. I, II, *Ann. of Math. (2)*. **63** (1956), 272–284, 430–436.
33. C. H. Dowker, Topology of metric complexes, *Amer. J. Math.* **74** (1952), 555–577.
34. R. W. Thomason, Homotopy colimits in the category of small categories, *Math. Proc. Cambridge Philos. Soc.* **85** (1979), 91–109. doi: 10.1017/S0305004100055535.
35. C. P. Rourke and B. J. Sanderson, *Introduction to piecewise-linear topology*. Springer-Verlag, New York, 1972.
36. J. R. Munkres, *Topology*, 2nd. edn. Prentice-Hall Inc., Englewood Cliffs, N.J., 2000.
37. A. Arhangel'skiĭ, Some types of factor mappings and the relations between classes of topological spaces, *Dokl. Akad. Nauk SSSR*. **153** (1963), 743–746.
38. E. Michael, Bi-quotient maps and Cartesian products of quotient maps, *Ann. Inst. Fourier (Grenoble)*. **18** (1968), 287–302 vii (1969).
39. J. H. C. Whitehead, Note on a theorem due to Borsuk, *Bull. Amer. Math. Soc.* **54** (1948), 1125–1132.
40. J. Dugundji, *Topology*. Allyn and Bacon Inc., Boston, Mass., 1978. ISBN 0-205-00271-4.
41. P. G. Goerss and J. F. Jardine, *Simplicial homotopy theory*. Modern Birkhäuser Classics, Birkhäuser Verlag, Basel, 2009. ISBN 978-3-0346-0188-7. doi: 10.1007/978-3-0346-0189-4. Reprint of the 1999 edition [MR1711612].
42. W. G. Dwyer and H.-W. Henn, *Homotopy theoretic methods in group cohomology*. Advanced Courses in Mathematics—CRM Barcelona, Birkhäuser Verlag, Basel, 2001. ISBN 3-7643-6605-2.
43. G. Friedman, Survey article: an elementary illustrated introduction to simplicial sets, *Rocky Mountain J. Math.* **42** (2012), 353–423. doi: 10.1216/RMJ-2012-42-2-353. URL `http://dx.doi.org/10.1216/RMJ-2012-42-2-353`.
44. G. Segal, Classifying spaces and spectral sequences, *Inst. Hautes Études Sci. Publ. Math.* **34** (1968), 105–112.

A. Generalities on quotient maps

In our definition of cell structures, we required the cell structure map $\varphi : D \to \overline{e}$ of a cell e to be a quotient map. In order to perform operations on cellular stratified spaces, such as taking products and subspaces, we need to understand basic properties of quotient maps.

A.1. *Definitions*

It is well-known that the quotient topology does not behave well with respect to certain operations of topological spaces, such as taking products and subspaces. We need to impose stronger conditions.

Definition A.1: A surjective continuous map $f : X \to Y$ is called *bi-quotient*, if, for any $y \in Y$ and any open covering $\mathcal{U}$ of $f^{-1}(y)$, there exists finitely many $U_1, \ldots, U_k \in \mathcal{U}$ such that $\bigcup_{i=1}^{k} f(U_i)$ contains a neighborhood of y in Y.

Another important class of maps are hereditarily quotient maps.

Definition A.2: A surjective continuous map $f : X \to Y$ is called *hereditarily quotient* if, for any $y \in Y$ and any neighborhood U of $f^{-1}(y)$, $f(U)$ is a neighborhood of y.

A.2. *Properties*

This quotient topology condition imposes some restrictions on the topology of $\overline{e}$, especially when e is closed. For example, $\overline{e}$ is metrizable for any closed cell e. A proof can be found in a book [27] by Lundell and Weingram. Their proof can be modified to obtain the following extension of this fact.

Lemma A.3: *Suppose $\varphi : D \to \overline{e} \subset X$ is an n-cell structure with $\varphi^{-1}(y)$ compact for each $y \in \overline{e}$. Then $\overline{e}$ is metrizable. In particular, it is Hausdorff and paracompact (Theorem 41.4 in* [36]*).*

Proof: For $y, y' \in \overline{e}$, define

$$\overline{d}(y, y') = \min \left\{ d(x, x') \,\middle|\, x \in \varphi^{-1}(y), x' \in \varphi^{-1}(y') \right\},$$

where d is the metric on D^n. By assumption, $\varphi^{-1}(y)$ and $\varphi^{-1}(y')$ are compact and $\overline{d}(y, y')$ is defined. The compactness of $\varphi^{-1}(y)$ and $\varphi^{-1}(y')$ also implies that $\overline{d}$ is a metric on $\overline{e}$.

Let us verify that the topology defined by $\overline{d}$ coincides with the quotient topology by φ. The continuity of φ with respect to the metric topologies on D and $\overline{e}$ implies that open subsets in the $\overline{d}$-metric topology are open in the quotient topology. Conversely let U be an open subset of $\overline{e}$ with respect to the quotient topology. We would like to show that, for each $y \in U$, there exists $\delta > 0$ such that the open disk $U_\delta(y; \overline{d})$ around y with radius δ with respect to the metric $\overline{d}$ is contained in U. Let δ be a Lebesgue number of the open covering $\{\varphi^{-1}(U)\}$ of the compact metric space $\varphi^{-1}(y)$. For $y' \in U_\delta(y; \overline{d})$, there exist $x \in \varphi^{-1}(y)$ and $x' \in \varphi^{-1}(y')$ such that $d(x, x') < \delta$. Thus $x' \in U_\delta(x; d) \subset \varphi^{-1}(U)$ by the definition of Lebesgue number, or $y' = \varphi(x') \in U$, and we have $U_\delta(y'; \overline{d}) \subset U$. □

Definition A.4: We say a cell structure $\varphi : D \to \overline{e}$ is *relatively compact* if $\varphi^{-1}(y)$ is compact for each $y \in \overline{e}$. We also say that the cell e is *relatively compact*.

In particular, when $\varphi : D \to \overline{e}$ is proper (i.e. closed and each $\varphi^{-1}(y)$ is compact), $\overline{e}$ is metrizable. On the other hand, the properness of φ implies that φ is a bi-quotient map.

It is straight forward to verify that a hereditarily quotient map can be restricted freely.

Lemma A.5: *Any hereditarily quotient map $f : X \to Y$ is a quotient map. More generally, for any subspace $A \subset Y$, the restriction $f|_{f^{-1}(A)} : f^{-1}(A) \to A$ is hereditarily quotient, hence a quotient map.*

Proof: Suppose f is hereditarily quotient. For a subset $U \subset Y$, suppose that $f^{-1}(U)$ is open in X. For a point $y \in U$, $f^{-1}(U)$ is a neighborhood of $f^{-1}(y)$. Since f is hereditarily quotient, $f(f^{-1}(U)) = U$ is a neighborhood of y in Y. Thus y is an interior point of U and it follows that U is an open subset of Y.

Since the definition of an hereditarily quotient map is local, $f|_{f^{-1}(A)}$ is hereditarily quotient for any $A \subset Y$. □

Remark A.6: See also Arhangel'skii's paper [37] for hereditarily quotient maps.

Lemma A.7: *Any bi-quotient map is hereditarily quotient. In particular, it is a quotient map.*

Proof: By definition. □

Michael [38] proved that bi-quotient maps are abundant.

Lemma A.8: *Any one of the following conditions implies that a map $f : X \to Y$ is bi-quotient:*

(1) f is open.
(2) f is proper.
(3) f is hereditarily quotient and the boundary $\partial f^{-1}(y)$ of each fiber is compact.

Proof: Proposition 3.2 in [38]. □

Recall that a product of quotient maps may not be a quotient map. There exists a space X and a quotient map $f : Y \to Z$ such that the product $1_X \times f : X \times Y \to X \times Z$ is not a quotient map. The following fact is a well-known result of J.H.C. Whitehead [39]:

Lemma A.9: *For a locally compact Hausdorff space X, $1_X \times f$ is a quotient map for any quotient map f.*

Unfortunately the domain of cell structures may not be locally compact.

Example A.10: $D^2 - \{(1,0)\}$ is locally compact, while $\mathrm{Int}(D^2) \cup \{(1,0)\}$ is not locally compact.

The domain D of an n-cell structure $\varphi : D \to \overline{e}$ is often a stratified subspace of D^n under a normal cell decomposition of D^n. In other words, D is obtained from D^n by removing cells. In the Example A.10, D^2 is regarded as a cell complex $D^2 = e^0 \cup e^1 \cup e^2$. $D^2 \setminus e^0$ is locally compact, while $D^2 \setminus e^1$ is not. More generally we have the following criterion of locally compact subspaces in a CW complex.

Proposition A.11: *Let X be a locally finite CW complex and A be a subcomplex, then $X \setminus A$ is locally compact.*

This is an immediate corollary to the following fact, which can be found, for example, in Chapter XI of Dugundji's book [40] as Theorem 6.5.

Lemma A.12: *Let X be a locally compact Hausdorff space. A subspace $A \subset X$ is locally compact if and only if there exist closed subsets $F_1, F_2 \subset X$ with $A = F_2 \setminus F_1$.*

Proof of Proposition A.11: Since X is locally finite, it is locally compact. The CW condition implies that A is closed in X. □

Let us go back to the discussion on products of quotient maps. The main motivation of Michael for introducing bi-quotient maps is that they behave well with respect to products.

Proposition A.13: *For any family of bi-quotient maps* $\{f_i : X_i \to Y_i\}_{i\in I}$, *the product*

$$\prod_{i\in I} f_i : \prod_{i\in I} X_i \longrightarrow \prod_{i\in I} Y_i$$

is a bi-quotient map.

Proof: Theorem 1.2 in [38]. □

The following property is also useful when we study cell structures.

Lemma A.14: *Let* $f : X \to Y$ *be a quotient map. Suppose that* Y *is first countable and Hausdorff and that, for each* $y \in Y$, $\partial f^{-1}(y)$ *is Lindelöf. Then* f *is bi-quotient.*

Proof: Proposition 3.3(d) in [38]. □

Corollary A.15: *Let* $\varphi : D \to \overline{e}$ *be a relatively compact cell. Then* φ *is bi-quotient.*

Proof: By Lemma A.3, $\overline{e}$ is first countable and Hausdorff. By assumption each fiber $\varphi^{-1}(y)$ is compact and so is the boundary $\partial\varphi^{-1}(y)$. The result follows from Lemma A.14. □

B. Simplicial topology

In this second appendix, we recall basic definitions and theorems in PL (Piecewise Linear) topology and simplicial homotopy theory used in these notes. Our references are

- the book [35] of Rourke and Sanderson for PL topology,
- the book [41] by Goerss and Jardine for simplicial homotopy theory.

B.1. *Simplicial complexes, simplicial sets, and simplicial spaces*

Let us fix notation and terminology for simplicial complexes first. Good references are Dwyer's monograph [42] and Friedman's survey article [43].

Definition B.1: For a set V, the power set of V is denoted by 2^V.

Definition B.2: Let V be a set. An *abstract simplicial complex* on V is a family of subsets $K \subset 2^V$ satisfying the following condition:

- $\sigma \in K$ and $\tau \subset \sigma$ imply $\tau \in K$.

K is called *finite* if V is a finite set.

Definition B.3: An *ordered simplicial complex* K is an abstract simplicial complex whose vertex set P is partially ordered in such a way that the induced ordering on each simplex is a total order.

An n-simplex $\sigma \in K$ with vertices $v_0 < \cdots < v_n$ is denoted by $\sigma = [v_0, \ldots, v_n]$.

There are several ways to define the geometric realization of an abstract simplicial complex.

Definition B.4: For an abstract simplicial complex K with vertex set V, define a space $\|K\|$ by

$$\|K\| = \left\{ f \in \mathrm{Map}^f(V, \mathbb{R}) \,\middle|\, \sum_{v \in \sigma} f(v) = 1, f(v) \geq 0, \sigma \in K \right\},$$

where $\mathrm{Map}^f(V, \mathbb{R})$ is the set of maps from V to $\mathbb{R}$ whose values are 0 except for a finite number of elements. It is equipped with the compact-open topology. The space $\|K\|$ is called the *geometric realization* of K.

Lemma B.5: *Suppose the vertex set V of an abstract simplicial complex K is finite. Choose an embedding*

$$i : V \hookrightarrow \mathbb{R}^N$$

for a sufficiently large N so that the $i(V)$ is affinely independent. Then we have a homeomorphism

$$\|K\| \cong \left\{ \sum_{v \in V} a_v i(v) \,\middle|\, \sum_{v \in \sigma} a_v = 1, a_v \geq 0, \sigma \in K \right\}.$$

Example B.6: Consider $2^V \setminus \{\emptyset\}$ for $V = \{0, \ldots, n\}$. This is an abstract simplicial complex. Then we have a homeomorphism

$$\|2^V \setminus \{\emptyset\}\| \cong \left\{ (t_0, \ldots, t_n) \in \mathbb{R}^{n+1} \,\middle|\, \sum_i t_i = 1, t_i \geq 0 \right\} = \Delta^n.$$

Δ^n is a convex polytope having $(n+1)$ codimension 1 faces. Each codimension 1 face can be realized as the image of the map

$$d^i : \Delta^{n-1} \longrightarrow \Delta^n$$

defined by

$$d^i(t_0, \ldots, t_{n-1}) = (t_0, \ldots, t_{i-1}, 0, t_i, \ldots, t_{n-1}).$$

We also have maps

$$s^i : \Delta^n \longrightarrow \Delta^{n-1}$$

defined by

$$s^i(t_0, \ldots, t_n) = (t_0, \ldots, t_i + t_{i+1}, t_{i+2}, \ldots, t_n).$$

For an ordered simplicial complex K, we may forget the ordering and apply the above construction. However, there is another construction.

Definition B.7: For an ordered simplicial complex K with vertex set V, let K_n be the set of n-simplices in K. Each element σ in K_n can be written as $\sigma = (v_0, \ldots, v_n)$ with $v_0 < \cdots < v_n$. Under such an expression, define

$$d_i(\sigma) = (v_0, \ldots, v_{i-1}, v_{i+1}, \ldots, v_n).$$

Define

$$\|K\| = \left(\coprod_n K_n \times \Delta^n \right) \Big/ \sim,$$

where the relation $\sim$ is generated by

$$(\sigma, d^i(\boldsymbol{t})) \sim (d_i(\sigma), \boldsymbol{t}).$$

This is called the *geometric realization* of K.

Lemma B.8: *For a finite ordered simplicial complex, the above two constructions of the geometric realization coincide.*

The above construction can be extended to simplicial sets and simplicial spaces.

Definition B.9: A *simplicial set* X consists of

- a sequence of sets $X_0, X_1, \ldots$,
- a family of maps $d_i : X_n \to X_{n-1}$ for $0 \le i \le n$,
- a family of maps $s_i : X_n \to X_{n+1}$ for $0 \le i \le n$

satisfying the following relations

$$\begin{cases} d_i \circ d_j = d_{j-1} \circ d_i, & i < j \\ d_i \circ s_j = s_{j-1} \circ d_i, & i < j \\ d_j \circ s_j = 1 = d_{j+1} \circ s_j & \\ d_i \circ s_j = s_j \circ d_{i-1}, & i > j+1 \\ s_i \circ s_j = s_{j+1} \circ s_i, & i \le j. \end{cases}$$

The maps d_i's and s_j's are called *face operators* and *degeneracy operators*, respectively.

When each X_n is a topological space and maps d_i, s_j are continuous, X is called a *simplicial space*.

Remark B.10: It is well known that defining a simplicial set X is equivalent to defining a functor

$$X : \Delta^{\mathrm{op}} \longrightarrow \mathbf{Sets},$$

where Δ is the full subcategory of the category of posets consisting of $[n] = \{0 < 1 < \cdots < n\}$ for $n = 0, 1, 2, \ldots$.

Example B.11: For a topological space X, define

$$S_n(X) = \mathrm{Map}(\Delta^n, X).$$

The operators d^i and s^i on Δ^n induce

$$d_i : S_n(X) \longrightarrow S_{n-1}(X)$$
$$s_i : S_{n-1}(X) \longrightarrow S_n(X).$$

When each $S_n(X)$ is equipped with the compact-open topology, these maps are continuous and we obtain a simplicial space $S(X)$. This is called the *singular simplicial space.*

Usually $S_n(X)$'s are merely regarded as sets and $S(X)$ is regarded as a simplicial set, in which case we denote it by $S_n^\delta(X)$.

Example B.12: Let K be an ordered simplicial complex on the vertex set V. Define

$$s(K)_n = \left\{ [\underbrace{v_0, \ldots, v_0}_{i_0}, \underbrace{v_1, \ldots, v_1}_{i_1}, \ldots, \underbrace{v_k, \ldots, v_k}_{i_k}] \;\middle|\; \sigma = [v_0, \ldots, v_k] \in K, \sum_{j=0}^{k} i_j = n \right\}.$$

Then the collection $s(K) = \{s(K)_n\}$ becomes a simplicial set. This is called the *simplicial set generated* by K.

Definition B.13: The *geometric realization* of a simplicial space X is defined by

$$|X| = \left(\coprod_n X_n \times \Delta^n\right)\Big/\sim$$

where the relation $\sim$ is generated by

$$(x, d^i(\boldsymbol{t})) \sim (d_i(x), \boldsymbol{t}),$$
$$(x, s^i(\boldsymbol{t})) \sim (s_i(x), \boldsymbol{t}).$$

The map induced by the evaluation maps

$$\Delta^n \times S_n(X) \longrightarrow X$$

is denoted by

$$\mathrm{ev} : |S(X)| \longrightarrow X.$$

Note that the geometric realization of an ordered simplicial complex is defined only by using face operators.

Definition B.14: A Δ-*set* X consists of

- a sequence of sets $X_0, X_1, \ldots$,
- a family of maps $d_i : X_n \to X_{n-1}$ for $0 \leq i \leq n$,

satisfying the following relations

$$d_i \circ d_j = d_{j-1} \circ d_i,$$

for $i < j$.

When each X_n is equipped with a topology under which the d_i's are continuous, X is called a Δ-*space*.

Remark B.15: A Δ-set X can be identified with a functor

$$X : \Delta_{\mathrm{inj}}^{\mathrm{op}} \longrightarrow \mathbf{Sets},$$

where Δ_{inj} is the subcategory of Δ consisting of injective maps. In particular, any simplicial set can be regarded as a Δ-set.

Definition B.16: The *geometric realization* of a Δ-space X is defined by

$$\|X\| = \left(\coprod_n X_n \times \Delta^n\right)\Big/_{\sim},$$

where the relation $\sim$ is generated by

$$(x, d^i(\boldsymbol{t})) \sim (d_i(x), \boldsymbol{t}).$$

Remark B.17: Note that any simplicial space X can be regarded as a Δ-space. However the geometric realization of X as a Δ-space, $\|X\|$, is much larger than that of X as a simplicial space. $\|X\|$ is often called the fat realization.

In order to study the homotopy type of simplicial complexes, and more generally, regular cell complexes, the notion of regular neighborhood is useful. Let us recall the definition.

Definition B.18: Let K be a cell complex. For $x \in K$, define

$$\mathrm{St}(x; K) = \bigcup_{x \in \overline{e}} e.$$

This is called the *open star* around x in K. For a subset $A \subset K$, define

$$\mathrm{St}(A; K) = \bigcup_{x \in A} \mathrm{St}(x; K).$$

When K is a simplicial complex and A is a subcomplex, $\mathrm{St}(A; K)$ is called the *regular neighborhood* of A in K.

The regular neighborhood of a subcomplex is often defined in terms of vertices.

Lemma B.19: *Let A be a subcomplex of a simplicial complex K. Then*

$$\mathrm{St}(A; K) = \bigcup_{v \in \mathrm{sk}_0(A)} \mathrm{St}(v; K).$$

B.2. *Locally cone-like spaces*

Let us first define the operation $*$ on subsets of a Euclidean space.

Definition B.20: For subspaces $P, Q \subset \mathbb{R}^n$, define

$$P * Q = \{(1-t)p + tq \mid p \in P, q \in Q, 0 \le t \le 1\}.$$

This is called the *convex sum* or *join* of P and Q

When P is a single point v, $v * Q$ is called the *cone* on Q with vertex v.

Remark B.21: When P and Q are closed and "in general position", $P * Q$ agrees with the join operation.

Definition B.22: We say a subspace $P \subset \mathbb{R}^n$ is *locally cone-like*, if, for any $a \in P$, there exists a compact subset $L \subset P$ such that the cone $a * L$ is a neighborhood of a.

Remark B.23: Locally cone-like spaces are called polyhedra in the Rourke-Sanderson book [35].

Example B.24: The half space $\mathbb{R}^n_+ = \{\boldsymbol{x} \in \mathbb{R}^n \mid x_n \geq 0\}$ is locally cone-like.

Example B.25: A vector (or affine) subspace of $\mathbb{R}^n$ is locally cone-like.

Example B.26: A convex polytope in $\mathbb{R}^n$ is locally cone-like.

Lemma B.27: *The class of locally cone-like spaces is closed under the following operations:*

- *finite intersections,*
- *finite products,*
- *locally finite unions.*

Corollary B.28: *Euclidean polyhedral complexes are locally cone-like.*

The following theorem characterizes compact locally cone-like spaces.

Theorem B.29: *Any compact locally cone-like space can be expressed as a union of finite number of simplices.*

B.3. *PL maps between polyhedral complexes*

In PL topology, we study triangulated spaces. Following Rourke and Sanderson, let us consider locally cone-like spaces. We also consider polyhedral complexes in a Euclidean space.

Definition B.30: Let P and Q be locally cone-like spaces. A map

$$f : P \longrightarrow Q$$

is said to be *piecewise-linear (PL)* if, for each $a \in P$, there exists a cone neighborhood $N = a * L$ such that

$$f(\lambda a + \mu x) = \lambda f(a) + \mu f(x)$$

for $x \in L$ and $\lambda, \mu \geq 0$ and $\lambda + \mu = 1$.

Lemma B.31: *PL maps are closed under the following operations:*

- *products,*
- *compositions,*
- *the cone construction.*

Another important property of PL maps is the extendability.

Lemma B.32: *Let P be a convex polytope and*

$$f : \mathrm{Int}P \longrightarrow \mathbb{R}^n$$

be a PL map. Then it has a PL extension

$$\tilde{f} : P \longrightarrow \mathbb{R}^n.$$

Theorem B.33: *Let K and L be Euclidean polyhedral complexes. For any PL map*

$$f : K \longrightarrow L,$$

there exist simplicial subdivisions K' and L' of K and L, respectively, such that the induced map

$$f : K' \longrightarrow L'$$

is simplicial.

Proof: By Theorem 2.14 in [35]. □

C. Topological categories

In this third appendix, we recall basics of topological categories. Our references are Segal's paper [44] and the article by Dwyer in [42].

C.1. *Topological acyclic categories*

Definition C.1: A *topological quiver* Q is a diagram of spaces of the form

$$s, t : Q_1 \longrightarrow Q_0.$$

For a topological quiver Q, define

$$N_n(Q) = \{(u_n, \ldots, u_1) \mid s(u_n) = t(u_{n-1}), \ldots, s(u_2) = t(u_1)\}.$$

An element of $N_n(Q)$ is called an *n-chain* of Q.

Definition C.2: A *topological category* C is a topological quiver equipped with two more maps

$$i : C_0 \longrightarrow C_1$$
$$\circ : N_2(C) \longrightarrow C_1$$

making the following diagrams commutative:

(1)

$$\begin{array}{ccc} N_3(C) & \xrightarrow{\circ\times 1} & N_2(C) \\ {\scriptstyle 1\times\circ}\downarrow & & \downarrow{\scriptstyle \circ} \\ N_2(C) & \xrightarrow[\circ]{} & C_1, \end{array}$$

(2)

$$\begin{array}{ccccc} C_1 & \xrightarrow{i\times 1} & N_2(C) & \xleftarrow{1\times i} & C_1 \\ & \diagdown\!\!\diagdown & \downarrow{\scriptstyle \circ} & \diagup\!\!\diagup & \\ & & C_1 & & \end{array}$$

Elements of C_0 are called *objects*. An element $u \in C_1$ with $s(u) = x$ and $t(u) = y$ is called a *morphism* from x to y and is denoted by $u : x \to y$. The subspace of morphisms from x to y is denoted by $C(x, y)$, i.e. $C(x, y) = s^{-1}(x) \cap t^{-1}(y)$. For $x \in C_0$, $i(x)$ is called the *identity morphism* on x and is denoted by $1_x : x \to x$.

When C_0 has the discrete topology, C is called a *top-enriched category.*

Definition C.3: An *acyclic topological category* is a top-enriched category C in which, for any pair of distinct objects $x, y \in C_0$, either $C(x, y)$ or $C(y, x)$ is empty and, for any object $x \in C_0$, $C(x, x)$ consists of the identity morphism.

Remark C.4: When the topology of C_0 is not discrete, we need to assume that C_1 decomposes into a disjoint union of $i(C_0)$ and its complement.

The following fact is well known.

Lemma C.5: *For an acyclic topoloical category C, define a relation $\leq$ on C_0 as follows:*

$x \leq y$ if and only if $C(x,y) \neq \emptyset$.

Then the relation $\leq$ is a partial order on C_0.

Definition C.6: For an acyclic topological category C, the poset $(C_0, \leq)$ is called the *underlying poset* of C and is denoted by $P(C)$. The canonical functor from C to $P(C)$ is denoted by

$$\pi_C : C \longrightarrow P(C).$$

In these notes, we are concerned with cellular structures.

Definition C.7: A topological category C is said to be *cellular* if both C_0 and C_1 have structures of cellular stratified spaces and the structure maps $s, t, i, \circ$ are morphisms of cellular stratified spaces.

Remark C.8: In the above definition, we assume that each $C(y,z) \times C(x,y)$ is a cellular stratified space under the product stratification. See Proposition 6.14 and discussions in §6.2 for conditions under which the bi-quotient assumption in Lemma C.17 is satisfied.

When C_0 is not discrete it is not easy to define cellularness, since the pullback of stratified spaces over a stratified space may not be a stratified space in general.

Functors between topological categories are always assumed to be continuous.

Definition C.9: A *continuous functor* f from a topological category C to another D is a pair $f = (f_0, f_1)$ of continuous map

$$f_0 : C_0 \longrightarrow D_0$$
$$f_1 : C_1 \longrightarrow D_1$$

that are compatible with all structure maps of topological category.

Topological categories defined in Definition C.2 are usually called small topological categories, in the sense that collections of objects and morphisms form sets. In these notes the only category which is not small is the category **Spaces** of topological spaces and continuous maps. The continuity of a functor to **Spaces** is defined as follows.

Definition C.10: Let C be a (small) top-enriched category. A functor $f : C \to$ **Spaces** is said to be *continuous* if, for each pair $x, y \in C_0$ of objects, the adjoint

$$C(x,y) \times f(x) \longrightarrow f(y)$$

to the map $f(x,y) : C(x,y) \to \mathrm{Map}(f(x), f(y))$ is continuous.

For a continuous functor $f : C \to \mathbf{Spaces}$, define $\mathrm{tot}(f) = \coprod_{x \in C_0} f(x)$. The canonical projection onto C_0 is denoted by $\pi_f : \mathrm{tot}(f) \to C_0$. Define $C_1 \square_{C_0} \mathrm{tot}(f)$ by the following pullback diagram

$$\begin{array}{ccc} C_1 \square_{C_0} \mathrm{tot}(f) & \xrightarrow{s \square 1} & \mathrm{tot}(f) \\ \downarrow & & \downarrow \pi_f \\ C_1 & \xrightarrow[s]{} & C_0. \end{array}$$

Then the *colimit* of f, $\mathrm{colim}_C\, f$, is defined by the following coequalizer diagram

$$C_1 \square_{C_0} \mathrm{tot}(f) \underset{\mu_f}{\overset{s\square 1}{\rightrightarrows}} \mathrm{tot}(f) \xrightarrow{p_f} \operatorname*{colim}_C f,$$

where μ_f is defined on each component by the adjoint $C(x,y) \times f(x) \longrightarrow f(y)$ to $f(x,y) : C(x,y) \to \mathrm{Map}(f(x), f(y))$.

C.2. *Nerves and classifying spaces*

One of the most fundamental facts is that the collection $N(X) = \{N_n(X)\}_{n \geq 0}$ defined in Definition C.1 forms a simplicial space.

Lemma C.11: *For a topological category X, we have*

$$N_n(X) = \mathrm{Funct}([n], X),$$

where the poset $[n] = \{0, \ldots, n\}$ is regarded as a topological category (with discrete topology). Thus $N(X)$ defines a functor

$$N(X) : \Delta^{\mathrm{op}} \longrightarrow \mathbf{Spaces}.$$

In other words, $N(X)$ is a simplicial space for any topological category X.

Definition C.12: The simplicial space $N(X)$ is called the *nerve* of X. The source and the target maps on X can be extended to

$$s, t : N_k(X) \longrightarrow X_0$$

by $s(\boldsymbol{f}) = f(0)$ and $t(\boldsymbol{f}) = f(k)$, respectively. These are also called the *source and target maps*.

Definition C.13: The geometric realization of $N(X)$ is called the *classifying space* of X and is denoted by BX.

When we form the geometric realization of a simplicial space, nondegenerate chains are essential.

Definition C.14: For a topological category X, define

$$\overline{N}_n(X) = N_n(X) \setminus \bigcup_i s_i(N_{n-1}(X)).$$

Elements of $\overline{N}_n(X)$ are called *nondegenerate* n-chains.

Lemma C.15: *When P is a poset regarded as a small category, $\overline{N}(P) = \{\overline{N}_n(P)\}$ is an ordered simplicial complex and we have*

$$BP = \left\|\overline{N}(P)\right\|.$$

More generally, we have the following description.

Lemma C.16: *When X is a topological acyclic category, the simplicial structure on $N(X)$ can be restricted to give a structure of Δ-space on $\overline{N}(X)$. Furthermore the composition*

$$\left\|\overline{N}(X)\right\| \longrightarrow \|N(X)\| \longrightarrow |N(X)| = BX$$

is a homeomorphism.

When an acyclic topological category C is cellular in the sense of Definition C.7, the classifying space BC has a canonical cell decomposition.

Lemma C.17: *Let C be an acyclic topological category in which each morphism space $C(x,y)$ is equipped with a structure of cellular stratified space whose cell structures are bi-quotient. Then the classifying space BC has a structure of cellular stratified space.*

In particular when all morphism spaces are cell complexes, the classifying space BC has a structure of cell complex.

Proof: Under the identification $BC \cong \left\|\overline{N}(C)\right\|$, any point in BC can be represented by a pair $(x,t) \in \overline{N}_k(C) \times \mathrm{Int}(\Delta^k)$ uniquely. In other words, we have a decomposition

$$BC = \coprod_{k=0}^{\infty} \overline{N}_k(C) \times \mathrm{Int}(\Delta^k)$$

as sets. The bi-quotient assumption on cell structures on $C(x,y)$ implies that $\overline{N}_k(C)$ has a structure of cellular stratified space. (See Definition 6.13 for product cellular stratifications.) Thus the above decomposition and the cellular stratification on each $\overline{N}_k(C)$ define a stratification on BC.

For each cell $\varphi : D \to \overline{e} \subset \overline{N}_k(C)$, let us denote the cell in BC corresponding to $e \times \mathrm{Int}(\Delta^k)$ by $e \times e^k$. The composition

$$D \times \Delta^k \xrightarrow{\varphi \times 1_{\Delta^k}} \overline{N}_k(C) \times \Delta^k \longrightarrow \|\overline{N}(C)\| = BC$$

defines a cell structure on $e \times e^k$. □